Making Geography Matter

What is the purpose of Geography? What do geographers study and why? How do they seek to shape the world they interrogate?

This book addresses these questions by examining the lives and works of individual geographers, both past and present. Like all disciplines, Geography is no more nor less than the collective endeavours of researchers and teachers operating in specific contexts. The contexts both shape, and are shaped by, these individuals. This book's biographical and autobiographical chapters transport readers to the times and places where geographers have sought to make Geography matter. The result is a more vivid, grounded understanding of the discipline than the many high-level surveys of geographic thought paradigms currently written for university students.

This book's accessible essays each conclude with a study task. *Making Geography Matter* is aimed at university students and their teachers who wish to understand the goals, history and evolving practice of Geography. It provides an alternative perspective – both concrete and engaging – to the many student-focussed texts that map out numerous 'isms and ologies'.

Noel Castree has worked at the universities of Manchester, Wollongong and Liverpool, and the University of Technology Sydney. He is managing editor of the journals *Progress in Human Geography* and *Environment and Planning F*. He is author of the books *What Future For the Earth?* (2026) and *Making Sense of Nature* (2013).

Trevor Barnes is Professor and Distinguished University Scholar at the Department of Geography at the University of British Columbia, Canada, where he has been since 1983. His research is in economic geography and on the post-war history of human geography. He is a Fellow of both the Royal Society of Canada and the British Academy.

Jennifer Salmond is Professor of Geography at the University of Auckland, New Zealand. She is a physical geographer whose research interests include urban meteorology, air pollution, climatology and critical physical geography.

Making Geography Matter

The Past and Present of a Changing Discipline

Edited by Noel Castree, Trevor Barnes and Jennifer Salmond

Routledge
Taylor & Francis Group
LONDON AND NEW YORK

Designed cover image: billyrenkl.com/fine-art/gleanings/
july-25-1838-and-thoughts-flow/

First published 2025
by Routledge
4 Park Square, Milton Park, Abingdon, Oxon OX14 4RN

and by Routledge
605 Third Avenue, New York, NY 10158

Routledge is an imprint of the Taylor & Francis Group, an informa business

British Library Cataloguing-in-Publication Data
A catalogue record for this book is available from the British Library

Library of Congress Cataloging-in-Publication Data
Names: Castree, Noel, 1968- editor. | Barnes, Trevor J.,
editor. | Salmond, Jennifer, editor.
Title: Making geography matter : the past and present of a changing
discipline / Edited by Noel Castree, Trevor Barnes and Jennifer Salmond.
Description: Abingdon, Oxon; New York, NY : Routledge, 2025. | Includes
bibliographical references and index.
Identifiers: LCCN 2024026665 (print) | LCCN 2024026666 (ebook) |
ISBN 9781032380506 (hardback) | ISBN 9781032380513 (paperback) |
ISBN 9781003343240 (ebook)
Subjects: LCSH: Geography--Study and teaching (Higher) | Geographers. |
Geography--Vocational guidance.
Classification: LCC G73 .M255 2025 (print) | LCC G73 (ebook) |
DDC 910.71/1--dc23/eng/20240629
LC record available at https://lccn.loc.gov/2024026665
LC ebook record available at https://lccn.loc.gov/2024026666

ISBN: 978-1-032-38050-6 (hbk)
ISBN: 978-1-032-38051-3 (pbk)
ISBN: 978-1-003-34324-0 (ebk)

DOI: 10.4324/9781003343240

Typeset in Sabon
by SPi Technologies India Pvt Ltd (Straive)

Contents

Illustrations

Figures

Tables

Box

Acknowledgements

The editors would like to thank the authors for their willingness to engage in a different kind of writing project and their enthusiasm for the material. The editors are enormously grateful for the many different tasks Margot Wilson and Ellie Buttle carried out in preparing the book manuscript for submission to the press. We also wish to thank the professional editorial team for their patience and assistance in compiling and producing this book.

Contributors

Trevor Barnes is Professor and Distinguished University Scholar at the Department of Geography at the University of British Columbia, Canada, where he has been since 1983. His research is in economic geography and on the post-war history of human geography. He is a Fellow of both the Royal Society of Canada and the British Academy.

Luke Bergmann is Associate Professor and Canada Research Chair in the Department of Geography at the University of British Columbia. His research interests include critical-computational and social-theoretic geographies.

Gary Brierley is Professor and Chair of Physical Geography in the School of Environment at the University of Auckland. Building on a background in Geography, Geomorphology and Sedimentology (UK, Canada), he co-developed the River Styles Framework as a tool to support coherent management applications (Australia). His recent work in Aotearoa New Zealand adopts a more-than-human lens that works at the interface of big data (digital rivers) and oral (local and indigenous) narratives in the quest to Find the Voice of the River and Let the River Speak.

Noel Castree has worked at the universities of Manchester, Wollongong and Liverpool, and the University of Technology Sydney. He's managing editor of the journals *Progress in Human Geography* and *Environment & Planning F*. He is author of the books *What Future For the Earth?* (2026) and *Making Sense of Nature* (2013).

Tim Cresswell is Ogilvie Professor of Geography at the University of Edinburgh. He is a cultural geographer by training, and the author or editor of a dozen books and over 100 articles on the role of space, place and mobility in social and cultural life. Cresswell is also a widely published poet with three collections – most recently *Plastiglomerate* (Penned in the Margins, 2020).

Veronica della Dora teaches Human Geography at Royal Holloway, University of London. Her research interests and publications span historical and cultural geography with a specific focus on landscape, sacred space and the geographical imagination. Her monographs include *Imagining Mount Athos: Visions of a Holy Place from Homer to World War II* (University of Virginia Press, 2011), *Landscape, Nature and the Sacred in Byzantium* (Cambridge University Press, 2016), *Mountain: Nature and Culture* (Reaktion, 2016) and *The Mantle of the Earth: Genealogies of a Geographical Metaphor* (University of Chicago Press, 2021)

Danny Dorling is a geographer. He is a patron of RoadPeace, Comprehensive Future, and Heeley City Farm. In his spare time, he makes sandcastles.

Federico Ferretti received his joint PhD at the Universities of Bologna and Paris 1 Panthéon-Sorbonne. After gaining research and teaching experience in France, Switzerland, Brazil and Ireland, he serves as a Professor of Geography at the University of Bologna. His research interests lie in the history and philosophy of geography, with a special interest in critical, radical and anarchist tendencies and in Latin American critical, feminist and decolonial geographies.

Rita Gardner's early career was as an academic geomorphologist in the University of London (1979–1996), with research interests in environmental change and soil conservation in South Asia and the Middle East. With a track record of promoting and supporting Geography in and beyond the academy, she was appointed (1996) the first woman Director of the Royal Geographical Society (RGS; founded in 1830), soon after it and the Institute of British Geographers (IBG) merged. She retired from the RGS-IBG in 2018, her work having been recognised with a CBE for services to geography and six honorary degrees, and has since been leading the work of the UK Academy of Social Sciences as its Chief Executive.

Katherine Gibson is a Professorial Research Fellow in the Institute for Culture and Society at Western Sydney University and co-founder of the Community Economies Institute, an international not-for-profit Non-University Education and Research Organization. She is an economic geographer with an international reputation for innovative research on economic transformation and over 30 years' experience of working with communities to build resilient economies. As J.K. Gibson-Graham, the collective authorial presence she shares with the late Julie Graham (Professor of Geography, University of Massachusetts Amherst), she has written many books and conducted action research projects.

Michael F. Goodchild is Emeritus Professor of Geography at the University of California, Santa Barbara. He received his BA degree from Cambridge University in Physics in 1965 and his PhD in geography from McMaster University in 1969. He was elected a member of the National Academy of Sciences and a Foreign Member of the Royal Society of Canada in 2002, a member of the American Academy of Arts and Sciences in 2006, and a Foreign Member of the Royal Society and Corresponding Fellow of the British Academy in 2010; and in 2007 he received the Prix Vautrin Lud. His research interests centre on geographic information science, spatial analysis and uncertainty in geographic data.

Emily Hayes is a Research Associate and Associate Lecturer in Geography and Social Anthropology at Oxford Brookes University. She has a PhD in Geography (University of Exeter, 2016); an MSc in Archaeology and Environmental Sciences (Paris 1 and 10, 2005); and a BA in Archaeology and Anthropology (University of Cambridge, 1999). She has published in *Early Popular Visual Culture*, the *Transactions of the British Institute of Geographers*, and the *Journal of Historical Geography* where she is also a Reviews Co-Editor. She is a committee member of the RGS-IBG History and Philosophy of Science Research Group.

Andrew Herod writes frequently upon issues of labour, globalization, and geographical scale. In 2023 he was awarded a John Simon Guggenheim Memorial Foundation Fellowship. He is a former local elected official, having served for 14 years in the government of Athens-Clarke County, Georgia, as a County Commissioner and Mayor Pro Tem.

Mike Hulme is Professor of Human Geography at the University of Cambridge, Director of Studies for geography at Pembroke College. His work illuminates the numerous ways in which the idea of climate change is deployed in public, political and scientific discourse and is the author of 12 books on climate change including *Climate Change*

– *Key Ideas in Geography* (Routledge, 2021) and *Why We Disagree About Climate Change* (CUP, 2009). From 2000 to 2007 Hulme was the Founding Director of the Tyndall Centre for Climate Change Research.

Sue Jackson is a cultural geographer who studies the interaction between Indigenous customary and state environmental governance systems. For the past 20 years, her research has focused on the meaning of water, and its symbolic significance and material value to both Indigenous and settler communities of Australia.

Innes M. Keighren is Professor of Historical Geography at Royal Holloway, University of London. He has research interests in geography's disciplinary and discursive histories, in book history, and in the history of science. He is author of *Bringing Geography to Book: Ellen Semple and the Reception of Geographical Knowledge* (2010) and co-author of *Travels into Print: Exploration, Writing, and Publishing with John Murray, 1773–1859* (2015).

Kent Mathewson is author and editor of articles and books on the history of Geography, ancient and traditional agriculture, cultural and historical geography. Topics include: Carl Sauer and Elisée Reclus, raised fields and ethnic groups. He has taught at various universities in the US and Latin America.

Charles J. Pattie is a Professor of Politics at the University of Sheffield. He researches the geography of elections and election campaigns.

Jamie Peck is Professor of Geography and Distinguished University Scholar at the University of British Columbia, Canada. With long-term research interests in economic geography, urban and regional studies, and institutional political economy, his recent publications include the monograph *Variegated Economies*. The Managing Editor of *EPA: Economy & Space*, Peck was the recipient of the 2023 Prix International de Géographie Vautrin Lud.

Geraldine Pratt is Professor of Geography and Canada Research Chair in Care Economies and Global Labour at the University of British Columbia, author of *Working Feminism and Families Apart*; co-author of *Gender, Work and Space* and *Film and Urban Space*; and co-editor of *The Global and the Intimate*. She co-authored with Caleb Johnston a site-specific testimonial play, *Nanay*, performed in Vancouver, Berlin, Edinburgh, Whitehorse and Manila. She is founding member of the UBC Center for Climate Justice.

Jennifer Salmond is Professor of Geography at The University of Auckland. She is a physical geographer whose research interests include urban meteorology, air pollution, climatology and critical physical geography.

Eric Sheppard is the Alexander von Humboldt Chair of Geography at The University of California, Los Angeles. His scholarship embraces geographical political economy and socio-spatial theory, currently focusing on uneven geographies of capitalist globalization, more-than-capitalist practices, southern urban theory, urban social movements and urban land transformations.

Jamie Woodward is Professor of Physical Geography at the University of Manchester. He is a geomorphologist with interests in rivers and how humans interact with them, from the ice age to the present day. He is a Fellow of the British Society for Geomorphology and Vice-President for Research and Higher Education at the Royal Geographical Society (2022–2025).

1 Making the 'geo' matter

The discipline and discourse of geography

Noel Castree, Trevor Barnes and Jennifer Salmond

Introduction

What is Geography about? What are its key discoveries and ideas? By what means do geographers interrogate the world? How has their thinking shifted over time and why? Students of Geography at universities are usually expected to develop an informed perspective on the aims and scope of their chosen discipline, and the major schools of thought within it (both past and present). Often, this perspective will emerge from a semester-long course of study, perhaps at the beginning or very end of their degree programme. This book has been conceived and written for them. As we'll explain in this introductory chapter, the aim is to offer a fresh and engaging approach to material that students often find dry, irrelevant and, if truth be told, rather boring.

Geography: Realities and their representation

Geographers study the what, how and why of where. They interrogate the world in three dimensions – four, if you account for continuity and change through time, and even five if the virtual realm created by modern technology counts. Their grand challenge is to illuminate the character and interrelations between phenomena, both human and non-human, on the variegated surface of our fast-changing planet. Thus, while many other academic disciplines are specialised – that is, concerned with particular slices of life – Geography aims to investigate everything, everywhere, all at once. There's seemingly no aspect of reality that's out of scope.

While Geography's ambition may appear impossibly large – maybe foolhardy, even hubristic to the point on absurdity – someone needs to do it. If Geography didn't exist, it would be necessary to invent it. After all, there's an enormous societal appetite for geographical knowledge and insight about the world around us. The popularity of books by former foreign affairs correspondents Robert Kaplan and Tim Marshall attest to this. Then there are hugely popular TV series and documentaries about natural hazards, wildlife, overseas travel, human impacts on the environment and spectacular landscapes – for instance, *Years of Living Dangerously* (Netflix), *Planet Earth* (BBC) and *Ningaloo Nyinggulu* (ABC). *National Geographic* magazine is as widely read as ever, while any number of daily news reports offer basic geography lessons each time events in places like Brazil, Gaza, Haiti, Liberia, Niger, Palestine, Singapore, Tuvalu and Ukraine are recounted.

The contemporary world is remarkably differentiated, despite over-blown claims sometimes made about 'the death of geography' – a spatial shrinking caused by ever faster, more efficient, cheaper communications and transportation networks. Geography

DOI: 10.4324/9781003343240-1

still matters: it makes a huge difference to your experience of life whether you live in Turin, Tuktuyuktuk, Tripoli, Townsville, Tokyo, Tobago, Tehran, Takapuna or Taubate. However, today's world is also tightly integrated, albeit unevenly: movements of people, money, information, goods, services, animal species, greenhouse gases, viruses and more besides make it ever harder to separate 'the local' from 'the global'. In the 21st century, the world's geographies are extraordinarily complex and fluid. An array of causes and consequences, both near and far, create geographies of difference and connectivity. Again, Geography matters – how could it not? – but in new ways. Recent claims that the Earth has entered 'the Anthropocene' imply that the entire physical geography of the last 11,500 years (the Holocene) is fast giving way to vast new geographies forged by humans, and as much by accident as design. The changes are rapid and already enormously consequential. At the time of writing, July 2023 has been declared the hottest month ever recorded. That's almost certain to have changed by the time you read these words.

Geographers generate ideas, theories, information, propositions, maps, images and narratives for a variety of purposes. These include describing, ordering and classifying; creating meaning and explaining the worlds around us; criticising aspects of reality with a view to improving them; identifying risks and threats, such as floods; predicting future trends and possibilities; protecting or restoring the natural environment; and suggesting how to promote positive social change based on values such as fairness or equality. How we choose to study and know our worlds (the tools we use to measure; the words, numbers, cartographs etc. we use to describe, explain, model and forecast) doesn't sit in intellectual isolation alongside (but just out of reach of) reality, isolated in the pages of textbooks, academics' offices or the silence of examination halls. Instead, our diagnostic and prognostic tools become integral to the ways we live and order our lives, the decisions we make and the futures we plan and try to create. Geography matters, but so does the way we choose to represent it, with 'reality' and how we depict it influencing each other ceaselessly.

It's an open question how well geographers of the past and present comprehend our multi-dimensional world, however good their intentions were or are. This book doesn't intend to judge. Instead, it introduces readers to the various approaches that geographers have developed – mostly, although not always, in good faith – to shed light on how and why the geographies they study matter. The focus, then, is on the practices of making sense of the geographical world. Without those geographical practices the realities they reveal would be unknown or understood quite differently. Our concern is with what we will call, in a loosely metaphorical way, Geography's 'grammar'. In linguistics, grammar is the set of rules within a given language that allows meaningful statements using words to be made. Without grammar there'd be only a cacophony of utterances, scribbles on a page or symbols on a screen. Geography's grammar comprises the overarching 'rules' or norms used to make meaningful statements about the geographical world. The 'rules' are unwritten, quite numerous and have evolved slowly over time. Geography's grammar is not the same as its compendious vocabulary, let alone the countless facts about the world's human and physical geography. Facts only become meaningful when there's an appropriate 'grammar' that allows the apprehension of social, physical and virtual reality by way of coherent statements and representations. Below, we explain how Geography's rich 'grammar' has taken the form of sometimes competing, sometimes overlapping 'projects' – that is, coherent collections of ideas and practices that form around a particular area of geographical study, created by particular individuals or groups of geographers, to serve a particular (professional) goal.

Both historically and now, geographers employed in universities and other institutions of research and higher learning compete for the attention of peers and students to represent the world's changing geographies. They advocate for, and debate, the merits of different approaches. This highlights a crucial feature common to all academic disciplines. Each geographer, in their own way, tries to make the world intelligible by employing ensembles of precepts, concepts, theories, models, hypotheses, categorisations and methods, all geared to various goals (e.g., truth-seeking or the pursuit of social justice or policy prescription). These ensembles are influenced, but not dictated, by the social, material or virtual realities they seek to depict, predict or evaluate. Those ensembles, and the decisions made around them, were (are) not inevitable. They didn't (don't) have to turn out as they did or do. There's no single road to the truth or even a singular truth. Researchers aren't omniscient, unbiased, objective surveyors of a reality out there awaiting proper discovery. However talented and credentialed, they're finite humans operating in specific contexts that both enable and constrain both what they do and what they can know. Their combined efforts don't beat a path towards some absolute truth. Instead, given a person's experiences and situation, given their specific biographies, they offer contestable and revisable insights that may, or may not, prove valuable for citizens, politicians, businesspeople, political activists and others as new challenges and opportunities arise in our dynamic world.

Making Geography Matter aims to show students how, why and to what ends the lived lives of different geographers have made the worlds they represent meaningful through their study. The verb 'to make' signals our intention to explore how geographers situated in different times and locations make the world 'speak'. That doesn't just happen automatically. It takes considerable effort, talent and preparation, as will be evident in the ensuing biographies and autobiographies of individual geographers. The rest of this introduction lays out the rationale for the book and for the particular approach that we and our contributors take. In *Making Geography Matter*, we want students to think about how knowledge is produced, prioritised, turned into action (or not), the purposes to which it is directed, its successes and failures, and how all of these can change through time. Who gets to speak about the world's geography, why and how? This edited collection, we hope, will both inform and inspire students interested in matters of G/geography, why they matter, how and for whom.

Making sense of Geography and geographers: A focus on people, their contexts and their 'projects'

Worldwide, at any one time, tens of thousands of university students are studying for certificates, diplomas or degrees in Geography and cognate subjects. They're taught about topics that include water resource management, foreign direct investment, involuntary migration, urban policy, climate change, economic growth regions, trading blocs, geopolitics and food supply chains. In some countries, such as the United Kingdom, students learn about the aims, history and achievements of Geography as a discipline, as well as key disciplinary debates and controversies in the past and present. Fundamentally, these modules and courses, taught in Anglo-American Geography departments, are about the various modalities of geographical research: that is, how geographical research is done and for what purpose, and then how it is turned into geographical knowledge. They ask about the means and ends of geographical knowledge.

When we editors were undergraduate students (depending on which one of us, anywhere from 30 to 50 years ago!) such modules and courses were compulsory in undergraduate Geography degrees. They still are in several departments. As a result, there are many textbooks in this area that offer broad surveys of human, environmental and physical geography (e.g., Cloke, Crang & Goodwin, 2013; Inkpen & Wilson, 2013; Johnston & Sidaway, 2015; Nayak & Jeffery, 2011). But Geography today is a large, dynamic, heterodox and porous discipline (which bleeds into multiple other fields). This means that textbooks (such as Tim Cresswell's *Geographic Thought* [2012], Kevin Cox's *Making Human Geography* [2014] and Pauline Couper's *A Student's Introduction to Geographical Thought* [2015]) are hard-pressed to survey properly all human or physical geography – never mind both, as well as their interstices. If you imagined David Livingstone were tasked with updating *The Geographical Tradition* (1993) – admittedly not a textbook, but a highly authoritative and large survey of the discipline's origins and evolution – it would be at least *twice* as long! And yet any attempted *brief* surveys of so vast an academic terrain would necessarily simplify and gloss, with much of the interesting and important detail that someone like Livingstone provides rinsed out. The resulting general but succinct treatments of topics like 'spatial science', humanistic geography, GIScience, post-colonialism and so on threaten to put students and teachers off for lack of specificity (see *Geography: A Very Short Introduction* by Herbert & Matthews, 2008). However, as noted, detailed treatments are now too big for the average degree student (or indeed lecturer!) to want (or be able) to read.

One solution is for some student-focussed overviews of Geography's history to treat human and physical geography separately. While certainly meritorious, these various books pose two significant challenges to students and their instructors. First, they present at a very high level of abstraction the varied 'isms and ologies' – so-called 'paradigms' – that have both unified and differentiated geographers within and across the 'divide' between human and physical geography since around 1950. Readers of these books grapple with broad summaries of sometimes exotic-sounding and esoteric approaches such as 'positivism', 'humanism', 'post-humanism', 'behaviouralism', 'Marxism', 'post-Marxism', 'feminism', 'queer theory', 'post-colonialism', 'non-representational theory' and many more besides. The synoptic approach ends up depersonalising Geography, making intellectual change appear like tidal currents, sweeping numerous geographers along willy-nilly.

The other challenge is that there are now *so many* isms and ologies. Geography is *extraordinarily* heterodox. Instructors and students can barely keep up. For example, in human geography the fairly recent 'more-than-representational', 'post-humanist', 'more-than-human' and 'affective' turns have added to a very long list of overarching approaches to investigate the geographical world. The sheer number of approaches threatens to overwhelm students and, worse, to dull their senses (a well as their instructors'). It's no surprise, for instance, that Ron Johnston's much used book *Geography and Geographers* evolved from an admittedly lengthy text when first published 40-plus years ago to a veritable door-stopper, co-authored with James Sidaway, ten years ago. If the book were further updated by Sidaway (Johnston, sadly, passed away in 2020), it would probably be over 600 pages long. Questions now arise about the wisdom of trying to offer 'total surveys' of Geography's evolving territory. It would be a full-time occupation.

Our book aims to step outside the traditional mould. It is explicitly *not* synoptic in its coverage and refrains from venturing another aerial view of disciplinary 'isms and ologies' and their various strands and intersections. Instead, we aim to do two things. One is to emphasise that, at base, the creation and evolution of new investigative approaches in Geography is, as in all disciplines, driven by sentient and engaged *individuals*: by real people in real times and in real places. In this book we home in on some of the key individuals who have sought to make Geography matter – both big G (the discipline) and small g (the insights provided by work conducted within the discipline). Our contributors seek to paint vivid pictures of lives lived and contributions made since Geography was instituted in Western universities in the late 19th century. There's a mix of biographical and autobiographical chapters. Our authors are looking, as it were, from the other end of the telescope, eschewing the larger view of, say, 'science', 'landscape study', 'applied geography' or 'critical human geography', to focus on the 'small' lives and endeavours of particular scholars. To call them small is not in any way to belittle their often large achievements and effects within the discipline. We believe that stories about individual geographers rather than aggregated stories about isms and ologies will resonate better with students, capturing their imagination. They promise to 'humanise', vitalise and 'socialise' Geography.

Aside from our focus on the lives and works of particular geographers, we also foreground what we call 'projects'. This is the book's second key point of difference compared to other texts. As noted, many of the works listed above highlight 'paradigms', 'isms' and 'ologies' – broadly, these are packages of ontological, epistemological, categorial and methodological beliefs-cum-practices that define how some practitioners together believe geographical research ought to be done. We take a different approach to how geographers have sought to shine a light on the world's changing, diverse and interlinked geographies. By 'projects' we mean a set of heterogeneous practices enacted by particular geographers in order to achieve their professional goals.

Projects are different 'bundles' of practices (i.e., modes of thinking and acting), some large and some more contained. They range from advocacy directed at potential or actual sponsors and patrons of the discipline, to formative research conducted within Geography (conceptual, empirical, applied), to efforts to translate that research into directly useable or understandable forms to which non-academics can relate. All disciplines are an amalgam of sometimes overlapping, sometimes antagonistic, sometimes relatively discrete projects that variously co-exist with or shape one another. Given its enormous span of subject matter, it's no surprise that Geography's projects have been many and varied, with some more ambitious than others. Our aim in *Making Geography Matter* is less to be exhaustive and more to reveal the richness of projects pursued by geographers – with varying success – at different times in various places.

We cover both human *and* physical geography, unlike most texts about the history and nature of Geography. We also cover both broader and narrower projects because both kinds of project matter. We favour the motif (signalled in the title of our book) of 'mattering' because, in different ways, each project involves geographers actively showing the importance of particular ideas, forms of evidence, activities or goals within or beyond the university world. Those who pioneered projects or who sought to redirect them have self-consciously attempted to recruit followers. During their careers, many geographers pursue more than one project, sometimes even reinventing themselves in the process by criticising their own earlier project (see Chapter 8 on David Harvey). There's been no consensus, especially since the 1970s, about how best to elucidate the 'what, how and why of where' or reveal what was (un)important to study, how and why. This is what makes Geography such a rich, dynamic intellectual ecosystem for those willing to move between niches and forge new connections.

The structure and content of this book

As already noted, the book comprises biographical and autobiographical chapters. These are organised into three intersecting parts. We describe these before focussing more fully on each chapter in turn. In the following section we focus on some key issues about how we arrived at our selection of geographers.

Part 1 considers the work of geographers who (have) devoted considerable energy to defining, instituting, evolving and seeking to secure the status of Geography as a university (and school) *discipline*. In our book they are personified by individuals from **Halford Mackinder** (1861–1947), who held the first continuing university appointment in the UK in Geography, in the late 19th century, to the first ever female director of the Royal Geographical Society-Institute of British Geographers (RGS-IBG), **Rita Gardner** (born 1955). Part 2 focusses on geographers who have sought to breathe intellectual life into the discipline at different points in time, in the process setting examples for others to follow in their research and university teaching. This is the largest part of the book and considers the careers of everyone from influential American geographer **Carl Sauer** to the present-day practitioners **Gary Brierley, Tim Cresswell, Andy Herod, Sue Jackson, Mike Hulme** and **Gerry Pratt**. While most of the people profiled in this book work or worked in university departments (or schools) of Geography, not all have or do. This is a useful reminder that geographical inquiry often occurs in other disciplines too, and sometimes outside universities altogether. For instance, Australian **Sue Jackson** (a contributor to this book) once worked in a federal government research organisation chock-full of Geography graduates.

If Part I is about efforts to show that having a discipline called Geography matters, and if Part II focusses on the efforts of certain people to show how geographical perspectives on the world matter in different registers, then Part III is about making these perspectives count outside the academic realm where they're created. This part covers the works during the first half of the 20th century of the late **Dudley Stamp** in the UK, the action research during the 1960s and 1970s conducted by **Bill Bunge** and collaborators in North America through to current efforts by **Danny Dorling, Kathy Gibson** and **Jamie Woodward** to render geographical insight and information beneficial in society. The work undertaken in the discipline of Geography feeds into a wider societal *discourse* about geography shaped by journalists, travel writers, documentarians, politicians, community leaders and many others. Part III explores efforts to link the discipline to the wider discourse.

What – to be more specific – are the 'projects' in question in the 20-plus chapters? **Emily Hayes** kicks things off with her examination of **Mackinder's** efforts to institute in British universities (and schools) a Geography discipline that was, at the time, far from being a secure piece of the academic furniture. **Innes Keighren** then explores one of Anglophone Geography's earliest thought-shapers, the American **Ellen Semple** (1863–1932). Her tireless work sought to provide an intellectual framework that bridged the study of society with that of the physical environment, so-called 'environmental determinism'. By the 1930s environmental determinism is already heavily discredited, leading to much soul-searching by Anglophone geographers. After 1945 things changed. A new generation sought to put Geography on a new intellectual footing, accenting the values and tools of science, and seeking to identify patterns and variances on the face of the Earth at different scales. British geographers **Richard Chorley** (physical, 1927–2002) and **Peter Haggett** (human, born 1933) were instrumental to this project, as **Trevor Barnes** shows. Part 1 ends with **Rita Gardner's** autobiographical chapter. Well over a century after Mackinder and others sought to make space for Geography in the research and educational system, Gardner recounts her considerable efforts to advance Geography in early 21st-century Britain through the RGS-IBG – the UK's chief professional body for geographers. The context she worked in was very different to Mackinder's, both externally and in terms of Geography's internal configuration. Her task was to advocate for the discipline on behalf of all those who in the UK identify with it.

Part II begins with **Kent Matthewson's** overview of **Carl Sauer's** (1889–1975) remarkable career. Early on, Sauer renounced Semple's environmental determinism, setting in its place a new subdiscipline, cultural geography, that was deeply historical and alive to human agency but still very much tied to the physical environment through the landscape form. Later he became a pioneering environmentalist, scathing of modernity's destructive impact on the face of the Earth. **Eric Sheppard** follows, recounting **David Harvey's** (born 1935) remarkably successful attempt to remake human geography as a politically left-wing discipline directed to the pursuit of social justice. Harvey's Marxism was a founding part of 'radical Geography', a now very large and diverse field better known as 'critical human geography'. Harvey was among the first radicals in Geography who launched far-reaching criticisms of both Geography and a world that, in his view, it largely failed to question. Around the same time as Harvey's path-breaking project was underway, Irish-American geographer **Anne Buttimer** (1938–2017) was among those seeking to 'humanise' what, at times, seemed like an anaemic, statistically focussed human geography that traded in only hollowed-out, paper-thin versions of real people. **Federico Ferretti** writes about Buttimer's remarkable career as a self-styled 'humanistic geographer'. With others, Buttimer's aim was to focus research on emotionally equipped, reflective human beings as they experienced particular places in their life-world and throughout their life-course.

In a mix of biography and autobiography, **Michael Goodchild** (born 1944) narrates how the Swiss-American geographer **Waldo Tolber** (1930–2018) and, later, Goodchild himself, took traditional cartography into the modern era via the project of Geographical Information Science (GIS). The ability to map digitally the concentration, dispersion, partitioning, relative position and patterning of various phenomena at different spatial and temporal scales remains an absolutely vital aspect of life, integral to all manner of organisations, policies and practices. GIS is today a powerful enterprise, contributing to everything from environmental management and urban planning to surveillance and national security. As with mapping, place has long been a central preoccupation of geographers. **Jamie Peck** demonstrates that the British geographer **Doreen Massey** (1944–2016) made place central to her work, offering strikingly novel conceptualisations of it. In particular, she was concerned to rethink place in an increasingly interconnected world, a 'global sense of place' as she called it, along with its political and ethical implications. Massey's contribution added to the already rich body of work in human geography on the nature and importance of place.

If place is a key concept in human geography, so is landscape in Geography as a whole. Reacting in part to Sauer's work, in the 1980s **Denis Cosgrove** (1948–2008) helped to pioneer a new understanding of the nature and significance of human landscapes, exemplified especially by his work in Renaissance Italy. **Veronica Della Dora** examines Cosgrove's motivations, propositions and insights, which have left an enduring mark on landscape study in Geography and cognate disciplines. Borders and boundaries – their nature, maintenance, exact position and consequences – are also a major focus in human geography (and not just among political geographers). **Charles Pattie** shows how the late **Ron Johnston** (1941–2020) made key contributions to demonstrating how and why borders and boundaries matter. Johnston contributed to Geography in so many ways, but key among them was his work on electoral districts in Australia, the UK and the USA. This work illustrates the value of a wider geographical project to understand socially created divisions at different geographical scales. The partitioning of land, water, the airwaves and even near space surrounding Earth is an elemental and consequential multi-scalar process. And scale is precisely what **Andy Herod** discusses in his autobiographical account of his career as a pioneering 'labour geographer'. Scale is a central concept across Geography as a whole, though it is variously employed and thus complex. Herod's sustained work has advanced understandings of scale as 'socially constructed' and deeply political rather than objectively given and merely technical. While other understandings of geographical scale exist, Herod's work has done much to demonstrate how and why scale matters.

The last five autobiographical chapters in Part II introduce readers to other important geographical projects. **Tim Cresswell** has spent much of his career examining forms of human mobility, how they arise and why they are of consequence. The movement of people, goods, information and so on between places and countries is an elemental preoccupation of human geographers, while physical geographers have long been interested in cycles, flows and processes that operate across scales. Cresswell's work has sparked new insights into liberty, social power and inequality, raising profoundly important questions about who gets to move when, where and for how long. **Geraldine Pratt** is also concerned with mobility, but in the context of the difference that the proximity (or concentration and co-location) of people makes. Her pioneering research into women's employment, women's domestic roles and their access to jobs (both near and far) contributes richly to a long-standing geographical project: to make clear how 'nearness' and 'farness' shape capabilities for action and ultimately for a good life. Pratt's work on gender roles, identities, social class and female migration reveals complex stories of the intertwining of geography with people's life opportunities and trajectories.

Gary Brierley then explains his evolution from a river scientist to a reflexive, socially aware and ethically conscious investigator who refuses to 'objectify' rivers or to seek some 'neutral' perspective on how they operate. His most recent research as a 'critical physical geographer' shows that it's difficult to separate the realities of the physical environment from the particular cognitive and normative values we bring to bear when making sense of the world. His project, pursued in two former settler colonies (Australia and now New Zealand), is to question the merits of a supposedly dispassionate scientific perspective. This reveals that physical geographers must take *their own* human geographies very seriously when considering how and with what effects they create knowledge about the material world. **Mike Hulme**, once a leading climate scientist, is very aware of this fact. His chapter explores how Geography – a discipline to which he returned around 15 years ago – usefully allows knowledge of the natural world and people to be combined in creative, often dissonant ways that ensure scientific insight is properly contextualised. Hulme resists any suggestion that the world can be seen through a single lens, however large and clear it may seem to be.

Finally, related to Brierley's chapter, **Sue Jackson's** Australia-based research has also focussed on water. A society-environment geographer, she was an early contributor to new thinking about natural resource management in countries with displaced, dispossessed and discriminated-against pre-colonial minorities. Her project is to make just social space for indigenous Australians so they're taken seriously as knowledge-holders and rights-bearers when sharing water and maintaining water quality. A long-term fieldworker who moves in and out of universities, Jackson's research grounded so-called 'post-colonial' and 'de-colonial' arguments in a range of complex local-scale geographical contexts. Her work, in part advocacy as much as research, gives voice to minority perspectives and highlights how dominant social groups mistake their own understandings for truth and 'common-sense'. There's more than one way to apprehend the realities of geography, both in Geography and the wider society.

Part III comprises two biographical and three autobiographical chapters examining how geographical insight has been made to matter in the wider world. **Trevor Barnes** explores British geographer **Dudley Stamp's** (1898–1966) effort to forge an applied geography; that is, his use of geographical knowledge instrumentally to solve present problems and shape the future. In Stamp's case, it was done by harnessing the power of the nation state and large international organisations. Then **Luke Bergmann and Trevor Barnes** detail the pioneering 'action research' pursued by **Bill Bunge** (1928–2013) and his collaborators in the late 1960s and early 1970s. Whereas Harvey sought intellectually to radicalise Geography, Bunge aspired to do so in practice, becoming a community activist working with marginalised black Americans in inner-city Detroit. Whereas Stamp was among those geographers who tailored their work to the agendas of national governments, Australian human geographer **Kathy Gibson** (with **Julie Graham** [1945–2010] and others) seeks to support communities working outside the mainstream of politics. Her project to make visible and support grounded 'alternative geographies' constitutes one of the most significant activist ventures within critical human geography during the last 25 years. It speaks of, and for, different ways of living on Earth during a period when capitalism and other systemic forces dominate most people's lives. The project seeks real and metaphorical spaces in the interstices where seeds of wider possible change are planted.

Less common are successful attempts to shape both public policy and public understanding. Given this, the efforts of **Jamie Woodward** (a British physical geographer) and **Danny Dorling** (a British quantitative human geographer interested in socio-geographical inequality) are notable. Woodward uses his credentials and experience as an accomplished physical geographer to conduct policy-relevant research into microplastics pollution in UK rivers. This project of 'speaking truth to power' has been highly successful. Since around 2015 Woodward and his co-workers have demonstrably affected political debates in Westminster (London) and raised big questions about whether private water companies effectively uphold the public interest in clean water. It's a project Woodward stumbled into, even though other geographers were already policy focussed. Finally, Dorling is among the few academic geographers to enjoy a public profile in the news media because he chooses to write books for a general audience about poverty, wealth and equality in the UK and beyond. His project to be a 'public geographer' in Britain is sustained and spirited. It exposes him to a sometimes hostile 'public sphere' (where reasoned debate is often eclipsed by sound bites and vitriol) and yields lessons about communicating research beyond the discipline of Geography. As with Woodward, Dorling's 'public geography' isn't a project he planned from the start, but rather one he ended up investing a lot into because he recognises that popular understandings can benefit from more expert ones.

Canonisation, hagiography, Whig histories and the evolution of projects

Having explained the rationale for and focus of the book, we will clarify a few important issues to avoid misunderstanding.

First, this is emphatically *not* a book of the 'greatest geographers'. There can be no roll-call of the 'best geographers ever' because there's no universal criterion of 'bestness' and no 'complete' geographer either. Everyone is different. Even for a given project there's no pre-eminent geographer, though clearly some have been very influential. While each of the geographers profiled in the book is, or was, highly accomplished, there are (or were) others pursuing similar projects across Anglophone Geography. Indeed, unlike some disciplines – such as Economics – where there's taken to be a set of widely recognised thought-shapers, Geography's heterodoxy prevents any easy canonisation of a select few. While the likes of Chorley, Haggett, Harvey and Massey may have been unusually formative in their efforts, there are so many interesting and influential others in Geography to whom we could have instead devoted chapters.

For those geographers whose work *is* profiled in this book we're careful not to idealise or exaggerate their efforts. We encourage readers always to be constructively critical. Relatedly, we note demographic features of the discipline that constrain, and still limit, its intellectual vibrancy and bandwidth. Until the 1990s, Anglophone Geography was largely male and overwhelmingly white. Even today, it's largely white, though thankfully that's changing as ethno-cultural diversity is increasing among the discipline's student and professional body. Happily, far more of its practitioners are women and non-binary including within some of the hitherto masculine sub-disciplinary bastions like economic geography or some branches of physical geography.

Without doubt, the demographics of Geography at any one time influence the sort of knowledge and insight geographers are able to produce. In the Austrian philosopher Ludwig Wittgenstein's famous maxim, 'the limits of our language are the limits of our world', with language reflecting wider cultural particularities. The more we diversify the membership of Geography's research and teaching community, the more expansive those limits become. We're better able to represent the questions, ideas and creativity found in the wider societies that all academic disciplines at some level seek to serve. A brilliant example is the project pursued by Bawaka Country in Australia. Starting over a decade ago, a group of academic geographers and indigenous people have together sought to be a voice of Country in order to reconfigure how Anglophone Geography researchers and students think about their own perspectives and obligations in our diverse but unjust world. Their work aims to constructively decolonise Geography (e.g., see Bawaka Country *et al.*, 2015).

Secondly, this book contains chapters about the lives and works of a number of geographers who are now dead – some a long time ago, others much more recently. This raises questions about the value accrued today from looking at the past. We certainly want to avoid a 'Whig history' of untrammelled progress, the result of the cumulative efforts of supposedly 'great geographers' who are to be venerated. Our aim, instead, is to shed light on different, often influential, context-specific efforts to advance the cause of big and small G geography. By examining these efforts, we can learn lessons about how geographical research, teaching and outreach is both a product and producer of the prevailing time-space conjunction. Geography is a discipline *fully immersed in the world*: it's affected by the world that it seeks in good faith to understand. As we look back to the efforts of geographers in the past, we discern the reasons, strategies and tactics they employed to advance Geography and geographical perspectives. Those pursuing projects today likewise employ reasons, strategies and tactics both to justify and 'sell' to others their projects. The projects discussed by the living geographers in this book – such as those of Sue Jackson and Jamie Woodward – aren't *intrinsically* 'better' than those pursued by, say, Ellen Semple or Dudley Stamp. Instead, we might judge them 'better' because our present values and circumstances make us sympathetic while at the same time disposing us to be critical of efforts in the past to make G/geography matter. The past often appears from the perspective of the present to be a 'foreign land' where people do things differently. Context matters enormously, making the value of what geographers do necessarily contingent (Stoddart, 1981).

Thirdly, what might seem like 'old' projects live on today in new and often vigorous ways. In other words, readers shouldn't think that the projects of, say, Sauer or Bunge began and ended with these individuals, their collaborators and their doctoral students. On the contrary, the projects evolved in both specific and general ways by all sorts of other human and physical geographers. Sauer's environmentalism is now almost obligatory across the entire discipline, and Bunge's activist-political sensibility is one of the defining features of contemporary human geography as a whole. There are also crossovers from human to physical geography. For example, the 'critical physical geography' of Rebecca Lave and fellow travellers is a case in point. Without Harvey's project to radicalise human geography 50 years ago, the well-springs of insight that eventually contributed to Brierley's evolution away from dispassionate, asocial 'science' simply wouldn't have existed. Yet Harvey never said much about physical geography, either then or now. This shows how one project bleeds into and alters others in unexpected, slow yet often fruitful ways.

How Geography is made to matter: A framework for understanding

The geographers profiled in this book made, and continue to make, G/geography matter in a range of registers and situations. While each of their stories is individual, using more general concepts it's possible to identify commonalities across the profiled geographers in how each made their work worthy of attention. These concepts will help readers to join the dots between otherwise very different people, both past and present. To repeat an earlier point, we take it for granted that "our discipline cannot be understood independently of the societ[ies] ... in which the practices of Geography are embedded" (Harvey, 1984: 1). But we also maintain that geographers' efforts aren't reducible to only the circumstances in which they work. There's a degree of autonomy or agency in play, supported by the institutionalised principle within the university called 'academic freedom' and embedded contractually in the awarding of academic tenure. Tenure means that academics who hold it cannot be removed from their positions except under extraordinary circumstances. It means that geographers can determine their own research and teaching activities without pressure from others, from Departmental Heads to Ministers of Higher Education

We can divide the factors influencing academics and the knowledge they produce into two kinds: *internal* and *external* (Table 1.1).

Internal factors stem from the institutional character of academic disciplines along with the university and individual departments in which they are housed. That internal institutional setting comes with specific rules and procedures, designated practices and expectations, normative values about what is appropriate (or not), and criteria of success or failure. In turn, through a process of socialisation, those rules, norms and practices shape individuals operating within the institution, both staff and students, shaping the geographical knowledge produced. It begins as an entering undergraduate, continues through post-graduate study, and is fully realised as members of staff, lecturers and pro fessors, who then enforce such standards on others as well as themselves. Disciplines discipline their practitioners.

Table 1.1 Academics and factors shaping their activities

External Factors		*Internal Factors*
Authoritarian Political Regime	*Democratic Political Regime*	*University Career*
Justification of ideology, e.g., Lysenkoism	Government institutions use budgets to generate knowledge and training of personnel to further interests of the nation, e.g., military funding of research in Cold War America	Responses to the university's rules for academic advancement
Exile or 'disappearance' of academic critics of the regime, e.g., Chilean geographers after the Augusto Pinochet military coup in 1973.	State auditing and evaluation of universities, e.g., the UK's periodic REF	Funding Time Peer review publication
Practical knowledge to support state projects, e.g., Chinese geographers used to help plan the Chinese state's One Belt, One Road initiative.	Opportunities exist for academics to use their research to shape public understanding or to key in to the goals of communities, businesses and other organisations	Colleagues/students within discipline and beyond Access to travel (money and time) Own interests, curiosity and creativity

Internal institutional factors influence geographical knowledge production in multiplicitous ways. They can be seemingly benign, a result of happenstance, such as the university you choose to attend, hearing a charismatic lecturer, selecting a particular graduate supervisor, going to a certain conference or workshop. Less benign is that these internal institutional factors can also be deployed as part of a larger strategy to advance an academic career. This is the argument of the political geographer, Peter Taylor (1976), who referenced the example of the shift in geographical knowledge from regional description to scientific rigour (the "quantitative revolution"). Taylor argued that, given the institutional rules of the academy around career advancement, each new generation of scholars has a vested self-interest in differentiating itself from the previous generation by adopting a very different intellectual approach that shows off their originality, skill, intellectual independence and drive, while also casting doubt on what went before. By so doing, the new generation is rewarded by quickly moving up the career ladder, while the older generation become has-beens. In Taylor's example, the quantitative "revolution" in geographical knowledge was carried out almost exclusively by young ambitious men in their twenties and thirties (including Taylor). They separated their 'new' Geography from the 'old' kind by aims, vocabulary, methods, philosophical justification, styles of presentation, presentation venues and journals. It made the old form of Geography look old-fashioned, anti-modern, anti-intellectual and irrelevant. Consequently, the "young Turks", as they were called, leapt ahead in their careers, by-passing the otherwise slow, grinding process of academic professional advancement. Key players like Peter Haggett and Brian Berry were full professors by their early thirties.

External factors are, like internal ones, numerous. They form the larger context outside the university that influences inside academic knowledge. Because they are contextual, such factors necessarily change over time but are also quite different according to place and spatial scale. Pressing global issues like, most recently, Covid, or biodiversity loss, or movements of refugees, or the continuing climate crisis brought about by global warming, clearly have shaped knowledge that geographers produce. Other times the issues are more local or regional, such as the urban-poverty crisis of American cities of the 1970s and 1980s, the AIDS epidemic in Africa during the 1990s and into the new millennium, and the contemporary recognition of the marginalisation of indigenous peoples in locations like Brazil, Mexico, Canada and Australia.

Reflections on the influence of external factors on geographical knowledge have often emphasised the formative role of the state and its political make-up and aims. At one extreme are authoritarian states. Under them, academics not only lose their academic freedom but occasionally lose their lives. Academics are sometimes compelled ideologically to support and justify the state and to contribute to practical projects of state maintenance. For example, following a military coup in Chile in 1973 led by General Augusto Pinochet (ironically a geographer), the Institute of Geography at the University of Chile was hollowed out. 38 of 47 geographers were removed. Their work was judged either in conflict with, or irrelevant to, the new authoritarian military regime. Two of the geographers 'disappeared' altogether. Consequently, after the coup, "the face of Chilean geography changed dramatically" (Caviedes 1991: 360). Those geographers who were kept specialised in military geography, or else did applied work and were technically proficient. A more contemporary example within another state authoritarian regime is the work of Chinese geographers at the Institute of Geographical Sciences and Natural Resources Research, Chinese Academy of Sciences. Many of them have been enlisted in the state project of the One Belt, One Road global infrastructure initiative that's a centrepiece of Xi Jinping's Chinese foreign policy.

Even in democratic, so-called 'free' societies, institutional power from central government influences knowledge that academics produce. Frequently steering knowledge is money, the funding of research. A classic example was the use of state funding in the United States during the Cold War to steer the production of particular kinds of academic knowledge. Much of that funding was distributed by the military such as the Office of Naval Research and national security agencies like the Central Intelligence Agency (CIA). They were primarily interested in knowledge that would facilitate and realise their interests. For example, large sums of money from the CIA and the US State Department funded so-called Area Studies programmes within American universities, both public and private. As interdisciplinary programmes that included geographers, Area Studies generated knowledge and expertise in regions which the US as a post-war superpower knew little about but wanted to know more – the Soviet Union and Eastern Europe, South-East Asia, the Middle East. University Area Studies programmes not only provided information to the state but also trained personnel for positions within it such as the diplomatic corps, the military and covert national security organisations.

Governments continue to use funding to influence the production of knowledge. Nationwide evaluations of public academic programmes, such as the former Research Assessment Exercise (RAE) in the UK (now called the REF) that occurs every five years or so, clearly influence not only the character of universities and academic life, but also the knowledge that's produced and its form. For example, the increasing weight given to "impact", one of the three criteria used by the REF to judge academic outputs, has meant social relevance and consequence have become important shapers of the research projects that British academics, including geographers, now undertake.

Conclusion

To recap, "Geography is ... about what is where and where is what; and why and when, and who and how" (Dorling & Lee, 2016: 6). If Geography were a house it would comprise an unusually large number of rooms, connected by very many stairways and corridors for those willing to move around, and with numerous windows and doors opening out to the wider world both near and far. We might think of the rooms – varying in size, shape and furnishings – as the projects pursued by geographers over time as they look at the world outside. There is much internal renovation of the house of Geography – walls are removed, basement rooms dug out and new passages created, but fresh partitions are also installed. Unlike conventional buildings, there's no architect and the 'house' far from consistent, inside or out – this despite many external factors bearing down on the discipline. Like unruly bees, geographers variously cooperate but then go off separately on their own travels. As people move in and out of the house – students, professional geographers and sometimes people from other disciplines – so they are shaped by, and shape, those parts of the structure in which they spend time. This is what makes the house such a vibrant place, more so now – arguably – than at any time in Anglophone Geography's approximately 140-year history.

Geography will never be a cohesive discipline, with practitioners united in their aims and practices. Geographical knowledge and insight will always matter in more than one way outside the university world. This is as it should be. The challenge is to ensure that Geography is an arena where the quest to understand the world is pursued openly, critically and with full recognition that there's no one final geographical truth awaiting discovery. Between an impossible unity of understanding and unhelpful Balkanisation, the metaphorical house of Geography is always full of life, unfinished and in the process of being made.

* * *

A note for instructors: How to use this book

Academics who devise lecture and seminar series about the aims, content, evolution and wider impacts of Geography will, we hope, find this book a useful resource. As noted in the introduction, in our experience students are more engaged by stories about the lives and works of people than by overviews of isms, ologies and paradigms. In our experience, when asked to write set essays or sit unseen written examinations, student responses often float around on high, misusing terms like 'ontology' and 'positivism' and lacking assurance in the complex thought-scapes of fields like feminist geography and qualitative GIS. That suggests a more targeted approach is advisable.

Our framing device of 'the project' allows readers to sample some of the people whose work is profiled in this book. But each chapter is also about more than the individuals in question. With some 20 chapters to choose from, instructors could home in on, say, seven or eight of the people and projects, possibly supplemented by lectures or seminars focussed on other geographers chosen by the instructor. The people-project focus allows specificity while opening out onto wider currents of geographical thought and practice. Each chapter ends with an essay task that instructors can use, or adapt to their own needs. Similar tasks can be replicated for other geographers and projects not covered in *Making Geography Matter*. Clearly, older geographers of influence tend to have a secondary literature written about them that increases options for student study.

By contrast, some leading living geographers may not (yet) have been the subject of much serious published commentary. In these cases, as we show at the end of Danny Dorling's chapter (among others), there are still effective ways to devise rich student assessment of both the person and their project.

Instructors may also want students to compare and contrast projects, using individual geographers as exemplars. For instance, the impact of Jamie Woodward and colleagues' research into microplastic pollution of British rivers arises from its scientific credentials as much as the project's specific empirical content. Compare this with Gary Brierley's 'more-than-science' approach to understanding and managing rivers in Australia and New Zealand. What are the respective aims, merits and weaknesses of the projects? And how does Brierley's critical physical geography compare with Chorley and Haggett's belief that Geography's entire subject matter can be analysed using a common scientific approach to detecting order and variety at the Earth's surface?

Instructors can focus on projects not considered here, drawing on one or more geographers. For instance, the modelling project (conceptual, physical or computational) has been really important across Geography since the 1960s; then the adaptation of statistical techniques to geographical questions and data-sets is another key project that's evolved in remarkable ways. We could go on. As a heuristic, it may help to bundle projects into those focussed on: (i) the proclaimed goals of Geography; (ii) the key dimensions of a 'geographical perspective' on reality or key geographic preoccupations (e.g., the concentration, co-location and dispersion of phenomena or the shaping of spatial form by socio-economic processes and vice versa); (iii) key investigative tools (e.g., hypothesis testing or theory formation); (iv) the content and objectives of geographic education; and (v) the communication and translation of geographic research outside universities.

Another approach instructors can take to supplement this book's contents is to identify interesting geographers and consider what project(s) they pursued, how and why. For instance, influential Anglophone geographers not considered in this book include Louise Amoore, Kay Anderson, Victor Baker, Mike Batty, Brian Berry, Piers Blaikie, Isaiah Bowman, Bruce Braun, Neil Brenner, Brett Christophers, Andrew Clark, Nigel Clark, Paul Cloke, Stuart Elden, Sarah Elwood, Kevin Grove, Stewart Fotheringham, William Garrison, Peter Gould, Derek Gregory, Julie Guthman, Torsten Hagerstrand, Peter Hall, Cole Harris, Gillian Hart, Richard Hartshorne, Lesley Head, Richie Howitt, Matt Huber, Peter Jackson, Mike Kirkby, Rob Kitchin, Owen Lattimore, Rebecca Lave, Billie Lee Turner, David Ley, Paul Longley, Jamie Lorimer, Geoff Mann, Mark Maslin, Linda McDowell, Donald Meinig, Don Mitchell, Gunnar Olsson, Tim O'Riordan, Jamie Peck, Jonathan Phillips, Allan Pred, Jenny Robinson, Paul Robbins, Andrew Sayer, Stanley Schumm, Eric Sheppard, Neil Smith, Susan Smith, Ed Soja, David Stoddart, Griffith Taylor, John Thornes, Nigel Thrift, Yi-Fu Tuan, William Warntz, Sarah Whatmore, Michael Watts, Gilbert White, Alan Wilson and Kathryn Yusoff - but there are so many others. And then there are people who advance geographical understanding (quite how well is open to question) having been trained in different fields – such as Robert Kaplan and Tim Marshall, or the British earth system scientist Tim Lenton (at Exeter University). These people could also be the focus of lectures, seminars and assessment tasks, with their projects set in the context of the wider societal discourse about (or of) geography that the discipline of Geography aims to influence.

In sum, we believe there are ways to make this book an active resource feeding into student assessments, so that their learning can be formally evaluated.

References

Bonnett, A. (2008) *What Is Geography?* (London: SAGE).
Caviedes, C. (1991) 'Contemporary geography in Chile: A story of development and contradictions', *The Professional Geographer* 43, 4: 359–362.
Cloke, P., Crang, P. & Goodwin, M. (2013) *Introducing Human Geographies* (London: Routledge).
Country, Bawaka et al. (2015) 'Working with and learning from Country: Decentring human authority', *Cultural Geographies* 22, 2: 269–283.
Couper, P. (2015) *A Student's Introduction to Geographical Thought* (London: SAGE).
Cox, K. (2014) *Making Human Geography.* (New York: Guilford Press).
Cresswell, T. (2012) *Geographic Thought* (Malden: Wiley-Blackwell).
Dorling, D. & Lee, C. (2016) *Geography* (London: Profile Books).
Harvey, D. (1984) 'On the history and present condition of Geography', *Professional Geographer* 36, 1: 1–11.
Herbert, D. & Matthews, J. (2008) *Geography: A Very Short Introduction* (Oxford: Oxford University Press).
Inkpen, R. & Wilson, G. (2013) *Science, Philosophy and Physical Geography*, 2nd ed. (London: Routledge).
Johnston, R. J. & Sidaway, J. (2015) *Geography and Geographers*, 7th ed. (London: Routledge).
Livingstone, D. (1993) *The Geographical Tradition* (Oxford: Blackwell).
Nayak, A. & Jeffery, A. (2011) *Geographical Thought* (London: Routledge.
Stoddart, D. (1981) *Geography, Ideology & Social Concern* (Oxford: Blackwell).
Taylor, P. J. (1976) 'An interpretation of the quantification debate in British Geography', *Transactions of the Institute of British Geographers* 1, 2: 129–142.

Part I

Making Geography

2 Absolute beginner?

Halford Mackinder and the popularization of geographical knowledge

Emily Hayes

Portrait of Halford Mackinder (1861–1947). Halford J. Mackinder, Maull & Fox, [Date unknown]. Royal Geographical Society-with Institute of British Geographers (RGS-IBG) Picture Library. Contact email: Images@RGS.ORG

Introduction

Halford Mackinder's 60-year career was intimately bound to the British Empire, and laced with imperial projects, ideals and interests. While Mackinder was certainly an imperialist, he was not just an imperialist and he denied being a political jingoist. Mackinder also constructively served the changing geographical educational project both

DOI: 10.4324/9781003343240-3

in thought and in practice. This includes ventures that only partially succeeded, or even failed. While I am interested in Mackinder's role in defining the discipline and shaping geographical pedagogical practice, I want to see these as the consequence of the sometimes competing, and sometimes co-operating, institutional, academic and socio-political interests and challenges that he faced throughout his life.

Section 1 unpacks Mackinder's family background, upbringing and university education. Section 2 discusses perhaps his most famous paper, his 1887 'On the scope and methods of geography', originally delivered as a lantern-lecture[1] at the Royal Geographical Society (RGS) in London. Founded in 1830, the RGS was established to promote and sustain geographical knowledge. Its original Fellows were primarily travellers, explorers and members of the armed forces, but over time a growing number were natural scientists like Mackinder. Section 3 discusses Mackinder's Oxford University Extension (UE) lectures around the UK and his plan for a dedicated geographical institute. Section 4 explores Mackinder's interest in the German theory of anthropogeography and his expedition to Mount Kenya in the 1890s. Section 5 is about his historical geography and geography textbooks. The final section summarizes Mackinder's geopolitical theorizations and political career.

The larger argument of the chapter is that Mackinder's primary concern was promoting a geographical point of view as well as developing and sustaining an international, rather than exclusively national, geographical community and tradition. This is sometimes forgotten by other accounts of Mackinder that treat him only as an imperialist, with his writings in geopolitics considered simply a means to that end. This chapter suggests there is more to Mackinder once we study his full life and biography.

Mackinder's childhood, family and educational background

Born in Gainsborough, a small town in England's northern Midlands, Mackinder's father was a country doctor and his mother a home-maker. His family were to some degree cosmopolitan and outward looking. Mackinder's father had studied in Paris and his practice was sufficiently successful for the family to afford a French governess. After attending a local grammar school, in 1874 Mackinder was sent to the newly founded public boarding school, Epsom College in Surrey, south of London. In the 1870s science education and dedicated science teachers were a rarity. Mackinder was privileged to be taught science at a time when such subjects were not yet integrated into all schools' curricula. He was taught botany by William Thysleton Dyer (1843–1928), a Natural Science graduate who later became director of Kew Gardens (Blouet 1987, 17). As a student, Mackinder enjoyed the outdoors, and was also an imaginative thinker. His prize-winning essays 'Geological Epsom' and 'A glimpse of A.D. 1950,' which envisioned a future world transformed by technological innovations, show already his early geographical interests (Hayes 2019, 3). A technology that caught his imagination was the magic lantern, a device that projected images onto a screen. As a teenager he had been struck by its use in a presentation by the natural philosopher, John Tyndall (1820–1893), in London at the Royal Institution, then a centre of experimental science and public lectures (Hayes 2019).

The lecture by Tyndall would have been very different from Mackinder's experiences at school. British education was dominated by the church and clergymen teachers. Geography was barely taught at any level. Nor was it viewed as a serious subject. In 1878 a teacher from Nottinghamshire described Geography as 'drill, drill, drill, to get it well done' and 'depending to a large extent upon the memories of children, as there are a large number of proper names (mountains, lakes, seas, rivers, towns, etc.) not previously known' (Wing 1959, quoted in Andrews 1986, 230). Teaching was typically dull and uninspiring and pedagogical materials were also limited.

That began to change as the profusion of geological, and later biological, knowledge challenged the biblical creation story and Christian theological notions of time (Stoddart 1966, 686). Also significant was the publication of Charles Darwin's (1809–1882) *On the Origin of Species by Means of Natural Selection, or The Preservation of Favoured Races in the Struggle for Life* (1859). It brought about 'an intellectual revolution', leading to 'an intellectual gulf between the younger generation and their parents' (Pease 1916, 17). Scientific practitioners became infamous as they challenged the authority of the church and "natural theologians", those who believed in some version of the biblical narrative. Followers of Darwin also sought to establish the value of science as a general approach, as a new and environmentally sensitive form of knowledge-making that should be used to refurbish teaching methods across the education spectrum, to make subjects relevant and exciting. Mackinder grew up and studied during this period of transition. By the late nineteenth century, the belief that science alone offered solutions to social problems, and an ethical basis for daily life, became widely accepted (McLeod 2000, 150). In turn, that affected the conception and practice of Geography to which Mackinder later contributed.

Towards the scope and methods of Geography and beyond

In 1880 Mackinder received a junior studentship to study Natural Science at the University of Oxford. He specialized in physiology, that is, human and non-human anatomy (Coones 1989), taking classes in what today is Oxford's Natural History Museum (Figure 2.1). In 1883 he received a first-class degree in Natural Science. Unusually, he had the chance to study other subjects. He also took a second-class degree in history, and the Burdett-Coutts scholarship to study geology. As an undergraduate, he was also a member of the Junior Science Club, which organized 'friendly discussion[s] of scientific and other topics' (Junior Science Club, Bodleian 1895, 19) and included lectures by famous scientists. These varied courses and activities shaped Mackinder's geographical outlook, including his belief that geographical education could positively transform society (Blouet 1987, 26).

Apart from an education in natural sciences that was increasingly influenced by Darwinian thought, Oxford also exposed Mackinder to a diverse range of philosophies and intellectual opinions and practices. For example, he encountered late eighteenth-century and early nineteenth-century German philosophical texts that had been recently translated into English (Mander 2011; Hayes 2021a, 7). The university also introduced him to socially progressive theorists and activists. For example, much discussed was the Christian, proto-socialist philosophy of the art critic and artist, John Ruskin (1819–1900). In the same vein, he was also exposed to the radical political activist William Morris (1834–1896), the founder of the Arts and Crafts movement, who extolled the virtues of 'constructive socialism' (Mulvaney 1985, 51). And he also attended lectures by the new Reader in Anthropology, Edward Burnett Tylor (1832–1917), who judged indigenous cultural artefacts as elements embedded within the evolution of human groups that ranged on what was then seen as a provisional theoretical continuum going from "savagery" to "civilization". Yet Tylor also helped develop the concept of "culture" and believed in the 'psychic unity' of humankind (Mann 2022, 159–160). The larger point, as Driver (2010, 150 and 154) suggests, is that during this period, when Mackinder was a student, socialism, science and imperialism were intellectually linked and entangled. As a young man, Mackinder had a front-row seat at one of the prime sites where this was unfolding.

Figure 2.1 The central court of the Natural History Museum, Oxford after it opened to the public. Illustrated London News, 6 October 1860. N, 2288 b.6, vol. 37, 310.

Mackinder, however, was not the first, or only, geographer in mid-1880s Oxford. John Frederick Heyes (1857–1937), a natural sciences graduate specializing in chemistry, was also active. Heyes had connections with the RGS before Mackinder, participating in the Society's Schools Medals Prize scheme. He had also authored polemical articles in student newspapers lamenting both politicians' lack of geographical knowledge and the absence of geographical teaching at Oxford. In 1887, Heyes even wrote a geographical manifesto, *Geosophy: An ideal with practical issues* (Hayes 2021a, 2021b) calling for a comparative cultural, religious, philosophical and ethical practice of Geography.

After graduation, Mackinder worked for the Oxford University Extension (UE) scheme. UE aimed to provide university education to working- and middle-class audiences of men and women who otherwise would be excluded. Mackinder's Extension work became for him a form of fieldwork and practical training. Little is known of Mackinder's UE lectures on physiography, political economy and geography. But he was a popular lecturer and incorporated into his geographical teaching Darwinian understandings of organisms' adaptation to their environments. The success of his extension lectures brought him to the attention of the RGS at a time when it was seeking to improve its scientific reputation.

In 1887, Mackinder 'as a young revolutionary' (1931, 323) was invited by the RGS to give a lecture (Figure 2.2). It was titled, 'On the scope and methods of geography' and illustrated by slide images projected from a magic lantern (Hayes 2018). Drawing on what he learned at Oxford, he defined Geography's scope and methods broadly, appealing to different understandings of the subject, and linking it to ideas from a number of adjacent fields such as anthropology, biology, geology and physics. Mackinder's argument was based on the Darwinian evolutionary understanding that 'communities of men' were 'units in the struggle for existence, more or less favoured by their several environments' (1887, 143). He also distinguished Geography from the discipline of Geology. The latter was a respected and authoritative science that was essential for providing evidence in support of the Darwinian theory of natural selection, showing that species adapted to environments over extended periods of time (Stoddart 1980, 199–200). '[T]he geologist looks at the present that he may interpret the past; the geographer looks at the past that he may interpret the present', Mackinder argued (1887, 146). Thus, Geology and Geography shared the same material base, but asked different questions and investigated them with different aims in view.

Figure 2.2 Image of the Royal Geographical Society, Saville Row. Edward Whymper (1840–1911), 1, Savile Row [the RGS], from a wood engraving, [1894] in H. R. Mill, The Record of the Royal Geographical Society, 1830–1930, London: Royal Geographical Society, 1930.

[Copyright RGS-IBG]

Geology had professionalized and specialized into sub-fields before Geography. It was geologists' study of landforms 'which had previously had widest popular appeal and least technical content that the geographers proposed to usurp' (Stoddart 1980, 196). However, unlike Geology, Mackinder argued that Geography could link arts and sciences. This was potentially a major contribution of the discipline. The "physical" and "political" should be connected, unified. For the former was the basis for understanding the latter. Mackinder contended that Geography was the discipline to do that; it could overcome the abyss between the natural sciences and humanities, which hitherto had 'upset the equilibrium of our culture' (Mackinder 1887, 145), referencing a wider public debate taking place in Europe and in the UK.

Mackinder relied heavily on the works and thoughts of others, especially on the writings of the polymath, and cousin of Charles Darwin, Francis Galton (1822–1911). One of the reformers of geographical education at the RGS, in 1886 Galton categorically stated the political value of teaching people how to 'sympathise with other races, and to regard them as kinsmen rather than aliens' (1886, 337–338). But he was also a proponent of social Darwinism, the idea that some classes and "races", a vaguely defined term, could be arrayed hierarchically from better to worse in terms of their ability to adapt to specific environments (climates, temperature zones, altitudes and subsistence opportunities). In parallel, Galton had a stronger grasp of developments in physical science and philosophy. For him Geography was an art rather than a science (Mackinder 1887, 165). Galton's views on language, perception, visualization and teaching methods fashioned the scope and methods of Mackinder's Geography.

So, while not wholly original, '[S]cope and methods' nevertheless opened academic space for university Geography between the well-established disciplines of geology and history. Mackinder argued that the 'new' Geography synthesized the mass of information gathered by both the human and physical sciences. He also sought to shift the perception of Geography as a subject without method and aim, by vividly recasting it as a science of laws with a social purpose. Moreover, Geography would be a science conveyed in accessible language and images. His vision was literally embodied in his RGS multimedia audio-visual lecture performance, which combined lantern projections and spoken word:

> Imagine our globe in a landless condition, composed ... of three concentric spheroids – atmosphere, hydrosphere, and lithosphere ... Next introduce the third set of world-wide forces – the inclination of the earth's axis to the plane of its orbit and the revolution of the earth round the sun.
>
> (1887, 155)

He continued to develop his concentric metaphor throughout his life. The distributions of various features on the Earth's surface and the causes of those distributions were Geography's subject matter (1887, 160). In addition, his use of magic lantern projections to illustrate his lecture was radical, though this was deemed in some quarters to be of dubious scientific and pedagogical value.

Following the success of the '[S]cope and methods' lecture, in 1887 Mackinder was appointed Reader in Geography at Oxford.[2] In contrast to the UK, Geography was already well established in European universities. The courses that he taught in his first year at Oxford had already been trialed on UE audiences. His human geography university lectures were attended primarily by historians, but those on physical geography attracted few science students (Blouet 1987, 2008).

The Oxford University Extension scheme and a geographical institute

As Reader, Mackinder remained a UE lecturer, committed to democratizing education by building on what was 'begun in the elementary schools' (1890, 39). Via the expanding railway network he travelled some 30,000 miles (Gilbert 1951, 28), developing his 'showman' skills (Freshfield 1932, 2).[3]

Between UE teaching commitments, during the 1890s Mackinder travelled and gave public lectures in the UK, the United States and Europe. His writings included published talks that explored the historical connections between European geographical traditions. These papers discussed environmental determinism and anthropogeography, drawing on the work of the German Friedrich Ratzel (1844–1904) and his American student, Ellen Churchill Semple (1863–1932) (see Keighren in this volume). Carlotta Santini (2017) has traced the vein of cultural and historical relativism in Ratzel's thought. Ian Klinke's work on Ratzel takes a different course by assessing Ratzel's life, travels and works in relation to Cameroonian historian, Achille Mbembe's (b. 1957) studies of colonial violence as necropolitics, the politics of death. Beneath the 'language of cosmopolitanism and civilizational progress' that flowed into Mackinder's work, Klinke (2023, 18) discerns Ratzel's obsession with death. Alongside a vision of genetically similar human beings, he shows how Ratzel's entrenched racialized Geography justified exploitation, inequality and violence. After visiting Germany in 1895, Mackinder hatched a project to create a geographical institute (1895, 376–378), a dedicated space for the discipline.[4] Housing specialized professors and staff, the institute would provide purpose-designed lecture rooms, a library, a map collection, physical models, diagrams, views, photographs, globes, and specimen collections and other teaching equipment (1895, 377).

The relationship between Mackinder and the spaces and instruments for geographical pedagogy requires elaboration. Turn-of-the-century classrooms often had 'whitewashed walls full of crumpled maps, worn-out charts, and other scraps of paper' (Dams, Depaepe and Simon 2014, 52). Few schools offered Geography classes or were suitably equipped to do so. Debates raged about teaching methods, with some believing that children's minds were at risk of degenerating through the provision of too many images, or materials that were too easy for them. In 1890, Mackinder insisted in a letter to Francis Galton that '[I]t is wrong to teach the broad matter of geography from anything but a globe' (1890, 1). To that end Mackinder developed what he called an epidiascope projective device (Figure 2.3). It would consist of a number of billiard ball-sized 3D globes depicting a range of topographical features. He believed the device would give teachers the 'power of turning the globe in any direction and causing the image to rotate correspondingly' (1890, 2). Though the device was not built, it showed Mackinder's continual concern for the teaching of Geography. In 1893, Mackinder was one of the founders of the Geographical Association, established to assist Geography teachers by sharing teaching materials, including lantern slides (Walford 2001, 63).

Anthropogeography and Mount Kenya

In the 1890s environmental determinism as promoted by Friedrich Ratzel (1899 [1882]) and Ellen Semple (1911) surfaced in Mackinder's work. He hoped that human geography would become 'anthropogeography or the geography of men'. Mackinder believed that the anthropogeographer was 'the most typical and complete of geographers'. Although

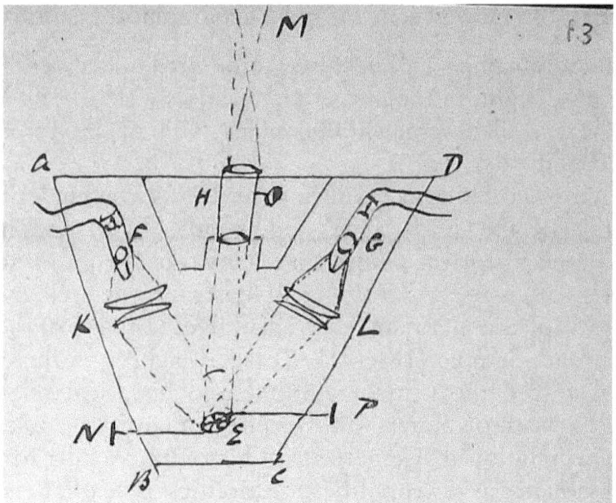

Figure 2.3 Sketch of Mackinder's projection device (1890).

Mackinder, Halford John to Francis Galton, 2 May 1890, p. 3, GALTON /3/3/13/5 in Galton Papers, UCL [Copyright Halford Mackinder estate].

he said that each "race" was shaped by its history, Mackinder believed over the long term in the power of environmental determinism: 'nature reasserts her supremacy' over a population. Moreover, he believed that anthropogeography's methods were 'essentially the same as those of physical geography'.[5] Human and physical phenomena were alike, he argued: 'The streams of commodities over the face of the earth, considered as an element in human environments, present many analogies to the currents of the ocean or the winds of the air' (Mackinder 1895, 375).

Mackinder had the chance to test Ratzel's anthropogeography in the field when in 1899 he led an RGS expedition to Kenya, then part of the British Protectorate of East Africa. The business of exploration was different from travel and different from the practices of academic Geography. Yet, each overlap and each pertain to the history of Geography. The aim of Mackinder's expedition was to prove that he could be both an explorer and an armchair geographer. The Kenya expedition was Mackinder's first extended fieldwork abroad and with peoples with whom he was unfamiliar. His previous travels clearly had not prepared him for this project, which was not only a failure in terms of geographical ideals and education, but resulted in tragic human consequences. His immediate objective was to climb the summit of Mount Kenya, which would be the first recorded European expedition to Africa's second-highest peak, which today is known locally in Kenya as Kirima Kiri Nyaga (Nyamanga, 2022). While the Kenya expedition therefore had symbolic value for the RGS, which was a stakeholder, it was poorly planned, early on exhausting its food supplies. Moreover, numerous specimens that were collected along the way were also lost, rendering the expedition of limited scientific value. Even worse, eight indigenous porters whose names were recorded were sentenced to corporal punishment, likely by Mackinder, and shot (Kearns 2021, 13). In light of this, and greater attention to postcolonial and decolonizing geographies, a lecture theatre in the Oxford School of Geography which once bore Mackinder's name was renamed (Oxford and Colonialism network, SGE, [2022]).

On his return from Kenya in 1899, Mackinder, together with the Scottish physical geographer, Andrew. J. Herbertson (1865–1915), established the Oxford School of Geography. The school admitted women, a decision which helped the geographical community to flourish. Significantly, summer schools were also open to both men and women, allowing teachers the chance to train in Geography.

Historical geography and Geography textbooks

After Mackinder returned to the UK he began energetically to promote a particular kind of British imperialism. Blouet calls Mackinder 'a visionary imperialist' (2008). By that he meant an aspiration for the Empire to become a league of democracies with common defence and economic policies. Mackinder advocated an imperial tariff system, and widespread economic development across the Empire. These measures were checks on the liberal desire for laissez-faire, unregulated trade although they also represented a continuation of imperial domination and exploitation (Mackinder 1919, 247).

During this period as a teacher of Geography, Mackinder further promoted the usefulness of photographic plates of a topographical and borderless globe which showed the relative remoteness of the British Isles and the continental landmasses. This was an important step in the development of his own theory of historical geopolitical sea, land and air powers. Such images had the advantage of conveying the underlying planetary topography rather than the political geographies of national and imperial borders. 1902 also saw the publication of the book series *The Regions of the World* edited by Mackinder. His own contribution, *Britain and the British Seas* (Figure 2.4) (1902), has been accused of racism. Yet, it should be noted that he used the term 'race' in numerous ways to denote nations, ethnicities, language groups and other identities. This complicates any facile generalizations about Mackinder's understanding of the term, especially when we bear in mind that the concept and term "ethnic" did not become widely used in English until after 1935 (Huxley and Haddon, 1936, 85). *Britain* is notable for drawing on what were then the latest understandings of the archaeological record and, in an era before detailed genetic analysis, morphological and cranial measurements that were deemed to be scientifically valuable/accurate within biological anthropology. Mackinder's book stressed the millennia-long connections between succeeding populations across the islands and continent of Europe rather than any kind of insular exclusivity. Harder to fathom was Mackinder's faith in Britain's moral good and its ability to lead and sustain an international federation of democracies.

Throughout, Mackinder remained convinced of the need to promote school Geography to secure the continued health of the discipline (Blouet 2008). He thought it essential that young children be exposed to, and study, nature at school (Mackinder 1904a). In 1906, Mackinder assumed the presidency of the Froebel Society, which promoted kindergarten teaching methods and encouraged elementary schools to prepare students for secondary education (Hayes 2019). The Colonial Office Visual Instruction Scheme, known as COVIC, was another school education project he promoted. He and the committee of imperial administrators, educational reformers and instructors hired Alfred Hugh Fisher (1867–1945), a writer and illustrator, to create a photographic and artistic record of the peoples and landscapes of the British Empire for use as educational materials. The aim was to foster mutual cultural understanding within the colonies, and a common imperial identity (Ryan 1994). From Fisher's fast moving, empire wide travels over three years, COVIC produced textbooks and sets of lantern slides numbering several hundred each (Ryan 1994; Moser 2019). Soon after, 'On thinking imperially' (1907) appeared. This text attempted to distinguish between xenophobic jingoism which served national

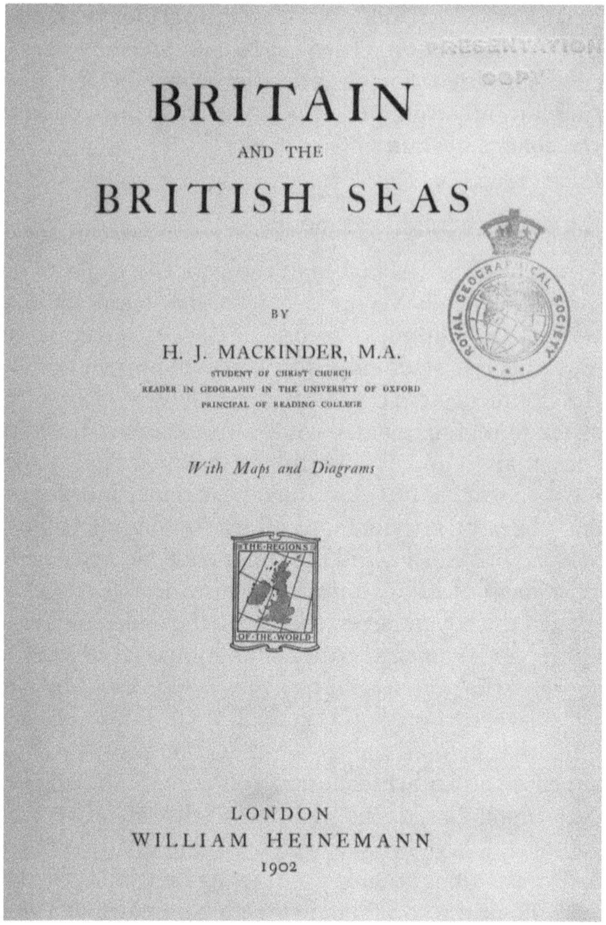
BRITAIN
AND THE
BRITISH SEAS

BY

H. J. MACKINDER, M.A.
STUDENT OF CHRIST CHURCH
READER IN GEOGRAPHY IN THE UNIVERSITY OF OXFORD
PRINCIPAL OF READING COLLEGE

With Maps and Diagrams

THE REGIONS OF THE WORLD

LONDON
WILLIAM HEINEMANN
1902

Figure 2.4 Image of title page, *Britain and the British Seas* (1902).
[Copyright RGS-IBG]

interests first, particularly British ones deemed by some to be superior to those of other peoples and countries, and what Mackinder presented as an enlightened form of imperialism that would encompass connected and federated mutually sustaining democracies.

Mackinder believed that 'no teaching of Geography really worthy of a university would be practicable' until students both at primary and high schools had a better grounding in Geography's basics (1921, 378). As Geography's status as a distinct subject was still uncertain, he continued to promote the concentric method that he first elaborated in 1887. It emphasized the interactions of the land, sea and air, and enabled pupils to 'visualize with a single grasp our whole world of varied scenes and incessant change'. As he elaborated, 'almost every great problem of today is a whole-world problem, and the comprehensive outlook must be cultivated' (1921, vi). To our contemporary eyes the value of this planetary perspective is vital. However, today we can see that Mackinder's emphasis on the physical form and phenomena of Earth to some extent occluded the many and varied historical, cultural and religious cosmological understandings held by humans.

The London School of Economics and Mackinder's entry into politics

In the new century Mackinder's professional activities proliferated. From its establishment in 1895, he delivered lectures at the London School of Economics and Political Science (the LSE). In 1900 he was appointed Reader in Economic Geography there. In 1902 he took up a full-time lectureship at London University, whilst also advising on the organization of a Geography department at the LSE. From 1903 to 1908 he was Director of the LSE as a whole and of the School of Geography. This move triggered his resignation in 1905 from his remaining Oxford commitments.

There was a longstanding criticism of politicians for their lack of geographical and cartographical knowledge (Hayes, 2021b). That applied also to the British military. Mackinder felt obliged to redress this lack and from 1906 to 1914 he ran an annual course for army officers (Blouet 2008). More generally, he was concerned to inform the wider public about the world. He wrote newspaper articles and gave public speeches about economic issues, the question of home rule in Ireland, imperial concerns, the lack of adequate public understanding of maps and geography, and class inequalities (Blouet 2008). Over time the range of his interests only broadened. Engaging with a range of visual media and the new medium of radio, his lectures and writings became ever more interdisciplinary, embracing such diverse topics as archaeological and historical knowledge, Einsteinian theories of relativity, European philosophy, and literature, poetry and plays. All this was despite the fact that within the academy the divide between sciences and the arts became only more entrenched.

This turn to wider interests corresponded with his move away from the academic discipline of Geography to a concern for on-the-ground change that he thought he could best participate in by entering national politics. After multiple failed attempts to be elected a Member of Parliament (MP) in England and Scotland, in January 1910 he won a seat for the Camlachie division of Glasgow, standing for the Conservative and Unionist Party. He occupied it until 1922 although he never became a minister. In 1919 he was appointed British High Commissioner to South Russia. His Tory (Conservative) government, however, would not follow his advice to provide support for the White Russian, anti-Bolshevik forces that continued to fight the new Russian Bolshevik government under Lenin. Even after ceasing to be an MP he worked as a government adviser promoting a geographical outlook. Whether he made a positive difference, though, remains unclear.

His influence was greater as an 'instigator of political geography' and a geopolitical theorist (Mayhew 2000, 772). Several of his books investigated how the interplay of physical and human geography fashioned patterns of politics and power. The textbook *Britain and the British Seas* (1902) examined the strategic and imperial implications of Britain's geographical position and resources. The book's latter part focussed on Britain as an imperial state, and specifically London as a global financial and power-brokering centre. His RGS lecture in 1904, 'The geographical pivot of history' (Mackinder 1904b), became enormously influential. It set out to 'exhibit human history as part of the life of the world organism' (Mackinder 1904b, 422). It proposed a fundamental geopolitical thesis that if one power, or alliance of powers, could gain control of Eurasia and use the resources of that region, it would dominate the world. He called the inland area of Eurasia the 'pivot area' (Figure 2.5), believing that newly laid rail networks in that region gave whichever power was controlling it the potential for world domination. He

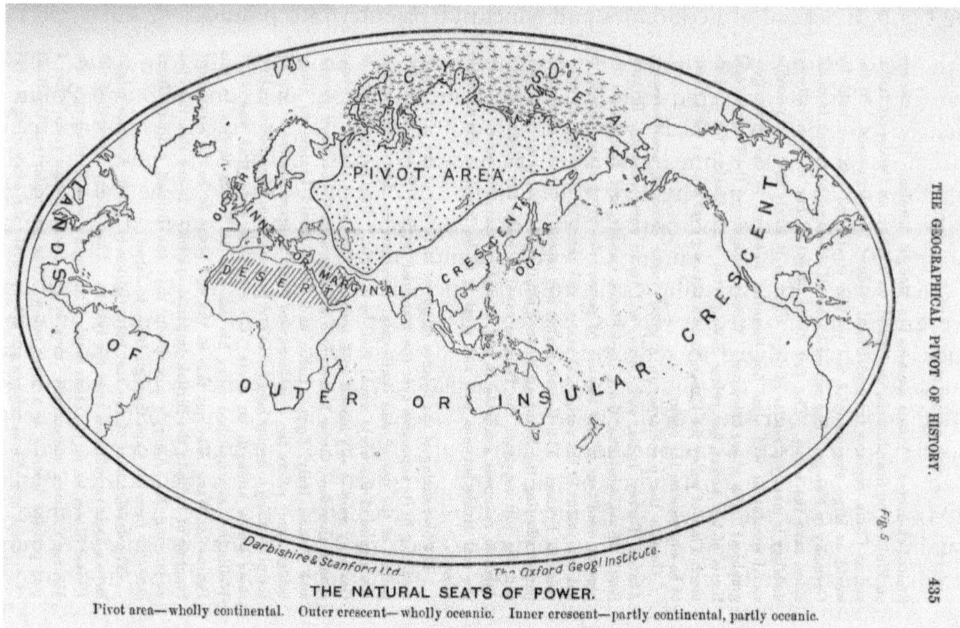

Figure 2.5 'The Natural seats of Power' from Mackinder, H. J. (1904) 'The geographical pivot of history', *The Geographical Journal* 23(4), 435. '

developed this hypothesis further as the 'Heartland' theory in *Democratic Ideals and Reality* (1919), coining the saying:

> Who rules East Europe commands the Heartland:
> Who rules the Heartland commands the World-Island:
> Who rules the World-Island commands the World.
>
> (Mackinder 1919, 106)

States were expanding territorially, he argued. Writing in 1919 after the Russian Revolution, the pivot area was recast as the World-Island's Heartland (Figure 2.6). The latter was relocated west from north-central Asia to the Black Sea area between the Mediterranean and the Arctic Ocean. In the Heartland area, Mackinder claimed, there was a risk of mass movements taking control of populations and state structures alike. Consequently, Mackinder favoured the federation of states encompassed by a closed system of empires. He advocated the independence of states from the Baltic Sea to the Caucasus mountains range. As he foresaw, Poland, Czechoslovakia and Hungary had the potential to serve as buffer states for Western European nations from the twin threats of Germany and Russia. Regrettably, his Heartland theory attracted attention, including from German geo-strategists such as Karl Hausshofer (1869–1946) who influenced Adolf Hitler's Third Reich territorial policy of *Lebensraum*, and later US and Russian Cold War (1947–1991) strategies (Ó Tuathail 1992, 115; Barnes and Minca 2013). With the recent invasion of Ukraine by Russian forces, the theory is again of topical salience.

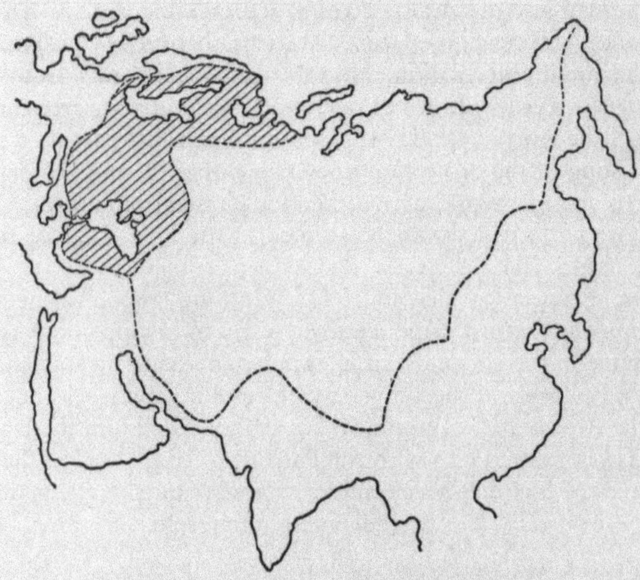

FIG. 24.—The Heartland, with the addition of the basins of the Black and Baltic Seas, and the uppermost (plateau) valleys of the Chinese and Indian rivers.

Figure 2.6 The extended area of 'The Heartland', in Halford Mackinder, *Democratic Ideals and Reality* (1919), p. 76.

The Heartland theory overshadowed other ideas put forward within *Democracy and Ideals* (1919). A key innovation turned on 'universal democracy' (Mackinder 1919, 7) with a concomitant critique of nineteenth-century nationalism (Mackinder 1919, 237). Though Mackinder has been described as a conservative, in this work he favoured the French revolution of 1789 and the Russian one of 1917 because in both cases unelected monarchs, a king and a czar respectively, were overthrown and replaced by more democratic forms of governance. While this work, like his other texts, is striking in its lack of attention to women, it does attend to class inequalities. A nation, he contended, 'cannot be independent if it is broken into classes and interests which are for ever seeking to range themselves for fighting purposes with the equivalent classes and interests of other nations' (1919, 253–254). Additionally, this work argued for devolution and a more balanced division of power and resources across a nation's regions (Mackinder 1919, 254). In this sense he was neither a conventional conservative nor an imperialist.

During the last decade of his life, Mackinder increasingly reflected on popular science and philosophy, but also the relationship of geographical theory and methods to these two. He believed that Geography was 'a concrete philosophy based on science, rather than a science comparable with any one of the analytical specialist sciences' (1937, 179). Its special contribution stemmed from its unique position of bridging both sciences and humanities allowing it to 'co-ordinate essential science with humane knowledge incommensurable with science' (1937, 181).

Geography, in Mackinder's final conceptualization, was recast as a study of global patterns of both human and physical processes, representing both an art and a philosophy (1942), a stance that prefigured earth systems theory. He proposed to abandon 'regions, except the initial home region, and substitute focal points, from which the visualizing and rationalising eye would sweep over gradually widening areas until it embraced the entire surface of the world' (1943, 71). This approach was part of 'a series of spherical studies... consisting of the projection upon the globe of a pattern depicting features and/or processes of a single world activity, either physical or human'. The 'integrated complex of all those patterns...would be studied in the final stage of regions growing from focal points' (1943, 71).

Mackinder was 86 when he died in 1947. While his appointment as a University Reader in Geography at Oxford came relatively early in his life and was hugely significant for the discipline in the UK, most of his academic honours came relatively late. He was 62 when made Professor at the University of London, and it was not until he was in his eighties that the American Geographical Society awarded him the Daly Gold Medal (1943) and the RGS honoured him with the Patron's Medal (1945). The lateness of his recognition reflected the fact that Mackinder was always outside the mainstream of geographical thought – ironically, this despite his founding role.

Conclusion: Taking stock of Mackinder's impact

Halford Mackinder straddled the boundary between different phases of geographical practice. He was at the very beginning of the promotion of Geography as a university academic subject, becoming a Reader in the "new" Geography at the University of Oxford. But he was also a participant in imperial exploration and its disturbing consequences, advocating for the subject's utility in military, political and colonial decision-making. Arguably, his importance in histories of Geography has been exaggerated. While Mackinder's aspirations to grow a sustainable community of geographers, and provide materials for beginners of all ages, are laudable, I have suggested in this chapter the considerable difficulties that he, along with the RGS and teachers, faced in those tasks. It makes for a less heroic but a more complicated and richer history of both him and the discipline.

This chapter reveals a young man's political engagement, his absorption of late nineteenth-century socio-political ideals, and a commitment to furthering and developing those ideals pragmatically. It has presented Mackinder's enthusiasm for innovating educational technologies, and his efforts to improve and sustain geographical education. Mackinder defined the ideal geographer 'as a Cartographer ... [who] would produce scholarly and graphic maps [...] as a teacher ... [who] would make maps speak' (1895, 376). How successfully he passed on his method to students and teachers who attended his courses is not known. What is clear, though, is how his approach to Geography was driven by his teaching experiences and the needs of learners, often beginners, of all ages.

Academic fame usually focusses on one or a limited set of achievements, with the more complex context of an individual's life and their wider set of contributions airbrushed out. This is the case for Mackinder, with the development of his ideas consequently remaining ill-understood. As a result, subsequent generations of geographers have sometimes believed that they were effecting revolutionary change when, in fact, they were just revisiting concepts that had previously been proposed by Mackinder. But in turn, Mackinder cannot be seen as the absolute beginner of academic Geography; his ideas were themselves adapted from several scientific practitioners both in the past and those who were contemporaries.

Mackinder was, for many, the epitome of British imperialism (Kearns 2010), an originator of geopolitics. While never an anti-imperialist, he never advocated a simple-minded version of imperialism either. In his forecasting and speculations Mackinder envisaged a globalized, integrated world. He showed that in the past the British Isles had been populated by peoples from elsewhere and colonized. He predicted that in the future as geopolitical blocks consolidated Britain would become united with Europe.[6] Rather than ruled from a single imperial centre, he aspired for the Empire to become a league of democracies with common economic and defence policies. By revising what we thought we knew, the history of Geography allows us to see Mackinder, and thus other geographers, while still often flawed, as a more complicated, less simple-minded figure: neither a model geographer to be revered nor a caricature to be reviled and banished from the histories that we write and teach. By revising what we thought we knew, the traditions of Geography can be remade to show how the history of Geography matters to the planetary imaginaries and understandings of environmental and Anthropocenic imperatives, but also the lack of diversity and attention to indigenous perspectives that our decolonializing processes (Radcliffe 2017; Esson et al. 2017) are reckoning with. Critical assessments of people like Mackinder have the power to inspire the creative, inclusive and plural geographical outlooks that we need to address these issues.

Notes

1 As discussed in more detail below, the magic lantern was an early version of the slide projector and an ancestor of today's PowerPoint.
2 The RGS and the common university fund at Oxford shared the costs of the Reader's annual salary of £300 (Blouet 2008).
3 He also became ambitious to extend the extension scheme. Michael Sadler and Mackinder delineated their plan for the Extension scheme in the 1890 publication, *University extension: has it a future?*
4 The Royal Institution in London may also have been a model. The Outlook Tower in Edinburgh, managed by the Scottish geographer Patrick Geddes as a space dedicated to geography materials and teaching from 1892, may have provided additional inspiration.
5 Yet, in parallel, for Mackinder human geography resulted from 'the conflict of two elements, the dynamic and the genetic' (Mackinder 1895, 375).
6 One way in which he envisaged the world was as an archipelago of continental islands.

Study task

This can be set as a term paper question or as part of a timed exam:

With a focus on the life of Halford Mackinder, consider how our understanding Geography's past can best be used in the present.

Starter reading list

Barnett, C. (1995) 'Awakening the dead: Who needs the history of geography?', *Transactions of the Institute of British Geographers* 20,4: 417–419.

Davies, A. (2023) 'Of elephants and discipline: For a recursive history of geography', *Dialogues in Human Geography* 14, 2: https://doi.org/10.1177/20438206231177064

Kearns, G. (2010) 'Geography, geopolitics and Empire', *Transactions of the Institute of British Geographers* 35, 2: 187–203.

Kearns, G. (2021) 'Topple the racists 2: Decolonising the space and the institutional memory of geography', *Geography* 106, 1: 4–15.

Keighren, I. M. et al. (2017) 'Teaching the history of geography: Current challenges and future directions', *Progress in Human Geography* 41, 2: 245–262.

Livingstone, D. (1992) *The Geographical Tradition*, chapter 1 (Oxford: Blackwell).

References

Andrews, H.F. (1986). 'A French View of Geography Teaching in Britain in 1871', *Geographical Journal* 152, pp. 225–231.

Barnes, T.B. & Minca, C. (2013). 'Nazi spatial theory: The dark geographies of Carl Schmitt and Walter Christaller,' *Annals of the Association of American Geographers* 103, pp. 669–687. 10.1080/00045608.2011.653732

Blouet, B.W. (1987). *Halford Mackinder: A Biography*. College Station: Texas A & M University Press.

Blouet, B.W. (2008). *Mackinder, Sir Halford John (1861–1947)*. Published online: 03 January 2008. 10.1093/ref:odnb/34760

Coones, P. (1989). 'The centenary of the Mackinder readership at Oxford', *The Geographical Journal* 155(1), pp. 13–23. http://www.jstor.org/stable/635377

Dams, K., Depaepe, M. & Simon, F. (2014) (first published 1999). 'Sneaking into school: Classroom history at work', in Grosvenor, I., Lawn, M., and Rousmaniere, K. (eds), *Silences and Images: The Social History of the Classroom*. New York: Peter Lang, pp. 15–46.

Darwin, C. (1859). *On the Origin of Species*. London: John Murray.

Driver, F. (2010). 'In search of the imperial map: Walter Crane and the image of empire,' *History Workshop Journal* 69, pp. 146–157.

Esson, J., Noxolo, P., Baxter, R., Daley, P. & Byron, M. (2017). 'The 2017 RGS-IBG chair's theme: Decolonising geographical knowledges, or reproducing coloniality?', *Area* 49(3), pp. 384–388. 10.1111/area.12371

Freshfield, D. to Mill, H.R. (2 February 1932). RGS-IBG collections, Hugh Robert Mill Collection, Box 3/Freshfield.

Galton, F. (1886). 'Opening remarks by the President', *The Journal of the Anthropological Institute of Great Britain and Ireland* 15, pp. 336–338.

Gilbert, E.W. (1951). 'Seven lamps of geography. An appreciation of The Teaching of Sir Halford J. Mackinder', *Geography* 36(1), pp. 21–43.

Hayes, E. (2018). 'Geographical light: The magic lantern, the reform of the Royal Geographical Society and the professionalization of geography c.1885–1894', *Journal of Historical Geography* 62, pp. 24–36 10.1016/j.jhg.2018.04.003

Hayes, E. (2019). 'Fashioned in the light of physics: The scope and methods of Halford Mackinder's geography', *British Journal for the History of Science* 52, pp. 569–594. 10.1017/S0007087419000475

Hayes, E. (2021a). 'Wordplay, world-play and interdisciplinary imaginations', *Transactions of the Institute of British Geographers* 46, pp. 510–524. 10.1111/tran.12432

Hayes, E. (2021b). 'On the plurality of words: the portmanteau word of geosophy and its persistence in the disciplinary baggage of geography', *Journal of Historical Geography* 73, pp. 52–61. 10.1016/j.jhg.2021.08.001

Huxley, J.S. & Haddon, A.C. (1936) [1935]. *We Europeans: A Survey of 'Racial' Problems*. London: Harper & Brothers.

Junior Science Club archives, Bodleian 1895,19, Oxford.

Kearns, G. (2010). 'Geography, geopolitics and empire,' *Transactions of the Institute of British* Geographers 35(2), pp. 187–203.

Kearns, G. (2021). 'Topple the racists 2: Decolonising the space and the institutional memory of geography', *Geography* 106(1), pp. 4–15. 10.1080/00167487.2020.1862575

Klinke, I. (2023). *Life, Earth, Colony: Friedrich Ratzel's Necropolitical Geography*. Ann Arbor: University of Michigan Press.

Mackinder, H.J. (1887). 'On the scope and methods of geography', *Proceedings of the Royal Geographical Society and Monthly Record of Geography* 9(3), pp. 141–174. http://www.jstor.org/stable/1801248

Mackinder, H.J. & Sadler, M.E. (1890). *University Extension: Has It a Future?* England: H. Frowde.

Mackinder, H. to Francis Galton. (2 May 1890). *Francis Galton Papers*, Wellcome Collection Library, London, pp. 1–4.

Mackinder, H.J. (1895). 'Modern geography, German and English', *The Geographical Journal* 6(4), pp. 367–379.

Mackinder, H.J. (1902). *Britain and the British Seas*. New York: D. Appleton and Company. https://archive.org/details/britainandbriti02mackgoog/page/n33/mode/2up?q=Eratosthenes

Mackinder, H.J. (1904a). 'The development of geography teaching out of nature study', *The Geographical Teacher* 2(5), pp. 191–197. http://www.jstor.org/stable/40553787

Mackinder, H.J. (1904b). 'The geographical pivot of history', *The Geographical Journal* 23(4), pp. 421–437. http://www.jstor.org/stable/1775498

Mackinder, H.J. (1907). 'On thinking imperially,' in Sadler, M. (ed.), *Lectures on Empire*. London: Privately printed.

Mackinder, H. (1919). *Democratic Ideals and Reality*. New York: National Defense University Press Publications, pp. xvii–xxii.

Mackinder, H. (1921). 'Geography as a pivotal subject in education', *The Geographical Journal* 57(5), pp. 376–384. http://www.jstor.org/stable/1780845

Mackinder, Halford J. (1931). 'The human habitat', *Scottish Geographical Magazine* 47, pp. 321–335, 10.1080/14702543108555217

Mackinder, H.J. (1937). 'The music of the spheres', *Proceedings of the Royal Philosophical Society* 63, pp. 170–181.

Mackinder, H. (1942). 'Geography, an art and a philosophy', *Geography* 27(4), pp. 122–130.

Mackinder, H. (1943). 'The development of geography' *Geography* 28(3), pp. 69–77.

Mander, W.J. (2011). *British Idealism: A History*. Oxford: Oxford University Press.

Mann, V. (2022). 'Mind and knowledge in the early thought of Franz Boas, 1887–1904', *History of the Human Sciences* 35, pp. 57–184. 10.1177/09526951221091137

Mayhew, R. (2000). 'Halford Mackinder's "new" political geography and the geographical tradition,' *Political Geography* 19(6), pp. 771–791.

McLeod, H. (2000). *Secularisation in Western Europe, 1848–1914*. Basingstoke and London: Macmillan Press LTD.

Moser, G. (2019). *Projecting Citizenship: Photography and Belonging in the British Empire*. University Park, PA: Penn State University Press.

Mulvaney, D.J. (1985). *'So much that is new': Baldwin Spencer, 1860–1929: A Biography*. Victoria: University of Melbourne University Press.

Nyamanga, P. (2022). 'The Kikuyu Community of Kenya'. https://artsandculture.google.com/story/the-kikuyu-community-of-kenya-national-museums-of-kenya/qQUR6HqkIjAUIQ?hl=en

Ó Tuathail, G. (1992). 'Putting Mackinder in his place', *Political Geography* 2, pp. 100–118.

Oxford and Colonialism Network. (2022). 'School of Geography and the Environment'. https://oxfordandcolonialism.web.ox.ac.uk/school-of-geography-and-the-environment

Pease, E.R. (1916). *The History of the Fabian Society*. London: A. C. Fifield.

Radcliffe, S.A. (2017). 'Decolonising geographical knowledges', *Transactions of the Institute of British Geographers* 42, pp. 329–333. 10.1111/tran.12195

Ratzel, F. (1899) [1882]. *Anthropogeographie*. Stuttgart: J. Engelhorn.

Ryan, J.R. (1994). 'Visualizing imperial geography: Halford Mackinder and the Colonial Office Visual Instruction Committee, 1902–11,' *Ecumene* 1, pp. 157–176.

Santini, C. (2017). 'At the origins of modern geography. The Oecumene: An anthropogeographical pattern', *History of European Ideas* 43(6), pp. 560–569.

Semple, E.C. (1911). *Influences of Geographical Environment on the Basis of Ratzel's System of Anthropo-Geography*. New York: Holt; London: Constable.

Stoddart, D.R. (1966). 'Darwin's impact on geography', *Annals of the Association of American Geographers* 56, pp. 683–698. https://www.jstor.org/stable/2561794

Stoddart, D.R. (1980). 'The RGS and the "new geography": Changing aims and changing roles in nineteenth century science', *Geographical Journal* 146, pp. 190–202. http://www.jstor.org/stable/632860

Walford, R. (2001). *Geography in British Schools, 1850–2000: Making a World of Difference*. London: Woburn Press.

Wing, D.W. (1959). 'A local school in the nineteenth century', *University of Nottingham Institute of Education Bulletin* 31, pp. 17–25.

3 Geography as the science of environmental influences

Ellen Semple and the search for disciplinary relevance

Innes M. Keighren

Ellen Semple (1863–1932) photographed in 1914, the year she was awarded the Cullum Geographical Medal.

Reproduced by permission of the University of Kentucky Special Collections Research Center, PA46M139.

DOI: 10.4324/9781003343240-4

Introduction

The discipline of Geography has always been the butt of jokes. On 2 June 1842, for example, the American newspaper *The Brooklyn Eagle* sought to amuse its readers with a joke poking fun at Geography's seemingly ill-defined and all-encompassing scope. "John what is Geography?", a teacher asks his student. "Geography is the history of everything but the sun, moon and stars, and the steam engine", John replies sardonically. It was a weak gag, but it would nevertheless have struck a chord with readers for whom Geography was an amorphous and apparently undisciplined discipline: seeming to be about everything, yet nothing in particular. John's mid-nineteenth-century vision of Geography was not far removed from the truth. Academic geographers themselves recognised that the discipline resisted straightforward definition. One anonymous 1843 contributor to *The Journal of the Royal Geographical Society of London*—then the discipline's flagship English-language periodical—agreed that "With respect to Geography...it's much more easy to say what it is not, than what it is" (p. 330). For many of its nineteenth-century practitioners, however, the value of Geography lay precisely in its breadth and in its ability to bring other disciplinary perspectives together in combination. Geography, the Victorian polymath Francis Galton noted in 1855, "links the scattered sciences together, and gives to each of them a meaning and a significance of which they are barren when they stand alone" (p. 81).

Relaxed as geographers may have been in the middle of the nineteenth century about their discipline's expansive and elusive nature, the increasing specialisation and professionalisation of academic knowledge production had a negative effect. During the second half of the century it provoked a crisis of sorts over Geography's legitimacy and status. No longer was "What is Geography?" the question that was most urgent and perplexing; it had been replaced by the altogether more critical query "Is Geography a science?". Answering that question, and addressing its implications for the discipline's methods and practice, became a preoccupation for those who were eager to make Geography matter—to show that it was a specialised field of knowledge, that it could make sense of the world in ways that were founded in observation and evidence, and that it was methodologically and theoretically distinct from those disciplines whose fields of study it overlapped. Between the late nineteenth and early twentieth centuries, making Geography matter was a project based on finding ways to bring the human and physical together under a single analytical umbrella. Among the many attempts to achieve this outcome, perhaps the most impactful, yet controversial, was that advanced by the American geographer Ellen Churchill Semple (1863–1932).

This chapter seeks to understand Ellen Semple's geographical project: her attempt to develop a theoretical and methodological approach to the study of Geography that would unite the discipline's human and physical concerns through the systematic examination of the role of environmental influences on human societies. At the same time, Semple wanted to demonstrate that Geography was legitimately a science—one capable of classification, generalisation, and prediction, not merely description. The chapter begins by examining Semple's early life and education for what they tell us about the social and intellectual contexts in which her geographical outlook took shape. In following Semple's career as an academic geographer, the chapter then considers how—through publication, fieldwork, and teaching—she sought to develop and promote her particular vision of Geography as the systematic study of environmental influence. Finally, the chapter examines how her ideas were received by other geographers, the impact those ideas had on the discipline, and how the legacy of her geographical project has been considered in the years since her death. Taken together, the chapter tells a story of how—at a particularly significant moment in its disciplinary history—Geography was made to matter by equipping it with an apparently objective and distinctive purpose. At the same time, in examining the reasons why Semple's approach was ultimately rejected, the chapter considers the toxic legacy of so-called 'environmental determinism' in the discipline.

The path to becoming a geographer

Semple was born in Louisville, Kentucky on 8 January 1863, amid the tumult and confusion of the American Civil War (Keighren, 2010). Her childhood was, notwithstanding this inauspicious context, a happy and privileged one. Her father, Alexander, was a successful merchant who had profited during the Civil War through trade in hardware and guns. Her mother, Emerin, was intellectually curious and encouraged Semple's education and that of her older sisters. Although her parents separated while she was still young, and Alexander died when Semple was just twelve, Semple's childhood passed largely "in an atmosphere of wealth and refinement" (Colby, 1933, p. 229). Under Emerin's direction, and benefiting from wider social changes that had slowly promoted women's access to education during the second half of the nineteenth century, Semple read widely, received private tuition, and attended some of Louisville's most prestigious girls' schools (Johnson, 2008). At the age of fifteen in 1878, Semple followed the example set by her older sister, Patricia, and enrolled at the prestigious women-only Vassar College in Poughkeepsie, New York. Although officially a year under age, Semple completed the daunting suite of entrance examinations without apparent difficulty and went on to distinguish herself during a four-year degree programme. Semple later attributed many of her achievements as a professional researcher to the training that she had received at Vassar, particularly in composition and rhetoric (Bingham, 1932). Her power of persuasion, in talk and in text, was important to the later dissemination of her geographical ideas.

Geography was not on the curriculum at Vassar and Semple's interest in the subject emerged only later through reading and private study in the years after her graduation. Her emerging preoccupation with Geography reflected a longer-standing interest and belief in the influence of the physical environment on human history and social development. While many nineteenth-century historians treated Geography only metaphorically—as the 'stage' on which the 'drama' of human history played out—Semple was convinced that the physical environment was more than an inert backdrop. In her view it was, in fact, centrally important to explaining the course of historical events. It is difficult to know precisely why this idea resonated so strongly with her, but it is possible that the proximity of the Civil War, which had showcased the significance of regional differences within the United States and highlighted the role of geographical factors in directing military strategy, had served to bring these considerations to the fore. Whatever the motivation, Semple spent the middle years of the 1880s trying to identify, variously in historical and geographical scholarship, an intellectual framework that she might use to explore and better understand the role of environmental influences on history and society. As she later recalled, this was a time during which she "began to scent the importance of geographical influences", but "struck no trail of a previous investigator that was reliable enough to follow" (quoted in Keighren, 2010, p. 14).

A turning point in Semple's intellectual journey came during a visit to Europe in 1887. In London, Semple met Durren J. H. Ward. He was an American scholar who had recently completed a doctoral degree at Leipzig University in Germany. Ward enthralled Semple with his account of life at Leipzig, particularly the memorable lectures of a charismatic professor there, Friedrich Ratzel. An intrigued Semple borrowed from Ward his copy of Ratzel's recent book, *Anthropogeographie* (1882). Over the next six months she read across Ratzel's wider body of scholarship. In Ratzel's work—attentive as it was to the links between the biological and the geographical—Semple found an immediate sense of intellectual connection and resolved to travel to Leipzig to further her study of Geography under him. Before doing so, she returned to Vassar where she completed a master's degree on the sociology and political economy of slavery and maintained a lively correspondence with Ward in which the pair discussed and dissected Ratzel's ideas (Adams, 2011; Keighren, 2010). Semple eventually embarked on her studies at Leipzig in the autumn of 1891. Although university regulations prevented women from formally enrolling or sitting examinations, Semple was allowed to join Ratzel's lectures and the pair quickly established what became an enduring friendship that they maintained through correspondence until Ratzel's death in 1904. What, though, did Semple learn from Ratzel and how did Ratzel's own geographical project help inform Semple's?

The influence of Friedrich Ratzel

Like Semple, Ratzel (shown in Figure 3.1) had not always been a geographer. His university training in the 1860s had been primarily in the fields of zoology and geology, and his intellectual outlook had been shaped strongly by the scholarly debates that followed the publication of Charles Darwin's *On the origin of species by means of natural selection* (1859). Darwin's theory of evolution by means of natural selection argued that all life on Earth was locked in a struggle for existence, exhibiting constant gradual change through the differential preservation of traits best adapted to particular environmental circumstances. This was a model that saw evolution, at least partly, as the response to external conditions or pressures—such as climate and competition for food and resources—but one whose implications for understanding human evolution in particular were deliberately understated in Darwin's book. Ratzel studied under one of Germany's leading proponents of Darwinian evolution, the zoologist Ernst Haeckel. He inherited from Haeckel the view that human adaptation to the environment could be understood, and addressed scientifically, as a matter of biological fact. Haeckel's understanding of human evolution was also, however, political: he believed that evolution had led to the development of several distinct human species, 'races' that could be arranged in a hierarchy of relative superiority and inferiority (Richards, 2018). Finally, for Haeckel and other scholars, Darwinian evolution and its associated metaphors of 'struggle' and 'favoured races' also seemed to offer a model to understand human social development—a field of enquiry that became known as social Darwinism and which was later used to lend pseudoscientific credibility to German Nazi ideology during the twentieth century (Hawkins, 1997; Giaccaria and Minca, 2016).

The combined influences of Darwinian evolution and social Darwinism eventually led Ratzel to believe that human societies and nation-states were fundamentally no different to biological organisms—that they were likewise locked in a perpetual struggle for

Figure 3.1 Friedrich Ratzel (1844–1904) photographed in 1886, the year before Semple first read his *Anthropogeographie* (1882).

Reproduced by permission of the University of Kentucky Special Collections Research Center, PA46M139.

existence and competition for resources, particularly geographical space (portions of land and sea), and could be studied in a similar way. This belief informed his 1882 book, *Anthropogeographie*—the text that so enthralled Semple when she read it in London. What Ratzel's book sought to do, among many other things, was to show how the development of human societies and cultures was a function of their relationship with the geographical environment. *Anthropogeographie* also set out—as Semple (1900a, p. 137) later explained in the gendered language of the time—"a comprehensive statement of the laws underlying the relationship of man to his environment" and particularly the influence of topography and climate. In Ratzel's work, Semple had found the intellectual framework she had been looking for, but it was not quite the statement of laws she later described.

Ratzel was conscious, as one historian of academic Geography has noted, that "the geographer cannot formulate laws expressed with mathematical precision, nor can he conduct laboratory experiments. The only experiments open to him are those conducted by nature" (Crone, 1970, p. 34). In this respect, *Anthropogeographie* was a statement of precepts about how the environmental study of human geography—anthropogeography—might proceed in the future, rather than a definitive presentation of laws established on the basis of data or observation in the field. In order for the validity of Ratzel's ideas to be demonstrated, they needed to be tested in the field—in Geography's disciplinary laboratory. Central to any such test would be to resolve the question of whether the environment merely influenced human society or was actually a determinant of it. This was not, in any sense, a new question, but one that reflected more than 2,000 years of speculation about the role played by the physical environment in shaping human life and society (Livingstone, 2024). But what was it that underpinned the belief that the environment had a role to play at all?

Understanding theories of environmental influences

Environmental influence, at least as far as Ratzel and Semple understood it, drew on certain historical, philosophical, and medical discourses that had been developed by ancient Greek and Roman scholars. Perhaps the most significant of these ideas concerned the apparent relationship between the physical environment and the human body and mind. Classical understandings of the human body were exemplified by the so-called humoural theory—the idea that health, mood, and temperament were determined by the relative balance or imbalance of four bodily fluids: blood, yellow bile, black bile, and phlegm. Ill health, as well as variations in temperament and personality, was seen to be the consequence of an imbalance in these fluids. Each of the humours was thought, moreover, to exist in relationship with one of the four fundamental elements that governed the natural world: earth, air, fire, and water, and their associated characteristics of cold, dry, hot, and wet (Keighren, 2015). The humours would vary with the seasons as temperatures changed, and the rains came and went, but geographical conditions—so it was believed—could also exert long-term influences on the humours that meant that whole populations could be shaped variously physically, mentally, and morally by the topography and climate to which they were exposed.

This way of thinking about Geography's influence on human populations—however fanciful and non-scientific it may seem today—was both enormously influential and long-standing. It fed into European scholars' damaging belief that certain physical environments promoted the development of human civilisation while others encouraged a descent into sloth and savagery. Well into the twentieth century—and despite considerable advances in medical knowledge—a belief in the relative superiority and inferiority of different human populations on the basis of their geographical location remained persistent and widespread. It was a belief that had also functioned as a moral rationale of sorts for European imperialism, as well as forming the basis of European scholarly constructions of 'race' as a (supposedly) scientific category (Livingstone, 2002). Even given the dominance of this bio-geographical understanding of human life and society, there remained considerable doubt and debate about the ability of populations, through social and economic organisation, to counteract and overcome the constraints of their environmental circumstances.

In Geography, there was widespread acceptance of the influence of the physical environment on humans, but disagreement over the extent of that influence. At one end of the spectrum of opinion (see Table 3.1) were scholars like Ratzel and Semple who, as proponents of environmentalism (later called environmental determinism), were convinced of the predominant influence of the physical environment. At the other end were those, like the French geographers Paul Vidal de la Blache and Jean Brunhes, who advocated a position later called 'possibilism'. This saw the physical environment as presenting a range of possibilities to which human populations could respond (Berdoulay, 2009). An intermediate position between the destiny of determinism and the free will of possibilism—'probabilism'—was briefly advanced in the middle of the twentieth century, by which point the debate over environmental influences in the discipline was beginning to fade in any case (Flowerdew, 2009).

Table 3.1 Three ways environmental influence was conceptualised during the twentieth century

Degree of influence	*← lesser influence*		*greater influence →*
Concept	*Possibilism*	*Probabilism*	*Determinism*
Argument	The physical environment imposes constraints on human societies, but populations retain free will and can adapt to those constraints and modify their environment through certain social and cultural practices.	The physical environment is a central factor in determining human life, but it is not the only one. While human responses to environmental constraints are limited, more than one 'destiny' exists.	The physical environment, including topography and climate, is the primary influence on human societies, shaping their social and cultural organisation in ways that are impossible to resist.
Proponents	Paul Vidal de la Blache Richard Hartshorne Carl Sauer	Oscar Spate	Ellsworth Huntington Ellen Churchill Semple Thomas Griffith Taylor

For all that Semple's interest in Ratzel's ideas reflected a genuine belief in their utility as a framework for investigation, her adherence to them also satisfied a more subjective and aesthetic sense that they were ideas that simultaneously enlivened the narrative of history and elevated the study of Geography above the level of the descriptive. This is a position she made explicit, years later, in a paper delivered at the prestigious Eighth International Geographic Congress in Washington, DC in 1904: "By the introduction of the human element", she told her audience, "geography is lifted out of the dull round of formal studies and the Earth becomes the setting of a great world drama. By the introduction of the geographic element, history becomes vitalized; through it now pulses the lifeblood of the people" (Semple, 1904, pp. 366–367). Semple's rhetorical flourish was characteristic of the way she would go on to make her case for the study of environmental influences. It also shows that her project to make Geography matter was at least partly rooted in a desire to showcase the subject's value for other disciplines, particularly history, and to distance it from pejorative associations with methodological dullness—the "deadly tedium of 'locate this' and 'bound that'" (White, 1977, p. 49)—that often characterised outsiders' understandings of what geographers did. In this sense, Semple's geographical project was predicated both on a claim that Geography could be studied in a scientific way and that it had the capacity to excite the imagination.

Bringing anthropogeography to the United States

Semple completed a second and final period of study in Leipzig in 1895 and on her return to the United States the following year her focus shifted to bringing Ratzel's ideas—anthropogeography in particular—to the attention of Anglophone geographers. Rather than attempt a straightforward translation of Ratzel's *Anthropogeographie*, however, Semple saw her task as altogether more nuanced. First, it would be necessary to test the application of anthropogeography in the field, since its validity would ultimately be seen by others to depend on how closely its precepts were supported by on-the-ground evidence. Second, the range of historical examples Ratzel had used to illustrate his claims would need to be expanded to include those more obviously familiar to Anglophone audiences. Third, the literary and rhetorical presentation of anthropogeography would need to be pitched in a way designed to appeal particularly to English-language readers (Keighren, 2010). Semple envisioned this, moreover, as a long-term project and it was one that she would not complete fully until the publication in 1911 of her most significant work, *Influences of geographic environment on the basis of Ratzel's system of anthropo-geography*.

Semple's project to bring anthropogeography to the United States began in earnest with the publication in 1897 of her first scholarly paper. Entitled "The influence of the Appalachian barrier upon colonial history", Semple's article sought to demonstrate Ratzel's approach with reference to a distinctively American topographical example: the Appalachian Mountains and the role they had played in the early colonial settlement of North America. In Semple's analysis, the Appalachians had served to limit westward expansion of the colonies and had thereby given to the colonists "a certain solidarity which they would not have otherwise possessed"—a unity of outlook that played to their advantage during the American Revolutionary War (Semple, 1897, p. 35). Semple's paper appeared in the first volume of a newly established periodical—the *Journal of School Geography*—one aimed at elementary- and secondary-school teachers of Geography. It was an audience with whom Semple was familiar. Upon graduating from Vassar, she had begun occasional teaching at girls' schools in Louisville and later, in 1893, she and her sister Patricia had established their own—the Semple Collegiate School—to prepare young women for college (Kleber, 2001). At a time when the discipline of Geography was less well established in America's universities, Semple's decision to prioritise the communication of her ideas to school teachers was a pragmatic one, reflecting the nationwide presence of Geography in the school curriculum, but it also mirrored her own sense about the value and importance of the subject.

By 1900, Semple had contributed a further five articles to the *Journal of School Geography*, each addressing a particular component of anthropogeography. This work served to promote a vision of the discipline as fundamentally human-centred and as one based on the articulation of general principles, rather than on the memorisation of abstract facts about geographical phenomena (Adams, 2011). Anthropogeography was central, Semple believed, in making students active rather than passive learners and encouraging interpretation rather than mere recollection: "every child should become an infant discovered on the cartographical page", she would later tell an audience at the Eight International Geographic Congress, "in order to acquire a self-constructed knowledge of every ocean, continent, and country as the basis for anthropogeographical deductions" (Semple, 1904, p. 368). Through her publications in the *Journal*, Semple was taking a bottom-up approach to the dissemination of her ideas and to her wider project of making Geography matter. At the same time, and despite maintaining a career-long interest in the teaching of Geography as a school subject, she was aware that her influence was unlikely to extend upwards, to Geography as it was being taught and practised at university level, without a major shift in her publication strategy and in the nature of the evidence she employed to illustrate her claims. Semple had long known that the validity of anthropogeography would depend upon its being tested in the field and, as the nineteenth century drew to a close, undertaking that test became her priority.

Testing anthropogeography as a field science

Semple's opportunity to put the principles of anthropogeography to the test of evidence came almost by accident. In 1899 she was invited to join a charitable project then being organised by the Kentucky Federation of Women's Clubs in the remote mountain town of Hazard in eastern Kentucky. Isolated by the Appalachian Mountains, Hazard was far from dominant modes of trade and communication, and its residents, though resilient and resourceful, experienced poverty and hardship that were unparalleled in the state. Hazard was a place, Semple later explained, "where the people are still living the frontier life of the backwoods, [and] where civilization is that of the eighteenth century" (Semple, 1901, p. 588). The project in which Semple was involved was one that drew on religious and class-based ideas of moral improvement. She sought to transplant some of the social norms and imagined sophistication of urban Louisville to the Kentucky Mountains. The upper-class women of Louisville saw themselves as missionaries among mountaineer "natives", and offered instruction in hygiene, sewing, cooking, and reading to people they considered to be living a backwards life (Semple, 1900b, p. 160). Semple, meanwhile, saw in Hazard and its population the effects of environmental influence in action.

The Kentucky mountaineers were, in Semple's analysis, direct descendants of the early Anglo-Saxon settlers of Virginia and North Carolina, preserved since then by their topographical isolation. Hazard was a cultural time capsule of sorts where "the conserving power of the mountains" had carried "a bit of the eighteenth century…into this strongly contrasting twentieth century" (Semple, 1901, p. 588). For Semple, 120 years of geographical and social separation had served to deteriorate the population's "naturally fine stock"—its Anglo-Saxon genetic inheritance—through "the effect of too close intermarriage, moonshine whiskey and wretchedly cooked food" (Semple, 1900b, p. 158). The Kentucky Mountains appeared to Semple to be a natural laboratory in which anthropogeography could be tested: a case study that would allow her to evidence, with reference to real-world data, the validity of her claims about the influence of the geographical environment. Semple consequently undertook a 350-mile horseback journey though the Kentucky Mountains in order to investigate the ways geography—and topography in particular—had shaped the habits, culture, and economy of the region's population. What anthropogeography predicted, and what Semple claimed to see in her investigations, was that the Kentucky mountaineers would have more in common—physiologically and socially—with genetically unrelated populations living in similar mountain environments than they would with genetically similar Anglo-Saxon populations living under different environmental circumstances.

When Semple published the results of her investigation in the Royal Geographical Society's prestigious *Geographical Journal* in 1901, she presented example after example that seemed to support her position: the too-close intermarriage of Kentucky mountaineers was just like that seen "in the upland valleys of Switzerland and other mountain countries where communication is difficult" (p. 592); the vernacular architecture of the mountaineers' homes was similar to the "village dwellings in Norway" (p. 597); the design and construction of farmers' barns served to "vividly recall the mountain architecture of some of the Alpine dwellings of Switzerland and Bavaria" (p. 597); and the intricate and careful quilt making of the mountain woman had aesthetic parallels "in the wood-carving of the Alps and Scandinavian mountains, the metal-work of the Caucuses, [and] the Cashmere shawls of the Himalayas" (p. 608). Wherever Semple looked, she saw a population that had been moulded by topography and geographical isolation and one that had cultural parallels across the globe in other similarly remote locations.

The publication of her paper on the Anglo-Saxons of the Kentucky Mountains was a turning point in Semple's geographical project—it brought her work to the attention of academic geographers internationally and demonstrated, at least for this one case study, that anthropogeography could be studied in the field. The paper cemented Semple's emerging professional reputation and proved influential throughout the first decade of the twentieth century. In 1910, the American Geographical Society took the unusual step of reprinting the paper in its own journal "in response to a repeated demand from students of Geography and Sociology for copies" (Semple, 1910, p. 561). The paper's positive reception encouraged Semple to collate some of her earlier publications on the influences of geography (the material environment) on history (the course of events through time) into a full-length monograph. It was published to critical praise in 1903 as *American history and its geographic conditions*.

The success of these publications opened professional doors for Semple. In December 1904, for example, she became a founding member of the newly established Association of American Geographers—one of only two women to be selected for inclusion in the Association's first cohort (Keighren, 2010). She would later go on to become its first woman president. Membership of the Association, which was then limited to the discipline's most prominent figures, gave Semple an important warrant of professional credibility and was undoubtedly central to her securing her first formal academic appointment at the University of Chicago, where a new graduate school of Geography had been established in 1903. At Chicago, which quickly emerged as a leading disciplinary centre for Geography, Semple taught to her interests, developing a course—"Some Principles of Anthropogeography"—that became a staple of the department's offering (Keighren, 2010).

Chicago served as a disciplinary incubator for American Geography in the first half of the twentieth century, producing a disproportionate percentage of the subject's PhD graduates. Many of them took Semple's course. Through her teaching, Semple was able to instill a career-long enthusiasm for anthropogeography in many of her students, some of whom went on to teach similar courses at other institutions. At the same time, teaching allowed Semple to revisit and refine the principles of anthropogeography, testing through presentation and discussion in the classroom the examples that best illustrated the role of geographical influences and the rhetorical strategies that might prove most effective in communicating her viewpoint. In much the same way that the Kentucky Mountains had served as a laboratory to test anthropography as a field science, the lecture theatres of the University of Chicago were laboratories in which Semple tested the arguments that underpinned her geographical project. Teaching anthropogeography gave Semple both the confidence and impetus to see through her long-held plan to communicate Ratzel's ideas more broadly, something she did in 1911 with the publication of the book with which she is today most often associated: *Influences of geographic environment on the basis of Ratzel's system of anthropo-geography*. While the book's publication marked a culmination of sorts for Semple's geographical project, its reception tells us something significant about the impact her views had on the discipline of Geography more broadly.

Tracing the reception of Semple's ideas

The reception of Semple's *Influences* was conditioned, at least partly, by a moment of acute disciplinary crisis. On 31 August 1911, the president of the Geographical Section of the British Association for the Advancement of Science, Sir Charles Close, delivered a devastating paper at the Association's annual meeting in Portsmouth. In it he lamented the current state of Geography (Close, 1911). Close, who was then Director General of the Ordnance Survey, took aim at a discipline he considered to be fundamentally unscientific—one that was inconsistent in its scope of investigation and in the methods it employed (Keighren, 2010). Close's argument was based on an analysis and classification of the papers that had been published in the Royal Geographical Society's *Geographical Journal* during the preceding five years. Close found that almost 60 per cent of the papers were devoted to reporting the results of travel and exploration, often in a descriptive and "humdrum" way (Close, 1911, p. 417). Too little of the work being done in Geography was, in Close's view, sufficiently rigorous or scientific. He argued that if Geography was to be recognised "by the world of science as an independent, indispensable, and definitive branch of knowledge", it must, therefore, "prove its independence and value by original, definitive, and, if possible, quantitative research" (Close, 1911, p. 409). Close's remarks were incendiary and provoked considerable soul-searching among professional geographers in Britain and North America about their validity and about how best they might be refuted (Keighren, 2010).

Read in the immediate aftermath of Close's presidential address, Semple's *Influences* was understood by many readers as a likely basis by which his accusation could be disproved. As one early reviewer of the book noted, *Influences* was "a satisfactory answer—if an answer were required by one of unbiased mind—to the charge lately made that Geography is not a science, but a hanger-on of other sciences, a picker-up of crumbs falling from their table" (Anonymous, 1912, p. 160). Close's address had, in this respect, served inadvertently to create a particularly receptive environment for Semple's geographical project. But the reception of her ideas was not limited solely to the reading of her book—it also hinged on Semple's willingness to present her views on the international lecture circuit.

Almost as soon as Semple had finished work on *Influences* in early 1911, she embarked on an 18-month round-the-world tour designed to further test her ideas about environmental influence in a range of different geographical contexts (Adams, 2014). She travelled extensively in Southeast Asia, India, and Europe, before she arrived in Britain in the summer of 1912 for an extended stay. In Britain, she lectured widely in London and across Scotland to public and academic audiences, often illustrating her talks with hand-coloured lantern-slide images from her recent travels, and taught at Oxford, where she contributed to Halford Mackinder's popular biennial summer school for Geography teachers (Keighren, 2008). Perhaps the most consequential of these engagements was the lecture she delivered in November 1912 at the Royal Geographical Society in London on the influences of Geography on Japanese agriculture (Semple, 1912).

At the time of her presentation, women were not permitted to join the Society's fellowship—a position that was becoming increasingly untenable as women routinely demonstrated their abilities as geographers in a range of academic and professional contexts. In retrospect, it is clear that Semple's presentation mattered not only for what it had to say about her wider geographical project, but also for what it meant for women's standing within the discipline. Semple's talk was chaired by the Society's president, George Curzon. Although he had been a longstanding opponent of women's admission to the Society, his views on what was often referred to as the 'Lady Question' were in a state of flux. He was confronted by Semple's achievements as a geographical traveller and author, and by the campaigning efforts of other women, particularly those associated with the newly established Geographical Circle of the women-only Lyceum Club, who regularly questioned the Society's exclusionary policies (Keighren, 2017). In his comments at the end of Semple's lecture, Curzon praised her "unusual power...of deducing from the phenomena of material existence large generalizations and scientific laws" (Curzon, 1912, p. 607). Combined with her extensive recent experience of geographical travel, and her significant record of publications, it was clear, Curzon (1912, p. 603) explained, that Semple was "evidently a master of the subject". The old objections to women's admittance to the fellowship crumbled in the face of evidence of their capabilities and achievements. Before the end of the year the Society balloted its fellows on women's admission, with Curzon at the fore in arguing the case: "women", he told the fellows, "have conducted explorations not inferior in adventurous courage or in scientific results to those achieved by men; they have made valuable additions to the literature of travel, and have been invited to lecture in our great Universities" (quoted in Keighren, 2010, p. 103). The vote passed in favour of the proposal, and women's admission to the Society followed in January 1913 (Evans, Keighren, and Maddrell, 2013).

The twin crises in British Geography in 1911 and 1912—Close's damaging address and the uncertain position of women in the discipline—set the context for the reception of Semple and her ideas, both there and in North America. The apparent, and accidental, topicality of Semple's geographical project largely conditioned the discipline's response to *Influences* and served to cement its effect on the way academic Geography imagined its purpose and future direction. The book, and its reception, secured Semple's professional reputation. In 1914 she was awarded the American Geographical Society's prestigious Cullum Geographical Medal, joining a roster of previously male-only recipients that included the polar explorers Fridtjof Nansen and Robert Falcon Scott, and the prominent geomorphologist William Morris Davis. Yet, within a generation, the discipline had come almost entirely to reject Semple's project and to see her as "the bogey-lady of a slightly silly concept that has now happily been abandoned and forgotten" (quoted in Keighren, 2010, p. 174). What had changed? Why had Semple's project come no longer to matter? How could the judgement of Semple's peers have changed so rapidly?

Conclusion

Of the various factors that contributed to the declining influence and eventual refutation of Semple's ideas, the most fundamental centred around the question of evidence and how closely Semple's deterministic explanations fitted with what could actually be seen in the real world. Over time, a growing disjuncture was apparent between the theory of anthropogeography and the observations that emerged from the study of human populations in a range of geographical environments. We see this, for example, in the early fieldwork experience of the influential cultural geographer Carl Sauer, who became the most prominent twentieth-century opponent of Semple's project (Solot, 1986; see the chapter by Kent Matthewson, this volume). Ironically, Sauer had completed his PhD at Chicago, and had come under the influence there of the environmentalism that pervaded the department during Semple's tenure. Sauer's doctoral research focused on the cultural geography of the Ozark Highlands in Missouri. Much of his initial interpretation of the region was filtered through the lens of environmental influence. Over time, however, Sauer realised that his observations did not fit with the theoretical precepts of Semple's model. The Ozark Highlands was a region defined by its topography and climate, yet the population exhibited considerable variety in cultural practices, like agriculture and architecture. Semple's model would have predicted a cultural uniformity imposed by shared geographical influences, not cultural diversity. What Sauer was seeing was evidence of earlier migration: the Ozark Highlands were populated by the descendants of migrant communities from Kentucky, New England, France, and Germany who had retained their social and cultural practices, despite the new geographical environment in which they were now living (Keighren, 2010). The realm of culture, Sauer saw, mattered just as much as did environmental influences.

In addition to the failure of Semple's model to adequately explain the combined significance of geographical and cultural factors in shaping human societies, the wider context of her work effectively became taboo in the second half of the twentieth century as a consequence of the extreme ends to which related geographical ideas were put in support of German Nazi ideology and ideas of racial superiority in the 1930s and 1940s (Keighren, 2020). By the time Semple died in 1932, however, the influence of her thinking was already on the wane, and she is generally remembered now as the prophet of an unsophisticated and rather naïve form of geographical explanation that emerged from, and implicitly reinforced, a twentieth-century intellectual culture defined by white supremacy.

The appeal of environmental and geographical explanations to the study of human history and geopolitics has never, however, entirely disappeared (Meyer and Guss, 2017). It is evident, today, in a range of popular books on neo-environmental determinism, like Jared Diamond's *Guns, germs, and steel* (1997), Robert Kaplan's *The revenge of geography* (2012), and Tim Marshall's *Prisoners of geography* (2015). Such works tend, in offering seemingly common-sense explanations of Geography's significance, to lack nuance. But understanding the role of the physical environmental in shaping patterns of human life and existence has arguably never been more urgent. We are residents now of the Anthropocene (a new, potentially threatening epoch in the Earth's long history) and our lives will be shaped by the global climate crisis in ways that we cannot yet know. In seeking explanations that can anticipate the social consequences of this future uncertainty, geographers and other scholars are engaging anew with the role of environmental influences, and finding novel and important ways to make Geography matter.

Study task

Variations on this question can be set on an unseen examination, an open-book examination or as a term-paper assignment:

To what extent does the socially uneven impact of the 'climate crisis' show that the study of environmental influence alone is insufficient?

Starter reading list

Adamson, George C. D., Matthew J. Hannaford, and Eleonora J. Rohland. (2018) 'Re-thinking the present: The role of a historical focus in climate change adaptation research', *Global Environmental Change* 48, pp. 195–205. j.gloenvcha.2017.12.003

Dalby, Simon. (2015) 'Environment: From determinism to the Anthropocence'. In John Agnew, Virginie Mamadouh, Anna J. Secor, and Joanne Sharp (eds), *The Wiley Blackwell Companion to Political Geography*. Chichester: John Wiley and Sons, pp. 451–461.

Fleetwood, Lachlan F. (2023) 'Histories of habitability from the Oikoumene to the Anthropocene', *WIREs Climate Chance* 14(5). e840, 10.1002/wcc.840

Hulme, Mike. (2011) 'Reducing the future to climate: A story of climate determinism and reductionism', *Osiris* 26(1), pp. 245–266. 10.1086/661274

Lunstrum, Elizabeth and Pablo S. Bose. (2022) 'Environmental displacement in the Anthropocene', *Annals of the American Association of Geographers* 112(3), pp. 644–653. 10.1080/24694452.2021.1995316

Millar, Susan W. S. and Don Mitchell. (2017) 'The tight dialectic: The Anthropocene and the capitalist production of nature', *Antipode* 49(S1), pp. 75–93. 10.1111/anti.12188

Radcliffe, Sarah A., Elizabeth E. Watson, Ian Simmons, Felipe Fernández-Armesto and Andrew Sluyter. (2010) 'Environmentalist thinking and/in geography', *Progress in Human Geography* 34(1), pp. 98–116. 10.1177/0309132509338749

References

Adams, E. E. (2011) *Ellen Churchill Semple and American geography in an era of imperialism.* PhD thesis. Williamsburg, VA: The College of William and Mary.

Adams, E. E. (2014) 'Colonial geographies, imperial romances: Travels in Japan with Ellen Churchill Semple and Fannie Caldwell Macaulay', *The Journal of the Gilded Age and Progressive Era*, 13(2), pp. 145–165.

Anonymous (2 June 1842) *The Brooklyn Eagle, and Kings Country Democrat*, 1(147), unpaginated.

Anonymous (1843) 'Elements of geography', *The Journal of the Royal Geographical Society of London*, 13, pp. 330–338. 10.2307/1798154

Anonymous (1912) 'Review of *Influences of geographic environment*, by Ellen Churchill Semple', *Scottish Geographical Magazine*, 28(3), pp. 160–162. 10.1080/14702541208554737

Berdoulay, V. (2009) 'Possibilism', in Kitchin, R. and Thrift, N. (eds) *International Encyclopedia of Human Geography*, vol. 9. Oxford: Elsevier, pp. 312–320. 10.1016/B978-008044910-4.00721-5

Bingham, M. T. (1932) 'Ellen Churchill Semple, geographer', *Vassar Quarterly*, 17(3), pp. 237–240.

Close, C. F. (1911) 'The position of geography', *The Geographical Journal*, 38(4), pp. 404–413. 10.2307/1778545

Colby, C. C. (1933) 'Ellen Churchill Semple', *Annals of the Association of American Geographers*, 23(4), pp. 229–240. 10.1080/00045603309357074

Crone, G. R. (1970) *Modern Geographers: An Outline of Progress since AD 1800.* 2nd ed. London: The Royal Geographical Society.

Curzon, G. (1912) 'Discussion', *The Geographical Journal*, 40(6), pp. 602–607. 10.2307/1778217

Darwin, C. (1859) *On the Origin of Species by Means of Natural Selection, or the Preservation of Favoured Races in the Struggle for Life.* London: John Murray.

Diamond, J. M. (1997) *Guns, Germs, and Steel: The Fates of Human Societies.* New York: W. W. Norton.

Evans, S. L., Keighren, I. M. and Maddrell, A. (2013) 'Coming of age? Marking the centenary of women's admission to the Royal Geographical Society', *The Geographical Journal*, 179(4), pp. 373–376.

Flowerdew, R. (2009) 'Probabilism', in Kitchin, R. and Thrift, N. (eds) *International Encyclopedia of Human Geography*, vol. 9. Oxford: Elsevier, pp. 448–450. 10.1016/B978-008044910-4.00729-X

Galton, F. (1855) 'Notes on modern geography', in *Cambridge Essays, Contributed by Members of the University*, vol. 1. London: John W. Parker and Son, pp. 79–122. https://hdl.handle.net/2027/umn.31951p00098504t

Giaccaria, P. and Minca, C. (eds.) (2016) *Hitler's Geographies: The Spatialities of the Third Reich.* Chicago: University of Chicago Press.

Hawkins, M. (1997) *Social Darwinism in European and American thought, 1860–1945: Nature as Model and Nature and Threat.* Cambridge: Cambridge University Press.

Johnson, J. M. (2008) *Southern Women at the Seven Sister Colleges: Feminist Values and Social Activism, 1875–1915.* Athens: University of Georgia Press.

Kaplan, R. D. (2012) *The Revenge of Geography: What the Map Tells Us about Coming Conflicts and the Battle against Fate.* New York: Random House.

Keighren, I. M. (2008) 'Giving voice to geography: Popular lectures and the diffusion of knowledge', *Scottish Geographical Journal*, 124(2–3), pp. 198–203. 10.1080/14702540802411782

Keighren, I. M. (2010) *Brining Geography to Book: Ellen Semple and the Reception of Geographical Knowledge*. London: I.B.Tauris.
Keighren, I. M. (2015) 'Environmental determinism', in Wright, J. D. (ed.) *International Encyclopedia of the Social and Behavioral Sciences*, vol. 7. 2nd ed. Oxford: Elsevier, pp. 720–725. 10.1016/B978-0-08-097086-8.72016-2
Keighren, I. M. (2017) '"A Royal Geographical Society for ladies": The Lyceum Club and women's geographical frontiers in Edwardian London', *The Professional Geographer*, 69(4), pp. 661–669. 10.1080/00330124.2017.1289773
Keighren, I. M. (2020) 'History and philosophy of geography III: The haunted, the reviled, and the plural', *Progress in Human Geography*, 44(1), pp. 160–167. 10.1177/0309132518818725
Kleber, J. E. (ed.) (2001) *The Encyclopedia of Louisville*. Lexington: The University Press of Kentucky.
Livingstone, D. N. (2002) 'Race, space and moral climatology: Notes toward a genealogy', *Journal of Historical Geography* 28(2), pp. 159–180. 10.1006/jhge.2001.0397
Livingstone, D. N. (2024) *The Empire of Climate: A History of an Idea*. Princeton: Princeton University Press.
Marshall, T. (2015) *Prisoners of Geography: Ten Maps that Tell You Everything You Need to Know About Global Politics*. London: Elliot & Thompson.
Meyer, W. B. and Guss, D. M. T. (2017) *Neo-environmental Determinism: Geographical Critiques*. Cham: Palgrave Macmillan.
Richards, R. J. (2018) 'Ernst Haeckel: A dream transformed', in Harman, O. and Dietrich, M. R. (eds) *Dreamers, Visionaries, and Revolutionaries in the Life Sciences*. Chicago: University of Chicago Press, pp. 35–50.
Semple, E. C. (1897) 'The influence of the Appalachian barrier upon colonial history', *The Journal of School Geography*, 1, pp. 33–41.
Semple, E. C. (1900a) 'Anthropogeographie', *The Annals of the American Academy of Political and Social Science*, 16(1), pp. 137–139. 10.1177/000271620001600110
Semple, E. C. (1900b) 'A new departure in social settlements', *The Annals of the American Academy of Political and Social Science*, 15(2), pp. 301–304. 10.1177/000271620001500211
Semple, E. C. (1901) 'The Anglo-Saxons of the Kentucky Mountains: A study in anthropogeography', *The Geographical Journal*, 17(6), pp. 588–623. 10.2307/1775213
Semple, E. C. (1903) *American History and Its Geographic Conditions*. Boston: Houghton, Mifflin and Company.
Semple, E. C. (1904) 'Emphasis upon anthropo-geography in schools', *The Journal of Geography* 3, pp. 366–374. 10.1080/00221340408985533
Semple, E. C. (1910) 'The Anglo-Saxons of the Kentucky Mountains: A study in anthropogeography', *Bulletin of the American Geographical Society* 42(8), pp. 561–594. 10.2307/199958
Semple, E. C. (1911) *Influences of Geographic Environment, on the Basis of Ratzel's System of Anthropo-geography*. New York: Henry Holt and Company.
Semple, E. C. (1912) 'Influence of geographical conditions upon Japanese agriculture', *The Geographical Journal*, 40(6), pp. 589–603. 10.2307/1778216
Solot, M. (1986) 'Carl Sauer and cultural evolution', *Annals of the Association of American Geographers*, 76(4), pp. 508–520. 10.1111/j.1467-8306.1986.tb00133.x
White, C. L. (1977) 'Ellen Churchill Semple'82: The dean of American geography', *Vassar Quarterly*, 74(1), pp. 49–50.

4 Keeping human and physical geography together

Richard Chorley and Peter Haggett's scientific turn*

Trevor Barnes

Richard Chorley in 1962.

From Stoddart, D. 1997. Richard J. Chorley and modern geomorphology. In D. R. Stoddart, Ed., *Process and form in geomorphology*. London: Routledge, pp. 383–399; p. 392.

Peter Haggett in 1984.

From A. D. Cliff, P. R. Gould, A. G. Hoare and N. J. Thrift, Eds., *Diffusing geography: Essays for Peter Haggett*. Oxford: Blackwell.

Introduction

Born between the two World Wars and only 20 miles (32 kms) apart in "the same corner of western Somerset" (a county in the West of England), Richard ("Dick") Chorley (1927–2002) and Peter Haggett (1933–) from the late 1950s fundamentally reconceived the geographical project (Dow 1982, 1). Before their intervention, Anglo-American

* I am extremely grateful to Tony Gattrel, Peter Haggett, and Olav Slaymaker for their enormously helpful comments and information that significantly improved the chapter.

DOI: 10.4324/9781003343240-5

human and physical geography were primarily descriptive, concerned with enumerating raw facts, prose-driven, about specific and concrete entities, illustrated by black and white photos of the phenomena depicted, and fixed geographically by using two-dimensional maps of various kinds and at various scales.

After the Second World War, the descriptivism of the older approach to Geography became increasingly out of step with other kindred disciplines – both social and earth sciences – which although concerned with similar substantive topics to Geography, drew instead on a scientific analysis directed at explanation. Geography seemed in comparison lightweight and naïve, continuing to peddle "mere description", compiling dull gazetteers of facts. Neil Smith (1989) likened the state of early post-War Geography to a museum-like existence, preserved under glass, unchanging, stuck in the past.

Chorley and Haggett were among the first to challenge the old geography, breaking open the museum cases, replacing their content with a "new geography". That new geography – their project – involved recasting human and physical geography as exact sciences akin to physics, chemistry, or the new biology of genetic analysis and DNA. Not that Chorley and Haggett were the only ones to do this within Anglo-American Geography – there was a similar, even stronger and more sweeping movement in North America (Barnes 2008). But the result was the same. Geography became a science of spatial relations: "spatial science". How geography looked on the page, how it was written, what skills were needed, what counted as good work or, its opposite, shoddy work, and who was deemed a capable geographer, were irrevocably altered.

While Chorley and Haggett believed their project made Geography intellectually respectable, putting disciplinary knowledge on a secure scientific footing, it also offered a solution to a problem long dogging the subject: the relationship between its often only loosely joined two halves, human and physical geography. If these two halves could not be joined, it would raise an existential question about Geography as a separate discipline. For Chorley and Haggett, though, that issue was fully resolved by the "new geography". Human and physical geography would be united by a single common scientific methodology giving the discipline "a monistic view" (Haggett and Stoddart 2009, 65).

A unified Geography had been an aspiration from the very birth of the academic subject. During this early period, the two parts of the discipline – human and physical – were often integrated by a crude environmental determinism; the idea that the physical world shaped the human world (see Chapter 3). As the American geographer Ellen Churchill Semple (1863–1935), one of the great early twentieth-century environmental determinists, lyrically put it in the opening of her *Influences of geographic environment*:

> Man is a product of the earth's surface. This means not merely that he is a child of the earth, dust of her dust; but that the earth has mothered him, fed him, set him tasks, directed his thoughts, confronted him with the difficulties that have strengthened his body and sharpened his wits, given him the problems of navigation or irrigation, and at the same time whispered hints of their solution. (Semple 1911, 1)

Lyrical though it might be, the underlying thesis of environmental determinism was at base coarse, unscientific, smacking of racism and imperial jingoism (Keighren, Chapter 3, labels it a "toxic legacy"). As a theoretical project environmental determinism couldn't be sustained, and it wasn't, floundering during the early interwar period. As a result, the disciplinary glue holding together human and physical geography was lost. It took another 30 years of so before there was another attempt, as Gregory and Castree (2011, lvii) write, "to jumpstart the stalled geographical experiment ... to unite human and physical geography". That was Chorley and Haggett's. The approach they advocated was very different from environmental determinism. Uniting Geography would be not a one-way causal arrow between the physical and human world, but a common approach, the scientific method. Human and physical geographers would be joined by collectively upholding logical rigour; theoretical explanation; mathematical formulations often embodied in abstract models; precise data measurement; hypothesis formulation and testing; and the use of quantitative and statistical techniques for analysis and verification.

The early years

Chorley and Haggett first met in early summer, 1958, at Steyning's Tea Rooms, Bridgwater, Somerset.[1] They had come to discuss their teaching as Demonstrators in their upcoming first-year undergraduate lab and field methods class at the Department of Geography, Cambridge University.[2] Haggett had joined the Department as a Demonstrator the year before, while Chorley was newly hired. The class ran the whole academic year, meeting for two hours every Monday, Wednesday, and Friday.[3] There was also a second-year graduate student assistant ("Student Demonstrator") who was writing a doctoral thesis on 19th-century Kent hop farming, David Harvey (see Chapter 8). Because Chorley and Haggett were both required by the Head of Department, Alfred Steers, to attend every class even though the other might take the lead on teaching, that gave, as Chorley (1995, 360) wrote, "some 120 hours ... for Peter and I to chat with one another". Those chats helped to forge a lasting intellectual partnership and deep friendship, marking the beginning of their quest to unify human and physical geography by a common scientific method. The trajectories that brought each of them to Steyning's Tea Room were similar but different.

Richard Chorley

Chorley was born in Minehead, Somerset, in 1927,[4] and attended the local grammar school. The Second World War ended five days before he turned 18 (the conscription age). He was still required to enlist for two years of national service, in his case joining the Corps of Royal Engineers (1946–1948). In 1948, he went up to Oxford to study Geography at Exeter College. His tutor for the entire three years was the iconoclastic geomorphologist, Robert Beckinsale (1902–1998), who was already beginning to break away from traditional descriptive physical geography and its domination by the so-called Davisian ("denudational chronology") model of landform erosion (Beckinsale 1997, 205).

The Davisian model was named after the Harvard geologist and geographer, William M. Davis (1850–1934), for work he carried out at the very end of the nineteenth century. It recognized three distinct descriptive stages in the evolving form of the physical landscape. After an abrupt period of land uplift, erosional forces set in, creating first, a "youthful" landscape of deep, narrow V-valleys, second and following, a "mature" landscape of broad U-valleys, and finally, ending as an "old" flattened landscape of lateral erosion, a peneplain. Beckinsale's scepticism of Davis' simple descriptive and teleological life-cycle model of geomorphological transformation influenced Chorley. He called

Davis' work "highly ambiguous [and] subjective" (Dow 1982, 2). Instead, Chorley aimed for scientific certainty and objectivity. Beckinsale knew where Chorley could find both.

Awarded a prestigious Fulbright Fellowship,[5] in late summer 1951 Chorley enrolled in the graduate geology programme at Columbia University, New York City, working with Arthur Strahler.[6] In an unpublished note, Chorley recalled his first encounter with his supervisor. Describing Strahler as a "tall, gangling, 34-year-old Midwesterner who reminded me of [the Hollywood actor] James Stewart", at that meeting, Strahler said:

> Well, I guess you might like to look at this!", [and he] handed me a copy of his erosional slope development paper [Strahler (1950)] and ushered me out. In a state of excitement I went to the Lions' Den, ordered myself a large chocolate malted with two scoops of ice cream, and settled down to read the paper. The milk shake rose about half the length of the straw—and remained there! I recall that the juke box was playing Glen Miller's 'String of Pearls'. (quoted by Haggett 2008, 349, fn. 5)

Strahler's 1950 paper was the first published direct formal assault on the Davisian model. While Strahler (1950, 674) acknowledged that "our understanding of erosional landforms is still in a stage of generalized verbal description", he recognized the future possibility of "relating quantitatively the effect of causative factors to [geomorphological] form characteristics".[7] By carefully measuring, drawing on formal statistical methods and fluid mechanics, and theoretically using the notion of equilibrium from von Bertalanffy's "general systems theory", Strahler (1992, 65–66) moved from, as he labelled it, a "pre-paradigm" "qualitative/descriptive geomorphology" to a scientific "quantitative/dynamic geomorphology".

No wonder Chorley stopped mid-suck. He had never read anything quite like it. Along with other students in his cohort supervised by Strahler, Chorley had, as he later put it, "stumbled into the control room of a scholarly revolution" (quoted by Schumm 2004, 672). Supported by large gobs of Cold War military money that funded students, Strahler's paradigm-busting scientific "quantitative/dynamic geomorphology" wasn't completely a bolt out of the blue.[8] But it was the first to pinpoint an enemy, Davis, along with his model of "denudational chronology" and concomitant historical-descriptive approach. Strahler offered a clear alternative, the scientific method, and later codified in his best-selling textbook, *Physical Geography*, which continued in new editions even after his death in 2002.

While continuing his doctoral research, Chorley became first an Instructor at Columbia's Geography Department (1952–1954), and later an Instructor in Geology at Brown University. The death of his father in 1957 caused him to return to England, however. A Hail-Mary letter and CV sent to Alfred Steers, Head of the Cambridge Department of Geography, miraculously netted Chorley an interview and the offer of a Demonstratorship beginning in Michaelmas term 1958.[9]

Peter Haggett

Peter Haggett attributed his early interest in Geography both to an inspiring teacher at his grammar school in Bridgewater (Dow 2000), and to a spinal cord injury sustained at age 16 in a rugby match. The latter resulted in him being "encased, Pharaoh-like, in a long head-to-toe plaster cast" for 3 months in a hospital bed (Haggett 1990, 1). By a system of pulleys, levers, and mirrors, he could still read. The "first serious book" brought to him while recuperating was Charles Cotton's *Geomorphology of New Zealand* (1942). He was captivated "by Cotton's diagrams of finely etched landforms", and his geophysical account of the causal processes that underlaid them (Haggett 1990, 2). He was entranced. He would be a geographer interested in spatial morphological structures and their scientific explanation.

In 1951, Haggett won a scholarship to St. Catharine's College (the "geography college") at Cambridge University. He originally intended to specialize in geomorphology, particularly glaciology. His tutor was the brilliant glaciologist, Vaughan Lewis, a Cambridge Mathematics graduate, who encouraged Haggett "to think mathematically about geographical problems" (Haggett 2008, 338). During his final undergraduate year, though, Haggett had an unlikely road-to-Damascus conversion reading the German location theorist August Lösch's just-translated monograph, *The Economics of Location* (1954). The book provided a general theoretical framework for representing the morphological structure of an urban system. The many geometric figures of urban morphology that littered Lösch's book reminded Haggett of the "snow and ice, its crystal structure" that he studied in glaciology (Dow 2000).[10] In Lösch's case, though, "they were crystal structures that were man-made" (Dow 2000). With that recognition, Haggett changed focus. Rather than identifying physical morphological structures like New Zealand's landforms or crystals of snow or ice, he concentrated on human-made geographical structures. As Haggett later reflected: "I looked at the socio-economic landscape in terms of its morphology, trying to see what its structural components were" (Browning 1982, 47). Following his training in physical geography under Lewis, Haggett translated that morphology into mathematics, especially geometry (the mathematics of spatial form). With the spatial morphology formalized, he drew on the scientific method to provide causal explanations.

In 1954, awarded top marks, Haggett stayed at Cambridge to write a PhD (although it was not given until the late 1960s). Even then, like Chorley, he believed that Geography must change, to move away from prose-based description to mathematical (scientific) explanation. Further, Haggett believed such a change could occur quickly, and be carried out by a new generation of scholars like him drawing on scientific methods. He'd already seen that kind of change at Cambridge when he was an undergraduate. In 1953 at Cambridge's Cavendish Laboratory, a 30-year-old Francis Crick, who was yet to complete his PhD, and a 24-year-old graduate student, James Watson, turned biology and the world upside down with their explanation of the structural form of the DNA molecule, relying on scientific analysis and data modelling (Haggett 2008).

In 1955, Haggett spent two years as an Assistant Lecturer at University College, London, where he jointly taught a class on economic geography. Bill Mead, his co-instructor, "encouraged [him] to introduce industrial location models", including their stark mathematical forms that were beginning to be formulated in Walter Isard's contemporaneous Regional Science (Haggett 2008, 339). In 1957, Haggett returned to Geography at Cambridge as a University Demonstrator.

The "Terrible Twins" of British Geography

While the geographer Derek Ford (2002) said being a Cambridge Demonstrator was only "about one step up from being a cleaner on the premises", for Chorley and Haggett the position enabled them to begin transforming British Geography into a unified spatial science. Both believed that to make that change first involved providing geographers with the necessary toolbox of spatial analytical methods. This had been the central topic of their conversation at Steyning's. To that end, their first-year Cambridge students were taught a double-distilled dose of quantitative techniques, "to do with statistical methods, matrices, set theory, trend surface analysis, and network analysis" (Chorley 1995, 361). It replaced the previous lab exercises based on "artsy cartography and the laborious construction of map projections" (Beckinsale 1997, 8). Quantitative methods and locational analysis had never been systematically taught before in British Geography Departments. But for Chorley and Haggett, the first step to fermenting a disciplinary revolution was for human and physical geography to share the same scientific model of explanation and set of analytical techniques.

In 1963, the year that both Chorley and Haggett were promoted to permanent Lecturers at Cambridge, a second pedagogical step in the revolution was taken. Chorley and Haggett organized a week-long residential summer school for sixth-form Geography teachers at Madingley Hall, a converted rambling Tudor mansion later fitted with tennis courts and a croquet lawn located 5km outside Cambridge. Sixth form was an extra two-year period of schooling that British students aged between 16 and 18 could elect to take to sit specialized exams, "A [Advanced] levels", which if successful allowed them to enter university. Chorley and Haggett believed that if they could convince sixth-form teachers to instruct their students in the new geography, it would boost demand at the university level for their approach. As Chorley (1995, 367) later wrote:

> we were convinced .... that the correct strategy was to attack points of critical vulnerability [within the British geographical establishment] and bypass the strong points. Accordingly, we identified a core of intellectually dissatisfied, very active and highly influential sixth-form teachers as our critical point of incursion into the existing system. It was clear that these teachers could be extremely influential in syllabus reform, that their pent-up professionalism could be rapidly unleashed in terms of their teaching and the changes in sixth-form teaching could rapidly diffuse up into the universities and down into the younger pupils.

Madingley Hall had been purchased just after World War II by Cambridge University's Board of Extra-Mural Studies (the arranger of extension classes; Haggett 2015). That Board was interested in especially equipping teachers with the skills to teach the so-called "new mathematics" being introduced into the curriculum of British schools. The Board saw what Chorley and Haggett were doing as complementary to that broader curricular transformation, and they were invited to offer a summer school.

In late July 1963, Chorley and Haggett's course, "Modern Geography", was given at Madingley to 28 "teachers and other serious students" to update them on "recent changes to their subject" made at university level (Haggett 2015, 6). The disciplinary changes reviewed, however, were of only one kind: those associated with Haggett and Chorley's new geography. The Madingley Hall summer school then ran annually for the next five years. Although the course title "meandered", its format remained constant (Haggett and Stoddart 2009, 69). It began with a review of the main developments in the field "with an emphasis on spatial modelling" and was followed by "practical classes" focused on hands-on techniques such as "morphometric analysis, graph theory, linear programming, [and] multivariate modelling" (Haggett 2015, 7). Staff were brought in from around the country to lecture on their specialty and paid a flat fee of five guineas (£5.25). The result was lots of converts. One schoolteacher said after the first summer school, "Madingley was a revelation of 'Road-to-Damascus' proportions" (quoted in Chorley 1995, 371).

Two books directly came out of Madingley, both edited by Chorley and Haggett: *Frontiers in Geographical Teaching* (1965) and *Models in Geography* (1967). Both volumes presented a vision of a scientific geography defined by abstraction, formal theoretical analysis, mathematical models, and statistical techniques. Traditional Geography, they argued in *Frontiers*, was "a Victorian academic structure every bit as solid and constricting as its architectural counterpart" (Haggett and Chorley 1965, 380). The house of geography, though, urgently needed remodelling and modernizing. That was just as true for physical as human geography. Both of Chorley's (1965a, b) chapters in *Frontiers* took direct aim at the Davisian model.

Models was even more sweeping in its ambition than *Frontiers*. Published both as an 816-page omnibus doorstopper as well as four shorter paperback "best-of volumes", it tried to demonstrate that the new geography could be applied to all sub-fields of the discipline. The tie that bound the 30-plus chapters was the idea of the model, defined by Haggett (1965, 19) two years earlier as an "idealized representation of reality". Model was code for joining abstract explanatory theory with analytical quantitative methods. Geographical modelling required not just one or the other but both. This was demonstrated in individual chapters that traversed the full spectrum of geographical concerns from hydrology to urban geography, from hierarchies of rivers to hierarchies of cities. Each chapter laid out with clinical precision a model-based version of a given geographical subfield, demonstrating its plausibility, utility, and relevance. As Reg Golledge (2006, 112, 110) said in a retrospective review, "For many of us *Models in Geography* became a bible" leaving a "lasting and significant impact on the discipline". It was a "textbook that moved generations" (Golledge 2006, 107). Pertinently for this chapter, *Models* offered a means to unify the discipline. As Golledge (2006, 111) again wrote, "*Models in Geography* clearly showed ... that Geography was as much a human science as a physical science".

Frontiers and particularly *Models* fundamentally shook the foundation of the older "Victorian" edifice of Geography. It introduced abstract theory, ideas of rigorous causal explanation, the careful collection of data, precise measurement, formal verification, and ideas from the philosophy of science. Chorley and Haggett (1967) in the introduction to *Models* invoked the work of the pre-eminent philosopher of science, Thomas Kuhn, on scientific paradigms, while former Student Demonstrator David Harvey (1963) in his chapter presented ideas from logical positivism. Moreover, both volumes made the case for a monistic discipline: one subject, one method. Inevitably, this rankled the old guard. That's why David Harvey joshingly dubbed Chorley and Haggett the "terrible twins" of British Geography (Haggett and Stoddart 2009, 74), and those who contributed to their edited volumes became known as "young Turks".[11] As Haggett (2008, 343) later wryly put it and in his usual overly modest way, "it was on the croquet lawns of Madingley that a very gentle and minor revolutionary ripple was set moving".

Chorley's general systems theory and Haggett's *Locational Analysis*

Apart from their joint work, Chorley and Haggett separately sought to remake their own corners of the discipline. For Chorley that was initially geomorphology, but later expanded to include climatology and hydrology. And for Haggett that was originally urban and economic geography, but later health geography, which ended up preoccupying him for the second half of his career. In Chorley's case there was no single standout work, but several, each additively contributing to his new "quantitative, model based paradigm" (Haggett 2002, 523). In Haggett's (1965) case, although his publications were prodigious, one towered over all others, his 1965 book, *Locational analysis in human geography*.

Richard Chorley

Chorley's enormous drive and energy allowed him to carry out multiple projects simultaneously, including, in his "spare time", a revered intellectual history of geomorphology, and continuing after his death (five fat volumes). Most germane here was his critique of traditional geomorphology and the "quantitative-theoretical alternative" he pursued following his New York City chocolate-malted epiphany (Haggett and Stoddart 2009, 71).

Despite Chorley's "chief adversary", Professor of Physical Geography at King's College, University of London, Sidney Wooldridge, saying about the new geomorphology, "I don't want it. I don't like it and I won't have it!", it came anyway (Slaymaker 2022, quoted 418). Early on Chorley made two significant kinds of contribution that hastened its arrival. The first was technical, providing both reviews and case studies that exemplified an assortment of statistical and modelling techniques for prosecuting a scientific geomorphology. Chorley's 1965a long review of applied statistical methods for geomorphologists especially "became a much-thumbed bible for graduate students and colleagues alike" (Haggett and Stoddart 2009, 71). The second was developing a larger methodological-cum-philosophical scientific framework for understanding the dynamic relationships among different physical geographical processes at different spatial scales: general systems theory. Originating with the Austrian biologist, Ludwig von Bertalanffy (1901–1972), general systems theory is about how parts of a larger system (conceived as "sub-systems") are directly or indirectly causally related, which then shape and change the whole. Chorley began corresponding with von Bertalanffy during the late 1950s, publishing his pioneering and influential paper, "Geomorphology and general systems theory" in 1962. It was followed by more expansive treatments that dealt with the full range of physical geographical processes as well as the ends of control and optimization. General systems theory typically took on the distinctive form of flow diagrams composed of different-sized shaded boxes, arrows of different lengths and directions, along with a generous scattering of mathematical signs and symbols. It is gently satirized along with Chorley at work in Figure 4.1.

Peter Haggett

Haggett's (1965, v) *Locational analysis in human geography* was based on his "much-thumbed and much-revised lecture notes" for his course, "Economic Geography – Studies in Location", which he offered to Cambridge undergraduates at 9 a.m. on Saturday mornings. In turn, those notes were typed up to become *Locational Analysis* (see Figure 4.2). Much of that volume's content derived from work carried out in urban and economic geography in North America. As Haggett said in an interview, "there was very little of my own research reported [in *Locational analysis*] .... [The book] rather was like a kind of Alistair Cooke commenting on the American scene" (Browning 1982, 47).[12]

Written with crystalline clarity, and lavishly illustrated with 162 separate figures, the genius of *Locational analysis* was finding a consistent theme that both held together its parts, and served as a paradigmatic alternative capable of relocating the discipline from its dated, creaky Victorian home to a new, smart, modern abode. That theme was geometrical (morphological) structure, which brought in its train abstract space, formal analysis, and a specialized mathematical vocabulary. The book became an academic blockbuster almost overnight. John Davey, an assistant editor at the book's publisher, Arnold,

Figure 4.1 Richard Chorley constructing one of his many works in the early 1960s, as viewed by a local wag.

Source: Courtesy of an anonymous contributor.

Figure 4.2 Peter Haggett (and son) typing *Locational Analysis in Human Geography* at his Somerset home, 1964.

Photo by Brenda Haggett with permission from Peter Haggett.

remembers it "selling a huge number of quantities in hardback at two guineas [£2.10] each, which is quite a high price ... and it was then reprinted every year".[13] There had never been a book quite like it before in British human geography. Maurice Yeates (1991, 301) said it was "the Grail of disciplinary intellectual coherence", while Dick Morrill (1991, 300) believed it "stood as a revelation and as a statement of faith in geography as a science". Tony Gatrell bought the book at age 17 on the recommendation of his sixth-form Geography teacher, Brian Fitzgerald, who had attended a Madingley summer school. Gatrell said he "knew the Preface by heart", and the book was his "pride and joy" (quoted in Thrift 1995, 383).

Reception and aftermath

It was not everyone's pride and joy, nor, more generally, was the new geography. Early on both Chorley and Haggett ran into flak from established British geographers. In November 1958, when Chorley was introduced to Sidney Wooldridge (1900–1963) as Strahler's former student, Wooldridge responded: "Next time you see Strahler tell him I spit at him!" (quoted in Beckinsale 1997, 6). Haggett was confronted by an even more imposing figure, Professor Sir Dudley Stamp (1898–1966), perhaps Britain's leading geographer of the 1940s and 1950s (see Chapter 18). In 1957 Stamp had already said in his Institute of British Geographers' (IBG) presidential speech, that he was "alarmed" by the "training" of geographers in "statistical methods" (Stamp 1957, 2). That criticism persisted in Stamp's comments on Haggett's (1964, 380) 1964 IBG conference presentation that employed multivariate regression. "Some of us might not like the look of this", Stamp said. It is like using "an enormous steam hammer ... [for] cracking nuts". The condemnation continued literally to the end in Stamp's last (posthumous) paper that he was scheduled to deliver at a 1966 Mexico City International Geographical Union Conference. He was going to say chidingly that *Locational Analysis* was "an introduction to a new Theology .... Quantification ... has become a religion to its devotees, its golden calf is the computer" (quoted in Stoddart 1997, 384). He never did, though. He died of a heart attack in the conference hotel lift before he got to the podium.

Despite these objections, maybe in part because of them, Chorley and Haggett's new geography gained pace rather than slackened. There was too much momentum both inside the discipline and outside. Inside, impetus came from: newly minted quantitatively trained undergraduates, with the Cambridge model replicated in other Geography departments; the Madingley summer schools that changed sixth-former expectations about Geography, which then carried over to the university; a generational shift as existing universities expanded and new ones were created resulting in large-scale hiring of "young Turks" (see fn. 12); and the publication of texts like *Frontiers*, *Models* and *Locational analysis* that excited students and became templates for new courses. Golledge (2006, 109), for example, said Barry Garner's chapter 9 in *Models* on urban geography "was comprehensive enough" for "it alone" to be the basis of an urban geography course that he "taught for many years at the Ohio State University".

Outside forces providing momentum turned especially on the generational move within the UK to break free of that country's early post-War stuffy traditionalism, which was rigid, class based, deferential, and deeply conventional – "Little Britain". In Geography that was embodied by figures such as Wooldridge and Stamp. Breaking free meant doing something completely different;[14] disobeying rules; thumbing noses at authority; embracing change and modernization. That's what the "terrible twins" and the "young Turks" were trying to do. As David Harvey (2002, 164), who was one of them, writes: "The modernization of Britain was firmly on the agenda, and a new structure of knowledge and power was needed to accomplish that task". There was "the idea", he later said, "that we could break out of tradition.... There was a modern geography waiting to be constructed and we were the ones who could do it".[15]

A modern Geography was constructed, and for a relatively brief period of a decade or so, the Chorley and Haggett vision of a monistic discipline began to be realized. From the early 1970s, however, that began to change with physical and human geography no longer coming together, but separating, moving apart. Physical geography continued its path, shifting ever further from Strahler's "qualitative/descriptive" mode to become increasingly a physical science defined by the scientific method – quantification, measurement, mathematical modelling, and theoretical explanation. In contrast, human geography practices, rather than narrowing, converging to a single scientific method, diverged and splintered, becoming instead ever more methodologically and philosophically pluralistic. That was represented from the early 1970s and into the 1980s and 1990s as a fusillade of "isms" – Marxism, humanism, feminism, critical realism, post-structuralism, post-colonialism, post-Marxism, and more recently, queer, gender, and critical race studies.

That unshackling of the intellectual trajectories of human and physical geography began with a series of reversals by former human geography proponents who had hitherto cleaved to the Chorley–Haggett line. The most renowned of the turncoats was David Harvey, the Student Demonstrator at the Cambridge first-year Geography class, who had been literally in the room where it happened. By 1972, however, influenced by Marxism, he wrote:

> [Geography's] quantitative revolution has run its course and diminishing marginal returns are apparently setting in as ... [it] serve[s] to tell us less and less about anything of great relevance.... There is a clear disparity between the sophisticated theoretical and methodological framework which we are using and our ability to say anything really meaningful about events as they unfold around us. ... In short, our paradigm is not coping well
>
> (Harvey 1972, 6)

The specific reasons for the rejection of the scientific method in human geography were multiple. It was partly political, and often framed as a lack of "relevance", Harvey's objection (see Chapter 8). It was partly analytical inconsistency, as in the Swedish geographer Gunnar Olsson's (1975) critique of spatial science's "model of models", spatial interaction theory. It was partly moral, as in David Ley and Marywn Samuels' (1978) humanist criticisms of the skeletal, de-contextualized, and ethically hollow model of humans and places deployed in spatial science (see Chapter 10). And partly it was philosophical and found in the British geographer Derek Gregory's (1978) use of European critical social theory to kill "the Gods of Positivism" that, he argued, held up the epistemological foundations of spatial science. Whatever the reason, the upshot was that from the early 1970s there was a shift in human geography away from spatial science. That didn't mean the scientific method wasn't still practised, and especially in particular subfields such as medical geography, into which Haggett moved, and even more so in Geographical Information Systems, tellingly relabelled Geographical Information Science (see Chapter 7). But it was no longer central within human geography, unlike in physical geography where it was the method.

Conclusion

Geographical ideas don't have a separate life of their own, falling from the heavens ready-made. Rather, they are utterly entangled, grounded, and embodied within the lives lived by the individuals who create and practise them. To fail to consider those lives when trying to understand the origin and effects of ideas is to engage in a form of "fetishism", that is, to presume ideas have no history, but come into the world fully formed.

The purpose of this chapter was to counter such fetishism. It was to show that history always matters, in this case, the life histories of the British geographers Dick Chorley and Peter Haggett, who promoted the idea that human and physical geography are united by a common scientific method. That idea did not abruptly arrive whole and complete, like a bolt from the blue, but emerged out of a hesitant, contested, and chance historical process set within the lives of both men. Ahead of time there were no guarantees that Geography would be defined by that idea, or that Chorley and Haggett would be involved. Many contingencies played out – from Haggett's schoolboy rugby accident to Chorley's first meeting with Arthur Strahler. To understand how those contingencies came together, how the world aligned temporarily at least in just the right way to make Geography what it became, it is necessary to practise some form of biography. As Hankins (1979, 5) writes, biography "gives us a way to tie together the parallel currents of history at the level where the events and ideas occur". Exactly. That was the intention of this chapter, to show geography-in-the-making rather than as ready-made.

Notes

1 Email from Peter Haggett to the author, December 23, 2022.

2 A Demonstrator was at the very bottom of the academic labour hierarchy on a fixed-term contract (5 years in the cases of Chorley and Haggett). Their job was instructing students in practical geographical techniques and skills. Above Demonstrators were Lecturers who were permanent staff and better paid. At the top were Professors or Chairs of which in the UK during the 1950s and 1960s there were typically only two or three in each Department.

3 Chorley and Haggett's course was for both human and physical geographers, and preliminary to specialization that occurred in the second year (email from Olav Slaymaker to the author, May 30, 2023).

4 Going to Chorley's obsession with numerical pattern, he rued British Summer Time, without which his birth would have conformed to the sequence: 3-9-27. Arriving in the world at 00.50, his birthday was disappointingly on September 4, 1927 (Beckinsale 1997, 3).

5 Beckinsale (1997, 5) says he was "fortunate" to win a Fulbright given that "two of his lowest marks" at Oxford were in his "special subject", landforms. For this reason, he was "debarred from pursuing ... post-graduate studies" in the UK (Beckinsale 1997, 5). Olav Slaymaker, a doctoral student of Chorley's, says "it was Beckinsale's letter of reference and his rehearsal of the general prejudice that existed in the Department of Geography towards Dick's iconoclastic ideas that saved the day" and won him the Fulbright (Email from Olav Slaymaker to the author, October 9, 2022).

6 Chorley's intention was to work with the geologist Armin K. Lobeck at Columbia, but Lobeck was ill when Chorley arrived and he switched to Strahler (Haggett and Stoddart 2009, 66).

7 Robert Horton's (1945) paper using the mathematics of topology to classify drainage networks was an important antecedent for Strahler's work in quantifying and theorizing geomorphologic processes.

8 Strahler's research, along with others working in the earth sciences, including Horton, benefitted enormously from Cold War military patronage. It resulted in tens of millions of dollars

being awarded for research, the significant expansion of the field, and the "create[ion of] a new intellectual map for these disciplines" (Doel 2003, 635).

9 Hearing that a geomorphologist had left the University of Sheffield, Chorley sent a letter and CV enquiring if there was a vacancy. As the story goes, because he had a spare stamp, on the off-chance he also sent the carbon copy of his CV and a new letter to Steers (Haggett and Stoddart 2009, 67). Michaelmas is the autumn term at Cambridge.

10 Haggett (2008, 338) wrote that when Lösch's book was translated, "it was Lewis – familiar as he was with solid-state arrays in physics – rather than economic geographers who saw its potential".

11 Haggett and Chorley (1989, xx) say that "the oldest contributor to *Models* was in his mid-thirties, almost all were still on temporary posts or had newly gained lectureships, none was a professor, none was well-known".

12 Alistair Cooke was a British-born BBC correspondent living in the United States. Every week between 1946 and 2004, he presented "Letter from America" on BBC radio, a 15-minute digest and interpretation of current American news and events for the British listening public.

13 John Davey interviewed by the author, Oxford, England, July 15, 2016.

14 "And now for something completely different" was a famous catchphrase from *Monty Python's Flying Circus*. While it had practical use in allowing that show to move from one improbable comedy sketch to another, it also reflected the belief among its performers, all, like Harvey, Oxbridge undergraduates during the 1950s, that Britain needed to change fundamentally. *Python* was part of a an early post-War male generational oppositional movement to a conservative, hide-bound and elite British culture and politics that included "angry-young-men" playwrights, e.g,, John Osborne, novelists, e,g,, Kingsley Amis, TV satirists, e.g., David Frost, poets, e.g., Philip Larkin, and musicians, e,g., Lonnie Donegan.

15 Email from David Harvey to the author, August 26, 2003.

Study task

Variations on this question can be set on an unseen examination, an open-book examination or as a term paper assignment.

Drawing on Richard Chorley and Peter Haggett's work, what are the benefits and drawbacks of using a common scientific method to study phenomena in both human and physical geography?

Starter reading list

Barnes, T. J. 2019. Pioneer: Peter Haggett. In P. Atkinson, S. Delamont, A. Cernat, J. W. Sakshaug and R. A. Williams, Eds, *SAGE research methods foundations*. London: Sage, pp. 1–13.

Chorley, Richard J. 1995. Haggett's Cambridge: 1957–66. In A. D. Cliff, P. R. Gould, A. G. Hoare and N. J. Thrift, Eds, *Diffusing geography: Essays for Peter Haggett*. Oxford: Blackwell, pp. 355–374.

Chorley, Richard J. 1971. The role and relations of physical geography. *Progress in Geography* 3, 87–109.

Gregory, Derek. 1978. *Ideology, science and human geography*, chapter. 1. London: Hutchinson, pp. 25–48.

Haggett, Peter and Chorley, Richard J. 1967. Model, paradigms and the new geography. In R. Chorley and P. Haggett, Eds, *Models in geography*. London: Methuen, pp. 19–41.

Kennedy, Barbara A. 1979. A naughty world. *Transactions of the Institute of British Geographers* 4(4), 550–558.

References

Barnes, T. J. 2008. Geography's *Underworld*: The military-industrial complex, mathematical modelling and the quantitative revolution. *Geoforum* 39(1), 3–16.

Beckinsale, R. 1997. Richard J. Chorley: A reformer with a cause. In D. R. Stoddart, Ed., *Process and form in geomorphology*. London: Routledge, pp. 3–14.

Browning, C. Ed. 1982. *Conversations with geographers: Career pathways and research styles.* Department of Geography occasional paper 16. Chapel Hill: University of North Carolina.
Chorley, R. J. 1965a. A re-evaluation of the geomorphic system of W.M. Davis. In R. J. Chorley and P. Haggett, Eds, *Frontiers in geographical teaching.* London: Methuen, pp. 21–38.
Chorley, R. J. 1965b. The application of quantitative methods to geomorphology. In R. J. Chorley and P. Haggett, Eds, *Frontiers in geographical teaching.* London: Methuen, pp. 147–163.
Chorley, R. J. 1995. Haggett's Cambridge: 1957–66. In A. D. Cliff, P. R. Gould, A. G. Hoare and N. J. Thrift, Eds, *Diffusing geography: Essays for Peter Haggett.* Oxford: Blackwell, pp. 355–374.
Chorley, R. J. and Haggett, P. Eds. 1965. *Frontiers in geographical teaching.* London: Methuen.
Chorley, R. J. and Haggett, P. Eds. 1967. *Models in geography.* London: Methuen.
Doel, R. E. 2003. Constituting the post war earth sciences: The military's influence on the environmental sciences in the USA after 1945. *Social Studies of Science* 33, 635–666.
Dow, M. W. 1982. Interview of Richard J. Chorley by Maynard Western Dow, Cambridge, UK. Online *Geographers on Film Transcriptions*, pp. 205–207. https://textarchive.ru/c-2977382-pall.html
Dow, M. W. 2000. *Geographers on film: Interview of Peter Haggett by Maynard Western Dow (68 min).* Pittsburgh, PA. Retrieved August 6, 2017, from https://www.youtube.com/watch?v=elSCKRTyl4E
Ford, D. 2002. GeomorpList posting. International Association of Geomorphologists, May 18, 2002. http://www.geomorph.org/sp/arch/item18.html
Golledge, R. G. 2006. Textbooks that moved generations. In R. J. Chorley and P. Haggett, Eds, 1967: *Models in geography.* London: Methuen. *Progress in Human Geography* 30(1), 107–113.
Gregory, D. 1978. *Ideology, science and human geography.* London: Hutchinson.
Gregory, D. and Castree, N. Eds. 2011. Editor's Introduction: Human geography. *Human geography, 5 volumes.* London: SAGE, pp. xxv–lxxix.
Haggett, P. 1964. Regional and local components in the distribution of forested areas in south-east Brazil: A multivariate approach. *Geographical Journal* 130, 365–378, and discussion, 378–380.
Haggett, P. 1965. *Locational analysis in human geography.* London: Arnold.
Haggett, P. 1990. *The geographer's art.* Oxford: Basil Blackwell.
Haggett, P. 1991. Author's response. *Progress in Human Geography* 15(3), 302.
Haggett, P. 2002. Obituary: Richard John Chorley, 1927–2002. *Transactions of the Institute of British Geographers* 27, 522–525.
Haggett, P. 2008. The local shape of revolution: Reflections on quantitative geography at Cambridge in the 1950s and 1960s. *Geographical Analysis* 40(3), 336–352.
Haggett, P. 2015. Madingley: Half-century reflections on a geographical experiment. *Geography* 100, 5–11.
Haggett, P, and Chorley, R. J. 1965. Frontier movements and the geographical tradition. In R. J. Chorley and P. Haggett, Eds, *Frontiers in geographical teaching.* London: Methuen, pp. 364–384.
Haggett, P. and Chorley, R. J. 1989. From Madingley to Oxford: A foreword to *Remodelling geography.* In B. Macmillan, Ed., *Remodelling geography.* Oxford: Blackwell, pp. xv–xx.
Haggett, P. and Stoddart, D. R. 2009. Richard J. Chorley 1927–2002. In H. Lorimer and C. J. Withers, Eds., *Geographers biobibliographical studies, volume 28.* London and New York: Continuum, pp. 65–88.
Hankings, T. L. 1979. In defence of biography: The use of biography in the history of science. *History of Science* 17, 1–16.
Harvey, D. 1963. Locational change in the Kentish hop industry and the analysis of land use patterns. *Transactions of the Institute of British Geographers* 33(2), 123–140.
Harvey, D. 1972. Revolutionary and counter-revolutionary theory in geography and the problem of ghetto formation. *Antipode* 4(2), 1–13.
Harvey, D. 2002. Memories and desires. In P. Gould and F. R. Pitts, Eds., *Geographical voices: Fourteen autobiographical essays.* Syracuse: Syracuse University Press, pp. 149–188.
Horton, R. 1945. Erosional development of streams and their drainage basins: Hydrophysical approach to quantitative morphology. *Bulletin of the Geological Society of America* 56, 275–370.
Ley, D. and Samuels, M., Eds. 1978. *Humanistic geography.* Chicago: Maroufa Press.
Morrill, R. L. 1991. Commentary 1. *Progress in Human Geography* 15(3), 300.
Olsson, G. 1975. *Birds in egg.* Michigan Geographical Publications, number 15. Ann Arbor: Department of Geography, University of Michigan.

Schumm, S. A. 2004. In memoriam: Arthur Newell Strahler (1918–2002). *Annals of the Association of American Geographers* 94, 671–673.

Semple, E. C. 1911. *Influences of geographic environment: On the basis of Ratzel's system of anthropo-geography*. New York: Henry Holy & Co.

Slaymaker, O. 2022. Epilogue: Personal reflection on geomorphology during 1965–2000 and geomorphological futures. In T. Burt, A. Goudie, and H. Viles, Eds, *History of the study of landforms. Volume 5: Geomorphology in the second half of the twentieth century*. Bath: The Geological Society Publishing House, pp. 415–428.

Smith, N. 1989. Geography as a museum: Private history and conservative idealism in *The Nature of Geography*. In N. Entrikin and S. D. Brunn, Eds, *Reflections on Richard Hartshorne's The Nature of Geography. Occasional publications of the Association of American Geographers*. Washington, DC: Association of American Geographers, pp. 91–120.

Stamp, L. D. 1957. Geographical agenda: A review of some tasks awaiting geographical attention. *Transactions and Papers of the Institute of British Geographers* 23(1), 1–17.

Stoddart, D. 1997. Richard J. Chorley and modern geomorphology. In D. R. Stoddart, Ed., *Process and form in geomorphology*. London: Routledge, pp. 383–399.

Strahler, A. N. 1950. Equilibrium theory of erosional slopes approached by frequency distribution analysis. *American Journal of Science* 248, 673–696, 800–814.

Strahler, A. N., 1992. Quantitative/dynamic geomorphology at Columbia 1945-60: A retrospective. *Progress in Physical Geography* 16 (1), 65–84.

Thrift, N. J. 1995. Peter Haggett's life in geography. In A. D. Cliff, P. R. Gould, A. G. Hoare and N. J. Thrift, Eds, *Diffusing geography: Essays for Peter Haggett*. Oxford: Blackwell, pp. 375–395.

Yeates, M. 1991. Commentary 2. *Progress in Human Geography* 15(3), 301.

5 Contemporary Geography

Advocating for a heterodox subject

Rita Gardner

Rita Gardner (1955–)

Introduction

For subjects to flourish they need institutions[1] whose purpose is to act as an inclusive 'home', focused on the discipline and its communities. These types of institutions are known as learned societies or professional bodies. They exist to promote and advance their discipline (learned societies), support and accredit its practitioners (professional bodies) and, when required, to defend the subject or profession and its communities. Many of the learned societies are charities, which requires them to be outward-facing organisations that support their disciplines for the public good. In contrast, professional bodies focus on supporting their members.

DOI: 10.4324/9781003343240-6

Such institutions need to be effective advocates, able to work collaboratively with their communities, inclusive and welcoming, and agile in understanding trends and averting risks. The primary risks to the 'health' of subjects in many countries are changes in government policy that potentially negatively affect if, when, and how the subject is taught at school; the training of teachers; and funding for teaching and/or research at university. Other risks may relate to decisions made by universities and competition for students from other disciplines. In the UK, unlike some other countries, such institutions also need ways of generating resources as they are not core funded by government. Common sources of income include membership subscriptions, publishing, and conferences, plus, for some, commercial activities and fundraising.

Geography is a large and complex discipline, with a wide range of methodologies and areas of study that, unusually, embrace aspects of humanities, social sciences, and physical sciences. Its UK learned society and professional body, the Royal Geographical Society (with The Institution of British Geographers) – more succinctly known as RGS-IBG – has an unusually long and complex history.

This chapter considers the development of the RGS-IBG between 1996 and 2018. This was the period of my tenure as Director. Using a small number of carefully chosen foci, it explores the strategic drivers and rationales for development, the changing political and regulatory contexts, and dives into a very small sample of the many new projects and enhanced activities that resulted. The examples demonstrate the developing breadth and depth of the Society's role and advocacy with a wide range of audiences. The RGS-IBG emerged as a powerful advocate and an inclusive 'home' for Geography, and its work helped to secure the subject's profile, its growth, and its recognition. Further details of RGS-IBG work can be read in the published Annual Reviews available online.

Background

The recent history of the organisation is arguably one of the most critical periods of change and growth since it was formed in 1830. Its purpose, set out in its Royal Charter of 1859, is "the advancement of Geographical Science". The charter was intended to give greater recognition to Geography as a discipline and the Society "a more permanent establishment among the scientific Institutions of the UK", many of which emerged in the Age of Enlightenment. It was among the first geographical societies in the world, with the Paris, Berlin, London, St Petersburg, and New York societies all being formed between 1821 and 1851. Today, 'geographical science' is interpreted broadly by the RGS-IBG to include the breadth of human, physical, and environmental geographies.

Between 1892 and 1933 two groups broke away from the RGS, with its good wishes, and formed independent organisations. In 1893 the Geographical Association (GA) was formed, owing in part to the RGS' opposition at the time to women members, as a specialist body for Geography teachers in England; it remains independent today. In 1933 a group of leading academic geographers formed the Institute of British Geographers (IBG). This was to meet the needs of the growing number of academic scholars in the 1920s and 1930s for publication outlets and research discussion meetings – needs that some felt were not being sufficiently met by the established societies of the RGS and GA (Steel, 1984).

The trigger for the developments I will be discussing below was the (re)merger of the RGS and the IBG in January 1995. Thus, this is also the story of forging a single 'new' merged institution from two organisations that had gone their own ways for some 60 years; each with a very different ethos, orthodoxy and, for the most part, memberships.

The journey towards a successful merger required the hard work, dedication, and passion of the RGS-IBG staff team, trustees, and committee members, and not least the geographical communities at large and the many volunteers. Strong leadership, underpinned by a shared vision, drive, and entrepreneurship was required to overcome challenges, raise funds, and identify and grasp opportunities. Above all, the RGS-IBG endeavour was planned and executed with the aim of bringing about lasting change for the benefit of Geography and geographers.

RGS-IBG in the making: approach and vision

The challenge at the outset of my time as Director was clear and daunting. Geography as a discipline lacked the influential, collective voice that is essential to successful promotion and safeguarding of any discipline in a democratic, lobby-based political setting and in a competitive world of higher education. Furthermore, the two organisations had merged on equal terms but differed fundamentally in several important ways:

- *Scale*: RGS with ca 35 paid staff, 10,000 Fellows, a listed 19th-century building in London; the IBG with 1.5 paid staff, ca 1,500 members and one rented office.
- *Membership*: IBG with a membership comprising the majority of UK academic geographers and postgraduate students; RGS's Fellowship drawn largely from interested lay persons, former Geography graduates, and expeditioners.
- *Tradition*: RGS with a great heritage of 19th- and 20th-century scientific exploration, a 'clubby' feel, strong connections with the UK establishment, and a history of entirely male directors; IBG with an egalitarian and inclusive ethos, pushing the boundaries of contemporary Geography, and a strong identity as the independent home of academic Geography in the UK.
- *Activities*: IBG with its world-leading annual conference, leading scholarly journals, and specialist research groups; RGS with its populist and well attended weekly Fellows' lectures, an internationally important historical collection of exploration and mapping, and a focus on expeditions through an expedition advisory centre and international, multi-disciplinary field research projects.
- *Connections*: IBG with its deep connections to the scholarly geographical world; RGS with connections through its Fellows to decision makers in government, the House of Lords, and high-profile explorers.

With difference came the potential for complementarity, but a new ethos had to be built and embraced by all, including a shared vision of the future which could be agreed and implemented. The merger agreement had started to embody some practical elements, but it had to go much further, finding an acceptable balance between (a) reassuring continuity for all members and (b) driving the change and modernisation that was needed to create the world-leading learned society that the subject deserved and which the merger offered the potential for. All this to be accomplished with sensitivity to the memberships, which together were the single most important source of expertise, income, and volunteer help.

We tackled this challenge by – in effect – creating a new organisation (RGS-IBG) that combined the best of both sets of existing activities and then grew innovatively to create a balanced, multi-dimensional learned society and professional body. From the outset we signalled a new ethos and approach, in being egalitarian, open, and inclusive, offering substantial public benefit, and putting the discipline of Geography at the forefront of all we did. The most important early task was to create the first (1998–2004) of five successive strategic plans under my leadership. We consulted widely and developed a plan for continuity and change, to nurture the new RGS-IBG, while signalling ambitions for the future.

The vision for how the RGS-IBG was to develop emerged steadily through the successive strategic plans. There was no 'blinding moment of truth', but I held strong strands of conviction throughout. I believed that to safeguard and champion the discipline strategically and effectively we needed to:

- Place academic Geography at the heart of the RGS-IBG. Research and the generation of new understandings and methodologies continually powers and renews disciplines, and knowledge from that flows to benefit other audiences, such as teachers, public, policy makers, and business.
- Embrace all the stages in the 'lifecycle' of geographers. A disciplinary community is a rich ecosystem, with individuals interacting with Geography at different life stages: as students at school and university; as professionals; as former Geography graduates – in roles of employers, parents, voting citizens, and retirees. The wider and larger the ecosystem we could embrace, the stronger and better connected the community, and the greater the resource available to the RGS-IBG – expertise, contacts, ambassadors, and money. And the better we could deliver public benefit.
- Ensure that Geography remained a single discipline enriched by its many facets. Geography is a lively, fissiparous subject, frequently generating new areas of study and with porous boundaries to other disciplines. The tensions to split have long been present, but there is enormous intellectual and applied strength, and distinctiveness, in having a discipline that can bring an integrated view of the world and a deep understanding to human–environment interactions, and a strong spatial perspective on the world.
- Develop and sustain breadth and balance across RGS-IBG activities and membership, including expedition activities. This was essential in order to sustain the merger, and important in its own right as the Society benefits greatly from being able to call upon high-profile members from a wide range of backgrounds – including business leaders, explorers, media figures, leading academics, and peers – to support its campaigning for Geography.
- Be bold and creative. To be an effective champion, an organisation needs to be trusted and respected for what it does. It also cannot afford to stand still. It needs to be able to 'think outside the box' to come up with effective ideas for projects at scale that meet identified needs and engage new audiences.

The conceptual framework of the RGS-IBG and its activities that emerged over those years (Figure 5.1) consisted of six pillars of activity for (largely) different audiences, supported by membership, improved facilities, access to collections, and enhanced income generation. We started with the two main pillars that the IBG and the RGS brought to the table: supporting research and higher education, and expeditions and fieldwork; together with the structural resources of a combined membership of ca12,000, 2 million historical items in the collections, and a worn, grand but forbidding-to-some, historical building.

Figure 5.1 The conceptual framework of the RGS-IBG.

We then built and sustained four additional pillars, of broadly similar strength, each of which also developed the membership. By 2006 substantial growth in all four was well underway: support for secondary education; advocacy and influencing of policy; public outreach and engagement; and developing a professional arm for working geographers.

The building blocks of the RGS-IBG's work that were initiated and grown over the 20-year period from 1998 to 2018 are illustrated in Figure 5.2. In addition to projects and activities in the six pillars, there was significant building and refurbishment work to provide the standard and scale of facilities needed (see Price Williams, 2012 for more information), new membership categories recognised different types of relationship, and we grew our income.

Change through heritage: 'Unlocking the Archives'

First, we needed to address our image and cement the merger in order to have a credible base and community support from which to develop. We used our most historic resources to bring about this change.

It culminated on June 8, 2004, when BBC Breakfast TV aired live from the RGS-IBG new buildings. That day we were officially 'unlocking our archives', giving full public and scholarly access to all 2 million items held in the Society's Collections for the first time ever. Recognised as the largest, and one of the most important, private geographical collections in the world, it comprised over a million maps, dating from 1472 to the present; 250,000 books; 500 m of shelved manuscripts, diaries, and archival papers; and 500,000 historical photographs, films, and paintings. It's an immeasurably rich, historical resource of people, places, exploration, and changing geographies, political contexts and perceptions over decades and spanning much of the world, from equatorial Africa to the Poles.

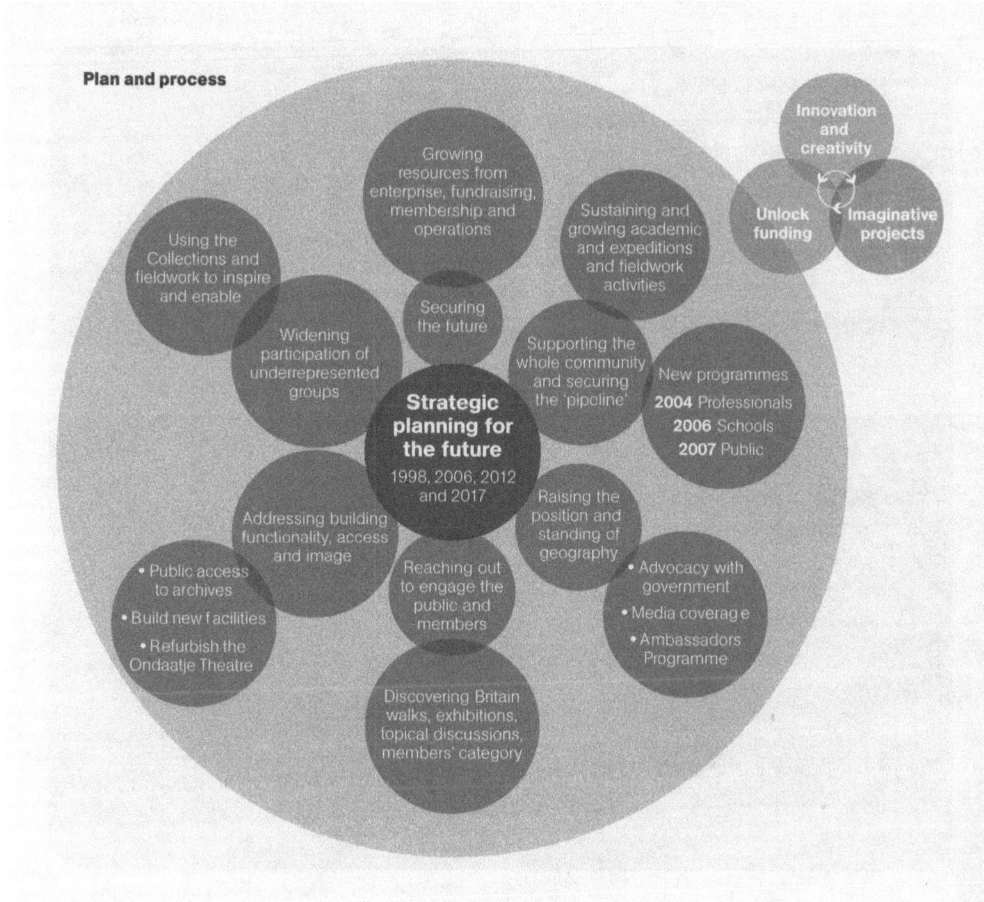

Figure 5.2 The building blocks of the four strategic plans between 1998 and 2017.

Source: Price Williams, 2017.

It is also an archive with deep, shared heritages and undeniably colonial roots. What to do with it had presented a significant challenge for a cash-strapped organisation largely funded by its members' subscriptions.

In 2001 the UK Heritage Lottery Fund approved a large grant of £5m towards the £7.24m project to 'Unlock the Archives' – one of the largest single projects in the Society's history. This followed three years of intense preparation of the funding bid involving heritage conservationists, architects, quantity surveyors, engineers, and many more specialists. And three years of sustained fundraising to meet the required matched funding of £2.24m.

The new contemporary buildings (Figure 5.3) housed state-of-the-art, temperature-controlled storage – a far cry from the multitude of hot offices, corridors, and dusty, cramped basement rooms that previously housed them. The eco-friendly Foyle Reading Room enabled public access, the glass pavilion provided much-needed exhibition space, and the Education Centre facilitated small conferences, meetings, and school class activities. The light, modern, and welcoming new RGS IBG entrance on Exhibition Road,

(a)

(b)

(c)

Figure 5.3 New and diverse public audiences and spaces for the Society's Collections: community research; Windrush Foundation exhibition launch; and the new entrance and exhibition pavilion, with the Foyle Reading Room facing into the garden below.

with its views of the gardens beyond, was designed as a very different experience from the traditional, dark entrance behind imposing wrought-iron gates on Kensington Gore. The architecture, by Studio Downie, won the Judges' Special Prize at the UK Construction Industry Awards in 2005.

However, this project was not just about properly conserving and giving public access to the archives, important though that was, nor just about having new, purpose-built exhibition, library, and education facilities dedicated for public use, even though they were in constant use. It was also the way we chose to kick-start the building of a modern image for the RGS-IBG and for public Geography (see Danny Dorling's chapter), to become visually part of the great cultural and learning quarter of Exhibition Road, and to work with diverse communities whose own heritage was represented in many different and often challenging and contested ways in the Collections. Opening the extraordinary resources to new and critical research and interpretation – as well as for enjoyment and learning – was a vital first stage in developing the RGS-IBG and in signalling a greater emphasis on being an outward-facing charity serving Geography and the public good in the 21st century.

As well as opening the archives, we were simultaneously, deliberately, and symbolically opening the RGS-IBG physically, intellectually, and visually to the world. We changed our image as an organisation overnight – aided by a new visual identity launched simultaneously – raised our profile, and gave Geography a modern 'home'. We also retained the vital support of all our members, the more traditional of whom welcomed the saving of the collections from a slow death from dust and heat and the retention of the historic feel of the 'old' building, while others identified with the revitalised image, contemporary spaces, and the symbolism of the new. 'Unlocking the Archives' had done so much more than simply find a solution to 'what do to with the Collections'. The shaping of the RGS-IBG had commenced.

Over the succeeding years, many wonderful exhibitions have drawn heavily on the collections and the new facilities, for example, the Enduring Eye exhibition (RGS-IBG, 2016). A number of ground-breaking projects have also worked in partnership with diaspora communities in the UK to explore and interpret shared heritages through the Collections, for example, the Crossing Continents: Connecting Communities programme (RGS-IBG, 2009). Scholarly research on the Collections, re-interpretation, and multidisciplinary collaboration have also flourished with the open access, for example, the Hidden Histories of Exploration project (Driver and Jones, 2009).

Boosting teaching and learning: the Action Plan for Geography

School pupils are the pipeline into UK higher education in Geography, in common with many other subjects. They underpin the size of the Geography university undergraduate body, and that directly affects the perceptions of university leaders, and the number of universities and academic staff that teach the subject.

It is fundamental, therefore, to the health and future of Geography – and to having trained geographers in the workforce – to sustain interest and uptake in secondary school and progression into university. One way of doing that is to influence how the subject is perceived in schools and to support and encourage teachers and pupils with, for example, great teaching resources and careers advice. Another way is to secure the subject's place in the school curriculum by working to influence government policy.

This and the following section explore two of the many ways in which the RGS-IBG has developed to support and influence both aspects in a sustained way, and what some of the outcomes were. While much of our work with schools was delivered independently, complementing the activities of the two Geography teachers' associations in the UK (GA and SAGT), some was collaborative. The RGS-IBG was building on a distinguished early RGS history in education. It was a major force in establishing Geography teaching in UK universities in the late 19th century (Mill, 1930), and in pressing for Geography to be included in the list of subjects to be offered in state secondary schools following the first Education Act in 1902. Through the second half of the 20th century RGS interests in education waned, and by 1994 its annual report simply noted that eight sixth-form lectures and young members' events had been held.

Figure 5.1 demonstrates the rationale for re invigorated action. In 1995 the numbers taking Geography GCSE in the UK stood at around 300,000. By 2011 they had fallen by about 40% to 180,737, despite an increase in total GCSE entries overall. The decline

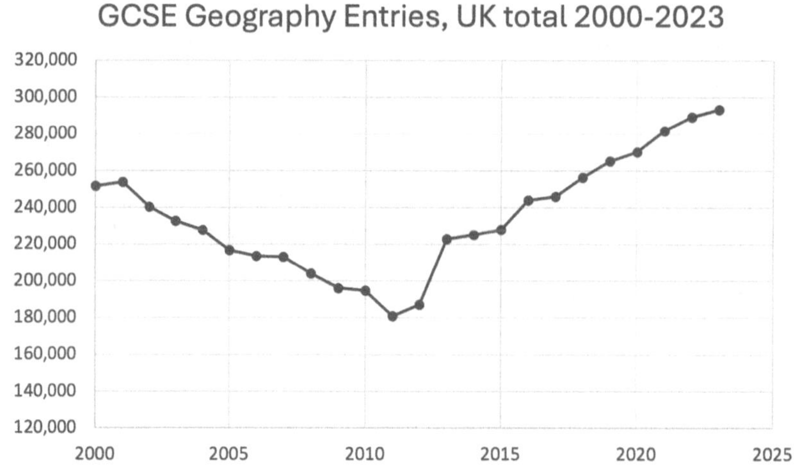

Figure 5.4 Entry numbers for Geography GCSE between 2000 and 2023; data from the Joint Council for Qualifications https://www.jcq.org.uk/examination-results/

commenced with the National Curriculum review of 1994, as a result of which Geography (and History) ceased to be compulsory study (see Walford, 2000). Several other reasons for prolonging the decline have also been suggested. Geography A-level suffered proportionately too, falling by 30% from 44,900 in 1998 to 31,226 in 2011. By the early 2000s it was clear that something needed to be done.

An introduction by RGS-IBG donor Sir Christopher Ondaatje to the Secretary of State for Education at the time, the Rt Hon Charles Clarke MP, at a parliamentary reception gave me the opportunity to request he establish a 'Focus Group for Geography' to parallel the group recently established for History by the government Department for Children, Schools and Families (DCSF). He agreed, asking us to coordinate the group, which was to be chaired by the Schools Minister. We brought together senior figures from business and industry, academia, government agencies, the media, teaching and teacher training, in a set of carefully orchestrated meetings and working groups over 18 months. In this time, we had to convince not one, but three successive Schools Ministers, of the importance of Geography to education and the need for Geography teaching to have a boost.

Our case centred on the need for young people – as citizens – to understand our rapidly changing world and some of the big challenges it faced; about the importance of fieldwork and 'learning in the real world'; the fact that Geography was the 'root' (and only) social science in the curriculum up to and including GCSE level; and about the skills and employability of geographers. In short, the arguments put forward in our advocacy pamphlet, 'Geography: An Education for Life' (RGS-IBG, 2001; Figure 5.5)

The outcome, in 2005, was an acceptance by DCSF of the need to boost Geography in schools, and a request for me as Director of RGS-IBG and David Lambert as CEO of the Geographical Association (GA) to produce an 'Action Plan for Geography' (APG). The plan was approved and granted £1m of government funding over five years (2006–2011). This marked, I believe, the first ever significant funding for school Geography from the government. That alone was a boost to teachers' morale and confidence and the activities furthered that.

Figure 5.5 'Geography: An Education for Life'. This pamphlet was the first public advocacy document written by the RGS-IBG, in 2001. Published in partnership with the Geographical Association, it was used widely in discussions with decision makers in government.

Source: RGS-IBG, 2001.

The APG comprised six main projects, three each to be led by the RGS-IBG and GA. The GA focused on pedagogy and the student voice, while the RGS-IBG established its Geography Ambassadors programme and provided curriculum resources for teachers at GCSE and A-Level. In 2010, for example, 1,000 undergraduate Geography Ambassadors engaged some 30,000 school pupils enthusiastically with the benefits of studying Geography. The RGS-IBG activities continue through to the present and are heavily used. The project led to the formation of an Education Department at the RGS-IBG, realising a key aim in our strategy, and that department has until recently introduced a string of innovative projects to support secondary Geography teachers, trainee teachers, and pupils.

Influencing policy: securing Geography in the English Baccalaureate

Education in the UK is highly politicised and thus liable to significant change after each national election, and particularly each time a new party is returned to power. The repeated changes can have significant consequences not just for schools, teachers, and pupils, but also for the curriculum and for individual subjects.

Under the New Labour government (1997–2010), and its school Quality and Curriculum Authority agency, the early secondary years (11–14 age) became a focus for skills and theme-based, integrated learning. Subjects and systematic subject knowledge were subordinated to this ethos and, some would argue, effectively de-prioritised. It has been suggested this may be another contributor to the falling numbers studying Geography GCSE.

With the election of the Conservative–Lib Dem Alliance in 2010, followed by successive Conservative governments, the ethos shifted dramatically and subjects were brought back to the fore under, in particular, the longstanding service of Nick Gibb MP as Minister of State for Schools (2010–2012; 2015–2021; 2022–). He favoured 'traditional subjects' and the identification of 'core' factual knowledge. At the heart of this approach was the creation of the English Baccalaureate (EBacc) at GCSE level (age 14–16). Although it was not officially mandatory for schools, it was given political teeth by its inclusion as a school performance reporting measure. It was *de facto* adopted by the majority of the 3,705 state secondary schools in England, Wales, and Northern Ireland (DfE, 2012).

At the outset, the big issue was the selection by Department for Education (DfE) of the subjects to be included in the EBacc from among the many core, foundation, and other optional subjects at GCSE. The DfE's rationale was for the EBacc to comprise a set of essential subjects that keep young people's options open for further study and future careers. These subjects would be studied by most pupils at GCSE level, leaving limited time for additional, optional choices for many pupils. The stakes were high in terms of sustaining the profiles and standing of individual disciplines in schools, and the pipeline of students through school and on into university. Stakes were equally high in terms of the nature of secondary education that young people received.

Nick Gibb had a high regard for learned societies and I met with him on a number of occasions. I remember the initial meeting to discuss the EBacc well. He expressed his view that History should definitely be included as the humanities subject, and that he was not convinced that Geography should be included as, in his view, it was often currently taught in a way that was too political, too focused on opinion and not sufficiently focused on 'facts'. He was a man who deplored 'topical issues' being taught without reference to a sound understanding of process and objective evidence. So, there was the challenge.

There is not space to go into the details of how Nick Gibb was persuaded to include Geography, and there were likely several influences, but the EBacc when announced included a requirement for pupils to study English (language and literature), Maths, two sciences, a modern foreign language, and either History or Geography. The RGS-IBG undoubtedly played a major role in successfully advocating for Geography's inclusion. This secured the future of the discipline at school, it enriched the EBacc with the understanding of our spatial, human, and physical worlds, and it also meant, importantly, that a social science subject was included in the EBacc. The decline in numbers taking Geography GCSE and A-level was reversed almost overnight and numbers grew steadily in the following years, regaining lost ground (Figure 5.4). The sustained growth in Geography in UK higher education in the years that followed may also owe much to this act of persuasion.

Subsequently the RGS-IBG went on to influence the major revisions by DfE to the content of the school Geography curriculum (age 5–16). Its traditional and content-rich approach was not welcomed by all teachers, but it served Geography well in ensuring its standing with key decision makers as a subject with rigour, and in achieving the end goal of reinforcing Geography in the EBacc.

These are two of the most impactful examples of our efforts in the RGS-IBG, between 2000 and 2018, to influence policy in a sustained manner to benefit Geography. Nothing could be more central to the Society's purpose of advancing Geography. We became active and respected influencers in many areas of school and higher education policy – curriculum, funding, and research – and earned the reputation of being highly effective advocates for our discipline.

Reaching public audiences: Discovering Britain

For many people their engagement with Geography starts and stops at school, and then remains fossilised. For others, Geography is equated with travel and 'other places' or seen through their children's homework projects. Yet, as other chapters in this book emphasise, over the past 60 years Geography as a discipline, and its methodologies and subject matter, have continuously evolved.

Why would we identify the public as an important audience for Geography and the RGS-IBG? For three reasons. First, in the knowledge that an understanding of Geography enriches everyone's capacity as community and global citizens. Second, the fact that people often influence the subject choices of their children at A-level and at university. Third, people may be active in settings where getting the Geography message across matters – from school governors to business leaders. Thus, we felt it important to attempt to find ways to excite the wider public about Geography today, its relevance to the challenges the world faces, the skills that are learned through it, and the excellent employability opportunities a degree in Geography opens up.

This may sound a simple task, but reaching people at scale is problematic. The Society had a track record in its longstanding regional lectures – established in 1988 – which still feature today. But we sought a much larger scale of outreach. For several years we encouraged the BBC to badge its so-clearly-geographical programmes as 'geography', much as they do with their history strand. But this was to no avail, even the notable BBC names close to the Society, including Paul Rose (a Society Vice-President) and Sir Michael Palin (Society President, 2009–2012), being unable to budge the BBC's worthy perceptions of the term 'geography'. Hence the labels 'environment, travel, and exploration' remained. There is nothing wrong in that, of course, but it's a missed opportunity to reach millions of people with 'geography'.

In a similar way to opening up work in other sectors, we started by devising and funding a single major programme of work to reach the UK public. As so often happens, ideas coalesced from a number of different avenues and in the spring of 2010, 'Britain from the Air' was born, in partnership with WeCommunic8, the street gallery experts. The idea was for a 100-panel street exhibition that drew people in with arresting arial images from across the UK, and about which we could tell engaging geographical stories. Between June and August 2010, we selected the images, identified four stories for each of the panels, and then at RGS-IBG I led the research and writing of 30,000 words of succinct, engaging, and accessible text. With a tight and fixed deadline to launch in Bath in early September, it was a manic summer. The street gallery exhibition (Figure 5.6) was launched by Michael Palin to national media acclaim.

Over the next nine years, the exhibition featured in the streets of ten major UK cities and the Shanghai Expo. We estimated that well over six million people viewed it, and judging from the extensive feedback, loved it. What pleased us most was the very wide demographic that was drawn in by the amazing images and then stopped to read, talk, and share their enjoyment. In a random survey of 250 viewers in Bath, 99% said they would recommend the exhibition to a friend. We had put Geography on the streets, for free, for everyone in ways that spoke to them. And our Society Fellows, students, and professional geographers were delighted to see it happen.

Figure 5.6 'Britain from the Air' street gallery (produced in association with WeCommunic8) in the centre of Bath.

Box 5.1 Quotes from exhibition viewers in Bath

"I've lived in Britain all my life, we grew up here and there's so many things I didn't know, that I'd never seen before…"

"There are things here you'd never pick up in normal life and it's presented so dramatically - things just stick in your mind…"

"It's brilliant that it's out where the public can see it and not tucked away in some museum somewhere. People walk past, have a look at it, talk about it, I think that's the way art should be…"

'Britain from the Air' was part of a bigger conceptualisation, 'Discovering Britain'. https://www.discoveringbritain.org/

This also included a programme of downloadable self-led walks, trails, and viewpoints, complete with stories, route details, and digital mapping. They were unique in being nationwide and highly geographical. By 2016 there were more than 350 activities on the website, developed by the project team at the Society and with the help of Fellows.

Following the decision to end this programme in 2019, the walks have found a new home in the Ordnance Survey walking routes.

Establishing Geography as a profession: Chartered Geographer

The traditional professions – for example law, medicine, accountancy – have long had regulated professional standards. Graduates are required to meet those standards in order to practise, and to maintain the professional accreditation they are mandated to undertake continuing professional development throughout their career.

During the later 20th century many of the science and environmental disciplines also introduced professional accreditation, to recognise expected standards of competency in the practice of those disciplines at work. These were optional rather than statutory, that is, people could practise without them, but it was in an individual's interest to become accredited. Examples include the chemists, geologists, engineers, and environmental scientists.

Few of the humanities and social sciences saw the need to pursue professional accreditations, not least because graduates work across a wide range of employment areas and tend to use generic higher skills rather than subject-specific ones. Notable exceptions include planners, librarians, and those working in business-related areas such as marketing and human resources.

In 2001 we extended the RGS-IBG into a professional body and established an accreditation for geographers, 'Chartered Geographer' (C.Geog.), a world first. Subsidiary specialist accreditations soon followed for GI, geomorphology, and teaching professionals.

Our rationale was five-fold.

1. We wanted to champion the fact that geographers bring a unique combination of skills, knowledge, and attributes to the workplace, including their geospatial skills, and to offer support and encouragement by creating a community of practising geographers. In doing so we aimed to raise the external recognition of Geography as a profession and of the value of geographers to employers.
2. With the rapid growth of the Geographic Information industry, well trained geographers were in demand. Their GIS technical skills combined with their knowledge of the underlying social and cultural processes that helped to explain the observed spatial distributions was a powerful combination. It was an appropriate time to start to recognise geographers' vital professional roles at the heart of this new industry.
3. We were conscious that some established chartered accreditations, especially for disciplines at the boundary with physical geography, were starting to attract geographers looking for recognition as accredited professionals to enhance their profile and employability.
4. Geography graduates were an important and large community with which the society wished to engage more directly and more fully.
5. Above all, we hoped there would be real benefit to the Chartered Geographers themselves in terms of their career progression, salary levels, and networking opportunities.

In establishing the scheme as a new venture, we adopted criteria and assessment processes consistent with those in many other professional accreditations. Initially fairly slow to take off, it steadily gained traction, and by 2018 the milestone of 1000 Chartered

Geographers on the active register was met. Today it is a vibrant community that holds regular meetings and networking events and links to an annual conference of geographers in government.

The C. Geog. community has become a force for action. Leading members were responsible, along with modest support from the RGS-IBG, for gaining Geography's inclusion as a recognised profession in government. This evolved from an informal cross-government geographers' network set up in 2015, with representation from the many government departments and agencies in Whitehall employing significant numbers of geographers, most notably in areas relating to the environment, regional development, planning, overseas development, GI and mapping, and international relations. In 2017 Geography was recognised as part of the science and technology profession, with the appointment of a Head of Geography Profession (David Wood), and by 2021 David had led it to become an analytical profession in its own right, sitting alongside the professions for economists, statisticians, and social researchers.

The fact that a professional accreditation (C.Geog) linked to the RGS-IBG already existed was important in this recognition. The government Geography profession currently has more than 3000 practising members and continues to work closely with the RGS-IBG Chartered Geographer scheme. The symbolic importance and practical value – in government and beyond – of having a recognised government profession for Geography and geographers is simply immense.

Sustaining merger strengths: research, expeditions, and membership

It was important to keep a balance between our six pillars of work as the RGS-IBG developed, and this included many incremental developments in our support for research, expeditions and fieldwork, and membership. Perhaps prime among them were the expansion of the Annual Conference for academic geographers and postgraduate researchers; the technical enhancement and growth in the Society's leading international scholarly journals; and the systematic expansion of the grants programme.

In 1995 the RGS-IBG awarded a total of £41,950 per year (equivalent to £76,752 in 2017 values) in small expedition grants, ranging in value from £100 to £1000 each, mostly to undergraduate university teams. In 2017 it awarded more than £200,000 per year in grants for field research projects, expeditions, and early career research across the breadth of Geography. Expanding the programme with significant new funding for small research grants for early career researchers was especially important, as existing sources were limited and yet the demand at a time of growth in Geography in higher education was substantial. Such small enabling grants can have a vital impact, disproportional to the size of the grant, in facilitating early career development post PhD. For many of the recipients it is their first research grant. Most of the additional funding came from the Society's membership as appeal donations, memorial and legacy funding.

There were some new major expedition, fieldwork, and field research grants and prizes too, kindly donated by Fellows and associated foundations. These included the Ralph Brown Award, the Neville Shulman Challenge Award, the Journey of a Lifetime Award jointly with BBC Radio 4, the Alexander Awards and, pledged in 2017, the Esmond B. Martin Royal Geographical Society Prize. The Expedition Advisory Centre also took on a high-profile role nationally in advising on field safety standards, including the new British Standard 8848 to guide overseas commercial expeditions.

One of the most rewarding innovations for me personally was the Learning and Leading programme (2008–2016). This used the transformative power of learning through Geography fieldwork and expeditions to provide life-changing opportunities and to enhance social mobility for young people from less advantaged backgrounds. Initiatives ranged from UK residential fieldwork (for A-Level students) to undergraduate 'fieldwork apprenticeships', and fieldwork training for teachers in deprived areas. Hundreds of young people and teachers facing challenging circumstances have benefited, and for many it was life-enhancing, building confidence, skills, and ambition.

More generally, there were undoubted operational challenges around generating income; co-ordinating a large volunteer team; upgrading a heritage building; and managing a staff team that grew from around 35 to 70. However, there were only two occasions when the merger of the RGS-IBG and its development came under significant pressure and challenge from parts of the Fellowship.

The first occasion was in 1996 when some former members of the IBG called a special general meeting to vote on the RGS-IBG association with the company Shell as a sponsor. Although the vote of Fellows was substantially in favour of retaining the sponsorship, the open, trustworthy, and honest way in which the vote was handled by all concerned left most Fellows willing to accept the majority vote (around 100 resigned). The Critical Geographers Forum was the most active voice in this debate (see Unwin, 1999 and other papers in that volume; and Gilbert, 2009). Shell's corporate sponsorship continued to support selected Society non-academic activities until 2014 after which, by mutual consent, it ceased.

The second was the decision to offer more research grants while reducing the number of Society-led, large field research programmes that had been characteristic of the RGS from the late 1960s to 2000. The careful decision was taken jointly by the Research and Higher Education, and Fieldwork and Expeditions Committees following an open review in 2003, and was endorsed by the Council as part of the 2005–2010 strategy.

A high-profile group of expeditioner Fellows of the former RGS were unhappy at the changes and could not be persuaded of the rationale for them. They mounted a forceful, public, and not wholly factually accurate, lobbying and media campaign in 2008 over what was essentially a private matter for the RGS-IBG and its Fellows. The vote of Fellows at the Special General Meeting called to resolve the issue was a majority (62%) in favour of the Society's position. However, it took several years to overcome a legacy of disharmony in what was otherwise a very harmonious and supportive Fellowship over the 22 years I led the RGS-IBG as Director.

In 2018, the membership stood around 16,000, 33% higher than at the merger in 1996. New categories of Member, Young Geographer, and Schools Member sat alongside that of Fellow. In 2017, approximately 3,000 volunteers supported our work; and, most importantly as a charity, an estimated 3.5 million engagements with Geography happened through the Society's work in one way or another (RGS-IBG, 2017).

Conclusion

The breadth and resilience of the RGS-IBG, built carefully and collectively over many years since 1995, with enormous support from Geography communities, has stood the test of time. The Society, guided by successive strategies and the hard work of staff, trustees, and numerous volunteers grew to serve geography and its many communities well.

The vision was for a world-leading institution, dedicated to advancing geographical science; sustaining the discipline and as a 'home' for its many communities; and as the means for everyone to discover, understand, and respect the world's people, places, and environments.

The following three published quotes summarise, for me, what the RGS-IBG had overcome and become by 2018:

> When I look at the successes of the RGS-IBG in representing Geography over the last decades, I reflect on the tentative beginnings of the merger. The cultural differences were extreme and there were many on both sides who doubted the wisdom of the venture.
>
> (Professor Emeritus David Sugden, President IBG and Vice-President RGS-IBG 1995–1996; quoted in Price Williams, 2017)

> The RGS-IBG is a most significant dimension of the institutional capacity of UK human geography and is unequalled in any other country. It is an extraordinary disciplinary resource.
>
> International Benchmarking Review of UK Human Geography (ESRC, 2013)

> Many contributors ...pointed to the effectiveness of the RGS-IBG in promoting its discipline at all levels and in all sectors with academic and professional activities, support for schools, an admirable programme of public engagement, and providing a single coherent voice, for example, to governments.
>
> Reflections on Archaeology (The British Academy, 2016)

This period was, however, just the beginning of a new chapter in the UK learned society that exists to advance geography. It builds on what went before and those who follow will continue to build. Neither geography nor the Society are bomb-proof and their resilience should never be taken as a given in a constantly changing world. Three aspects are, in my view, vital to the future of both:

First: being an effective, passionate, and knowledgeable advocate for geography. This requires constant awareness of evolving government and other decision-making agendas likely to affect Geography and geographers, and the ability to be agile in acting quickly to mobilise relevant networks and engage or intervene in the most appropriate, timely, and effective way. The future health of geography depends on this above all else. Having pro-active, passionate, and knowledgeable geographers within the RGS-IBG senior staff team is a vital component. Having an organisation that is proud of its discipline, and the community of geographers actively behind it, are essential too.

Second: sustaining finance. The Society has sound financial foundations and good reserves, grown over the years expressly to manage the risks of unexpected challenges, such as the recent Covid-19 pandemic when income fell. However, both income sources and reserves need careful stewardship. Additionally, successful fundraising will continue to be needed if the Society and its breadth of activities are to continue to flourish. Between 1998 and 2017 we raised £40 million, over and above the core income from membership subscriptions, operations, and enterprise activities. This was essential in enabling a step change in our support for, and promotion of, Geography and geographers and for us to advocate successfully on many fronts.

Third: retaining breadth and balance across the RGS-IBG. Few learned societies, and only the very largest, have the capacity, like the RGS-IBG, to engage and support their disciplines across the full breadth of stages and communities. There are many advantages in doing this, not least the ability to understand the big picture; the potential for innovative cross-overs between different communities of interest; to attract a large membership to help fund the organisation; and the ability to avoid polarisation.

It is for my successors, future trustees, and the community of Fellows, to determine how the Society proceeds in delivering its purpose of advancing Geography as the learned society and professional body.

Note

1 An institution is an organisation, establishment, foundation, society, or the like, devoted to the promotion of a particular cause or programme, especially one of a public, educational or charitable character (Dictionary.com).

Study task

Challenge

Imagine you are working within the geography learned society or equivalent in your country. Unexpectedly, government funding for geography research is suddenly under threat. You are being asked to research and make a case for continued funding, drawing on the beneficial impacts of geographical research beyond universities.

The case needs to be researched and evidenced, synthesized and presented.

- For the synthesis think carefully how you might structure the case and take note of the political, social and economic context in which it is being made.
- The presentation of the case could take one of several forms, for example a written report, a PowerPoint presentation, or a video.
- Some resources, relating to the UK, that you may find useful as evidence for your case are linked below. Your Geography Department or tutors may also be able to suggest a selection of individual research papers to read or particular journals to consult; and your national geographical society may be a good place for resources too.

Resources

- The SHAPE of Impact. A report published in January 2024 by the British Academy and the Academy of Social Sciences that analyses the impact of research across all the social sciences, arts and humanities disciplines (including Geography). This is based on nearly 4000 impact case studies submitted to the UK research excellence framework assessment in 2021 (REF 2021).
- The interactive SHAPE of Impact dashboard published with the report above, which enables summaries and searches to be made of the impact case studies submitted to the Geography and Environmental Studies review panel of REF 2021 (UoA 14).
- The Geography and Environmental Studies overview report found within the Panel C (social sciences) report for REF 2021.
- The Royal Geographical Society (with IBG) examples of impact.
- Many university department websites feature the impact of the research undertaken by their staff. For example, Durham University, Newcastle University, Oxford University, Southampton University among very many others.

- Research studies that have won Impact Prize Awards from the Economic and Social Research Council of UKRI (a government funding agency), that span all the disciplines in the social sciences and which are awarded each year.
- Information on the social sciences and humanities in the European Union's Horizon Europe research funding programme.

References

DfE (Department for Education), 2012, Education & Training Statistics for the United Kingdom, 38 p.

Driver, F. and Jones, L., 2009, *Hidden Histories of Exploration*. Royal Holloway University of London, 64p (ISBN: 9781905846306).

ESRC, 2013, *International Benchmarking Review of Human Geography*. ESRC, Swindon, UK.

Gilbert, D. 2009 Time to Shell Out? Reflections on RGS and corporate sponsorship. *ACME 8* s(3), 521–529.

Joint Council for Qualifications, n.d., https://www.jcq.org.uk/examination-results/

Mill, H.R., 1930, *The Record of the Royal Geographical Society, 1830–1930*. Royal Geographical Society, London, 288 p.

Price Williams, J., 2017, The Voice of Geography, RGS-IBG, 38 p.

Price Williams, J., 2012, The Home of Geography, RGS-IBG, 38 p.

RGS-IBG, 2001, Geography: An Education for Life, 8 p.

RGS-IBG, 2009, Annual Review, 32 p.

RGS-IBG, 2016, Annual Review, 32 p.

RGS-IBG, 2017, Annual Review, 32 p.

Steel, R.W. 1984, *The Institute of British Geographers: The First Fifty Years*. Institute of British Geographers, 170 p. (ISBN: 0901989363).

The British Academy, 2016, *Reflections on Archaeology*. British Academy, 52 p.

Unwin, T. 1999, Editorial. *Ethics, Place and Environment*, 2(2), 133–139.

Walford, R. 2000, *Geography in British Schools, 1885–2000: Making a World of Difference*. Taylor & Francis, 272 p. (ISBN: 9780713002072).

Part II

Making geographical knowledge

6 Landscape and environmental change

Carl Sauer on land and life

Kent Mathewson

Book cover of *Carl Sauer on Culture and Landscape*, LSU Press, including a photo of Carl O. Sauer (1889–1975). With permission from LSU Press.

Introduction

Carl Ortwin Sauer (1889–1975) occupied a key, if somewhat contrary, role in 20th-century American Geography from the 1925 publication of his first major disciplinary statement (Sauer 1925) to his death in 1975 (Leighly 1976). Sauer's 50-year span of professional prominence was neatly hinged at the mid-century mark. By 1950, much of the world accepted *TIME* magazine publisher Henry Luce's declaration that the 20th century was "the American Century." Of course, in retrospect this was short-sighted, a

DOI: 10.4324/9781003343240-8

claim that historians refer to as presentist, because it presumed that a difficult past necessarily gave way to a supposedly glorious present. In step with these times, mid-century American geographers celebrated the 50th anniversary of the Association of American Geographers with the publication of *American Geography: Inventory and Prospect* (James and Jones 1954). The book surveyed 22 of Geography's subfields. Preston E. James (1954) authored the overview and introductory chapter, placing regional studies as Geography's core organizing principle. The purpose of Geography was to describe the unique character of regions through a synthesis of their multitudinous constitutive elements, human and physical.

By 1954, Carl Sauer, was the acknowledged doyen of American historical geographers. Nonetheless, he declined the invitation to write the historical geography chapter in *Inventory and Prospect*. This was because from 1925, when Sauer moved from Ann Arbor to Berkeley, he increasingly distanced himself from the institutional cores of American academic geography in the Midwest and East Coast. The rest of his career (and life) was spent constructing, or perhaps more aptly cultivating, a highly distinctive geography, largely alternative, and at times oppositional, to the practice and theory of mainstream American Geography represented by *Inventory and Prospect*. The purpose of this chapter, by tracing his life, is to chart Sauer's geographical "project," or better, "nested projects" (it was multifaceted). While Sauer was the author and instigator of each of his projects, he enlisted collaborators and attracted followers that helped bring them to fruition, successfully impacting American geography.

Sauer's projects: an overview

Chronologically, the first of his projects was initiated while he was still in graduate school. He began to read widely in the German and French geographical literature, introducing the associated concepts and methods, especially those involving culture and landscape, into American Geography. His second project was to make field work and field courses core to university geographic curricula. He did this during his tenure at the University of Michigan (1915–1923), establishing the Kentucky field station, promoting the Michigan State Land Survey, and initiating the annual Midwestern geographer's field methods conference. Both of these projects were foundational for his third project: a three-decade-long program of field research in Mexico. His emphasis was Indigenous culture, particularly in the colonial and pre-colonial periods, with a focus on agricultural systems, plant domestication, and pre-Columbian demography. This entire complex of Sauer's cultural-historical research interests – field work, indigeneity, pre-colonial history, environmental relations, set in Mexico – was largely foreign to American Geography and geographers at that time. Sauer's fourth project was discrediting and dislodging environmental determinism from American Geography. By the 1920s "environmentalism" had achieved nearly paradigmatic status in American Geography. Sauer's 1925 "Morphology of Landscape" demonstrated its poverty and replaced it with human cultural agency as the driving force in landscape change.

All of these projects came together in his fifth major project and possibly the most important in the history of American Geography: the creation and building of an academic "school," the Berkeley School. Initially this was not by design, but Sauer soon saw that the individual and institutional components were at hand to build such a construct. From hiring German geographers as faculty members, to attracting field-seasoned graduate students from other disciplines, to forging links with Berkeley faculty in anthropology, history, and botany with Latin American interests, and distancing the evolving collective from mainstream American Geography, a school emerged. A hallmark of this school was the historical orientation, making the "fourth dimension," or time, central.

Sauer's final and capstone project, which marked his final decades, was raising both academic and public awareness of modernity's environmental depredations. From the late 1930s on, in print and in person, Sauer spoke out against the "destructive exploitation" of natural environments and the peoples that lived in relative harmony with them. In these regards, he was prescient. His "Man's Role" symposium in 1954 (Thomas 1956) was unprecedented and it preceded the emergence of the environmental movement by a decade or two.

In almost all of these projects, Sauer accomplished what he set out to do. In the case of his advocacy of sound and informed environmental stewardship, he was well ahead of the pack, but did not reach a mass audience as did some of the later figures like Rachel Carson. But among American geographers, Sauer stands alone in his distinctive record of projects conceived, carried out, and accomplished.

One of the key conceits of mid-20th-century America's manufactured consensual mentality was the notion of an ethnic "American" homogeneity, a unitary national culture and a national destiny that promised global pre-eminence. From the start, Sauer's formative years and social milieu in an ethnic and religious enclave in rural Missouri put him on a life-long course that was often at odds with convention (Sauer 1963). His scholarly work countered stock national conceits, and in more focused ways, he eschewed mainstream currents in various academic fields, particularly American Geography. Sauer's life and works can be viewed in roughly decadal segments (Leighly 1976). While most of his projects emerged and developed as his professional career progressed, many were also cross-linked to early familial, cultural, and locational influences (Williams 2014).

Early childhood: 1889–1899

Born on Christmas Eve, 1889, Sauer spent his early years in Warrenton, Missouri. Sixty miles west of St. Louis, on US Highway 40 – "America's Mainstreet" – and fifteen miles north of the Missouri River and the Ozark uplands, Warrenton was almost a caricature of a turn-of-the-century Midwestern town. Although Warrenton's population was less than 1000, it had almost a full range of services, including a small liberal arts college. Central Wesleyan College was founded by German Methodist ministers after the Civil War, initially as a school for war orphans, but soon expanded as a college of liberal arts. Instruction was in both English and German. Though clearly Midwestern in location, Warrenton was not entirely "All-American." It was at the center of Missouri's "Rhineland," home to many German immigrants, including a number of political refugees after the failed European revolutions of 1848. Many of the political refugees were university educated, distinct from other immigrants, or the more rooted Americans settling on the farming frontier. Sauer's father, William, was trained as a teacher and musician. He emigrated to the US in 1865, and accepted a position at Central Wesleyan in 1866, becoming one of its first faculty members. He was the college organist, botanist, and taught French. In 1878 he married Rosetta Vosholl, daughter of German immigrants.

Growing up, Sauer especially loved botanizing with his father and learning about the area's physical geography. The elder Sauer was a devotee of Goethe's *Naturphilosophie* which called for a direct engagement with nature, rather than testing hypotheses and dealing in abstractions. Carl Sauer's professional practice in the field followed this Goethean approach (Kenzer 1987), calling for immersion with the natural phenomena studied. In Sauer's case, this was the landscape (Gade 2011b).

Intensive schooling: 1899–1908

Enormously talented, Sauer skipped grades and completed his elementary, secondary, *and* undergraduate education by the time he was 18 years old, four years before the normal completion age. This was later mirrored in his professional career – from Instructor to Full Professor in only seven years. When he turned nine, William Sauer decided that Carl needed more intensive schooling than Warrenton could provide. With the help of friends and relatives in Germany, secondary schools were scouted, and in 1899 the family moved to Calw, a town in the Northern Black Forest region of Baden-Württemberg, Germany. Carl was enrolled in the local *Gymnasium* (secondary school). It was run along military lines with an emphasis on German history and geography. Despite his extreme aptitude, the *Gymnasium* was an ordeal, souring him on directed student learning, likely contributing to his later anti-institutional, anti-bureaucratic attitudes. However, he very much enjoyed the school field trips and being outside in nature. After two years in Calw, the family moved back to Warrenton. In 1906, at sixteen, Sauer enrolled in Central Wesleyan, finishing his Bachelor of Arts degree in two years. In his preliminary exams for graduation, he received a remarkable overall score of 98.25%. In his second and final year, he was effectively elevated to a lecturer position, and taught French, geology, geometry, and history.

Graduate school years: 1908–1915

In the autumn of 1908, at eighteen, the age when most students start undergraduate study, Sauer left home for graduate school. He chose Northwestern University, with its Methodist affiliation, studying in the Department of Geology, known for its field orientation (Kenzer 1985). Geology had become a popular university concentration in the US following the oil discoveries after 1900. Sauer's interest in the subject, however, was not that but in landforms and physiography. In 1909, he transferred to the new Department of Geography at the University of Chicago, believing that discipline would better meet his interests. Rollin Salisbury, perhaps the foremost physiographer in the US, chaired the Chicago department. Other faculty included Harlan Barrows and Ellen Churchill Semple, historical geographers, Paul Goode, cartographer, and Wellington Jones, fellow student and instructor who had field experience in South America. Sauer also took classes outside the department, particularly Henry Cowles' plant ecology course with its field excursions in the Indiana dunes and Chicago's marshes. Later Sauer always encouraged students to take courses outside Geography, especially scientific field-oriented offerings, along with anthropology and/or history. At Chicago, Sauer's intellectual potential was quickly recognized. In the summer of his first year, Salisbury commissioned from him a study of the Upper Illinois River region (Sauer 1916). The next summer he carried out a study of the Starved Rock subregion, with a detailed account of Native American occupancy (1918). This was one the first comprehensive and sympathetic accounts of Native American habitation (Williams 2014, 45). It foreshadowed his major contributions to Indigenous studies of the Americas (Denevan 1996).

During his second year of graduate studies, though, he began to wonder if he'd made the right decision. He disliked the competitiveness of universities and big institutions. He was also not keen on large cities, instead favoring small-town life. This became another hallmark of Sauer's later scholarship. It promoted the provincial, was anti-modernist, at odds with "progress" and urban-industrial civilization. After his third year, in July 1912, he took a year-and-a-half break from the university, working for the map maker, Rand McNally. Even though on "sabbatical," he continued to read widely, especially the European geographical literature. It was also during this period that he became disaffected with the environmental determinism that he had been taught by Barrows and Semple, the idea that humans are shaped by physical geography (see Chapter 3 by Innes Keighren). That rejection laid the foundation, a decade later, for his 1925 major disciplinary statement which disavowed the environmental determinist paradigm within American Geography (Sauer 1925).

But before he went back to graduate school, he had one piece of personal business to attend to. In December 1913, he married his long-time fiancée, Lorena Schowengerdt. Consequently, it was not until fall 1914 that Sauer finally began his doctoral field work on the Ozark Highlands, Missouri. He spent three months criss-crossing the region, by train, wagon, but mostly tramping on foot and riding on muleback, covering up to 30 miles a day. He got to know the region, its people, and its geography first-hand. The practice of "pedestrian geography," survey by foot, became yet another hallmark of Sauer's research method and the Berkeley School that he later founded. Returning to Chicago in early 1915, he spent the next year writing up the dissertation, defending it on December 15, 1915 (Sauer 1915).

Michigan years: 1916–1923

PhD in hand, Sauer accepted an instructorship to teach in Michigan's newly formed Department of Geology and Geography. He was the sole geographer. Much of his teaching involved field methods along with running the summer field camp. It was a hands-on and immersion-in-the-landscape approach to geographical instruction. Sauer was also heavily involved in off-campus activities. He was called on to survey Michigan's "cut over" lands and propose new land uses. To do so involved establishing the Michigan Land Economic Survey in 1922, which employed a similar approach to the one he used in his doctoral research in the Ozarks. He also carried out a regional field survey for the Kentucky Geological Survey. This resulted in his *Pennyroyal* monograph (1927a) that, as Karan (1983) has suggested, prefigured Sauer's turn to cultural and historical geography, and his main sub-disciplinary focus for the rest of his career.

Although Sauer accomplished much in Michigan and rose through the ranks meteorically – from Instructor to Full Professor in seven years (1915–1922), half the normal time – still he was not satisfied. He was weighed down by heavy state service obligations, teaching responsibilities on topics that did not especially interest him, and intellectually provincial Midwestern colleagues. For these reasons, in 1923 he accepted the Chair in Geography at Berkeley. The move opened up a whole new world and provided the platform for his future projects including: introducing the German concept of landscape or *Landschaft*; launching cultural geography (also largely a German import) (Sauer 1927b); building the "Berkeley School;" and making Latin America a major arena for cultural-historical geographic research.

Building Berkeley: 1923–1940

The Department of Geography at Berkeley dates to 1898, and was the first US Geography department. Sauer had turned down a position there in 1918, but he accepted the second offer made five years later, provided it came with two additional positions. One was for his Michigan student John Leighly to cover physical geography and cartography, and the other was for someone to teach commercial geography. At the time, Sauer had little confidence in American trained geographers and so many of his later departmental recruits were from Europe. American physical geographers generally followed William Morris Davis' cyclical stage theories. This was the idea that after new land was created following a period of uplift, erosional forces would ensue, producing in turn a "youthful," then "mature" and then "old-age" landscape. Sauer believed such an extrapolation from a simplistic life-cycle metaphor was naïve and should be replaced by grounded field work and empirical study. Likewise, environmental determinism, which he had seen up close in Chicago in lectures by Semple and Barrows. Its central thesis – nature determines human behavior and thought – was hopelessly simplistic, Sauer thought. Moreover, it led to dangerous regressive politics, cultural imperialism, and worse. Initially, Sauer challenged the Davisian dogma through physiographic study of California landforms. Turf battles with geologists, however, soon turned him instead toward the study of human cultural and historical landscapes in Southern California and Mexico. Cultural and historical landscape study, both in field and archive, was to become Sauer's overarching life-long project and his antidote to environmental determinism.

His formal assault on environmental determinism began with his 1925 Berkeley discussion paper, "Morphology of Landscape" (1925). Sauer would later say that this work had only the modest aim of trying to clarify the discipline's role within the university, and to stake a claim to "landscape" as geography's subject of study. In actuality, Sauer drew on sources from continental philosophy and geography to make the case against environmental determinism (May 2003). It was a *tour de force*, impressively learned, challengingly complex, and effectively argued. It was unprecedented in American geography (Penn and Lukermann 2003). The effect reverberated far beyond Berkeley. From 1925 on, the lure of environmental determinism lessened, until by 1950 it was largely eliminated. Instead, Sauer's program of study of cultural and historical landscape change increasingly attracted adherents, becoming the main mode of the Berkeley School. While the Ozarks and southern Kentucky originally attracted Sauer because they were living windows on the past (Mathewson 2004), the study of Mexico and Latin America for him became far richer, with a cultural heritage of millennial time depths. More generally, Sauer was an antimodernist, rooted in an abiding respect for traditional rural life, with an antipathy toward modern bureaucratic, urban commercial society (Mathewson 1987). His politics were conservative, not in the sense of pro-market right-wing, but as a belief in the conservation of premodern lifeways and landscapes. Here, Mexico was his "passage to premodernity."

While the original intent of his California move may not have been to form an academic "school," before long, it was clear Sauer was creating one (Speth 1981). This was another "project." Academic schools are similar to "invisible colleges," or informal networks of scholars working in related areas and topics, but schools often emerge around a charismatic figure with a broadly cohesive research agenda and a supportive institutional base. A body of published literature by the members gives it material expression. Less often, a programmatic statement provides direction (Crane 1972). In the case of the Berkeley School, all these criteria were met. Sauer has been widely characterized as charismatic. Whether that charisma was innate or not, Sauer's conscious distancing himself from mainstream American Geography, and forging an alternative/oppositional enterprise, added to it. It took a few years for Sauer and his followers to settle into a broadly cohesive research agenda, but once in place it endured to the present, albeit with updates and modifications. The same can be said for a body of published literature. The department already had the monograph series *University of California Publications in Geography*, and Sauer launched the *Ibero-Americana* series in 1932 with colleagues in anthropology and history (Parsons 1996a). These served as Berkeley School outlets for dissertations and theses, along with faculty research results. Finally, the University of California gave Sauer considerable latitude and support for his efforts throughout his career.

Even though Sauer disavowed ever setting about to establish a "school," by 1930 the Berkeley School began to emerge (Duncan 2000). In terms of graduate-level pedagogy and training, he kept a light hand on the tiller. He trusted students to find their own way in the field, and to largely choose topics and course work as desired. Lack of prior training in Geography was considered no impediment, and perhaps even preferable. Course work outside the department was encouraged. Anthropology, history, geology, botany, forestry, and agronomy were regular adjuncts, but lesser connected fields were also tapped. This multidisciplinary approach to graduate training and scholarly research set the Berkeley program apart from its counterparts and fed into the formation of the school.

Once Sauer set his sights on Mexico, a succession of graduate students followed him. The first cadre who wrote dissertations on Mexico included Fred Kniffen (1930) on the Colorado Delta, Peveril Meigs (1932) on Missions in Baja California, and Donald Brand's (1933) historical geography of NW Chihuahua. They were followed by: Henry Bruman (1940) on ancient alcohol areas; Dan Stanislawski's (1944) historical geography of Michoacán; Robert West (1946) on colonial mining; Brigham Arnold (1954) on early human occupancy in Baja California; Homer Aschmann (1954) on aboriginal ecology and demography, Baja California; and a dozen years after Sauer retired, H. Leonard Sawatsky (1967) on Mennonite colonization. Almost a quarter of the dissertations then, were on Mexican topics. The other Latin American dissertations were: Parsons (1948) on Antioquian colonization, Colombia; Gordon (1954) on the ecology of the Sinú region, Colombia; McBride (1940) on the Native economy of SW Guatemala; Johannessen (1959) on interior savannas of Honduras; Wagner (1953) on the cultural geography of Nicoya, Costa Rica; Duran (1953) on the physical and cultural geography of the Cayman Islands; Bowman (1941) on soil erosion in Puerto Rico; Innis (1958) on human ecology in Jamaica; Merrill (1957) on historical geography of St. Kitts and Nevis; and Edwards (1962) on aboriginal watercraft of western South America. Clearly the core domain of the Berkeley School was Latin America. All but two of these dissertators went on to careers in academia, but only a few produced "branches" on the Sauer Tree that went beyond the second generation (Brown and Mathewson 1999, Mathewson et al. 2020). As of 2020 the Parsons line has been by far the most prolific, with 20 main Latin Americanist branches and almost 200 "twigs" or doctoral descendants extending into the sixth generation.

Between 1925 and 1950, Sauer made 21 research trips to Mexico. During the winter and spring of 1942, he traveled the length of western South America under the auspices of the Rockefeller Foundation, surveying future field sites and scouting out local geographical institutions and scholars. During the summers of 1950 and 1952 he led students on field observations in the Caribbean, visiting seven islands and coastal Venezuela under the auspices of the Office of Naval Research (ONR). The ONR was a key source of funding for the Berkeley School's graduate research program. It supported most of the field work in the Caribbean as well as many of the studies in mainland Latin America (West 1979). In 1956 Sauer spent time in Costa Rica after attending the International Congress of Americanists in San Jose. A decade later he revisited Costa Rica to participate in the Organization of Tropical Studies (OTS) summer field course and took a side trip to see Guatemala.

Although the Berkeley School's primary focus was Mexico, and later Latin America more broadly, other topics were entertained. Of the 37 dissertations that Sauer oversaw (1927–1967), almost half (18) were on non-Latin American/Caribbean topics. Eleven of these were set in North America. Several of the dissertations on US topics were what the Germans termed *Heimatskunde*, or a study of the student's home place or region. Five dissertations were on topics outside the Western Hemisphere including New Zealand, Southeast Asia, Ethiopia, Morocco, and the Seringeti Savanna, East Africa. Almost all of the 37 dissertations have an historical orientation, whether focusing on cultural or physical geographic topics. The entire Sauer Tree has yet to be fully charted, but the total "progeny," counting just the holders of doctoral dissertations in the Sauer line (Latin Americanists and others) totals over 1000 (Mathewson et al. 2020). If the holders of only master's theses supervised by Sauerian progeny are included, the total might exceed 1500. It is highly unlikely that any other geographer can count as many offspring.

By 1940, at age 50, Sauer had largely accomplished his main aims. Drawing on the field-oriented natural sciences, particularly geology and ecology, he helped establish field studies as routine practice in human geography. That his erstwhile Midwestern colleagues had pursued micro-regional studies of limited intellectual content no longer concerned Sauer. He directed his field program and practice south of the border to cultures and histories of millennial rather than decadal time depths. Geographers were now viewing environmental determinism in the rearview mirror and looking to other explanatory modes. Sauer offered the Berkeley School approach to landscape study through cultural and historical lenses as one alternative. He put culture on the table with his "Morphology of Landscape" (1925) and his prospectus, "Recent Developments in Cultural Geography" (1927). His 1940 presidential address to the Association of American Geographers, "Foreword to Historical Geography" (1941) foretold his future work (Figure 6.1). Increasingly Sauer pushed back the timeline in his explorations, through archaeology but also imaginative speculation. He plumbed the depths of Mesoamerican cultural origins, with particular interest in plant domestication, and then on back into the Pleistocene to chart the first witnesses to the New World's pristine landscapes. In his post-retirement years (1957–1975) he speculated on topics such as "Seashore – Primitive Home of Man?" (1962) and "Fire and Early Man" (1961). His ideas have often since proven accurate (Dickson 2003).

Just as his "Morphology of Landscape" paper was a rebuke to environmental determinism, his 1940 presidential address was, in part, a retort to Richard Hartshorne's (1939) influential work, *The Nature of Geography*. Subtitled "A Critical Survey of Current Thought in the Light of the Past," Hartshorne's book sought to put an ahistorical regional geography at the methodological core of the discipline. Regions were defined by the accumulation of objective geographical facts, a form of philosophical positivism. For Sauer, Hartshorne's marginalization of both historical and physical geography, along with his positivist presentism, were diametrically (and dialectically) opposed to Sauer's conceptions and practice of Geography. For Sauer, a historical perspective was

Figure 6.1 Book cover painting by Antonio Sotomayor, prominent Bay Area artist, entitled "Carl Sauer and Field Party, Sonora, 29/30" (1932). Bottle of Old Carl "The Original Sauer Mash" commissioned for 1990 Golden Jubilee of Carl Sauer's Presidency of the AAG and annual meeting at LSU, Baton Rouge, LA. 1940.

fundamental, physical geography was foundational, and positivism in all its manifestations – philosophical, methodological, and political – were antithetical if not outright anathema (Entrikin 1984). Sauer dismissingly referred to "The Nature" as Hartshorne's "late resumé." (1941). He correctly saw it as legitimizing the recent regional work of American geographers, especially his former Midwestern colleagues, which tended to be descriptive, ahistorical studies of contemporary and largely homogeneous spaces.

In full stride: 1940–1957

By the onset of WWII, Sauer had established his Berkeley department and his research program as one of American Geography's key nodes, but one that charted an independent course (Parsons 1979). Not only was it separated by half a continent from the other centers, increasingly concentrated in the Midwest, but Sauer, that "Great God beyond the Sierras" as Hartshorne sarcastically referred to him, continued to keep his distance. And he increasingly distanced himself from the present. From the 1940s onward he explored the "farther reaches of human time," to borrow from John Leighly's section heading in *Land and Life* (Sauer 1965). Sauer explored the Pleistocene peopling of the Americas (1944, 1948, 1957). At the same time he continued his studies of human relations with plants and animals, especially domestication (1936, 1938, 1947, 1958). The capstone was a series of five lectures, the Bowman Lectures, delivered at the American Geographical Association in 1952 (Figure 6.2). Published as *Agricultural Origins and Dispersals* (1952), it was

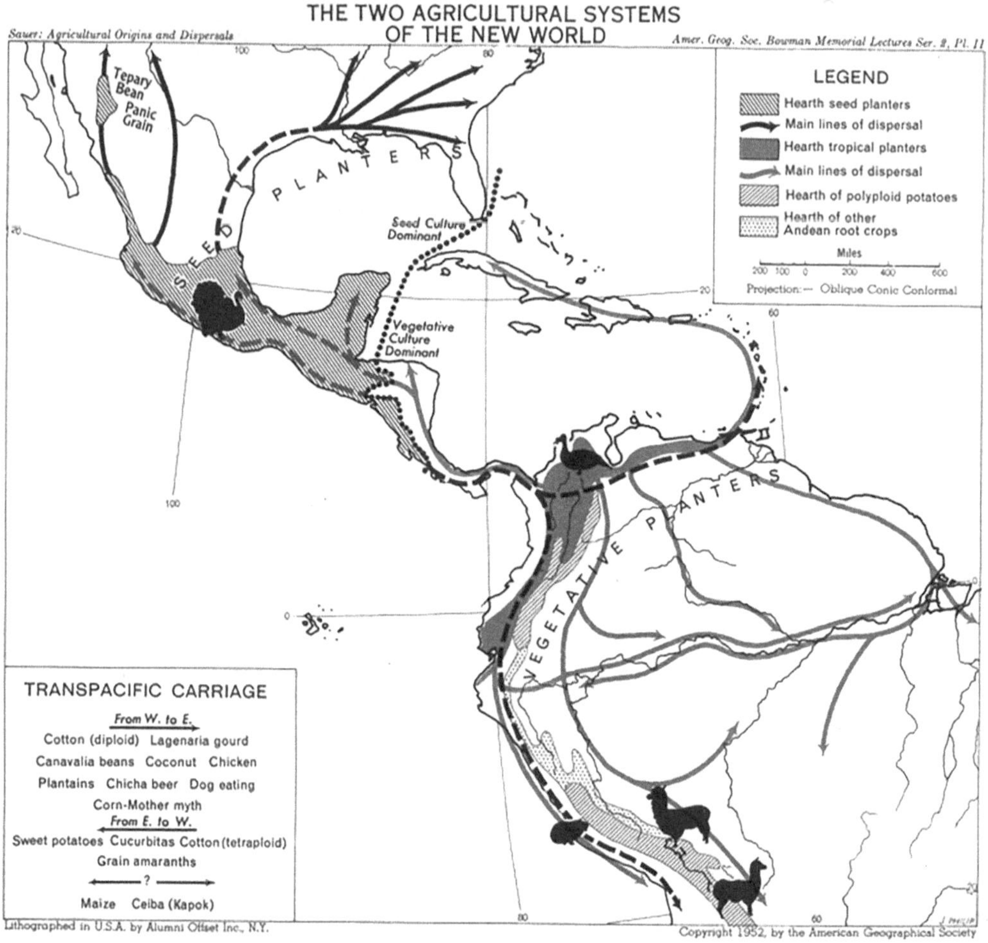

Figure 6.2 "The Two Agricultural Systems of the New World," American Geographical Society.
From *Agricultural Origins and Dispersals*. The Bowman Lectures, Series 2, Part II, American Geographical Society, 1952.

speculative, global in reach, spanning all human time, and prescient. Sauer posited that horticulture (the tending of individual and diverse plants) preceded agriculture (tending toward monoculture). This first emerged in Southeast Asia rather than Southwest Asia (the Levant). Sauer suggested root crops rather than seed crops as the initial domesticates. Root crops can be reproduced asexually (cloning), whereas seed crops must be reproduced sexually, requiring more selective effort. He argued that this initial "culture hearth" of domestication enjoyed diverse environments, and the innovators were "progressive fisherfolk" whose rivers and lakes provided abundant protein. Given ample leisure and sustenance, these tropical "Mesolithic fisherfolk" could experiment with plants for both food and material culture items useful for fishing, such as cordage and piscicides. Subsequent research, however, still favors seed cultivation and the Levant as the contexts for the earliest culture hearth of plant domestication (Harris 2002). In Sauer's retrovision, leisure and abundance rather than necessity and scarcity were the fostering conditions and forces.

Despite his frequent jeremiads (in print and in correspondence) about the shortcomings of modernity – centralization and bureaucratization, academic regimentation, positivism in social science, paucity of curiosity – Sauer was engaged throughout his career in multiple service activities beyond his own scholarly pursuits. He initiated the annual field conferences of Midwestern geographers (1923–1940), and Henry Wallace, US Secretary of Agriculture, appointed him to a special USDA advisory committee. The move to Berkeley in 1923 brought him relief from both the extra-curricular work in Michigan and its oppressive atmosphere. California and Berkeley were a tonic for him. He said, "I never went back to Ann Arbor and its raw fall weather without the feeling of a man going to a prison sentence..." (Williams 2014, 52). But during the 1930s, in part because of the Depression, he was involved in a number of off-campus initiatives. During this decade he served on the Science Advisory Board of the National Resource Council and undertook a lengthy expedition in the southern Piedmont to survey soil erosion. He was also a prominent consultant for the Social Science Research Council, the Advisory Council of Fairfield Osborn's Conservation Foundation, and later the ONR (Office of Naval Research).

In 1931 Sauer received a Guggenheim Fellowship for his Mexican field work. This was the beginning of a long and fruitful association with the Foundation. Through his association with the Guggenheim Foundation, Sauer came into contact with the heads of the Rockefeller and Carnegie Foundations. For someone who professed to be a provincial, and borderline anti-institutionalist, Sauer's growing national and international articulations were indeed impressive. In line with President Roosevelt's Good Neighbor Policy, in 1942 the Rockefeller Foundation sent Sauer on a six-month expedition to western South America to scout local scholarly talent and pursue his own research in economic botany (West 1982). He was also sent to Mexico to investigate Rockefeller-funded research that spawned the Green Revolution (Jennings 1988). Sauer authored stinging memos warning against replacing the peasant mode of production with modernized and highly capitalized farming. He was largely ignored.

Retirement

By the 1950s the Berkeley School was solidly in place, and Sauer's own reach and prominence well transcended Geography's disciplinary bounds. In October 1953 yet another foundation came calling for his expertise. The Wenner-Gren Foundation for Anthropological Research invited him to chair a planned symposium on human impact on the earth's surface (Williams 1987). Sauer readily accepted and was given license to select some 70 participants. The resultant roster included a number of geographers along with geoscientists, ecologists, anthropologists, historians, and others including luminaries such as Karl Wittfogel and Pierre Teilhard de Chardin. Urbanist Lewis Mumford and zoologist Marston Bates co-chaired the event with Sauer. The symposium was held in June 1955, in Princeton, NJ. The monumental *Man's Role in Changing the Face of the Earth* (Thomas 1956) followed. Praise and even celebrity followed, and Sauer was increasingly seen outside of Geography as its "doyen" (Mikesell 1987). He formally retired in 1957 but remained active until his death in July 1975 at age 85, a month after his wife Lorena's passing. In retirement he turned to historical geography, which he once referred to as "the apple of my eye" (Williams 1983). His subjects were human deep history, Geography's history, and a quartet of books: on the Caribbean, *The Early Spanish Main* (1966) (Figure 6.3); the North Atlantic in the late Middle Ages and early colonial

Figure 6.3 The Early Spanish Main, University of California Press, with permission.

period, *Northern Mists* (1968); *Sixteenth Century North America: The Land and People as Seen by the Europeans* (1971); and the posthumous *Seventeenth Century North America* (1980), in which the continent is seen through the eyes of Europeans. Fittingly, the last of his publications to appear while he was still alive was "The Fourth Dimension of Geography" (1974), in which he sketched the history of academic geography in the US, locating his own and the Berkeley School's place and time within this unfolding tableau. The paper ends with an admonition and warning: humans must recognize that they are not masters of an unlimited environment, and their continued technological intervention in the physical world will produce a crisis of survival for both humans and their earthly co-inhabitants.

Coda

One might legitimately ask why contemporary geographers, or especially students of Geography, should find Carl Sauer and his career worthy of study. After all, he was famously at odds with the present, and staunchly pessimistic about the future, especially modernity's stewardship of the natural world, and its relations with peoples not yet fully inducted into it. Sauer was not without his critics (Mathewson 2009). Prominent among them were the already mentioned Richard Hartshorne (1959), Peter Gould (1979), James Duncan (1980), Alan Pred (1983), and David Stoddart (1997). Quasi-critics include James Blaut (1980), Don Mitchell (2003), and Richard Peet (2003), who found Sauer's work commendable, but theoretically weak from a Marxist perspective. Price and Lewis (1993) countered charges that the Berkeley School was outmoded or had been off course from the start. But of Sauer's prescience, there should be no question. Two or three generations on, many of Sauer's insights have become accepted wisdom. Among them, three stand out. Foremost is his critique of the destructive effects of modernity on the environment producing something like ecocide (Speth 1977a). His criticisms helped lay the basis for an environmental geography (Mathewson 2011). Second, his advocacy of cultural diversity validated the value of preserving and protecting the inherited knowledge of Indigenous peoples and their peasant kindred (Lovell 2003). Third, his critique of presentism countered the view of history as cyclical or universal. Sauer held that to understand the present, one must see it as contingent, and non-recurrent – "an accretion of the past" (Speth 1977b).

Sauer's conception of Geography's charge in the scholarly division of labor was summed up in his "Foreword to Historical Geography," (1941) delivered as his presidential address to the annual meeting of the Association of American Geographers at Louisiana State University, December 1940. Sauer held that human geography must make "cultural processes the basis of its thinking and observation." Further, it must encompass "all the earth in all the time of human existence," to build a "retrospective science, which... acquires an ability to look ahead." These convictions were ably and amply demonstrated in his life's work, and the multiple projects which he undertook: to make field work a fundamental part of American Geography's pedagogy and practice; to build a distinctive cultural-historical-oriented program at Berkeley; to put the concept of landscape at the core of Geography's focus; to extend the temporal bounds of Geography's scope to humanity's beginnings; to broaden American Geography's purview to include the wider world, especially Latin America and the Global South; and perhaps most enduringly, to engender concern for the well-being of earth's life and landscapes in all their forms. As Sauer was fond of reminding and admonishing: it is humanity's preeminent project to "pass on a good earth."

Study task

This is best set as a term paper:

Whether by conscious design, or serendipity, the Berkeley School, led by Carl Sauer, created a distinctive "brand" of geography. Discuss the origins and development (through time) of this collectivity, its distinguishing traits, its prominent practitioners, and its notable publications.

Starter reading list

Denevan, William M. and Kent Mathewson (Eds). 2009. *Carl Sauer on Culture and Landscape: Readings and commentary*. Baton Rouge: LSU Press.

Harris, David R. 2002. The farther reaches of human time: Retrospect on Carl Sauer as prehistorian. *Geographical Review* 92: 526–544.

Kenzer, Martin S. 1987. Tracking Sauer across sour terrain. *Annals of the Association of American Geographers* 77: 469–474.

Leighly, John. 1976. Carl Ortwin Sauer, 1889–1975. *Annals of the Association of American Geographers* 66: 337–348.

Mathewson, K., A. L. Allen, A. Grismore, M. Lagos, J. R. Simms and B. Spencer. 2020. The Sauer Tree in time and place. *Journal of Latin American Geography* 19: 84–97.

Parsons, James J. 1979. The Later Sauer Years. *Annals of the Association of American Geographers* 69: 9–17.

Sauer, Carl O. 1941. Foreword to historical geography. *Annals of the Association of American Geographers* 31 (1): 1–24.

Speth, William W. 1981. Berkeley geography, 1923–33. In *The origins of academic geography in the United States*, ed. B.W. Blouet, 221–244. Hamden, CT: Archon Books.

Williams, Michael. 1983. 'The apple of my eye': Carl Sauer and historical geography. *Journal of Historical Geography* 9: 1–28.

Williams, Michael. 2014. *To pass on a good life: The life and work of Carl O. Sauer*. Charlottesville: University of Virginia Press.

References

Arnold, B.A. 1954. *Land forms and early human occupation of the Laguna Seca Chapala Area, Baja California, Mexico*. University of California, Berkeley [unpublished doctoral dissertation]

Aschmann, H.H. 1954. *The ecology, demography, and fate of the Indians of the Central Desert of Baja California*. University of California, Berkeley [unpublished doctoral dissertation]

Barnes, T. J. 2016. The odd couple: Richard Hartshorne and William Bunge. *The Canadian Geographer* 60 (4): 458–465.

Blaut, J.M. 1980. A radical critique of cultural geography. *Antipode* 12: 25–30.

Bookchin, M. 1971. *Post-Scarcity anarchism*. Berkeley, CA: Ramparts Press.

Bowman, 1941. *Soil rosion in Puerto Rico*. University of California, Berkeley [unpublished doctoral dissertation]

Brand, D.D. 1933. *The historical geography of Northwestern Chihuahua*. University of California, Berkeley [unpublished doctoral dissertation]

Brown, S.C. and K. Mathewson 1999. Sauer's descent? Or Berkeley roots forever? *Yearbook of the Association of Pacific Coast Geographers* 61: 137–157.

Bruman, H.J. 1940. *Aboriginal drink areas in New Spain*. University of California, Berkeley [unpublished doctoral dissertation]

Clark, A.H. 1954. Historical geography. In *American geography: Inventory and Prospect*, eds James, P.E. and C.F. Jones, 70–105. Syracuse, NY: Syracuse University Press.

Crane, D. 1972. *Invisible colleges. Diffusion of knowledge in scientific communities*. Chicago: University of Chicago Press.

Denevan, W.M. 1996. Carl Sauer and Native American population size. *Geographical Review* 86: 385–397.

Denevan, W.M. and K. Mathewson, eds. 2009. *Carl Sauer on culture and landscape: Readings and commentaries*. Baton Rouge: Louisiana State University Press.

Dickson, D.B. 2003. Origins and dispersals: Carl Sauer's impact on anthropological explanations of the development of food production. In *Culture, land, and legacy: Perspectives on Carl O. Sauer and Berkeley School Geography*, eds. K. Mathewson and M.S. Kenzer, 117–134. Baton Rouge: Louisiana State University Geoscience Publications.

Duncan, J.S. 1980. The superorganic in American cultural geography. *Annals of the Association of American Geographers* 70: 181–198.

Duncan, J.S. 2000. Berkeley school. In *The dictionary of human geography*, 4th ed., eds R.J. Johnston, D. Gregory, G. Pratt, and M. Watts, 45–46. Malden, MA: Blackwells.

Duran, E., Jr. 1953. *A physical and cultural geography of the Cayman Islands*. University of California, Berkeley [unpublished doctoral dissertation]

Edwards, C.R. 1962. *Aboriginal watercraft of Western South America: Distribution, history, and problems of origin*. University of California, Berkeley [unpublished doctoral dissertation]

Entrikin, J.N. 1984. Carl O. Sauer, Philosopher in spite of himself. *Geographical Review* 74: 387–408.

Gade, D.W. 2011a. Curious descendants: Sauer's web of scholarly influences. In *Curiosity, inquiry and the geographical imagination*, ed. D.W. Gade, 115–140. New York: Peter Lang.

Gade, D.W. 2011b. *Anima curiositas*: Carl Sauer. In *Curiosity, inquiry and the geographical imagination*, ed. D.W. Gade, 97–114. New York: Peter Lang.

Gordon, B.L. 1954. *Human geography and ecology in the Sinú Country of Colombia*. University of California, Berkeley [unpublished doctoral dissertation]
Gould, P. 1979. Geography 1957–1977: The Augean Period. *Annals of the Association of American Geographers* 69: 139–151.
Harris, D.R. 2002. The farther reaches of human time: Retrospect on Carl Sauer as prehistorian. *Geographical Review* 92: 526–544.
Hartshorne, R. 1939. *The nature of geography: A critical survey of current thought in the light of the past*. Lancaster, PA: Association of American Geographers.
Hartshorne, R. 1959. *Perspective on the nature of geography*. Chicago: Rand McNally.
Innis, D.Q. 1958. *Human ecology in Jamaica*. University of California, Berkeley [unpublished doctoral dissertation]
James, P.E. 1954. Introduction: The field of geography. In *American geography: Inventory and prospect*, eds P.E. James and C.F. Jones, 2–18. Syracuse, NY: Syracuse University Press.
James, P.E. 1983. The University of Michigan Field Station at Mill Springs, Kentucky, and field studies in American geography. In *The evolution of geographic thought in America*, eds W.A. Bladen and P.P. Karan, 59–85. Dubuque, IA: Kendall/Hunt.
James, P.E. and C.F. Jones 1954. *American geography: Inventory and prospect*. Syracuse, NY: Syracuse University Press.
James, P.E. and C. Mather 1977. The role of periodic field conferences in the development of geographical ideas in the United States. *Geographical Review* 67: 446–461.
Jennings, B.H. 1988. *Foundations of international agricultural research: Science and politics in Mexican agriculture*. Boulder, CO: Westview Press.
Johannessen, C.L. 1959. *The geography of the savannas of Interior Honduras*. University of California, Berkeley [unpublished doctoral dissertation]
Karan, P.P. 1983. Regional studies in Kentucky and American geography. In *The evolution of geographic thought in America: A Kentucky root*, eds. W.A. Bladen and P.P. Karan, 87–111. Dubuque, IA: Kendall/Hunt.
Kenzer, M.S. 1985. Carl O. Sauer: Nascent human geographer at Northwestern. *California Geographer* 25: 1–11.
Kenzer, M.S. 1987. Like father, like son: William Albert and Carl Ortwin Sauer. In *Carl O. Sauer: A tribute*, ed. M.S. Kenzer, 40–65. Corvallis: Oregon State University Press.
Kniffen, F.B. 1930. *The Colorado Delta*. University of California, Berkeley [unpublished doctoral dissertation]
Leighly, J. 1976. Carl Ortwin Sauer, 1889–1975. *Annals of the Association of American Geographers* 66: 337–348.
Leighly, J. 1979. Drifting into geography in the twenties. *Annals of the Association of American Geographers* 69: 4–9.
Lovell, W.G. 2003. "A first book in geography": Carl Sauer and the creation of *Man in Nature*. In *Culture, land, and legacy: Perspectives on Carl O. Sauer and Berkeley School Geography*, eds K. Mathewson and M.S. Kenzer, 323–338. Baton Rouge: Louisiana State University Geoscience Publications.
Mathewson, K. 1987. Sauer south by southwest: Antimodernism and the austral impulse. In *Carl O. Sauer: A tribute*, ed. M.S. Kenzer, 90–111. Corvallis: Oregon State University Press.
Mathewson, K. 2004. Sauer and the South: A deferred agenda. In *The role of the South in the making of American geography*, eds J. Wheeler and S. Brunn, 353–363. Columbia, MD: Bellwether Press.
Mathewson, K. 2009. Carl Sauer and his critics. In *Carl Sauer on culture and landscape: Readings and commentaries*, eds. W.M. Denevan and K. Mathewson, 9–28. Baton Rouge: Louisiana State University Press.
Mathewson, K. 2011. Sauer's Berkeley school legacy: Foundation for an emergent environmental geography? In *Geografía y Ambiente en América Latina*, eds G. Bocco, P.S. Urquijo and A. Vieyra, 51–82. Morelia, MI/MX: UNAM/CIGA.
Mathewson, K. and M.S. Kenzer, eds. 2003. *Culture, land, and legacy: Perspectives on Carl O. Sauer and Berkeley School Geography*. Baton Rouge: Louisiana State University Geoscience Publications.
Mathewson, K., A.L. Allen, A. Grismore, M. Lagos, J.R. Simms and B. Spencer. 2020. The Sauer Tree in time and place. *Journal of Latin American Geography* 19 (1): 84–97.

McBride, F.W. 1940. *Native economy of Southwestern Guatemala and its natural background*. University of California, Berkeley [unpublished doctoral dissertation]

Meigs, P., III. 1932. *Historical geography of the Dominican mission frontier*, Baja California. University of California, Berkeley [unpublished doctoral dissertation]

Merrill, G.C. 1957. *The historical geography of St. Kitts and Nevis, British West Indies*. University of California, Berkeley [unpublished doctoral dissertation]

Mikesell, M.W. 1987. Sauer and "Sauerology": A student's perspective. In *Carl O. Sauer: A tribute*, ed. M.S. Kenzer, 144–150. Corvallis: Oregon State University Press.

Mitchell, D. 2003. Cultural landscapes: Just landscapes or landscapes of justice? *Progress in Human Geography* 27: 787–796.

Parsons, J.J. 1948. *Antioqueño colonization in Western Colombia*. University of California, Berkeley [unpublished doctoral dissertation]

Parsons, J.J. 1979. The later Sauer years. *Annals of the Association of American Geographers* 69: 9–17.

Parsons, J.J. 1996a. Carl Sauer's vision of an Institute for Latin American Studies. *Geographical Review* 86: 377–384.

Parsons, J.J. 1996b. Mr. Sauer and the writers. *Geographical Review* 86: 22–41.

Penn, M. and F. Lukermann. 2003. Chorology and landscape: An internalist reading of "The Morphology of Landscape." In *Culture, land, and legacy: Perspectives on Carl O. Sauer and Berkeley School Geography*, eds K. Mathewson and M.S. Kenzer, 253–259. Baton Rouge: Louisiana State University Geoscience Publications.

Peet, R. 2003. Social relations: The missing dimension in Carl Sauer's theorization. In *Culture, land, and legacy: Perspectives on Carl O. Sauer and Berkeley school geography*, eds. K. Mathewson and M.S. Kenzer, 315–321. Baton Rouge, LA: Geoscience Publications.

Pred, A. 1983. From here and now to there and then: Some notes on diffusions, defusions, and disillusions. In *Recollections of a revolution: Geography as spatial science*, eds. Mark Billinge, Derek Gregory, and Ron Martin, 86–95. New York: St. Martin's Press.

Price, M. and M. Lewis. 1993. The reinvention of cultural geography. *Annals of the Association of American Geographers* 83: 1–17.

Sauer, C.O. 1915. The geography of the Ozark Highland of Missouri. PhD dissertation, University of Chicago.

Sauer, C.O. 1916. *Geography of the Upper Illinois Valley and history of development*. Illinois State Geological Survey, Bulletin No. 27.

Sauer, C.O. 1918. Part I, Geography. In *Starved Rock State Park and its environs*, eds., C.O. Sauer, G.H. Cady, and H.C. Cowles, 1–83. Geographic Society of Chicago, Bulletin No. 6.

Sauer, C.O. 1925. The morphology of landscape. *University of California Publications in Geography* 2(2): 19–53.

Sauer, C.O. 1927a. *Geography of the Pennyroyal*. Kentucky geological survey, ser. 6, vol. 25.

Sauer, C.O. 1927b. Recent developments in cultural geography. In *Recent developments in the social sciences*, ed. E.C. Hayes, 154–212. Philadelphia: Lippincott.

Sauer, C.O. 1938. Theme of plant and animal destruction in economic history. *Journal of Farm Economics* 20: 765–775.

Sauer, C.O. 1952. *Agricultural origins and dispersals*. Bowman memorial lectures, series 2. New York: American Geographical Society.

Sauer, C.O. 1963. Homestead and community on the Middle Border. In *Land use policy in the United States*, ed. H.W. Ottoson, 65–85. Lincoln: University of Nebraska Press.

Sauer, C.O. 1966. *The early Spanish Main*. Berkeley: University of California Press.

Sauer, C.O. 1968. *Northern mists*. Berkeley: University of California Press.

Sauer, C.O. 1971. *Sixteenth century North America: The land and the people as seen by the Europeans*. Berkeley: University of California Press.

Sauer, C.O. 1980. *Seventeenth century North America*. Berkeley, CA: Turtle Island Foundation.

Sauer, C.O. 1981. *Selected essays, 1963–1975*. Bob Callahan, ed. Berkeley, CA: Turtle Island Foundation.

Speth, W.W. 1977a. Carl Ortwin Sauer on destructive exploitation. *Biological Conservation* 11: 145–160.

Speth, W.W. 1977b. Historicism: The disciplinary world view of Carl O. Sauer. In *Carl O. Sauer: A tribute*, ed. M.S. Kenzer, 11–39. Corvallis: Oregon State University Press.

Speth, W.W. 1981. Berkeley geography, 1923–33. In *The origins of academic geography in the United States*, ed. B.W. Blouet, 221–244. Hamden, CT: Archon Books.
Stanislawski, D. 1944. *Historical geography of Michoacán*. University of California, Berkeley [unpublished doctoral dissertation]
Stoddart, D.R. 1997. Carl Sauer: Geomorphologist. In *Process and form in geomorphology*, ed. David Stoddart, 340–379. London: Routledge.
Sawatsky, H.L. 1967. *Mennonite colonization in Mexico: A study in the survival of a traditionalist society*. University of California, Berkeley [unpublished doctoral dissertation]
Thomas, W.L. 1956. *Man's role in changing the face of the earth*. Chicago: University of Chicago Press.
Wallach, B. 1999. Commentary: Will Carl Sauer make it across that Great Bridge to the next millennium? *Yearbook of the Association of Pacific Coast Geographers* 61: 129–136.
Wagner, P.L. 1953. *Nicoya: Historical geography of a Central American lowland community*. University of California, Berkeley [unpublished doctoral dissertation]
West, R.C. 1946. *The mining community in Northern New Spain*: The Parral Mining District. University of California, Berkeley [unpublished doctoral dissertation]
West, R.C. 1979. *Carl Sauer's field work in Latin America*. Ann Arbor, MI: Microfilms International.
West, R.C., ed. 1982. *Andean reflections: Letters from Carl O. Sauer while on a South American trip under a grant from the Rockefeller Foundation*, 1942. Dellplain Latin American Studies, no. 11. Boulder, CO: Westview Press.
Whittlesey, D. 1954. The regional concept and the regional method. In *American geography: Inventory and Prospect*, eds James, P.E. and C.F. Jones, 19–69. Syracuse, NY: Syracuse University Press.
Williams, M. 2014. *To Pass on a Good Earth: The life and work of Carl O. Sauer*. Charlottesville, VA: University of Virginia Press.
Williams, M. 1983. 'The apple of my eye': Carl Sauer and historical geography. *Journal of Historical Geography* 9: 1–28.
Williams, M. 1987. Carl Sauer and *Man's Role in Changing the Face of the Earth*. *Geographical Review* 77: 218–231.

7 From mapping to GIScience

A sixty-year project

Michael F. Goodchild

Waldo Tobler (1930–2018) in his thirties, circa 1965.

Source: Trevor Barnes

Michael Goodchild (1944–) in his thirties, circa 1975.

Source: author

Introduction

The discipline of Geography was in a state of flux in the 1950s. Harvard had closed its Geography department in 1948 (Smith, 1987), and although the practice of geography had thrived during World War II, when geographic knowledge and the technologies of mapping and aerial photography had been critical to Allied success, the end of the war and the return to a peacetime economy threatened to deprive Geography of much of its funding and stature. But a reversal of fortunes came quickly with the advent of the Cold War and the Soviet testing of the atomic bomb in 1949. By 1960 remote sensing and the development of intercontinental ballistic missiles had given new importance to image interpretation and the science of geodesy.[1] Digital computers became available for

DOI: 10.4324/9781003343240-9

applications that went far beyond their early uses in codebreaking and nuclear engineering. New civilian applications were emerging, though it would be a decade before the launch of the Landsat program prompted civilian uses of remote sensing, and three decades before the Global Positioning System (GPS) became available for non-military uses.[2]

Some of the most interesting early developments in the discipline around 1960 occurred at the University of Washington in Seattle, where a group of graduate students began to explore some new scientific possibilities under the influence of a young professor, William Garrison (1924–2015), as well as others. Those possibilities included: quantitative analysis, empowered by digital computers and new sources of digital data; theory-building as part of an exact and rigorous scientific vision of human geography; and the construction of geographic databases in support of urban planning. Several of the group went on to highly influential and distinguished careers as academic geographers. One of the most prominent was Waldo Tobler (1930–2018), a Swiss-American who entered Washington's PhD program after undergraduate work in Geography at the University of British Columbia. Waldo had been born in Portland, Oregon, where his father was a Swiss consular employee. He had joined the US Army in Europe in 1949 and had been exposed to German maps and aerial photographs as part of his work as an intelligence analyst. We do not know why he subsequently made the decision to become a geographer, and why he was recruited to the University of Washington. We do know, however, that although mild-mannered, modest, and courteous, his subsequent influence on the discipline of Geography was profound, reshaping the field, fundamentally altering its practices, and reconfiguring its constitutive elements.

Background

The 1960s were the heyday of what became known as the quantitative revolution in Geography (see Chapter 4). Anyone with mathematical skills, and for a period the University of Washington was the place to have acquired them, was in high demand. Consequently, Waldo had little difficulty obtaining a position at the University of Michigan in Ann Arbor, where his career flourished.[3] There he was a founding member of the Michigan Interuniversity Community of Mathematical Geographers (MICMOG). That community comprised quantitative and theoretical geographers drawn from three Michigan universities—Michigan State University in East Lansing, University of Michigan in Ann Arbor, and Wayne State University in Detroit. Members would meet monthly in New Brighton, MI, to present papers and to discuss theoretical geography. True to form, New Brighton was chosen because it was the mathematically derived most accessible point among the three universities. One evening in the late 1960s MICMOG was addressed by a young and iconoclastic doctoral student from McMaster University in Hamilton, Ontario. That was me, and it was how I first met Waldo.

I grew up in England, in the counties of Somerset and Devon. I entered the undergraduate program in physics at Cambridge in 1962, spending many weekends exploring the caves of the UK. Employment prospects in the UK were meager at that time. I was therefore attracted by an opportunity to turn my recreational interest in exploring caves into cave research as a graduate student at the Geography Department at McMaster University in Canada under the supervision of Derek Ford.

The circumstances of that first meeting might have been quite awkward. As a doctoral student I was inspired by Gerard Rushton, then a young faculty member at McMaster, to explore recent developments in quantitative human geography. Central place theory (Christaller, 1933) is a theory of the spatial distribution of different-sized settlements. It predicts that in an area of uniform population density, the location of settlements will take the spatial form of a regular hexagonal grid. Extensive efforts to find hexagons in settlement patterns have generally failed, though. One reason was that the assumption of uniform population density was not met in the real world. Arthur Getis applied central place theory to the location of supermarkets in the city of Tacoma, Washington (Getis, 1963), and tried to allow for population density difference. To do so, he based his argument on Waldo's work on the science of stretching and distorting maps (that is, the science of map projections). Getis suggested that even in an area of non-uniform population density, such as the suburbs of Tacoma, it was possible using different projections to distort the map so that population density appeared uniform. With that adjustment, one could then test whether the predicted hexagons of central place theory were present. The paper I presented at that MICMOG meeting argued that Getis's logic was flawed, however. Distorting the map distorted all the distances on the map. In turn, that undermined other assumptions central place theory made to derive the hexagonal settlement pattern, e.g., that consumers would always go to the nearest settlement that offered the good they wanted to buy. Consequently, even after making Getis's adjustment, one would still not expect to find a hexagonal grid. Further, I also showed that the observed locations of supermarkets in Getis's study were no closer to a hexagonal grid than randomly located points would have been. At one point Waldo interrupted me: "You seem to be enjoying yourself."

This chapter is about the growth of digital mapping and the emergence of geographic information science. It's a project that has occupied and transformed the work of many geographers. My focus is on two of them—Waldo Tobler and myself, Michael Goodchild (Figure 7.1)—and the roles we have played, alongside many others. Although a decade apart in age, both of us have found ourselves working within the discipline of Geography at a time of marvelous and rapidly expanding opportunities, brought about in part by the

Figure 7.1 Tobler and Goodchild, UC Santa Barbara, circa 2000.

Source: author.

advent of digital computers and their use to implement powerful techniques of mapping and quantitative analysis. In part, also, the discipline's focus on the geographic domain—that is, the world around us from human to global scales—offered a rich, rewarding, and largely untapped landscape for our field, for the rapid transition of geographical science into practical applications, and in understanding and representing societal impacts. Both of us were able to draw on a rich and rigorous background. In my case it was from an undergraduate degree in physics and in Waldo's case from the Swiss tradition of excellence in cartography. Although our first meeting was somewhat serendipitous, our subsequent careers steadily converged, both geographically and intellectually, in the field that we now term geographic information science.

Tobler's *Decennium Mirabilis*

While the four papers of Einstein's *annus mirabilis* appeared in quick succession in *Annalen der Physik* in 1905, Tobler's early burst of fundamental activity took rather longer, and occupied much of his period at Michigan in the late 1960s and early 1970s. Nevertheless, it was remarkable, as was Einstein's, in the range of topics on which he made significant advances.

The groundwork for this key period was laid by a paper published in 1966, while Tobler was still a graduate student and studying the mathematics of map projections (Tobler, 1966). In the paper he analyzed the Portolan charts created in the 13th century by Mediterranean navigators. He showed that they closely approximated an oblique Mercator projection, the method of flattening the Earth developed 300 years later by the cartographer Geradus Mercator (1512–1594) that allowed navigators to sail a loxodrome—a course of constant bearing that would also follow a straight line on the map. Map projections remained one of Tobler's passions for the rest of his life.

Another of his significant papers (Tobler and Wineburg, 1971) focused on spatial interaction models, which drew on an analogy between human interaction over space and Newton's law of gravitation (e.g., Stewart, 1948). A hoard of ancient Cuneiform-written tablets discovered in Cappadocia (Turkey) in 1925 contained the names of 119 early towns; in some cases their locations known. Tobler and Wineburg counted the number of times a pair of cities were both mentioned on the same tablet, and thus estimated the distances between pairs of towns by drawing on the distance-decay principles of spatial interaction modeling. In this way they were able to estimate the unknown locations. The paper remains an outstanding example of the use of geographic theory in prediction, and of an early application of computing to a cartographic problem.

While there were many other papers that indicated the richness and diversity of Tobler's early work, his 1973 paper in *Geographical Analysis* on choropleth maps (Tobler, 1973) provided an excellent illustration of the constant tension between art and science in cartography. It was also an early signal of the debates that would emerge over geographic information science in the 1990s. When maps were made by hand, cartographers were limited in the kinds of symbols they could generate to represent quantitative data. Choropleth maps show data by varying the shading or color applied to the reporting zones, such as counties. As a result, cartographers were forced to sort data into a limited number of intervals or classes so that each class could be assigned a distinct color or shading scheme. The problem of setting those class intervals became a major concern for cartographers. How they were set determined the message communicated by the choropleth map (Monmonier, 2018). But a computer-drawn map has no such limitations. Tobler's paper showed how the whole question of choropleth interval selection could be avoided by using the computer to generate a continuous variation of shading or color. To Tobler's scientific mind the loss of information in reducing data to a set of classes was unacceptable. Yet, Tobler's paper gained little traction because it challenged the role of the cartographer in controlling the message communicated through the map. Even today, it is difficult to find citations, or implementations of his ideas in widely used software.

Although each of these three papers were landmark contributions in their own right, it's somewhat odd that the lasting idea most strongly associated with Tobler's name was found in his 1970 paper (Tobler, 1970) that discussed potential methods for interpolating mapped data. In the course of it he wrote: "I invoke the first law of geography: everything is related to everything else, but nearby things are more related than distant things" (p. 236). While the statement might be better described as a tendency or principle rather than a law as that term is understood in the physical sciences, it's nevertheless useful as a basis for the detection of interesting anomalies: say, locations where a value is high but surprisingly surrounded by low values, and vice versa (Figure 7.2). It is formalized and operationalized in the science of geostatistics (Goovaerts, 1997) and well known in relation to time rather than space.

The statement attracted little attention at the time, but by the early 1990s it became enshrined as a fundamental principle of the emerging science of geographic information, where it is known as Tobler's First Law of Geography (Sui, 2004). It's the basis for regional geography because it underlies the ability to partition the geographic world into areas of similar characteristics. It is also the basis for spatial interpolation because it allows missing values to be interpolated from nearby measurements. And it's the basis of

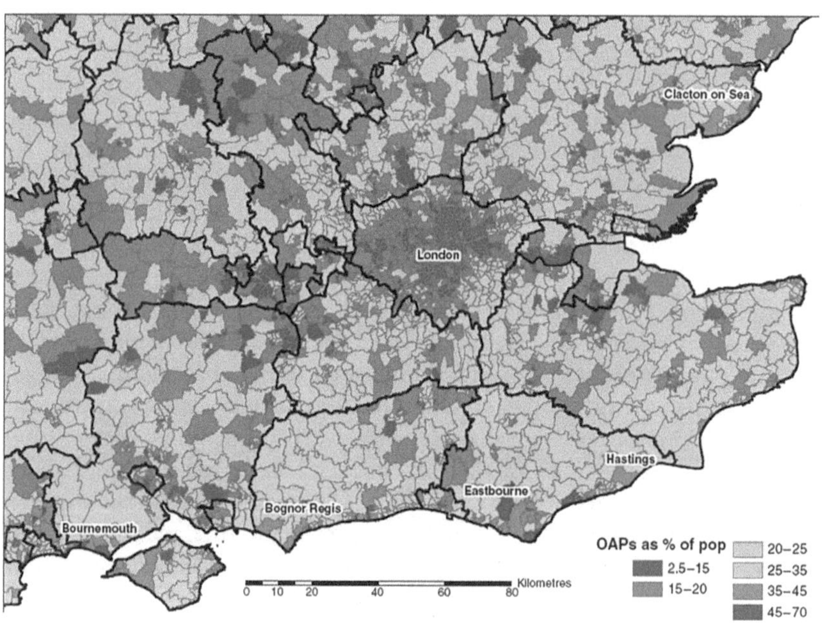

Figure 7.2 Tobler's First Law in practice: percentage of "old-age pensioners" in south-east England. Note the general absence of OAPs in London and to a lesser extent in its commuter belt, and the pockets of OAPs along the coast.

Source: author.

contour mapping, which relies on elevations changing slowly across space to create visualizations of topography.

In short, the early part of Waldo's career was the basis for the scientific approach to geography that was long the dream of many geographers, such as Schaefer (1953) and Bunge (1966) (see Chapter 20). It differed in two important respects, however: it was computational, relying on the power of the digital computer to capture, represent, and allow manipulation of geographic information; and its discoveries were largely about methods and the forms that characterize the Earth's surface, rather than dynamic processes that continuously modify those forms. In these ways, it set an early stage for the emergence of a science of geographic information (GIScience) as a significant part of the discipline of Geography.

The emergence of GIS

Although the development of early computers in the 1950s was motivated largely by nuclear engineering and codebreaking, by the late 1960s several projects around the world were beginning to demonstrate the value of computers in geographic research. In 1964 Roger Tomlinson (Tomlinson and Toomey, 1999) persuaded the Canadian federal government to invest in what he termed a geographic information system (GIS). In the late 1950s the Canadian government's Canada Land Inventory began creating large numbers of detailed maps of rural land use and potential land use. Tomlinson argued that only a computer-based approach could yield the vast number of measurements of land that were the primary focus of the inventory. IBM was contracted to build the Canada Geographic Information System (CGIS), and many new algorithms and computer peripherals were developed to operationalize what in the mid-1960s was an enormously challenging project at the extreme edge of computing capabilities (Goodchild, 2018).

In 1970 and 1972 Tomlinson convened two international meetings, both under the auspices of the International Geographical Union's Commission on Geographical Data Sensing and Processing. Their purpose was to exchange information about projects using computing technology for geographic research (Tomlinson, 1972). In addition to GIS, projects also included the automation of map production and editing, the processing of large quantities of remotely sensed data, and the geographic operations of the decennial census. The GIS vision was to perform many basic operations on geographic data, that is, data about *where*. To be realized the system needed to recognize one or more standard data formats and be programmed to execute the kinds of operations that Tobler was finding useful in his scientific approach to geographic research.

Two sharply different approaches were pursued in the years following the two meetings. A raster GIS was in principle easier to develop as it is based on dividing a map into a layer formed as an array of rectangular cells with a fixed resolution. GIS became the collection of such layers. This approach suited the remote sensing specialists because rasters are almost always the basis of digital representation of images. The second approach, a vector GIS, represents geographic data as a collection of the points, lines, and areas shown on maps. A point possesses geographic coordinates, a line was a series of points connected with straight-line segments (loosely described as a polyline), and an area was represented as a polygon defined by the coordinates of its vertices in clockwise or anticlockwise order. Various claims were made in the 1970s favoring either raster or vector approaches.

By 1980 a significant commercial software industry developed around the idea of a GIS. Applications were found in the management of land and natural resources; in the analysis and modeling of traffic flows; in the management of cadastral data, that is, data on property boundaries; in the operations of the decennial census; and in the production of maps. Many of today's more sophisticated applications of geographic data stem from this period, including mapping and wayfinding apps. Departments of Geography were introducing courses in GIS and training their students for a rapidly developing labor market.

GIS as a research tool

Many groups around the world were quick to jump on the potential of GIS as a tool supporting research. GIS was seen as analogous to the packages of statistical software that had become indispensable in universities beginning in the 1960s. Initially researchers wrote much of the software themselves, mostly in FORTRAN given its dominance as a language for the research community, and shared their codes freely. Harvard's Laboratory for Computer Graphics and Spatial Analysis became a major source of such software, and Waldo, I, and many others had our own collections of software to perform what Waldo called analytical cartography. But during the 1980s many of these activities found themselves losing in competition with the rapidly growing power and functionality of commercial products based on vector databases, most notably Esri's ARC/INFO.[4] It was seen as desirable to base teaching on the kinds of products that students would encounter in the workplace rather than the rather idiosyncratic and unreliable codes that originated with each instructor.

But a tension developed during this period between the commercial interests of the software industry and the demands of science. Tomlinson based the conceptual design of CGIS on the map, and on the accurate capturing of the features shown on the map. But this left open the question of the accuracy of the map itself. As Korzybski (1933) wrote, "The map is not the territory." Moreover, many of the maps that populated CGIS—maps of land cover, suitability for agriculture, etc.—were non-replicable in the sense that any two experts charged with creating a map of the same area would likely not produce the same map. Because of the uncertainties of mapping, there could be many different maps of the same area, with each one equally likely to be the truth. In other words, a map was the outcome of a stochastic error process. The positions of boundaries would be different among different maps, as would be their number of points, lines, and areas. Peter Burrough, an Anglo-Dutch professor of soil science, was one of the first to raise the issue of uncertainty and the first to publish in English a substantial and scientific book on the issue (Burrough, 1986). Clearly GIS raised some significant concerns as a tool for science.

A growing research establishment

By 1985 the developing focus on basic GIS issues led to an interest in building a funded research establishment for GIS. In the US Ron Abler led the move for a National Center for Geographic Information and Analysis (NCGIA), funded by the National Science Foundation (NSF) with five proposed research activities (Abler, 1987):

- new modes and methods of spatial analysis,
- a general theory of spatial relationships,
- artificial intelligence and expert systems in GIS,
- visualization, and
- social, economic, and institutional issues.

Although the term "spatial data handling" was central to the early promotion of GIS by Tomlinson and others, it is clear from this list that the "handling" part had matured into a much richer and challenging set of topics, many of which captured issues that had intrigued geographers for decades. On the other side of the Atlantic the UK funded several Regional Research Laboratories to develop national expertise in the application of GIS in a wide range of domains, and other centers were established in many other parts of the world. Abler's NCGIA was funded by NSF in 1988, with the award going to a consortium led by the University of California, Santa Barbara (UCSB); the State University of New York at Buffalo; and the University of Maine. Waldo had moved to UCSB in the late 1970s following the closure of the Geography department at Michigan, and I had moved from the University of Western Ontario to UCSB in 1988, during the preparation of the winning proposal. Thus, we both became part of the leadership of NCGIA, with eight years of funding to pursue topics of mutual interest.

NCGIA research was organized around research initiatives, each one initiated by an international meeting of about 35 people with interests in the topic (a "specialist meeting"), followed by a two-year period of research, followed by publications and a series of conference presentations on results. The structure was intended to engage as many colleagues as possible in a collaborative but distributed venture. The first such initiative began in December 1988 on the topic of Accuracy in Spatial Databases. It led to numerous efforts around the world and an edited collection of papers (Goodchild and Gopal, 1989). By 1996 at the end of the core NCGIA funding there were over 20 such initiatives, and since that time many other specialist meetings have been held on cutting-edge topics.

The last of the five research bullets listed above stimulated work on the social context of GIS. Initially these concerned such topics as the economics of geographic data production and barriers to the sharing of data, but they broadened significantly following a series of commentaries on the societal implications of GIS that appeared in the late 1980s and early 1990s and led indirectly to the emergence of GIScience.

The social critiques of GIS

By 1988 the growth of GIS, along with the founding of NCGIA, was attracting attention. That came not only from researchers interested in exploiting the new research tool, but from critical social theorists interested in reflecting on its societal impacts. One of the first of these commentaries was from Brian Harley, who began to apply a critical lens to cartography (Harley, 2001). He saw maps as a product not only of supposedly objective information, but also of the social and political agenda of the map maker. Wood (1992) captured this in the memorable phrase: "Whose agenda is in your glove compartment?" The glove compartment of cars stored not only gloves but paper maps, often freely distributed by gas stations and automobile clubs. Coupled with the concerns expressed earlier about non-replicability, this critique invited GIS researchers to consider the degree to which their data were social constructions, and all that that implied in the tensions between subjectivity and the scientific method.

Another line of argument came from the apparent tendency of GIS databases to simplify the world (Figure 7.3): with lines representing what were in fact continuous gradations; with a tendency to force the world into distinct but less than rigorously defined classes; with the preponderance of the vertical, "god's-eye" view of the geographic world; and with a general tendency to de-humanize the people occupying the areas mapped. Yet another critique saw GIS as invasive, surveillant, and endangering personal privacy. All

Figure 7.3 GIS simplifying the world: a soil map of part of southern California. Areas shown with the same color are not in fact identical; the boundaries between patches are actually approximations to continuous gradation; and two experts would not produce identical maps.

Source: SSURGO, US National Cooperative Soil Survey. https://databasin.org/datasets/028d6dc1c4084aeb96099355da5bc84a/.

these issues and more were articulated in 1995 in an important edited book on the societal implications of GIS (Pickles, 1995). Finally, many saw GIS as mechanical, non-intellectual, merely following procedure, and asked what justified its teaching in a university, and more specifically within a Geography department (Jordan, 1988).

In response, NCGIA proposed a new research initiative to address these critiques, known as Initiative 19. It held the customary specialist meeting, in March 1996 in South Haven, MN, stimulated research in several novel areas, and led to numerous publications including an edited volume (Nyerges, Couclelis, and McMaster, 2011).

The emergence of GIScience

By 1991 it had become clear to the leadership of NCGIA that a deliberate effort was needed to establish the intellectual basis for GIS, and for its place in academia. David Simonett, the director and Principal Investigator in the early years of NCGIA, had long regretted the lack of recognized scientific principles in his field of remote sensing. He hoped that the same mistake would not be made with GIS. I made two international conference presentations in that year outlining what I termed spatial information science, and in 1992 published a paper titled "Geographic information science" in the *International Journal of Geographical Information Systems*, thinking that "geographic" was more appropriate than "spatial," and as a deliberate play on the GIS acronym. The paper identified a research agenda for GIS that asked questions of fundamental significance to computer science, cognitive science, statistics, geography, and many other fields. In subsequent years many GIS journals changed their names from "systems" to "science," academic programs in GIScience were inaugurated, the University Consortium for Geographic Information Science was established in the US, and GIScience specialists were elected to academies around the world, led by Waldo's election to the US National Academy of Sciences in 1982.

The effects of the social critique were also profound. Following the Initiative 19 specialist meeting and the publication of the Pickles volume, many new efforts were aimed at making GIS more human-centric. They included a new field, public-participation GIS, dealing with more equitable access and functions aimed at the needs of individuals and communities; the concept of neogeography (Turner, 2006) as a recognition that through technological advances and increased affordability, the average citizen could now be both a consumer and producer of geographic technology; and a new recognition that people think about the geographic world in terms of named places rather than as coordinates (Goodchild, 2015). Research was also aimed at moving GIS away from its traditional European-based concepts of land ownership and mapping (see for example the work of Turk, Mark, and Stea, 2011, on Indigenous map-making).

GIScience today

I retired from UCSB in 2012 and Waldo's long and productive life came to a close in 2018. The analytical cartography and computational geography that Waldo envisioned in the 1960s had more than exceeded his expectations, growing into a significant part not only of Geography but of the entire academic world. Growing fields such as spatial history, geoarchaeology, spatial epidemiology, and business analytics applied GIS technology and GIScience in novel areas.

To end this chapter, it's appropriate to discuss a new phase of reflection on GIScience, driven in this case by issues of ethics. In 2020, the American Association of Geographers (AAG), in collaboration with UCSB, inaugurated the GeoEthics project in response to growing concerns about the privacy of individual locations and the use of GIS and related technologies for surveillance. It was also driven in part by both national and international concerns about diversity, equity, and inclusion, and a steady loss of openness and transparency in science. The project's organizing committee is chaired by me and Gary Langham, the Executive Director of the AAG. Locational privacy had surfaced during the critiques of GIS in the 1990s (Curry, 1998), but by 2020 smartphone apps were capturing locations and a new industry was aggregating and selling that information to market-research companies, raising a new level of concern (Valentino-DeVries et al., 2018).

The project held a summit in Santa Barbara in late June 2022, and released a report in the form of a white paper in October (https://www.aag.org/locational-information-and-the-public-interest-white-paper-2022/). The paper raises many issues and concerns, not only about locational privacy but also about the repurposing of GIS software for unacceptable applications, for generating false or misleading inferences from GIS analysis, and for ignoring the ever-present uncertainty in analyses and predictions using GIS. As with any powerful technology, it is all too easy to turn GIS tools to purposes that are malicious, subversive, or dishonest.

GIScience today might be defined as a form of critical thinking, in direct contrast to earlier claims that GIS was little more than the rote pushing of buttons. While the computer is capable of doing powerful things with geographic data, it is the user, not the computer, who must be constantly aware of ethical issues, of the ways in which geography is only approximately represented in a geographic database, of the many fallacies that may be encountered in making inferences from geographic analysis, of the need for openness and replicability, and of the need to include a variety of human perspectives on a problem. The field has been through many phases in its 60-year evolution, some dominated by simplistic enthusiasm and some by well-deserved critique. It has emerged as an important contribution to the academic enterprise and the endless search for human knowledge about our geographic world.

Acknowledgments

I am grateful to the editors and to Patrick McHaffie for comments on earlier drafts.

Notes

1 Geodesy is the mathematical science of the shape and area of Earth.
2 Initiated in 1972, Landsat is a US government program for acquiring satellite imagery of Earth.
3 As I will make clear below. I was both a long-time colleague and friend of Waldo Tobler and for that reason I will refer to him by his first name.
4 Esri, headquartered in Redlands, California, with an annual revenue of over $1bn, is the world's leading multinational corporation producing GIS software. ArcGIS is the name of one of their best-known premier software products.

Study task

The following question can be used in an exam or as a set essay question at undergraduate or Master's level:

Have Tobler and Goodchild (and many others) succeeded in providing an intellectual foundation for GIS, or was Jordan correct in asserting in 1988 that GIS was "non-intellectual expertise?"

Starter reading list

Goodchild, M.F. (1992) Geographical information science. *International Journal of Geographical Information Systems* 6(1): 31–45.
Goodchild, M.F. (2018) Reimagining the history of GIS. *Annals of GIS* 24(1): 1–8. DOI: 10.1080/19475683.2018.1424737
Goodchild, M.F. (2019) Geography and GIScience: An evolving relationship. *Canadian Geographer* 63(4): 530–539. DOI: 10.1111/cag.12554
Jordan, T. (1988) The intellectual core: President's Column. *Newsletter of the Association of American Geographers* 23: 1.
Longley, P.A., M.F. Goodchild, D.J. Maguire, and D.W. Rhind (2015) *Geographic Information Science and Systems*, 4th edn. Hoboken: Wiley.
Mark, D.M. (2003) Geographic information science: Defining the field. In M. Duckham, M.F. Goodchild, and M.F. Worboys, eds, *Foundations of Geographic Information Science*, pp. 3–18. New York: Taylor & Francis.
Wright, D.J., M.F. Goodchild, and J.D. Proctor (1997) Demystifying the persistent ambiguity of GIS as 'tool' versus 'science'. *Annals of the Association of American Geographers* 87(2): 346–362.

References

Abler, R.F. (1987) The National Science Foundation National Center for Geographic Information and Analysis. *International Journal of Geographical Information Systems* 1(4): 303–326. DOI: 10.1080/02693798708927819
Bunge, W. (1966) *Theoretical Geography*. Lund: Gleerup.
Burrough, P.A. (1986) *Principles of Geographical Information Systems for Land Resources Assessment*. Oxford: Clarendon Press.

Christaller, W. (1933) *Die Zentralen Orte in Süddeutschland*. Jena: Fischer. English translation from the German original by L. W. Baskin (1966) *Central places in Southern Germany*. Englewood Cliffs, NJ: Prentice Hall.
Curry, M. (1998) *Digital Places: Living with Geographic Information Technologies*. New York: Routledge.
Getis, A. (1963) The determination of the location of retail activities with the use of a map transformation. *Economic Geography* 39(1): 14–22. DOI:10.2307/142492
Goodchild, M.F. and S. Gopal, editors (1989) *Accuracy of Spatial Databases*. Basingstoke: Taylor & Francis.
Goodchild, M.F. (2015) Space, place and health. *Annals of GIS* 21(2): 97–100. 10.1080/19475683.2015.1007895
Goodchild, M.F. (2018) Reimagining the history of GIS. *Annals of GIS* 24(1): 1–8. 10.1080/19475683.2018.1424737
Goovaerts, P. (1997) *Geostatistics for Natural Resources Evaluation*. New York: Oxford University Press.
Harley, J.B. (2001) *The New Nature of Maps: Essays in the History of Cartography*. Baltimore: Johns Hopkins University Press.
Jordan, T. (1988) The intellectual core: President's Column. *Newsletter of the Association of American Geographers* 23: 1.
Korzybski, A. (1933) Science and sanity: An introduction to non-Aristotelian systems and general semantics. *The International Non-Aristotelian Library Pub. Co.*, pp. 747–761.
Monmonier, M. (2018) *How to Lie with Maps*. Chicago: University of Chicago Press.
Nyerges, T., H. Couclelis, and R. McMaster, eds (2011) *The SAGE Handbook of GIS and Society*. La Jolla: SAGE.
Pickles, J., ed. (1995) *Ground Truth: The Social Implications of Geographic Information Systems*. New York: Guilford.
Schaefer, F.K. (1953) Exceptionalism in geography: A methodological examination. *Annals of the Association of American Geographers* 43(3): 226–249. DOI: 10.2307/2560876
Smith, N. (1987) "Academic War over the Field of Geography": The elimination of geography at Harvard, 1947–1951. *Annals of the Association of American Geographers* 77(2): 155–172. DOI: 10.1111/j.1467-8306.1987.tb00151.x
Stewart, J.Q. (1948) Demographic gravitation: Evidence and applications. *Sociometry* 11(1/2): 31–58. DOI: 10.2307/2785468
Sui, D.Z. (2004) Tobler's first law of geography: a big idea for a small world. *Annals of the Association of American Geographers* 94: 269–277. DOI: 10.1111/j.1467-8306.2004.09402003.x
Tobler, W.R. (1966) Medieval distortions: The projections of ancient maps. *Annals of the Association of American Geographers* 56(2): 351–360. DOI: 10.1111/j.1467-8306.1966.tb00562.x
Tobler, W.R. (1970) A computer movie simulating urban growth in the Detroit region. *Economic Geography* 46(supplement): 234–240. DOI: 10.2307/143141
Tobler, W.R. (1973) Choropleth maps without class intervals? *Geographical Analysis* 5(3): 262–265. DOI: 10.1111/j.1538-4632.1973.tb01012.x
Tobler, W.R. and S. Wineburg (1971) A Cappadocian speculation. *Nature* 231: 39–41. DOI: 10.1038/231039a0
Tomlinson, R.F. (1972) *Geographical Data Handling*. Ottawa: International Geographical Union, Commission on Geographical Data Sensing and Processing.
Tomlinson, R.F. and M.A.G. Toomey (1999) GIS and LIS in Canada. Chapter 15 in G. McGrath and L. Sebert, editors, *Mapping a Northern Land: The Survey of Canada 1947–1994*. Montreal: McGill–Queen's University Press.
Turk, A.G., D.M. Mark, and D. Stea (2011) Ethnophysiography. In D.M. Mark, A.G. Turk, N. Burenhult, and D. Stea, editors, *Landscape in Language: Transdisciplinary Perspectives*, pp. 25–45. Amsterdam: John Benjamins Publishing.
Turner, A. (2006) *Introduction to Neogeography*. Sebastopol: O'Reilly Media.
Valentino-DeVries, J., N. Singer, M.H. Keller, and A. Krolik (2018) Your apps know where you were last night, and they're not keeping it secret. *New York Times*. December 10. https://www.nytimes.com/interactive/2018/12/10/business/location-data-privacy-apps.html?searchResultPosition=1 (Accessed November 17, 2022)
Wood, D. (1992) *The Power of Maps*. New York: Guilford Press.

8 Radicalizing geography

The case of David Harvey's Marxism

Eric Sheppard

David Harvey (1935–).

Source: https://www.gc.cuny.edu/people/david-harvey

Introduction

If you were to run into David Harvey informally, say at a Geography conference, your first impressions might be unprepossessing. Compact in stature, quietly spoken, and not one to draw attention to himself in social settings, you'd have little idea that you're talking to the world's most highly cited geographer. But a transformation happens when he steps up to speak: he projects an infectious energy and is capable of mesmerizing any audience. One of the discipline's most gifted orators, you watch him carefully develop his argument without notes in a seemingly off-the-cuff way that betrays his meticulous preparation. If you haven't experienced this, check out his personal website: davidharvey.org.

DOI: 10.4324/9781003343240-10

When he speaks Harvey favors a red shirt, symbolizing his identity as a radical and socialist. The word radical comes from the Latin *radix*, or root. Harvey has played a major role in uprooting Geography through two disciplinary revolutions: so-called 'spatial science' in the 1960s but particularly left-wing geography since the early 1970s (once called 'radical' geography and since around the 1990s 'critical' geography). His formative role in the latter revolution culminated in the international radical/critical geography project, which has profoundly shaped both human and nature/society geography, and now even physical geography (Lave, Biermann, & Lane, 2018).

This chapter introduces this remarkable man to you. Thinking geographically, I trace how Harvey's thinking co-evolved through the places where he has plied his trade, with particular attention to how moving from the UK to the US triggered his transformation from spatial science to radical geography (Figure 8.1).[1] As you'll see, his 1960s passion for eliminating socio-spatial inequality has not diminished, but his view as to what's necessary to achieve this has, rather radically. I then place him within the radical/critical geography project that he helped catalyze, noting the diversity of geographers it's brought into the field, and its effect on the kinds of questions Anglophone geographers have come to ask. Finally, I summarize why Harvey's rendition of geography matters.

Becoming a geographer in the UK

David Harvey grew up in a working-class family in Gillingham in southeastern England, a small town whose employment was dominated by Chatham Dockyard. Like so many geographers, he's rarely stood still: a Geography class on the US at the age of 12 led him to dream of traveling to the US and around the world, even though this "seemed hopeless.... We were too poor, and it was all too far away" (Harvey, 2006b, p. 184). Yet he was a precocious student. The British school system provided weekly Geography classes from the age of six, and by the time he was 16 he knew he wanted to become a geographer. He spent his last three secondary school years specializing in the subject, passed the entrance exam and interview, and was accepted to study Geography at Cambridge University—something that Britain's post-1945 state educational system made possible for a bright working-class boy.

Cambridge was the UK's leading Geography Department, dominated by regional geographers. Its redoubtable and somewhat domineering head of department, the geomorphologist J. Alfred Steers, was known for having walked the entire British coast. Harvey's undergraduate and PhD theses were classical regional geographies, exploring aspects of what made Kent distinctive in human and physical terms, with a focus on hop farming. He found studying at Cambridge a transformational experience. Socially, encountering the (largely male) elites who dominated Cambridge increased his awareness of class: "Call it class envy, prejudice, or war, but Cambridge taught me about class in a way that I had not earlier experienced" (Harvey, 2006b, p. 186). Intellectually, Cambridge exposed him to people pioneering a new quantitative geography examining how humans organized the world spatially. As an incoming PhD student he became a laboratory assistant in the first quantitative geography course taught in the UK, by Peter Haggett and Richard Chorley (see Chapter 4). This awoke in him an interest in the philosophy of science: on what basis, Harvey wondered, can humans construct reliable knowledge about the world?

This was reinforced when Harvey spent 1960–1961 at Uppsala University in Sweden on a research grant. He found himself sharing a room with the similarly philosophically inclined Swedish geographer Gunnar Olsson—a pioneering quantitative geographer at that time, known for his subsequent philosophical critiques of spatial science. After they got to talking, "I dropped the research project on Swedish demographics and just hung out all over Sweden,... retooling myself with all sorts of ideas and prospects for undertaking new kinds of research armed with different philosophical foundations and methods" (Harvey, 2006b, p. 186). His PhD in hand, in 1962 Harvey was appointed an assistant lecturer at Bristol University. When Haggett also arrived in 1966, as Professor, he found himself in a department devoting itself (unlike Cambridge) to Geography's quantitative revolution. This inspired his first book, *Explanation in Geography* (Harvey, 1969), which became the bible for how to practice the new quantitative scientific geography.[2]

A post-World War II optimism ruled at Bristol (and cognate centers of calculation for spatial science; Barnes, 2004), with the belief that 'positivist' social science could produce value-free 'objective' knowledge. The philosophy of science variously known as positivism or logical empiricism set out procedures to develop incontrovertible explanations of observed phenomena. *Explanation in Geography* was the discipline's most careful presentation of how geographers should do this. The proponents of this new geography believed that spatial science could not only explain the world, but also create a better and more equitable world. But the world was changing: the May 1968 student-led revolt in France catalyzed a sexual and musical revolution in such places as San Francisco and Woodstock, also inspiring anti-Vietnam War and anti-nuclear armaments demonstrations on both sides of the Atlantic (the latter prominently involving logical empiricism's leading philosopher, Bertrand Russell).[3] Further, the positivist vision of making a better world through value-free scientific geography was an impossible dream (Curry, 1985; Sheppard, 2014). It turned out that scientific applications tended to dovetail with some people's value-based agendas, thereby excluding others without much recourse.

To Baltimore, and to Marx

Harvey moved to the US in 1969, recruited to an Associate Professorship at Johns Hopkins University in Baltimore by the chair of its new Department of Geography and Environmental Engineering M. Gordon (Reds) Wolman (1924–2010), a fluvial geomorphologist practicing spatial science (Figure 8.1). If Harvey had barely noticed the cultural and political revolutions while in the UK, Baltimore—with its urban riots, Black Power movement, and Vietnam war protests—unsettled his thinking. Attracted to "a city deeply troubled by social unrest and impoverishment... I wanted to put my skills to work to try to deal with urban issues" (Harvey 2006b, p. 187). "My move to Baltimore was all about these things; I had done the methodological side. Now I needed to do social justice."[4] This emergent concern for social issues led him to question the scientific paradigm for explaining the world that he'd devoted the best part of a decade to. If the world is unjust, how can we change it?

In 1970 Harvey had begun reading the work of the influential revolutionary thinker Karl Marx (1818–1883): "We turned to Marx to see if there was anything there that could enlighten us, and most of us concluded that there was, even though none of us (that includes me) considered then ourselves 'Marxist'" (Harvey, 2016a). It was an empirical study of the Baltimore housing market that convinced him about Marx. Wolman had secured Harvey a US Federal Housing Authority grant to research the Baltimore housing market with his PhD advisee Lata Chatterjee. Trying to make sense of their data (Chatterjee, Harvey, & Klugman, 1974; Harvey & Chatterjee, 1974; Harvey,

Figure 8.1 Gordon 'Reds' Wolman.

Source: The Franklin Institute, with permission.

Chatterjee, Wolman, Klugman, & Newman, 1972), Harvey turned to Marx's idea that the use value of a house can differ markedly from its exchange value. He also was inspired by an essay on housing by the German political philosopher, collaborator with Marx and stockbroker Friedrich Engels (1820–1895), in which he argued that capitalist housing policy did not solve the housing problem—it just moved it around. It was not just Harvey who found this compelling: so did the Vice President of Chase Manhattan Bank when presented with their analysis (Harvey, 2016b, p. 35; Sheppard & Barnes, 2019a, pp. 192–193)!

Harvey's second book, *Social Justice and the City* (Harvey, 1973), narrates his intellectual conversion from positivism to Marxism. The book is a set of essays written in the early 1970s, divided into two parts at the suggestion of his editor John Davey: 'Liberal Formations' and 'Socialist Formations.' Liberal Formations refers to the explanatory framework he had advanced in *Explanation*. Although the first half of *Social Justice* is written in this tradition, Harvey is critical of four dualisms that he argues underlie this approach to urban studies. These separate space from society, economic production from income distribution, the city from the rural, and theory from method.

In Socialist Formations, Harvey (influenced by having read Marx) works to deconstruct these dualisms. In Chapter 1 he shows how while 'liberal' scholarship can take space seriously (Harvey conceptualized three kinds of space in *Explanation* absolute, relative, and relational—retaining these throughout his career), it is treated as a separate domain from that of social processes.[5] In Chapter 5 he argues that spatial and social processes are deeply intertwined and mutually constitutive: "The question 'what is space' is…replaced by the question 'how is it that different human practices create and make use of distinctive conceptualizations of space'" (Harvey, 1973, pp. 13 14). Conceptualizing spatial and social processes as mutually constitutive is now common practice.

Chapters 2 and 3 explore how societal wealth should be redistributed to reduce inequality and advance social and spatial justice. Here, wealth comes first and redistribution follows. In Chapter 5, however, he argues that these, too, are mutually constitutive. Like Marx, he argues that capitalist production produces inequality (some people and places become wealthy by impoverishing others), implying that the redistribution of wealth cannot be separated from its production. Chapters 1–3 treat the city as a distinct spatial form, to be analyzed separately from rural areas. Yet Chapter 6 conceptualizes cities as also shaped by the ways in which they are (often unevenly) connected with their hinterlands and the rest of the world. He argues that the wealth accumulating in cities is extracted in part from hinterlands and peripheral regions, subsequently dubbing this 'uneven geographical development.' Theory is presented as a separate value-free domain of analysis in Chapters 1–3, to be tested against our observations of the world. But in Chapter 4 he calls theorizing the observed world 'status quo theorizing,' arguing that theory must be revolutionary if we want to change the world.[6]

As the American philosopher Richard Bernstein (1976, p. 106) put it: "There has been a lack of critical self-consciousness among mainstream social scientists that the admonition to … study the way things are, is not so much a scientific imperative as a dubious moral imperative that has pernicious consequences in limiting human imagination and political and social possibilities." He meant that if we restrict ourselves to explaining the world as it is, this diverts attention from the more vital task of creating a better world. Harvey was just one of a group of spatial science adherents (including Olsson, Bill Bunge, and Richard Peet) becoming dissatisfied with the limits of empirical explanation, who abandoned spatial science for something more radical (Sheppard, 1995).

In his next book (which he still regards as one of his two most important), *The Limits to Capital* (Harvey, 1982), Harvey set out to spatialize Marx' classic three-volume analysis of capitalism *Das Kapital* (Marx, 1967 [1867], 1972a [1867–1896], 1972b [1885]). Redirecting his penchant for close philosophical reading to Marx, he summarized Marx's theorization of the capitalist mode of production and capitalist crisis, critically interrogating Marx's occasional references to space, and asked how attending to the capitalist *space* economy extends and challenges Marx's thinking.[7] *Limits* became the base from which the vast majority of Marxist economic geography sallied forth—required reading within the discipline despite its density of argument. Key concepts developed therein, including geographical uneven development, the spatial fix and class-monopoly land rent, are widely used in and beyond geography.

With this theoretical framework in place, Harvey returned to the substantive topic of *Social Justice* in two more books (Harvey, 1985a, 1985b). *The Urbanization of Capital* used the theoretical framework developed in *Limits* to conceptualize the political economy of a capitalist city. *Consciousness and the Urban Experience* examined issues of urban consciousness and class conflict, focusing heavily on Baron Hausmann's rebuilding of Paris between the February Revolution of 1848 and the 1871 Paris Commune. Harvey had spent a 1974–1975 sabbatical year in Paris shortly after finishing *Social Justice*. He had hoped to "sit at the feet, as it were" (Harvey, personal communication) of French Marxists who had come to prominence after 1968, only to be dismissed as an ignorant Johnny-come-lately American. A geographer, he instead spent much of the year exploring the city and its archives, continuing as he returned regularly to Paris (where he bought a flat). "The Paris study was always in the back of my mind as I was writing *Limits* just as *Limits* was in my mind while exploring what happened in Second Empire Paris" (Harvey, personal communication). This eventually culminated in what he regards as his other important book, *Paris, Capital of Modernity* (Harvey, 2003b).

Back to the UK

In 1987 Harvey returned to the UK as the Halford Mackinder Professor of Geography at Oxford University (Cambridge University's great rival). But there is a geography to Geography; unlike the strength of geographical political economy in North America, cultural and feminist geography were much more influential in the UK, with a focus on gender (influenced by Linda McDowell and Doreen Massey) and representation, and economic geography was focusing more on place specificity than capitalism *tout court*. Finding himself subject to feminist (Deutsche, 1991; Massey, 1991) and other critiques of his 'muscular' Marxism (Harvey, 1987; Saunders & Williams, 1986), Harvey set about writing *The Condition of Postmodernity* (Harvey, 1989).[8] One of his most heavily cited books, to his surprise as he found it "incredibly easy to write" (Harvey, 2006b, p. 189), Harvey's Marxist take on postmodernism dwarfed *Limits* and *Paris* in popularity. He attributes this impact to deploying a Marxism that takes seriously urbanization, spatio-temporality, place, culture, and environmental change; it certainly brought Geography to the attention of other critical social scientists and humanists.

Harvey was also uncomfortable with Oxford as a place. His Argentinian wife, an oceanographer, found it stultifying and he was also unsettled (Harvey, 2006b, p. 189):

> Oxford reminded me of my origins. The question of England and its relation to what had been Empire was still all over the place. I was forcefully reminded of what I had long ago rebelled against. The smugness of much of Oxford repelled me, and I could not understand how the Left in Britain, within geography and without, was taking such wishy-washy positions in relation to Thatcherism.

In 1993, he jumped at the opportunity to return to Baltimore.

Back to Baltimore, and on to New York City

Returning to his old haunts in Baltimore, Harvey experienced how you can never step into the same river twice. Reds Wolman had retired, and the department was now dominated by engineers. Channeling his frustration with the department but also with the way of the world at that time, and convinced that Marx was more relevant than ever, Harvey set about reintroducing him to academic critics and other interested readers.[9] After a life-threatening heart operation in 1996, he thought: "If I had something more to say then I had better say it as soon as possible" (Harvey, personal communication). He had written six monographs in the preceding 27 years but produces 17 in the subsequent 27. In their survey of Harvey's life and work, Castree *et al.* (2022, p. 38) call this his "public turn," with three interweaving strands of thinking: seeking pathways toward a better (non-capitalist) world; applying spatialized Marxist analysis to contemporary globalizing neoliberal capitalism; and popularizing Marx's economic writings. This turn was reinforced by his final relocation to New York City in 2001, joining Cindi Katz and his former doctoral student Neil Smith in the City University of New York (CUNY) Graduate School's Anthropology Department (Figure 8.2), and followed soon thereafter by Ruth Wilson Gilmore—Smith's former PhD advisee. There, he found a large group of students

Figure 8.2 Neil Smith.

Photo courtesy of Cindi Katz

passionate about Marxism in and beyond CUNY and was developing an interdisciplinary reputation as a foremost Anglophone interpreter of Marx—including incredibly popular online lectures on *Das Kapital* (https://davidharvey.org/). Notwithstanding Smith's untimely death, he remains at home in the Big Apple, inhabiting the very core of global capitalism.

Responding to critiques that his political imaginary is overly class-focused, masculinist, and overlooks the moral imperatives of our relationship with the more-than-human world, in *Justice, Nature and the Geography of Difference* (Harvey, 1996), *Spaces of Hope* (Harvey, 2000), and *Cosmopolitanism and the Geographies of Freedom* (Harvey, 2009) Harvey set about laying a philosophical foundation for creating a non-capitalist world. *Justice*, which he sees as his most geographical book (personal communication), develops what he had termed historical geographical materialism: "a political project that sees the transition from capitalism to socialism in historical-geographical terms" (Harvey, 1984, p. 10). Whereas *Justice* focuses on environmental movements and ethics, *Spaces* engages with European utopian thought and *Cosmopolitanism* with cosmopolitan philosophy since Immanuel Kant. A key organizing idea is the notion of possible worlds: that the world we live in—the one that empiricism seeks to understand—is just one of many possibilities.

Yet realizing non-capitalist alternatives, Harvey argues, requires overcoming what he calls militant particularism—the many, diverse, and localized critiques of capitalism emerging from peoples' specific place-based and/or bodily experiences of it, reflecting their distinct socio-spatial positionalities (Sheppard, 2002). But how do you transcend such particularities to create a global movement capable of replacing capitalism, without falling into the trap of proposing a universal template that has the effect of presenting one such critique (particularly Western positionalities) as superior to the others? In his view this requires dialectical geographical analysis. Instead of conceptualizing various possible worlds, places, bodily identities, and nature, as isolated and pre-given objects, dialectical thinking prioritizes their inter-relations and how these shape—and are shaped by—their co-evolution. This returns him to how he had theorized space already in *Explanation*: space should be conceptualized relationally, which means that what happens in any one place reflects how it is connected with, and positioned within, the rest of the world (cf. Massey, 2005). At the same time, he insists that political economy must be central to any such political analysis (his materialism: Harvey, 2020).[10] For him, cities are key places from which to launch anti-capitalist struggle (Harvey, 2012).

The second post-1996 strand of thinking applies Harvey's spatialized Marxian theory to analyze contemporary capitalist neoliberalism and globalization. In *The 'New' Imperialism* and *A Brief History of Neoliberalism* (Harvey, 2003a, 2006a), his most popular books after *Condition of Postmodernity*, he analyzes how the current era of capitalism has been dominated by American geoeconomics and geopolitics, providing a geographical analysis of how this gave birth to a neoliberal, anti-government era of capitalism that sees free markets and free trade as the key to prosperity for all people and places willing to play by these rules. Harvey argues that capitalism is constitutionally capable of overcoming crises by absorbing its outsides (nature, bodies, non-capitalist economic practices), converting them into fuel for capitalist expansion—a process he influentially dubbed accumulation by dispossession.[11]

The third strand is an increasingly intensive engagement with Marx's own thinking. The includes guides to *Das Kapital* and *Die Grundrisse* (Harvey, 2018, 2022), but also his geographical take on Marx's theorization of capitalism (Harvey, 2010, 2014, 2017). He focuses on the complex ways in which capitalism faces a variety of internal crises (already theorized in *Limits*) and external barriers, continually finding ways to transcend these (e.g., https://www.utube.com/watch?v=qOP2V_np2c0). In *Seventeen Contradictions and the End of Capitalism*, Harvey (2014) classifies these crises into 'foundational' (essential features of capitalism), 'moving' (constantly shifting contradictions that provide much of the dynamic force behind capital's historical and geographical evolution), and 'dangerous' (potentially fatal for capital and/or humanity: compound growth, capital's destructive relation with nature, and human revolution). In his view, only these last three—external barriers trenchant enough to escape accumulation by dispossession—could possibly terminate capitalism as a mode of production.

A radical geographer

Harvey has consistently sought to uproot disciplinary orthodoxy. In the 1960s this took the form of helping lead the quantitative revolution against descriptive regional geography. Once in Baltimore, it took the form of helping lead the revolution against what was by now the spatial science establishment (Harvey, 1972). Having turned to Marx, Harvey became radical in the political sense of advocating for an anti-capitalist political and social revolution. He saw his role as destabilizing the capitalist status quo by subjecting it to relentless criticism. *Limits* had exposed how the capitalist mode of production consistently reproduces socio-spatial inequality, not prosperity for all, with the corollary that eliminating inequality requires eliminating capitalism. The latter half of his career has been dedicated to prosecuting anti-capitalist alternatives.

In Baltimore, cultivating close connections between Johns Hopkins and Clark University (where Richard Peet had taken over publication of the journal *Antipode: A Journal of Radical Geography*, founded by David Stea and Ben Wisner in 1969), Harvey became a prominent member of an emergent group of self-styled radical geographers. Seeking to get a nascent radical geography off the ground within the US, Peet (1940–)

organized a session on American poverty at the 1971 Association of American Geographers (AAG) meeting in Boston, inviting Harvey to present what would become the most radical chapter in *Social Justice*: 'Revolutionary and counter revolutionary theory in geography and the problem of ghetto formation' (Harvey, 1972). He did so with panache. As Peet recalled:

> the room was full, like hundreds of people, you know.... I remember David said he had ten copies of his paper. It was the first mimeographed version of *Social Justice*, and he said he had ten copies, and it was like a dogfight to get them. Maybe 70 people rushed the stage at the end and grabbed these things. I thought, "God, we've arrived."[12]

Adopting the role of radical intellect, Harvey's acumen and charisma had transformative effects on what was to become the radical/critical geography project. He wrote a series of articles for early issues of *Antipode*; trained such influential second-generation radical geographers as Neil Smith and Dick Walker; attended many of the North American meetings organized by radical geographers—whether within the AAG or by the Union of Socialist Geographers (Figure 8.3); and took time to meet with and inspire young geographers interested in radical geography. As his stature and popularity grew, his papers and plenary sessions at AAG annual meetings became standing-room only, attracting the increasingly global audience of geographers attending these meetings. While he had slept on the street to protect the Black Panther Party's Baltimore office after Fred Hampton was executed in in 1969, which must have been an amazing experience for a young British academic newly arrived in the US, he was not one to man the barricades (Sheppard & Barnes, 2019a). As he recalls, Neil Smith and others "dragged me into political activities" (Harvey, 2006b, p. 188). Rather, he saw his role as developing a theoretical framework to ground radical geography as a serious academic endeavor, spreading radical geography through his teaching.

After *Limits* was republished and publicized in 1992 by the prominent progressive London publisher Verso Press, coupled with the influence of *Condition of Postmodernity*, Harvey's reputation as a radical scholar transcended Geography's bounds. Now widely seen as perhaps the most influential Marxist theorist in the Anglophone academy, his books and talks actively promote anti-capitalism and socialism to an interdisciplinary audience. His attendance at Geography meetings has become less frequent (if still avidly anticipated) as has his presence in Geography journals, but he remains a geographer at heart whose critiques and visions of worlds beyond capitalism remain closely informed by thinking geographically (Sheppard, 2015): taking seriously both socio-spatial theory and nature-society relations.

The radical/critical geography project

While David Harvey is indelibly associated with radical geography, it would be a mistake to equate the one with the other. Radical geographers were practicing their craft before that 1971 meeting and have always been more-than-Marxist. Radical geographers included the likes of the Russian anarcho-communist Pyotr Kropotkin (1842–1921), the French anarchist Élysée Reclus (1830–1905), and the German and American Sinologists Karl Wittfogel (1896–1988) and Owen Lattimore (1900–1989). Mary Arizona (Zonia) Baber (1862–1956), co-founder of the Chicago Geographical Society, promoted antiracism and supported Puerto Rican suffragists, and the Black geographer Thelma Glass at Alabama State University was one of the original Civil Rights activists (Peake & Sheppard, 2014).

Figure 8.3 Founding meeting of the Union of Socialist Geographers (Toronto, May 1974). David Harvey in front row, third from left.

Source: Copyright held by Colm Regan (colmregan@gmail.com), who gave us permission to use it in a paper published in the open-access journal *ACME* (Peake, L. and Sheppard, E. (2014) The emergence of radical/critical geography within North America. *ACME: An International Journal for Critical Geographies* 13(2): 305–327). It was republished in: Barnes, T.J. and Sheppard, E. (2019), *Spatial Histories of Radical Geography: North America and beyond*. New York City: John Wiley & Sons.

Bill Bunge (1928–2013) was a communist before he became a quantitative, and then radical, geographer (Bunge, 1966, 1971), Jim Blaut (1927–2000), David Stea (1936–) and Clark Akatiff marched on the Pentagon in 1967, and Bunge and Akatiff had organized radical geography sessions at AAG meetings in 1968 and 1970 (Akatiff, 1974, 2016).

As originally conceived, radical geography also was not Marxist. Reflecting on the founding of the journal *Antipode*, Ben Wisner recalls: "The key to *Antipode*'s origin is the term 'radical.' We were groping for root causes of the problems, contradictions, inconsistencies, and hypocrisies with which we had grown up in the 1940s and 1950s… Systematic study of Marxism and other critical disciplines was still ahead for most of us." (https://onlinelibrary.wiley.com/page/journal/14678330/homepage/editor_s_past_reflections.htm, accessed December 17, 2022). Those creating radical geography as a project were thus a diverse bunch, as evidenced in the early issues of *Antipode* (Huber, Knudson, & Tapp, 2019). They included feminists, anarchists, anti-racists, and environmentalists, and women and scholars of color were active.

Transitioning from loose collectives of like-minded geographers to a radical geography project requires intellectual entrepreneurs, however, as well as sympathetic places where momentum could be cultivated (cf. Barnes, 2004) and publication outlets. Peet's editorship of *Antipode*, in collaboration with Harvey's group at Johns Hopkins, accelerated and shaped this transition. *Antipode* and radical geography came to be dominated by white, mostly male Marxist geographers often with larger-than-life personalities (Peet, Bunge, Blaut, Harvey, Ed Soja (1940–2015)), narrowing its intellectual range.

Yet other nodes in North America were more diverse. The Union of Socialist Geographers, founded in Toronto in 1974, connected multiple sites in Canada and the US where radical geography was getting off the ground in its own ways, stretching to London, Sydney, Dublin, and Copenhagen, and producing its own *USG Newsletter* until its demise in 1982.[13] The USG supported a variety of intellectual traditions, embracing both intellectual debate and community and political action (Lauria et al., 2019; Peake, 2019). Bill Bunge and Gwendolyn Warren's Detroit Geographical Expedition, reproduced in Vancouver, Toronto and Sydney, practiced an on-the-ground radical geography seeking to inform and empower disadvantaged and racially diverse urban neighborhoods (Warren, Katz, & Heynen, 2019). Questions of 'race,' now so central to critical geography, were being prosecuted by such Black geographers as Harold Rose (1930–2016) and Donald Deskins (1932–2013) (Kobayashi, 2019). Radical environmental geographers Larry Wolf (1921–2016) and Wilbur Zelinsky (1921–2013) founded the Socially and Ecologically Responsible Geographers (SERGE) in the US, editing a third self-published radical geography journal, *Transition*, until its demise in 1986.

When Dick Peet handed over *Antipode* to a professional publisher (UK-based Basil Blackwell) and new editors in 1986, those taking over (Joe Doherty and myself) set about (re)diversifying' the journal's intellectual orientation, actively seeking submissions from anarchist, socialist, feminist, and environmental geographers.[14] Nevertheless, by the 1990s the rise of feminist and then post-structural geographies (Dixon & Jones III, 1998; Murdoch, 2005; Rose, 1993) had triggered the critiques that Harvey (1989, 1997, 1998, 1999) has felt compelled to respond to.

These critiques were also advanced against radical geography itself, now discursively equated with Marxist geography. Those seeking to escape this trap reframed the project as *critical* geography. The International Critical Geography Group, founded in 1996, has organized eight conferences around the world since 1997 (https://internationalcriticalgeography.org/conferences/, accessed December 17, 2002). The East Asian Regional Conference in Alternative Geography (EARCAG) held its first conference in 1997, with ten others since. Cementing this intellectual and discursive shift, a fourth open-access journal was created in 2002: *ACME: An International Journal of Critical Geography*. Its founding editors Pamela Moss, Lawrence Berg, and Caroline Desbiens (2002, p. 3) described its mission as follows: "We are interested in work done from a variety of critical and radical perspectives, as for example, anarchist, anti-racist, environmentalist, feminist, Marxist, postcolonial, poststructuralist, queer, situationist, and socialist." Use of the word 'critical' has received pushback, with mainstream scholars arguing that any good scholarship is critical, but it references a specific body of scholarship founded in 1930s Germany by the likes of Max Horkheimer, Theodor Adorno, Walter Benjamin, and Herbert Marcuse: the heterodox Marxist Frankfurt School now known as Critical Theory (Best, Bonefeld, & Chris, 2018). While departing substantially from Marx, rejecting capitalism and 'scientific' socialism, they retained Marx's concern for social and political structures and human emancipation.

Commenting on reformulation of radical as critical geography, Castree (2000) argued that it reflects a shift away from activism and towards the academy, within which geographers associating themselves with either term had by now assumed powerful disciplinary positions: editing influential mainstream journals and elected as presidents of major disciplinary associations. Self-styled critical geographers shared with radical geographers the belief that creating a truly equitable alternative to contemporary largely capitalist socio-political systems requires tearing these up by the roots. Yet they disagree on how to do this and on what the alternatives should look like. Stark philosophical and intellectual differences still occasionally blow up amongst radical/critical geographers, with internecine fights running the risk of endangering what is essentially a shared project (of one no and many yeses). Yet radical/critical approaches have become quite hegemonic within Anglophone human geography and political ecology. The radical/critical geography project is even gaining serious attention from physical geographers (Lave et al., 2018).

Diverse critical geographies have emerged also around the world, often locally created rather than diffusing from Anglophone powerhouses (Barnes & Sheppard, 2019; Berg, Best, Gilmartin, & Larsen, 2021). As Berg et al. note (2021, pp. 2–4), definitions of critical vary widely depending on the place where the term is used, and there can be stark disagreements even when thinking from the same place (for Japan, compare Mizuoka, 2019; and Nakashima, Fukuda, & Haraguchi, 2021). All this may seem a long way from the paths forged by Harvey, both geographically and intellectually. However, as Castree (2000, pp. 957–958) put it:

> without the efforts of the radical geographers of the 1970s and 1980s, the many young geographers who've fuelled the prodigious growth of the disciplinary Left in the 1990s would not have been able to nail their professional and personal colors to the Leftist mast so readily or so willingly.

Conclusion

As a project, radical/critical geography has always exceeded the Marxist framing pioneered by Harvey. Nevertheless, his influential early and ongoing contributions were key to its establishment. Radical/critical geography has profoundly reshaped the Anglophone discipline, from making sense of the world as it is to asking how we can create better—less unequal, more emancipatory and sustainable—possible worlds. This matters a great deal for the discipline, particularly as radical/critical thinking concatenates beyond Anglophone Geography. But it also matters in other ways. It has dragged Geography out of the academy, as geographers prioritize working with communities to address their problems (updating the Detroit Geographical Expedition vision – see Chapter 20). It's also brought Geography attention and respect from other social science and humanities disciplines, whose critical scholars increasingly appreciate the significance of thinking geographically for critiquing and changing a world that many now find deeply problematic. Geography also is influencing Anglophone universities and colleges, where geographers are often prominent in seeking to replace the 'neoliberalization' of academic norms. To retain this edge, the radical/critical geography project must relentlessly be willing to periodically renew itself through self-critique—something that its success can make difficult (Sheppard & Barnes, 2019b). Nevertheless, Geography has come to matter in ways profoundly shaped by Harvey and the project that bloomed in conversation with him.

Notes

1 One of the most fundamental ways in which geography matters is that the places we inhabit shape how we make sense of the world, even at a time when the internet and smartphones to some degree 'disembed' us from place (Barnes, 2004; Livingstone, 2003).
2 Bristol was the only UK department to offer me a place to study Geography in 1969 (despite my almost failing grade in Geography). These practitioners of the 'new geography' were only interested in my science grades. We were required to arrive a week early for a crash course in statistics. Harvey had left, but *Explanation* was our bible.
3 Harvey recalls that he was so focused on finishing *Explanation* and submitting it to the publisher that "I had hardly noticed what was happening. I felt sort of idiotic" (Harvey, 2006b, p. 187).
4 Interview with David Harvey, 2013.
5 His characterization of spatial science as separating space from society is a caricature; much work within that tradition used such social metrics as transport costs to parameterize space.
6 This triggered an animated response from the leading spatial scientist at that time when published in *Antipode*: Brian Berry (1972).
7 While Harvey made a huge intellectual transition after arriving in Baltimore, one thing remained constant: grounding his thinking in a close reading of European philosophy. Indeed, the case could be made that he made reading philosophy fashionable for up-and-coming geographers: Humanist geographers seeking an alternative to quantitative and Marxist geography began turning to phenomenology (Relph, 1976; Tuan, 1977), feminist geographers engaged with feminist philosophers (Rose, 1993), and 'post-prefixed' cultural geographers drew heavily on French philosophy.
8 He found these critiques somewhat unfair, given his already close relationship with such American feminists as Nancy Hartsock and Donna Haraway (personal communication; Harvey & Haraway, 1995).
9 The implosion of the Soviet world was presented as a moment of capitalist triumphalism—the end of history according to Francis Fukuyama (1992)—where Marx was seen as passé.
10 These are very difficult, deeply philosophical books, and Harvey's thinking here has been extensively criticized as too dismissive of particularism and of the politics of representation, too Marxist, and too abstract.
11 This somewhat pessimistic image reinforces Harvey's belief that capitalism must be contested globally as it cannot be undermined from below. This Godzilla-like view of capitalism, overcoming all barriers, has been criticized, also by fellow Marxian geographers, for failing to acknowledge the potential for localized more-then-capitalist economic practices to escape capitalism's grasp (Gibson-Graham, 1996, 2006; Sheppard, 2019).
12 Peet, R. Interview with Trevor Barnes. Worcester, MA, May 2002.
13 https://antipodeonline.org/supplementary-material/the-union-of-socialist-geographers-newsletter-1975-1983/, accessed December 17, 2022.
14 https://antipodeonine.org/2017/06/28/usg-newsletter-archive/, accessed December 13, 2022.

Study task

Variations on this question can be set on an unseen examination, an open book examination or as a term-paper assignment:

Taking David Harvey's turn to Marxism as an example, consider how and why it's important to infuse human geography with a radical (or critical) political sensibility.

Starter reading list

Blomley, N. (2006) 'Uncritical critical geography?,' *Progress in Human Geography* 30, 1: 87–94.

Blomley, N. (2007) 'Critical geography: Anger and hope,' *Progress in Human Geography* 31, 1: 53–65.

Bonnet, A. (2014) 'Critique,' in R. Lee et al. (eds) *The SAGE Handbook of Human Geography* (SAGE: London), volume 1, pp. XX.
Harvey, D. (1972) 'Revolutionary and counter revolutionary theory in geography and the problem of ghetto formation,' *Antipode* 6, 2: 1–13.
Harvey, D. (1974) 'Population, resources and the ideology of science,' *Economic Geography* 50, 3: 256–277.
Olson, E. & Sayer, A. (2009) 'Radical geography and its critical standpoints: Embracing the normative', *Antipode* 41, 1: 180–198.

References

Akatiff, C. (1974). The march on the Pentagon. *Annals of the Association of American Geographers*, *64*(1), 26–33.
Akatiff, C. (2016). Roots of radical geography. *Yearbook of the Association of Pacific Coast Geographers*, *78*(1), 258–278.
Barnes, T. J. (2004). Placing ideas, genius loci, heterotopia and geography's quantitative revolution. *Progress in Human Geography*, *28*(5), 1–31.
Barnes, T. J., & Sheppard, E. (Eds). (2019). *Spatial Histories of Radical Geography: North America and beyond*. New York City: John Wiley & Sons.
Berg, L. D., Best, U., Gilmartin, M., & Larsen, H. G. (Eds.). (2021). *Placing Critical Geography: Historical Geographies of Critical Geography*. London: Routledge.
Bernstein, R. J. (1976). *The Restructuring of Social and Political Theory*. Philadelphia: University of Pennsylvania Press.
Berry, B. J. (1972). Revolutionary and counter-revolutionary theory in Geography: A ghetto commentary. *Antipode*, *4*(2), 31–32.
Best, B., Bonefeld, W., & Chris, O. (Eds). (2018). *The Sage Handbook of Frankfurt School Critical Theory*. Thousand Oaks, CA: SAGE.
Bunge, W. (1966). *Theoretical Geography*. Lund: C.W.K. Gleerup.
Bunge, W. (1971). *Fitzgerald: Geography of a Revolution*. Cambridge, MA: Schenkman.
Castree, N. (2000). Professionalisation, activism, and the university: Whither 'critical geography'? *Environment and Planning A*, *32*(6), 955–970.
Castree, N., Charnock, G., & Christophers, B. (2022). *David Harvey: A Critical Introduction to His Thought*. London: Taylor & Francis.
Chatterjee, L., Harvey, D., & Klugman, L. (1974). *FHA Policies and the Baltimore Housing Market*. Baltimore, MD: Center for Metropolitan Planning and Research, Johns Hopkins University.
Curry, M. (1985). On rationality: Contemporary geography and the search for the foolproof method. *Geoforum*, *16*, 109–118.
Deutsche, R. (1991). Boys Town. *Environment and Planning D: Society and Space*, *9*(1), 5–30.
Dixon, D., & Jones III, J. P. (1998). My dinner with Derrida, or spatial analysis and poststructuralism do lunch. *Environment and Planning A*, *30*(2), 247–260.
Fukuyama, F. (1992). *The End of History and the Last Man*. London: Penguin.
Gibson-Graham, J. K. (1996). *The End of Capitalism (as we knew it)*. Oxford, UK: Blackwell.
Gibson-Graham, J. K. (2006). *A Postcapitalist Politics*. Minneapolis: University of Minnesota Press.
Harvey, D. (1969). *Explanation in Geography*. London: Edward Arnold.
Harvey, D. (1972). Revolutionary and counter revolutionary theory in geography and the problem of ghetto formation. *Antipode*, *6*(2), 1–13.
Harvey, D. (1973). *Social Justice and the City*. London: E. Arnold.
Harvey, D. (1982). *The Limits to Capital*. Oxford: Basil Blackwell.
Harvey, D. (1984). On the history and present condition of geography: An historical materialist manifesto. *Professional Geographer*, *36*(1), 1–11.
Harvey, D. (1985a). *Consciousness and the Urban Experience*. Oxford: Basil Blackwell.
Harvey, D. (1985b). *The Urbanization of Capital*. Oxford: Basil Blackwell.
Harvey, D. (1987). Three myths in search of a reality in urban studies. *Environment and Planning D: Society and Space*, *5*(4), 367–376.
Harvey, D. (1989). *The Condition of Postmodernity*. Oxford: Basil Blackwell.
Harvey, D. (1996). *Justice, Nature and the Geography of Difference*. Oxford: Basil Blackwell.

Harvey, D. (1997). Contested cities: social process and spatial form. In N. Jewson & S. MacGregor (Eds), *Transforming Cities*. 19–27. London: Routledge.
Harvey, D. (1998). The body as an accumulation strategy. *Environment and Planning D: Society and Space*, *16*(4), 401–421.
Harvey, D. (1999). On fatal flaws and fatal attractions. *Progress in Human Geography*, *23*(4), 557–562.
Harvey, D. (2000). *Spaces of Hope*. Berkeley: University of California Press.
Harvey, D. (2003a). *The New Imperialism*. Oxford, UK: Oxford University Press.
Harvey, D. (2003b). *Paris, Capital of Modernity*. London: Routledge.
Harvey, D. (2006a). *A Brief History of Neoliberalism*. Oxford, UK: Oxford University Press.
Harvey, D. (2006b). Memories and desires. In S. Aitken & G. Valentine (Eds), *Approaches to Human Geography*, (pp. 184–190). Thousand Oaks, CA: SAGE.
Harvey, D. (2009). *Cosmopolitanism and the Geographies of Freedom*. New York City: Columbia University Press.
Harvey, D. (2010). *The Enigma of Capital*. Oxford, UK: Oxford University Press.
Harvey, D. (2012). *Rebel Cities: From the Right to the City to the Urban Revolution*. London: Verso.
Harvey, D. (2014). *Seventeen Contradictions and the End of Capitalism*. Oxford, UK: Oxford University Press.
Harvey, D. (2016a). Books interview: David Harvey. Retrieved from https://www.timeshighereducation.com/books/interview-david-harvey-city-university-of-new-york-graduate-school
Harvey, D. (2016b). *The Ways of the World*. London: Profile Books.
Harvey, D. (2017). *Marx, Capital, and the Madness of Economic Reason*. Oxford: Oxford University Press.
Harvey, D. (2018). *A Companion to Marx's Capital*. The complete edition. London: Verso Books.
Harvey, D. (2020). *The Anti-capitalist Chronicles*. London: Pluto Press.
Harvey, D. (2022). *A Companion to Marx's Grundrisse*. London: Verso Books.
Harvey, D., & Chatterjee, L. (1974). Absolute rent and the structuring of space by governmental and financial institutions. *Antipode*, *6*(1), 22–36.
Harvey, D., Chatterjee, L., Wolman, M. G., Klugman, L., & Newman, J. S. (1972). *The Housing Market and Code Enforcement in Baltimore*. Baltimore, MD: Baltimore Urban Observatory.
Harvey, D., & Haraway, D. (1995). Nature, politics, and possibilities: A debate and discussion with David Harvey and Donna Haraway. *Environment & Planning D: Society & Space*, *13*(5), 507–527.
Huber, M. T., Knudson, C., & Tapp, R. (2019). Radical paradoxes: The making of Antipode at Clark University. In T. J. Barnes & E. Sheppard (Eds), *Spatial Histories of Radical Geography: North America and beyond* (pp. 87–115). Hoboken, NJ: Wiley Blackwell.
Kobayashi, A. (2019). Issues of "Race" and Early Radical Geography: Our invisible proponents. In T. J. Barnes & E. Sheppard (Eds), *Spatial Histories of Radical Geography: North America and beyond* (pp. 37–58). Hoboken, NJ: Wiley Blackwell.
Lauria, M., Higgins, B., Bouman, M. J., Mathewson, K., Barnes, T. J., & Sheppard, E. (2019). Radical geography in the Midwest. In T. J. Barnes & E. Sheppard (Eds), *Spatial Histories of Radical Geography: North America and beyond* (pp. 247–271). Hoboken, NJ: Wiley Blackwell.
Lave, R., Biermann, C., & Lane, S. N. (Eds). (2018). *The Palgrave Handbook of Critical Physical Geography*. Cham, Switzerland: Palgrave Macmillan.
Livingstone, D. N. (2003). *Putting Science in Its Place: Geographies of Scientific Knowledge*. Chicago, IL: University of Chicago Press.
Marx, K. (1967 [1867]). *Capital: A Critique of Political Economy, Vol. 1* (S. Moore & E. Aveling, Trans.). New York: International Publishers.
Marx, K. (1972a [1867–1896]). *Capital, Volume* 2. Penguin: Harmondsworth.
Marx, K. (1972b [1885]). *Capital, Volume 3* (Vol. 3). Penguin: Harmondsworth.
Massey, D. (1991). Flexible sexism. *Environment and Planning D: Society and Space*, *9*(1), 31–57.
Massey, D. (2005). *For Space*. London: SAGE.
Mizuoka, F. (2019). Japan: The Yada Faction versus North American Radical Geography. In T. J. Barnes & E. Sheppard (Eds), *Spatial Histories of Radical Geography: North America and Beyond* (pp. 301–314). Oxford, UK: Wiley Blackwell.

Moss, P., Berg, L. D., & Desbiens, C. (2002). The political economy of publishing in geography. *ACME: An International Journal for Critical Geographies*, *1*(1), 1–7.
Murdoch, J. (2005). *Post-structural Geography: A Guide to Relational Space*. Thousand Oaks, CA: Sage Publications.
Nakashima, K., Fukuda, T., & Haraguchi, T. (2021). Critical geographies in Japan: A diverse history of critical inquiry. In L. Berg, U. Best, M. Gilmartin, & H. G. Larsen (Eds), *Placing Critical Geographies* (pp. 95–117). London: Routledge.
Peake, L. (2019). The life and times of the Union of Socialist Geographers. In T. J. Barnes & E. Sheppard (Eds), *Spatial Histories of Radical Geography: North America and beyond* (pp. 149–181). Hoboken, NJ: Wiley Blackwell.
Peake, L., & Sheppard, E. (2014). The emergence of radical/critical geography within North America. *ACME: An International Journal for Critical Geographies*, *13*(2), 305–327.
Relph, E. (1976). *Place and Placelessness*. London: Pion.
Rose, G. (1993). *Feminism & Geography: The Limits of Geographical Knowledge*. Minneapolis: University of Minnesota Press.
Saunders, P., & Williams, P. (1986). The new conservatism: some thoughts on recent and future developments in urban studies. *Environment and Planning D: Society and Space*, *4*, 393–399.
Sheppard, E. (1995). Dissenting from Spatial Analysis. *Urban Geography*, *16*(4), 283–303.
Sheppard, E. (2002). The spaces and times of globalization: Place, scale, networks, and positionality. *Economic Geography*, *78*(3), 307–330.
Sheppard, E. (2014). We have never been positivist. *Urban Geography*, *35*(5), 636–644.
Sheppard, E. (2015). Thinking geographically: Globalizing capitalism, and beyond. *Annals of the Association of American Geographers*, *105*(6), 1113–1134.
Sheppard, E. (2019). Globalizing capitalism's raggedy fringes: Thinking through Jakarta. *Area Development and Policy*, *4*(1), 1–27.
Sheppard, E., & Barnes, T. J. (2019a). Baltimore as Truth Spot: David Harvey, Johns Hopkins, and urban activism. In T. J. Barnes & E. Sheppard (Eds), *Spatial Histories of Radical Geography: North America and beyond* (pp. 183–209). Hoboken, NJ: Wiley Blackwell.
Sheppard, E., & Barnes, T. J. (2019b). Conclusion. In T. J. Barnes & E. Sheppard (Eds), *Spatial Histories of Radical Geography: North America and beyond* (pp. 371–387). Hoboken, NJ: Wiley Blackwell.
Tuan, Y.-F. (1977). *Space and Place: The Perspective of Experience*. Minneapolis: University of Minnesota Press.
Warren, G. C., Katz, C., & Heynen, N. (2019). Myths, cults, memories, and revisions in radical geographic history: revisiting the Detroit Geographical Expedition and Institute. In T. J. Barnes & E. Sheppard (Eds), *Spatial Hstories of Radical Geography: North America and beyond* (pp. 59–85). Hoboken, NJ: Wiley Blackwell.

9 Open horizons

Doreen Massey's geographies

Jamie Peck

Doreen Barbara Massey, 1944–2016. 'For the future to be open, space must be open too' (Massey, 2005: 12).

Source: Wikimedia Commons

Introduction: radical geographer

The mobilizing slogan 'geography matters' is more closely associated with Doreen Massey than with anyone else in Geography, past or present (Massey and Allen, 1984). It was the title of a mid-1980s course text that she developed at the Open University (OU), an institution established under Prime Minister Harold Wilson's Labour government of the 1960s with an ambitious remit to widen working-class access to higher education. The OU was the base for Massey's academic career. (Retiring from teaching in 2009, she served as an especially active professor emeritus until her untimely death, at the age of

DOI: 10.4324/9781003343240-11

72, in 2016.) The fact that Doreen Massey stayed at the OU itself tells a story, both about *that* place and about *place* more generally. Despite receiving plenty of lucrative offers to move to other universities, and to join the jet-setting academic world for which she never cared, Massey remained at the OU. This reflected her commitments to the university's unique mission, as 'the ideal place to teach and to do research because of its social project and because of the way in which it really encourages us as academics to speak beyond the academy as well as within' (Massey, 2009: np). At the OU, Massey was in her element. As she once reflected, 'I loved every minute of it'.

For long-time friend and colleague John Clarke, an abiding memory of Massey was of her sitting in the cafeteria at the OU,

> engaged in animated conversation with several people – happily, imaginatively, creatively and critically. She's one of the few people I've known who managed to combine being critical with being enabling ... And these were always conversations with something at stake.
>
> (Clarke, 2016: 358)

Places, for Massey, were where people and projects came together, the site of encounters and conversations. She demonstrated this intersectional, interactive understanding of place arguments by theoretical and philosophical means, but also through everyday illustration and almost conversational exposition. One of Massey's rare gifts was a talent for explaining complex ideas in straightforward and unpretentious language. An ordinary day at the OU campus, for example, became an opportunity to reanimate 'a whole cartography of correspondence, ongoing conversations, pick[ing] up where I left off', after which she would commute back to her home in north London, 'check the post, the telephone messages, find out "what's been happening here" while I've been away' (Massey, 2005: 130). To travel between places, she continued, is to 'move between collections of trajectories and to reinsert yourself in the ones to which you relate ... Places not as points or areas on maps, but as integrations of space and time'.

Geography matters. On the face of it, the phrase can sound almost meek. It could be (mis)taken for a modest invitation to wallow in the ephemera of disciplinary stuff, like some A–Z compendium of scholarly artifacts, local curiosities, and 'matters' of geography. (But it was not that.) In explanatory terms, the petition that geography merely 'matters' could be heard as a polite addendum, as a nod towards the role of secondary details or mediating influences, perhaps making a difference at the margin. (It was not that either.) Such minimalist interpretations of the phrase radically underestimate Geography in general and Massey's project in particular. Her project breathed new life into some of Geography's staple concerns, with the uniqueness of places in a world of irreducible difference. This was a disruptive intervention, indeed a radicalizing one, founded on an injunction to think and act 'relationally'. To do so is not just to pay attention to connections, for example between local circumstances and so-called 'global forces', or between complex outcomes and 'general' theories, but to *hold them together*, as Massey liked to say. The general and the particular were like opposite sides of the same coin, each shaping and conditioning the other. A *relational* view of place, it follows, is about connections, confluences, and catalytic effects. It's not about obstacles and barriers, or stuff getting in the way. For Massey, places were boundless and open. They were always 'under construction', and saturated with power and politics, but also sites of generative interaction, renewal, and social potential. Geographical futures were made, not pre-determined. The horizon itself was an open one, both for social action and for the *geographical imagination.*

Doreen Massey radicalized Geography while at the same time recuperating and reimagining some of its signature concerns with place and uniqueness. As an adroit and creative social theorist, she made innovative contributions to political economy, feminism, and poststructuralism – but never in a compartmentalized way, as separate conversations. And she also proselytized for Geography, popularizing new ways of thinking place and space for the critical social sciences, while serving as one of the discipline's most effective public advocates. This means that it is both difficult and misleading to concertina Massey's story into an abbreviated academic biography. Fortunately, we have the benefit of several full-spectrum assessments of Massey's life and work, plus a collection of revealing interviews.[1] Building on these resources, the present chapter seeks to add to the conversation by focusing on the 'organic' character of Massey's project – and the constitutive connections between the professional, the political, and the personal – especially during its formative decades. This calls attention to the multifaceted but also deeply interconnected nature of her project, one that changed Geography.

Space invader

Doreen Barbara Massey came from a public-housing estate in Wythenshawe, an 'ordinary place' south of Manchester (Massey, 2001a; Figure 9.1). Born just before the end of the Second World War, she was the product of a working-class upbringing. Although never one for neat-and-tidy 'origin stories', this is where Massey came from, in both class and geographical terms, and it never ceased to matter to her. From an early age, she suffered the debilitating effects of osteogenic disease, effectively growing up with the newly established National Health Service, of which she would remain a stalwart supporter throughout her life. Academically gifted, Massey won a scholarship to Oxford,

Figure 9.1 Wythenshawe home, photographed by Doreen Massey.

Source: Borden et al., *The Unknown City*, Figure 28.2, Copyright 2000 Massachusetts Institute of Technology, by permission of The MIT Press an experience that quickly turned her, she relished saying, into 'a raving socialist' (quoted in *The Daily Telegraph*, 2016: 33). The Geography that she encountered in the mid-1960s was deeply conservative, if superficially apolitical. Oxford's School of Geography was an old boys' club in both a literal and metaphorical sense.[2] The department, she later recalled, had 'not yet taken on board *Models in Geography*', and the project of modernizing the field according to the scientific and systematic principles championed by Richard Chorley and Peter Haggett (see Chapter 4).[3] In fact, Massey had her own reasons to be skeptical of the discipline's incipient embrace of quantitative modeling and mathematical formalism, continuing to find value in Geography's enduring concerns with particularity and not-so-easily codified diversity, even as she bridled at the stuffiness and social detachment of the Oxford dons. 'I both hated it and loved it', she later said (quoted in Freytag and Hoyler, 1999: 83). Academically successful but out of place there, she had felt like a 'space invader'.

On graduating from Oxford in 1966, Massey promptly turned her back on at least that version of academic life. She moved to London, initially to an unfulfilling job in the private sector, before landing a position at the Centre for Environmental Studies (CES), a research institute newly established by the Wilson government with a remit to rethink approaches to urban and regional development. Very much a product of these times, CES had its own vision of modernization – expansive, progressive, scientific (see Barnes, 2018). There she joined an interdisciplinary team under the leadership of physicist-turned-geographer Alan Wilson, who was charged with advancing the problem-solving mission of CES through the application of the latest analytical techniques. Once again somewhat out of place, but now in a different way, Massey was put to work 'on employment distributions, linear programming, and the intoxicating logical intricacies of how to split the basic from the service sector' of an evolving economy. While she clearly enjoyed the challenge of tackling these elemental questions of 'applied' economic geography, her memories of the time were just as much about the experience of being in London in the swinging sixties, and a world that 'revolved around music and politics', Velvet Underground records, and anti-Vietnam war protests. It was a time of liberation and political struggle, pent-up desires for change, and the elusive promise of modernization. It was a 'mixture of repression and hope for a different future'.[4]

CES placed an abiding faith in the power of science, expert knowledge, and the promise of progressive reform. Massey (1993c: 11–12) was far from convinced by all of this, especially when it came to 'the assuredness of expertise, the apparent precision of quantification, ... the triumphalism of technical-experts-armed-with-mathematical models'. But this was also 'an amazing bunch of people', Massey remembering CES as a 'really good formative experience' (quoted in Freytag and Hoyler, 1999: 85). But if this was formative, what would prove to be transformative was the year that Massey spent, on leave from CES, as a Master's student at the Mecca of regional science, the University of Pennsylvania. She was there for advanced training in mathematical economics and orthodox location theory, the inherited tools of the trade at CES. More consequentially, Massey's time at Penn included an unplanned encounter with the work of Louis Althusser, in an elective class taken in the French department. Here was an approach to 'structuralist' social theory that was dynamic and multifaceted rather than deterministic and mechanical. It eschewed rigid forms of 'foundationalism' in favor of a more open-ended and anti-essentialist perspective. This was an important moment in Massey's complex but nevertheless generative relationship with Marxism, with which she would remain in life-long dialog, while never feeling comfortable with the label 'Marxist'.[5]

These Althusserian strands in Massey's thinking took a while to gestate, but the first order of business on returning to CES was to open up the space for a different kind of future. Her first academic paper was a root-and-branch critique of neoclassical location theory, the incumbent paradigm of rationally acting firms in competitive space that had been allied with a reformist orthodoxy in regional policymaking. Massey contended that 'consideration of "the spatial element" is essential to all effective economic analysis ... What are emerging as "locational problems", whether intra-urban, interregional or international, are the spatial manifestations of the contradictions of capitalism' (Massey, 1973: 38–39). While this might have been a little racy for some at CES, by this time the organization was pursuing a more radical program of critically oriented research and policy advocacy, positioned on the left flank of Labour Party thinking.

Massey's first book followed in 1978, a 'straightforward Marxist analysis' of capital and land, written with Alejandrina Catalano, but it was her extended collaboration with CES colleague Richard Meegan that established a template for a programmatic *alternative* to industrial location theory.[6] Truly original, *The Anatomy of Job Loss* set out to understand processes of employment decline in relation *both* to the big picture of 'wider economic and political circumstances' *and* to the nitty-gritty of 'what is actually going on in production, in the factory and the office' (Massey and Meegan, 1982: vii). Initiated by a grant from the Callaghan government, under the Left-leaning minister of the environment, Peter Shore, Massey and Meegan had been 'unpacking' the process of job loss at the level of industrial sectors, firms, and localities. This close attention to the causes, dynamics, and consequences of economic change, across dozens of sites and sectors, would epitomize what later became known as the 'restructuring' approach.[7]

The later stages of Massey's work on the job-loss project were completed with funding from a research fellowship that she secured from the Social Science Research Council (SSRC). Margaret Thatcher's Conservatives had swept to power in 1979, whereupon each minister of state was instructed to initiate significant expenditure cuts, the symbolic politics of which would include some 'sacrificial lambs' (Posner, 2002: 3). CES and the SSRC were both targeted. The Education Secretary, Sir Keith Joseph, who questioned the very idea of social *science*, took the view that the SSRC 'must go', launching an immediate review. In the end, the SSRC survived (with a new name and remit), but CES was not so lucky. As the preface to *The Anatomy of Job Loss* wryly noted, the authors' concern with 'the subject matter of this book was greatly increased on the closure of CES during [its] writing up' (Massey and Meegan, 1982: viii), both being made redundant. Ultimately, CES was wound up not because it had failed, but 'because it [had been] so successful' (McLoughlin, 1983: 17).

Massey transferred her SSHRC research fellowship to the London School of Economics, punctuated by a stint at Berkeley. Neither a regular nor an especially enthusiastic visitor to the United States, Massey's few months at Berkeley were once again transformative. What Richard Walker likened to 'a Vienna circle moment' cemented a radical course correction in the field of industrial-cum-economic geography. Graduate students there at the time recalled Massey's arrival with the 'energy of a whirling dervish' and a 'punk' attitude to established hierarchies, especially those of class and gender: this was a 'gardener's daughter who goes to Oxford', Michael Storper remembered, 'and she *lets you know it*, right, from the beginning'.[8] Rejecting the conservative-theory culture of neoclassical economics and the timidity of mainstream policymaking, Massey's take on 'restructuring' was disruptive in both methodological and political terms. A far cry from the idealized economy of neoclassical theory, it was about strikes and campaigns against factory closures, about alternative economic strategies and programs of 'restructuring for labor'.[9]

Spatial divisions

Published in 1984, Massey's *Spatial Divisions of Labour* is rated as one of the books of that decade with an indisputable claim to 'classic' status.[10] Research programs concerned with labor, locality, class, and gender have been substantially reshaped in its wake. It may be a little corny to say that this was the book that Massey was destined to write, but there is truth in her later reflection that, as someone 'from the North-west [of England she had] lived with, through and ... in combat with regional inequality since my childhood', the issues raised by which were never merely 'academic' (Massey, in Featherstone et al., 2013: 257). Programmatically, the book represented the capstone statement of her work on industrial restructuring, but it was also a decisive moment in Massey's extended dialog with Marxism, particularly 'one of the things that [she had] most taken from Marxism over the years [which was] thinking relationally' (Massey et al., 2009: 405). This required close attention to methodology and midlevel theory development, coupled with investigations at a similar analytical pitch to *The Anatomy of Job Loss*, foregrounding the complex nexus of 'industry' and 'region', and the intricacies of employment change as 'experienced by labour' (Massey and Meegan, 1982: 17). In retrospect, the book also serves as a prototype for a kind of conjunctural analysis formulated in expressly geographical terms, as a mode of *necessarily* contextual theorizing,[11] while at the same time being very much a product of a (conjunctural) historical moment – the deindustrialization and crisis-driven restructuring of the British economy, especially across its so-called 'problem regions', and the rise of Thatcherite neoliberalism.[12]

Working titles for the 1984 book convey a sense of what was at stake. It might have been called *Space and Class*, and thankfully not the cumbersome *Industrial Location and the Economy: Considerations on Space and Class*. Operationalizing the idea of the spatial division of labor as a midlevel conceptual framework, the book advanced an original take on spatialized class analysis, probing deeply into gender relations, government policies, and localized politics. Michael Storper (1986: 456) read it as an attempt to 'marry social scientific geography to the older regionalist tradition ... through the lens of political economy', in the process boldly ascribing *theoretical* salience to the question of regional differentiation. Geography therefore mattered in the book in more than an empirical or contingent sense, '[s]ocial change and spatial change [being] inextricably interdependent' (Urry, 1986: 278). If treatments of space in the austere tradition of neoclassical location theory were basically trivial, being reduced to the bare-bones questions of distance, proximity, and the sparse contours of costs and prices, *Spatial Divisions* presented a more richly granular understanding of *place*, as a site of historically embedded political cultures, industrial-relations traditions, gendered employment norms, and much more. When Massey talked about 'space', it was never just a container for social processes, or a setting for socioeconomic life. Here, space is 'much more than a mere stage on which other actors perform, or even a structure which constrains the operation of other social forces', Cochrane (1987: 356) explains; instead, it is 'an active participant in its own right', recursively shaping the fortunes of industries, the competitive strategies and employment policies of firms, and the counter-strategies of workers, social movements, and unions.

An object lesson in relational analysis, *Spatial Divisions* demonstrated that what Massey called the 'spatial structures of production' were not pregiven by the imperatives of capital, nor predictable outcomes of profit-driven accumulation, understood in terms of abstract laws of motion. In the language of the time, infused as it was by lessons derived from critical realism,[13] spatial structures and local outcomes could not be 'read off' from such structural conditions and general tendencies. Rather, they represented a socially complex, historically sedimented, and geographically uneven terrain, albeit a constantly moving and contested one. This shaped both the form and path taken by 'restructuring'. Consequently, Massey's project was not about simply complexifying more generalized (or more abstract) accounts of socioeconomic transformation, adding empirical details and 'local color'. It entailed the development of situated, mediated, and contextualized accounts of how these processes of socioeconomic transformation actually worked, in different places, for different industries, and for different groups of workers.

> The evolution of different kinds of spatial structure, their establishment, maintenance and eventual collapse and change, are not simply determined by the characteristics of the labour process, the requirements of accumulation, the stages of the mode of production, or even the demands of capital. None of these things in themselves 'result in' specific spatial forms. Spatial forms are established, reinforced, combated and changed through political and economic strategies and battles on the part of managers, workers and political representatives.
>
> (Massey, 1984b: 85)

By the same token, the uneven development of economic relations was not a predictable outcome of the seesawing laws of motion of capital accumulation, but was also shaped, reciprocally, by conditions on the ground, the localized balance of social forces and struggles, and the contingencies associated with 'highly differentiated and unique outcomes' (Massey, 1984b: 48). For Massey, these were never secondary or residual matters, nor details warranting attention only after the abstract theorizing was done, as if 'theory' itself were somehow 'above' such conditions, impervious and unmoved. Her search was for 'something more dynamic', beyond the foreclosures of structural theorizing. It was conditions on the ground, and the particularities and politics of their interpretation, that defined the terrain for exploration, exposition, and ultimately explanation – all the way out to the very meaning (and framing) of 'the economic'.[14]

Massey had taken from Althusser the insight that 'everything [is] always a product of what went before it' (Massey, in Featherstone et al., 2013: 254), a statement with implications not only for place-specific processes of historical conditioning but also for the multiplicity of combinatorial effects, political responses, and future trajectories that might follow from these (profoundly) 'mediated' circumstances – nothing being structurally pregiven. In *Spatial Divisions* these arguments were worked out in relation to the (differentiated) experiences of restructuring across various regions of the UK economy, yielding a rationale for conceptualizing and excavating the distinctive historical development of regions, places, and localities. 'Local' circumstances really mattered here, especially the roles performed, cumulatively and in combination, by localities in successive spatial divisions of labor.[15] Regional problems, in this sense, are not reducible to proximate regional causes, just as the so-called 'success' of regions cannot be put down to internal, endogenous factors alone, such as entrepreneurial qualities, superior technologies, better institutions, or inherently more efficient models of management. Massey's distinctive counter-argument was that 'you shouldn't just think about regions in terms of them being successes and failures [by virtue of] their own behaviour, [such] that they are to be blamed or congratulated on their economic failure or success' (Massey, in Featherstone et al., 2013: 257).

Massey was critical not only of neoclassical location theory, which stripped down the meaning of space and place to considerations of cost and accessibility, but also 'those structural Marxists – unnamed – who have seen geographical space as little more than a blank sheet on which the logical necessities of capital find their empirical expression' (Cochrane, 1987: 356). She refused to 'look at the world as if it were merely the predetermined product of a set of [capitalist] laws and tendencies' (Massey, 1984b: 6). Hers was an 'organic' treatment of uneven geographical development, generated relationally 'from below' (or within), through ongoing and inherently variable processes of interaction and (re)combination, in contrast to the more 'genetic' understanding of uneven development associated with those 'structural Marxists', such as David Harvey and Neil Smith, for whom such 'seesawing' tendencies tend necessarily to accompany the dynamics of capital accumulation, valorization, and devaluation.[16] Constructively, these alternate takes on uneven geographical development can be seen as complementary – the *yin* of the structural-cum-systemic alternating with the *yang* of the particular-cum-contingent. For his part, Smith was inclined to meet Massey half way. Harvey, on the other hand, was having none of it, complaining that 'Massey is so anxious to deny structuralist leanings or that the "logic of capitalist development" has any explanatory power in local settings that all theorising disappears beneath a mass of contingent labour-management relations in place', the result being 'laden down with a rhetoric of contingency, place, and the specificity of history'.[17]

The rift only widened through the 1980s and into the 1990s, later to take on various proxy forms, and although the personal and professional relations between Massey and Harvey were subsequently repaired, the methodological fracture never really was. In truth, there was (and remains) unfinished business on both sides. Massey's project received sympathetic but telling critique for its rather weakly principled 'micro-structuralism', for its largely absent theory of transformation (rules), and for sleight-of-hand slippages between richly documented case-study evidence and more plenary or placeholder concepts like 'spatial structures'.[18] If, for Massey, the various individual (and intricate) pieces of what amounts to a jigsaw-puzzle conception of uneven development ultimately fail to add up to a bigger-picture pattern (amenable to more synthetic theory claims), Harvey is vulnerable to the opposite charge. For Harvey, the mosaic is much more schematically drawn, his version of the jigsaw puzzle comprising just a few (almost interchangeable) pieces.

There are dangers, of course, in stereotyping these positions, which in explanatory terms represent opposite sides of the same coin. Massey never wanted to do away with more genetic, abstract theories, maintaining that 'the aim must be to *use* general theory not to reject it', with the crucial proviso that 'to use it means to use it in specific contexts – for there are no "general" ones, in that sense, to use it on' (Massey, 1989: 695, original emphasis). Her understanding of 'general' theory, however, could never be boiled down to a list of abstract tendencies, such as those associated with the logic of profit-seeking accumulation or the 'requirements' of the capitalist labor process; it was always more

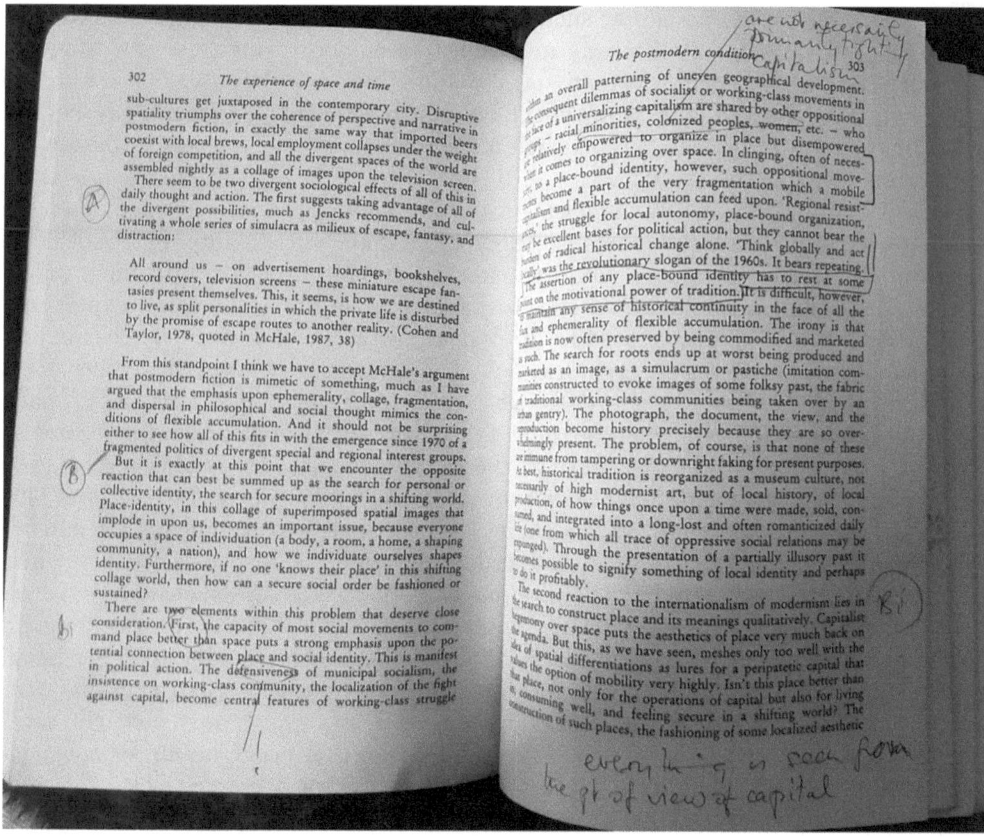

Figure 9.2 Massey's marginalia on her personal copy of Harvey's *The Condition of Postmodernity* : … 'everything is seen from the point of view of capital'.

Source: photograph courtesy of Mark Graham, with permission

conjunctural, more contextually specific and intersectionally complex, more contradictory and contestable.[19] Neither rejecting Marxism, nor buying the full package, she saw herself 'enriching and complexifying', and instead of being an 'archival Marxist', she wanted to 'bring what Marx had to say into the situation we have today' (quoted in Ijams et al., 1994: 107). As she wrote, clearly perplexed, in the margins of her personal copy of David Harvey's (1992) *The Condition of Postmodernity*, which she would later subject to withering critique (Massey, 1991b), 'everything [here] is seen from the point of view of capital' (see Figure 9.2).

Left turns

Ken Livingstone, the socialist politician who as 'Red Ken' served as Leader of the Greater London Council (GLC) from 1981 until its abolition at the hands of the Thatcher government in 1986, and later as a somewhat pinker Ken, as Mayor of London from 2000 to 2008, had known Massey from Labour Party circles since the mid-1970s. Speaking at an event at the Royal Geographical Society to mark her retirement, Livingstone recalled that, after his election as Leader of the GLC, 'Doreen was one of the first people I turned to for some input into our industry and employment policy' (Livingstone, 2013: xv), appointing her to the agency charged with realizing an alternative economic program, the Greater London Enterprise Board (GLEB). Around this time, Massey had been interviewed for the Chair of Geography at the OU, but there had been an inordinate delay before the formal job offer materialized. This led Massey's friends, Livingstone among them, to speculate that something might be amiss (after all, Massey had never completed a PhD, 'she had spiky multi-coloured hair [and] didn't look like the typical professor', which might be interpreted as a lack of gravitas). In fact, Thatcher's Education Secretary, Keith Joseph, had recently complained, when touring a university campus, that he was 'very worried about the coming politicisation of Geography' (Livingstone, 2013: xv). On top of this, the Conservative government's very public hostility to Livingstone's GLC and its allies was implacable – an attitude fulsomely reciprocated on the Left.

The early 1980s GLC was in the vanguard of experiments with municipal socialism.[20] The 600-page *London Industrial Strategy* detailed a program of progressive interventions across key sectors of the metropolitan economy – from instrument engineering and furniture manufacture to software, childcare, and transportation, from domestic work and cultural industries to defense conversation, energy, and retailing – in a way that owed a lot to Massey's politically attuned, sector-by-sector approach. The report's signature theme was 'restructuring for labor', a local government program of socialist demonstration projects and targeted financial assistance, which was positioned in opposition to the dominant model, 'restructuring for capital', enabled by market forces with the support of the Thatcher government. If Thatcherite neoliberalism had repudiated Keynesianism from the Right, the GLC offered a critique of the limitations of traditional Keynesianism from the Left, Michael Rustin observing in his evaluation of the *London Industrial Strategy* that, '[t]he view-point of capital is barely reflected in the document'. With an annual budget of £30–40 million, GLEB was the agency charged with delivering this politically (and indeed practically) ambitious strategy, eventually investing in around 100 enterprises with the creation of around 3,000 jobs. Yet if this represented a large-scale financial and political commitment from the perspective of local government, it was but a 'tiny' intervention relative to the scale of the London economy.[21]

Understood as a 'transitional' supply-side strategy, en route to a more comprehensive (and more than local) program of socialist transformation, GLEB was (over)promoted as 'an alternative model which works' (GLC, 1985: 46), far ahead of its mixed and modest results. Confronted by daunting problems of unemployment, job loss, inequality, and industrial adaptation, GLEB had been charged with the task of improvising, in real time and on a case-by-case basis, a radical and yet feasible experiment in 'socialist micro-economics', executed in but also *against* the capitalist marketplace, armed only with the limited operational, legal, and financial powers of local government. In this constrained context, GLEB had to balance concerns of commercial viability and financial responsibility with the social(ist) goals of supporting employment equity, trade-union rights, workplace democratization and enterprise planning, cooperative ownership, and socially useful production – in the face of a relentlessly hostile national government.

On the GLEB board, Massey was involved in the nuts and bolts of these decisions (not to say, judgments), which ranged from practical problems of alternative accounting to debates over theoretical and political principle. Hers was one of the more sober voices around the table when it came to the then fashionable 'tendency to "idealize" flexible manufacturing' as the locus of progressive intervention (Palmer, 1986: 123; cf. Murray, 1985). The issues involved here were immediate and practical, but they were also strategic and even theoretical. 'We had serious plans', she said of GLEB, which

> very definitely involved thinking through some major issues – if there was postFordism or neoFordism, what possibilities they offered for economic intervention at a local scale, etc – and that was very challenging ... [F]or many years we had all been saying 'the problems aren't problems of the city, they are problems of capitalism and the city is on the receiving end', and here we are sitting in the city trying to convince ourselves we could do something about it, which is quite contradictory ... GLEB was trying to intervene in a local economy not to help capital but to help labour ... [but] I'm still not sure to what extent capitalism can become kinder, more civilised.
>
> (Massey, in Freytag and Hoyler, 1999: 89–90)

For all of GLEB's achievements in elevating the level of debate and as a demonstration project for progressive local interventionism, its own limitations were exposed even prior to the abolition of the GLC. Massey (1987a: 29) ultimately concluded that GLEB had been 'unable to change the shape of the local economy', nor the logics of valuation that invariably privileged commercial considerations over social goals.

Going back to her time at CES, Massey had served on national-level policy committees for the Labour Party, although her faith in these efforts began to wane by the mid-1980s, as the party's leadership trimmed towards the center. She had concluded that the lack of meaningful progress on progressive policy change had nothing to do with gaps in knowledge, or limited understanding of the issues, but instead represented a failure of political will and commitment. The stakes, in retrospect, could hardly have been greater. The historic miners' strike of 1984–1985 ended in a crushing defeat for the union, on the heels of which Thatcher 's Conservatives drove through legislation to abolish the GLC along with the metropolitan tier of government across the country.[22] Social, regional, and class divides were widening, and in some cases fracturing, including in the most unequal region of them all, London and the south-east. Livingstone's GLC had dramatized the situation by posting a huge sign above County Hall (and directly across the River Thames from the Houses of Parliament) displaying the city's unprecedented unemployment count. London had its prosperous parts of course, but this was a city that, 'figured in the popular and political imagination, primarily in terms of the collapse of the old docks, the decline of manufacturing industry, and the classic problems, here writ large, of inner urban areas', although just as importantly it was 'a place of political contestation' (Massey, 2011: 5). When in 1986 the ESRC invited a small group of academics to consider 'horizons and opportunities in social science research', Massey's (1987b: 268) advice was emphatic: 'A first thing geographers could be doing ... is participating even more actively in public and policy debate'.

Modeling her own version of 'public Geography' (see Chapter 23), Massey was a regular contributor to *Marxism Today* through the 1980s (see Figure 9.3). Almost unimaginably today, this was a high-circulation magazine, available on every British high street, while also serving as the theoretical journal of the broad-left wing of the Communist Party of Great Britain.[23] Under the editorship of Martin Jacques, with the support of prominent allies like Stuart Hall and Eric Hobsbawm, the magazine became the leading venue for debate around the cultural and political meaning of Thatcherism. One of Massey's many contributions included a sobering analysis of the re-election of

Figure 9.3 'Labour's other task [is] how to win support for progressive politics amongst the new and growing sections of society' (Massey, 1983: 11).

Source: Barry Amiel and Norman Melburn Trust, Create Commons Attribution 2.5

Margaret Thatcher's Conservatives in 1983, mapping 'the political geography of a divided Britain'.[24]

> The Labour Party is left with two clear geographical bases – the regional heartlands and the inner areas of the big cities. Both are in economic decline. But they are very different politically: the old Labourism dominates the regions, the new alliances are growing in some cities. Weaving these together is a real challenge. They are different strengths, and Thatcher has already declared her intention of attacking both.
>
> (Massey, 1983: 19)

Following the abolition of the GLC in 1986, what became known as Ariel Road Group was an informal grouping of heavyweight political thinkers – among them Ken Livingstone, Stuart Hall, Hilary Wainwright, Michael Rustin, Michael Ward, Vella Pilay, and Robin Murray – gathering regularly in Massey's modest flat in Kilburn for what were remembered as 'very intense debates'. Besides earning a minor place in the annals of *Wall Street Journal* conspiracy theories (Champion, 2000), more enduring outcomes of the Ariel Road gatherings included the creation of the interdisciplinary journal *Soundings*, first published in 1995, and the germinal idea for what would be the last project of Massey's always engaged life, the Kilburn Manifesto. Launched in the aftermath of the global financial crisis of 2008, this was a programmatic effort to rethink, side by side, critical theory and political strategy.[25]

Conclusion: placing Massey

A succinct biography that Doreen Massey wrote, towards the end of her time at the OU, encapsulated her project this way: 'She has researched for many years issues such as space, place, cities and globalisation, especially trying to work between problems of conceptualisation and issues of immediate politics' (Massey, 2008: 497). Massey did so as a quintessential geographer, but also as a tirelessly transformative one. For her, both geography and Geography mattered in ways that were catalytic rather than constraining, opening new horizons rather than imposing limits. As a creative theorist and committed teacher, it mattered in much more than a disciplinary way. It also mattered in her pioneering roles as a political activist and public intellectual. Massey was never one for sniping from the sidelines, nor did she shy away from the long slog of working for progressive change through party politics, social movements, and even local government agencies, each twist and turn of which involved, she remembered, 're-negotiating the nature of one's involvement and responsibility' (Massey, 2008: 495). Doreen Massey was enormously influential, but not in a self-important or self-righteous way. She was an inspiring writer, an original thinker, and an infectious, magnetic speaker. Once self-describing as a 'nonstandard academic' (Ijams et al., 1994), she would joke that, at five foot one tall, she sometimes could barely see over the podium in public talks. Yet her personality and presence could fill the largest of venues. As her friend Hilary Wainwright (2018: 367) wrote of Massey, 'She radiated political energy and humanity, sparkling with a cheeky wit'.

Massey was once invited by a national newspaper to write about the single material thing that they most valued in their home. Looking around her modest and simply furnished flat in Kilburn, North London (Figure 9.4), she decided to choose her window, her window on the world. For Massey, this was certainly not, however, about looking down on the world from above, in detached contemplation. On the contrary, it symbolized her openness to the world, and to everyday worlds in constant change. From her window she could see planes on the flight path to and from Heathrow, as well as out onto the ceaseless interactions that made up life on the street. 'If it is now recognized that people have

Figure 9.4 Doreen Massey at home in Kilburn, 2010.

Source: photograph courtesy of Abel Albet i Mas, with permission multiple identities then the same point can be made in relation to places', she once wrote, 'multiple identities [that] can either be a source of richness or a source of conflict, or both' (Massey, 1991a: 28). Place, from this perspective, is not singular or static, neither is it coterminous with some introverted and homogenous notion of 'community'. As a domain of interactions and connections, it's always open, continuously under construction. Place is 'always in the process of being made', Massey (2005: 9) insisted, 'never finished; never closed'.

Notes

1 Biographical interviews can be found at Ijams et al. (1994), Freytag and Hoyler (1999), Massey et al. (2009), and Featherstone et al. (2013). For book-length assessments, see Featherstone and Painter (2013) and Werner et al. (2018). For more concise treatments, see Callard (2011), Castree (2016), Clarke (2016), Meegan (2017), Lee (2016, 2018), and Werner (2024). For edited collections of Massey's writings, see Christophers et al. (2018) and Featherstone and Kelliher (2022b). On Massey's political project, see Massey (2008), Livingstone (2013), Marriott et al. (2017), Featherstone and Kelliher (2022a), and Hatherley (2022).

2 Not coincidentally, Massey's feminist politics are also traced back to this period, even though they did not explicitly animate her theoretical project until later (Massey, 1994, 1989; Featherstone et al., 2013). The feminism of Massey's generation, she explained, 'was about much more general liberation', it was about 'a new society, a new way of being, a new way of organising things ... a bigger political struggle' (Massey, in Featherstone and Painter, 2013: 260, 258). It was also about a different kind of personal politics, in the workplace and beyond. The acknowledgments page of her first book, with Alejandrina Catalano, captured a sense of this: 'Neither of us has a "wife" to thank for solace and self-effacement' (Massey and Catalano, 1978: np).

3 I draw below on Massey's (1993c) reflections on the 1960s. See also Barnes (2018), Dear (2018), and McDowell (2018). On the Geography of Chorley and Haggett, see Barnes (this volume).

4 Massey (1993c: 10). Massey considered herself 'a child of [two] generations of social democratic interventionism' (Freytag and Hoyler, 1999: 84). First, there was the postwar world shaped by the 1945 Education Act, the National Health Service, and the welfare state. A quarter-century later came the apogee of Keynesian modernization, from the 1960s through to the crises of the 1970s, prelude to the neoliberal era. 'I grew up thinking of public sector housing as something wonderful', she once wrote, 'I really regret the passing of securities that that place used to offer, of – you know, I'm the welfare state generation. And I needed it, because I have quite a lot of problems with my health ... I needed the health service. And I had state education ... [There] were enormous securities about all that' (Massey, 2000: 141).

5 Massey would quite often say that she had found it 'very, very difficult to count [herself] as a Marxist' or to 'buy in to Marxism' (Massey et al., 2009: 403), while on other occasions invoking the Marxist *we*, for instance, 'As Marxists, we claim ...' (Massey, 1989: 694). If this was a club, it was one that she was disinclined to join, but from which she did not wish to be excluded. In fact, Massey spent a career resisting labels, refusing to be categorized or to stay in a prescribed academic 'lane'. As she once said, combining a rhetorical question with an appropriately relational response, 'If someone were to ask if I were a Marxist or not, my answer would entirely depend on who it was. If it was a total fundamentalist, I'd say I don't give a damn. If it was somebody on the right, I'd say yes' (quoted in Ijams et al., 1994: 107).

6 Massey and Catalano (1978); Massey and Meegan (1978, 1982).

7 On the restructuring approach, see Lovering (1989) and Meegan (2017).

8 Quoted in Peck and Barnes (2019). Prior to Massey's visit, there had reportedly been 'virtually no discussion of feminism' in the Geography and Planning departments at Berkeley, while the bandwidth of political debate had also been narrow (Gertler et al., 2018: 218). Graduate students there at the time included Judith Carney, Susan Christopherson, Meric Gertler, Amy Glasmeier, Susanna Hecht, Flavia Martinelli, AnnaLee Saxenian, Erica Schoenberger, and Michael Storper.

9 Massey's visit had 'an immense impact' on the graduate students and young faculty at Berkeley, 'showing a different path forward' (Marc Weiss, quoted in Peck and Barnes, 2019).

10 See Martin et al. (1993), Callard (2011), and Werner (2024).

11 For discussions of Massey's methodology in relation to conjunctural analysis, see Hall and Massey (2010), Clarke (2018), Hart (2018), and Peck and Ebner (2019).

12 The underdevelopment of these 'problem regions' of the United Kingdom, in the northern and western peripheries, has conventionally been pegged – in both orthodox theory and mainstream regional policy – to the supposed deficits of the regions themselves (say, in skills, capital stock, or institutional arrangements). Massey (1979) challenged this conventional wisdom, memorably asking 'in what sense' it represented a '*regional* problem'. Her later work substantiated the counter-claim that the country's *real* problem region was actually its 'most successful', London and the South East, the extreme concentrations of wealth and power within which, alongside poverty and insecurity, made it also the UK's most unequal (see Allen et al., 1998; Amin et al., 2000; Massey, 2007b).

13 The leading figure here was Andrew Sayer, a friend and interlocutor of Massey's since the 1970s (see Sayer, (1982, 1984); Peck, 2022). Massey's own commitments to critical realism were oblique rather than explicit – being mediated through her own tacit methodology, approach, and terminology – but they were meaningful nonetheless (see Lovering, 1989; Massey, 1985, 1995b: 306, 312–316, 322; 1989: 696; Peck et al., 2018).

14 See Allen and Massey (1988), Massey (1995a, 2013).

15 This argument was foreshadowed in Massey's landmark article, 'In what sense a regional problem?' Here she argued that 'The geographical distribution of economic activity which results from the evolution of a new form of division of labour will be overlaid on, *and combined with*, the pattern produced in previous periods by different forms of division of labour. This combination of successive layers will produce effects which themselves vary over space, thus giving rise to a new form and spatial distribution of inequality in the conditions of production, as a basis for the next "round" of investment. "The economy" of any given local area will thus be a complex result of the combination of its succession of roles within the series of wider, national and international, spatial divisions of labour' (Massey, 1979: 234–235).

16 Neil Smith (2008: 206), for example, saw uneven development as 'social inequality blazoned onto the geographical landscape'. On Massey's conception of uneven development, see Peck et al. (2022) and Peck (2023).

17 Harvey (1987: 369, 373), see also Smith (1986). Harvey might well concur with Julie Graham's (1998: 942) later assessment that, in 'hindsight [*Spatial Divisions*] can be seen as a founding text in the emerging tradition of poststructuralist economic geography', albeit while declining to see this as a compliment.

18 For sympathetic critiques, see Warde (1985), Cochrane (1987), and Massey's (1995b) response.

19 Massey was never content to account for complexities and particularities by beating them into compliance on the altar of 'general theory', but neither was she prepared to retreat from theoretical explanation itself, drifting into empiricism and *ad hoc* description. 'The fundamental methodological problem', she once wrote, is how to 'keep a grip on the generality of events, the wider processes lying behind them, without losing sight of the individuality of the form of their occurrence' (Massey, 1984a: 9).

20 On the municipal socialism of the time, see Ward (1983), Boddy and Fudge (1984), Murray (1985), and Cochrane (1988).

21 Rustin (1986: 77), Palmer (1986: 119); see also Cochrane (1986), Newman (1986).

22 On the politics of this tumultuous period, see Campbell (1984), O'Leary (1987), Hall and Jacques (1989), Beckett and Hencke (2009), and Hatherley (2020). On Massey's position and politics, see Massey et al. (1984), Massey and Wainwright (1985), Massey (2007a), Livingstone (2013), Hatherley (2022), and the collection assembled by David Featherstone, one of Massey's former students, and Diarmaid Kelliher (2022a).

23 See Callinicos (1985) and Shock (2020). On *Marxism Today*'s approach to Thatcherite neoliberalism, see Hall and Jacques (1989).

24 It is notable that one of the most compelling treatments of Massey's 'global' understanding of place was also published in *Marxism Today*, rather than an academic journal, as also was her influential work on the political geography of the labor movement (Massey and Miles, 1984; Massey, 1991a).

25 On *Soundings*, see Rustin et al. (2000) and Massey (2015). On the Kilburn Manifesto, see Hall et al. (2015) and Peck et al. (2014).

Study task

Variations on this question can be set on an unseen examination, an open-book examination, or as a term-paper assignment:

Doreen Massey's approach to understanding place is both analytically and politically superior to older understandings of 'the local' in Geography. Discuss.

(Note that, in the United States, the efforts of the late Allan Pred could also make for a good focus, instead of Massey's work).

Starter reading list

Featherstone, D. and Painter, J. (2013) 'Introduction: "There is no point of departure": The Many Trajectories of Doreen Massey.' In D. Featherstone and J. Painter (Eds.) *Spatial politics: Essays for Doreen Massey*. Oxford: Wiley-Blackwell, pp. 1–18.
Hart, G. (2018) 'Becoming a geographer: Massey moments in a spatial education.' In M. Werner, J. Peck, R. Lave, and B. Christophers (Eds), *Doreen Massey: Critical Dialogues*. Newcastle-upon-Tyne: Agenda, pp. 75–89.
Massey, D. (1991) 'A global sense of place', *Marxism Today*, June, pp. 24–29.
Massey, D. (1992) 'A place called home.' *New Formations*, 17(3), pp. 3–15.
Massey, D. (1993a) 'Power-geometry and a progressive sense of place.' In J. Bird, B. Curtis, T. Putnam and L. Tickner (Eds.) *Mapping the Futures*. London: Routledge, pp. 60–70.
Massey, D. (1993b) 'Questions of locality.' *Geography*, 78(2), pp. 142–149.
Meegan, R. (2017) 'Doreen Massey (1944–2016) 'A geographer who really mattered'. *Regional Studies*, 51(9), pp. 1285–1296.
Peck, J., Werner, M., Lave, R., and Christophers, B. (2018) 'Out of place: Doreen Massey, radical geographer'. In M. Werner, J. Peck, R. Lave, and B. Christophers (Eds), *Doreen Massey: Critical Dialogues*. Newcastle-upon-Tyne: Agenda, pp. 1–38.

References

Allen, J. and Massey, D. (eds) (1988) *The economy in question*. London: SAGE.
Allen, J., Massey, D. and Cochrane, A. (1998) *Rethinking the region: Spaces of neo-liberalism*. London: Routledge.
Amin, A., Massey, D. and Thrift, N. (2000) *Cites for the many, not the few*. Bristol: Policy Press.
Barnes, T. J. (2018) 'Her dark past', in Werner, M., Peck, J., Lave, R. and Christophers, B. (eds) *Doreen Massey: Critical dialogues*. Newcastle upon Tyne: Agenda Publishing, pp. 53–64.
Beckett, F. and Hencke, D. (2009) *Marching to the fault line: The miners' strike and the battle for industrial Britain*. London: Constable and Robinson.
Boddy, M. and Fudge, C. (eds) (1984) *Local socialism*. London: Macmillan.
Callard, F. (2011) 'Doreen Massey', in Hubbard, P. and Kitchin, R. (eds) *Key thinkers on space and place*. London: SAGE, pp. 299–306.
Callinicos, A. (1985) 'The politics of *Marxism Today*', *International Socialism*, 2(29), pp. 128–168.
Campbell, B. (1984) *Wigan Pier Revisited*. London: Virago.
Castree, N. (2016) 'A tribute to Doreen Massey (3 January 1944–11 March 2016)', *Progress in Human Geography*, 40(5), pp. 585–592.
Champion, M. (2000) 'Livingstone's economic intentions don't worry majority of critics', *Wall Street Journal*, 25 April, https://www.wsj.com/articles/SB956608791961303399
Christophers, B., Lave, R., Peck J. and Werner, M. (eds) (2018) *The Doreen Massey Reader*. Newcastle upon Tyne: Agenda Publishing.
Clarke, J. (2016) 'Doreen Massey (1944–2016): Making geography matter', *Cultural Studies*, 30(3), pp. 357–361.
Clarke, J. (2018) 'Finding place in the conjuncture: A dialogue with Doreen', in Werner, M., Peck, J., Lave, R. and Christophers, B. (eds) *Doreen Massey: Critical dialogues*. Newcastle upon Tyne: Agenda Publishing, pp. 201–213.

Cochrane, A. (1986) 'What's in a strategy? The London Industrial Strategy and municipal socialism', *Capital and Class*, 28, pp. 187–193.

Cochrane, A. (1987) 'What a difference the place makes: The new structuralism of locality', *Antipode*, 19(3), pp 354–363.

Cochrane, A. (1988) 'In and against the market? The development of socialist economic strategies in Britain, 1981–1986', *Policy and Politics*, 16(3), pp. 159–168.

Daily Telegraph (2016) 'Doreen Massey; geographer who examined how places are "socially constructed"', *Daily Telegraph*, March 21, p. 33.

Dear, M. (2018) 'Trainspotting in Bethlehem', in Werner, M., Peck, J., Lave, R. and Christophers, B. (eds) *Doreen Massey: Critical dialogues*. Newcastle upon Tyne: Agenda Publishing, pp. 65–74.

Featherstone, D. and Kelliher, D. (eds) (2022a) *Doreen Massey: Selected political writings*. London: Lawrence Wishart.

Featherstone, D. and Kelliher, D. (2022b) 'Introduction', in Featherstone, D. and Kelliher, D. (eds) *Doreen Massey: Selected political writings*. London: Lawrence Wishart, pp. 1–18.

Featherstone, D. and Painter, J. (eds) (2013) *Spatial politics: Essays for Doreen Massey*. Oxford: Wiley-Blackwell.

Featherstone, D., Bond, S. and Painter, J. (2013) '"Stories so far": A conversation with Doreen Massey', in Featherstone, D. and Painter, J. (eds) *Spatial politics: Essays for Doreen Massey*. Oxford: Wiley-Blackwell, pp. 253–266.

Freytag, T. and Hoyler, M. (1999) '"I feel as if I've been able to reinvent myself" – a biographical interview with Doreen Massey', in Massey, D., *Power geometries and the politics of space-time*. Heidelberg: Department of Geography, University of Heidelberg, pp. 83–95.

Gertler, M. S., Thomas, M. and Glasmeier, A. (2018) 'In Memoriam: Susan Christopherson (1947–2016)', *Cambridge Journal of Regions, Economy and Society*, 11(1), pp. 211–219.

Graham, J. (1998) 'Spatial divisions of labour: Social structures and the geography of production', *Environment and Planning A*, 30(5), pp. 942–943.

GLC [Greater London Council] (1985) *London Industrial Strategy*. London: GLC.

Hall, S. and Jacques, M. (1989) *New Times*. London: Lawrence and Wishart.

Hall, S. and Massey, D. (2010) 'Interpreting the crisis', *Soundings*, 44, pp. 57–71.

Hall, S., Massey, D. and Rustin, M. (eds) (2015) *After neo-liberalism? The Kilburn manifesto*. London: Lawrence and Wishart.

Hart, G. (2018) 'Becoming a geographer: Massey moments in a spatial education', in Werner, M., Peck, J., Lave, R. and Christophers, B. (eds) *Doreen Massey: Critical dialogues*. Newcastle upon Tyne: Agenda Publishing, pp. 75–88.

Hatherley, O. (2020) *Red metropolis: Socialism and the government of London*. London: Repeater.

Hatherley, O. (2022) 'Doreen Massey's radical 1980s', *Tribune Magazine*, June 3, 2022. https://tribunemag.co.uk/2022/06/doreen-massey-essays-1980s-history-socialism

Harvey, D. (1987) 'Three myths in search of a reality in urban studies', *Environment and Planning D: Society and Space*, 5(4), pp. 367–376.

Harvey, D. (1992) 'Postmodern morality plays', *Antipode*, 24(4), pp. 300–326.

Ijams, B. W., Popke, J. and Urch, K. (1994) 'Gender, space and the academy: An interview with Doreen Massey, The Open University', *disclosure*, 4: article 9. https://uknowledge.uky.edu/disclosure/vol4/iss1/ (Last accessed June 30, 2022).

Lee, R. (2016) 'Obituary: Doreen Massey', *Geographical Journal*, 182(3), pp. 311–312.

Lee, R. (2018) 'Doreen Barbara Massey', *Bibliographic memoirs of Fellows of the British Academy*, 17, pp. 145–178.

Livingstone, K. (2013) 'Foreword', in Featherstone, D. and Painter, J. (eds) *Spatial politics: Essays for Doreen Massey*. Oxford: Wiley-Blackwell, pp. xiv–xviii.

Lovering, J. (1989) 'The restructuring debate', in Peet, R. and Thrift, N. (eds) *New models in geography*, volume one. London: Unwin Hyman, pp. 198–223.

Marriott, J., Campbell, B, Cottam, H., Mouffe, C., Kalandides, A. and the Milton Keynes postgrads (2017) 'Memories of Doreen Massey', *Soundings*, 65, pp. 114–124.

Martin, R., Markusen, A. and Massey, D. (1993) 'Classics in human geography revisited: Massey, D. 1984: Spatial divisions of labour', *Progress in Human Geography*, 17(1), pp. 69–72.

Massey, D. (1973) 'Towards a critique of industrial location theory', *Antipode*, 5(3), pp. 33–39.

Massey, D. (1979) 'In what sense a regional problem?' *Regional Studies*, 13(2), pp. 233–243.

Massey, D. (1983) 'The contours of victory … dimensions of defeat', *Marxism Today*, June, pp. 16–19.

Massey, D. (1984a) 'Introduction: Geography matters', in Massey, D. and Allen, J. (eds) *Geography matters: A reader*. Cambridge: Cambridge University Press, pp. 1–11.
Massey, D. (1984b) *Spatial divisions of labour: Social structures and the geography of production*. Basingstoke: Macmillan.
Massey, D. (1985) 'New directions in space', in Gregory, D. and Urry, J. (eds) *Social relations and spatial structures*. London: Macmillan, pp. 9–19.
Massey, D. (1987a) 'Equal opportunities: The GLEB experience', in Cochrane, A. (ed.) *Developing local economic strategies*. Milton Keynes: Open University Press, pp. 23–34.
Massey, D. (1987b) 'Thoughts on a train going North', *Area*, 19(3), pp. 268–269.
Massey, D. (1989) 'Reflections on the debate: Thoughts on feminism, Marxism and theory', *Environment and Planning A*, 21(5), pp. 692–697.
Massey, D. (1991a) 'A global sense of place', *Marxism Today*, June, pp. 24–29.
Massey, D. (1991b) 'Flexible sexism', *Environment and Planning D: Society and Space* 9(1), pp. 31–57.
Massey, D. (1993c) 'The different sides of the sixties', *Environment and Planning A*, 25, 1_Suppl, pp. 10–13.
Massey, D. (1994) *Space, place and gender*. Cambridge: Polity Press.
Massey, D. (1995a) 'Masculinity, dualisms and high technology', *Transactions of the Institute of British Geographers*, 20(4), 487–499.
Massey, D. (1995b) 'Reflections on debates over a decade', in Massey, D., *Spatial divisions of labour: Social structures and the geography of production*, 2nd edition. Basingstoke: Macmillan, pp. 296–354.
Massey, D. (2000) 'Understanding cities', *City*, 4(1), pp. 135–144.
Massey, D. (2001a) 'Geography on the agenda', *Progress in Human Geography*, 25(1), pp. 5–17.
Massey, D. (2001b) 'Living in Wythenshawe', in Borden, I., Kerr, J. and Rendell, J. with Pivaro, A. (eds) *The unknown city: Contesting architecture and social space*. Cambridge, MA: MIT Press, pp. 458–475.
Massey, D. (2005) *For space*. London: SAGE.
Massey, D. (2007a) 'The world we're in: An interview with Ken Livingstone', *Soundings*, 36, pp. 11–25.
Massey, D. (2007b) *World city*. Cambridge: Polity.
Massey, D. (2008) 'When theory meets politics', *Antipode*, 40(3) pp. 494–497.
Massey, D. (2009) 'Space, place and politics', *OpenLearn*, 7 August 2009. https://www.open.edu/open.earn/society-politics-law/geography/doreen-massey-space-place-and-politics
Massey, D. (2011) 'A counterhegemonic relationality of place', in McCann, E. and Ward, K. (eds) *Mobile urbanism: Cities and policymaking in the global age*. Minneapolis: University of Minnesota Press, pp. 1–14.
Massey, D. (2013) 'Vocabularies of the economy', *Soundings*, 54, pp. 9–22.
Massey, D. (2015) 'Exhilarating times', *Soundings*, 61, pp. 4–16.
Massey, D. and Allen, J. (eds) (1984) *Geography matters: A reader*. Cambridge: Cambridge University Press.
Massey, D. and Catalano, A. (1978) *Capital and land: Landownership by capital in Great Britain*. London: Edward Arnold.
Massey, D., Human Geography Research Group, Bond, S. and Featherstone, D. (2009) 'The possibilities of a politics of place beyond place? A conversation with Doreen Massey', *Scottish Geographical Journal*, 125(3–4), pp. 401–420.
Massey, D. and Meegan, R. A. (1978) 'Industrial restructuring versus the cities', *Urban Studies*, 15(3), pp. 273–288.
Massey, D. and Meegan, R. (1982) *The anatomy of job loss: The how, why and where of employment decline*. London: Methuen.
Massey, D. and Miles, N. (1984) 'Mapping out the unions', *Marxism Today*, 28(5), pp. 19–23.
Massey, D., Segal, L. and Wainwright, H. (1984) 'And now for the good news', in Curran, J. (ed.) *Future of the left*, pp. 211–227.
Massey, D. and Wainwright, H. (1985) 'Beyond the coalfields: The work of the Miners' Support Groups', in Beynon, H. (ed.) *Digging deeper: Issues in the miners' strike*. London: Verso, pp. 149–168.
McDowell, L. (2018) 'North and south: Spatial divisions in a life lived geographically', in Werner, M., Peck, J., Lave, R. and Christophers, B. (eds) *Doreen Massey: Critical dialogues*. Newcastle upon Tyne: Agenda Publishing, pp. 41–52.

McLoughlin, J. B. (1983) 'Planning research and education in Australia', *Urban Policy and Research*, 1(4), pp. 16–21.
Meegan, R. (2017) 'Doreen Massey (1944–2016): A geographer who really mattered', *Regional Studies*, 51(9), pp. 1285–1296.
Murray, R. (1985) 'Benetton Britain: The new economic order', *Marxism Today*, November, pp. 28–32.
Newman, I. (1986) 'Greater London Enterprise Board: Vision and reality', *Local Economy*, 1(2), pp. 57–67.
O'Leary, B. (1987) 'Why was the GLC abolished?' *International Journal of Urban and Regional Research*, 11(2), pp. 193–217.
Palmer, J. (1986) 'Municipal enterprise and popular planning', *New Left Review*, I(159), pp. 117–124.
Peck, J. (2022) 'Hard work: Restructuring, realism, and regions', in Calder, G. and Sanghera, B. (eds) *Ethics, economy and social science: Dialogues with Andrew Sayer*. London: Routledge, pp. 139–152.
Peck, J. (2023) *Variegated economies*. Oxford: Oxford University Press.
Peck, J. and Barnes, T. J. (2019) 'Berkeley in-between: Radicalizing economic geography', in Barnes, T. J. and Sheppard, E. (eds) *Spatial histories of radical geography: North America and beyond*. Oxford: Wiley-Blackwell, pp. 211–246.
Peck, J., Werner, M. and Jones, M. (2022) 'A dialogue on uneven development: a distinctly regional problem', *Regional Studies*, 57(7), pp. 1392–1403.
Peck, J. and Ebner, N. (2019) 'Doreen Massey', in Atkinson, P. A., Delamont, S. and Williams, M. (eds) *The Sage encyclopedia of research methods*. London: SAGE, pp. 1–8.
Peck, J., Massey, D., Gibson, K. and Lawson, V. (2014) 'Symposium – The Kilburn Manifesto: After neoliberalism?' *Environment and Planning A*, 46(9), pp. 2033–2049.
Peck, J., Werner, M., Lave, R. and Christophers, B. (2018) 'Out of place: Doreen Massey, radical geographer', in Werner, M., Peck, J., Lave, R. and Christophers, B. (eds) *Doreen Massey: Critical dialogues*. Newcastle upon Tyne: Agenda Publishing, pp. 1–38.
Posner, M. (2002) 'Social sciences under attack in the UK (1981–1983)', *La revue pour l'histoire du CNRS*, 7.
Rustin, M. (1986) 'Lessons of the London Industrial Strategy', *New Left Review*, I(155), pp. 75–84.
Rustin, M., Davison, S. and Gilbert, J. (2000) 'The New Left and its legacies', *Soundings*, 74, pp. 136–163.
Sayer, A. (1982) 'Explanation in economic geography: Abstraction versus generalization', *Progress in Human Geography*, 6(1), pp. 68–88.
Sayer, A. (1984) *Method in Social Science*. London: Hutchinson.
Shock, M. (2020) *Renewing left-wing ideas in late Twentieth-Century Britain: Marxism Today, c. 1977–1994*, DPhil thesis, Oxford University.
Smith, N. (1986) 'Spatial divisions of labour: social structures and the geography of production', *Geographical Review*, 76(3), pp. 350–352.
Smith, N. (2008) *Uneven Development*, 3rd edition. Athens: University of Georgia Press.
Storper, M. (1986) 'Spatial divisions of labour: Social structures and the geography of production', *Progress in Human Geography*, 10(3), pp. 313–468.
Urry, J. (1986) Making space for space. *Economy and Society*, 15(2), pp. 273–280.
Wainwright, H. (2018) 'Epilogue: "How we will miss that chuckle": My friend, Doreen Massey', in Werner, M., Peck, J., Lave, R. and Christophers, B. (eds) *Doreen Massey: Critical dialogues*. Newcastle upon Tyne: Agenda Publishing, pp. 367–370.
Ward, M. (1983) 'Labour's capital gains: The GLC experience', *Marxism Today*, December, pp. 24–29.
Warde, A. (1985) 'Spatial change, politics and the division of labour', in Gregory, D. and Urry, J. (eds) *Social relations and spatial structures*. London: Macmillan, pp. 190–212.
Werner, M. (2024) 'Doreen Massey', in Gilmartin, M., Hubbard, P., Kitchin, R. and Roberts, S. (eds) *Key thinkers on space and place*, 2nd edition. pp. 245–252. London: SAGE.
Werner, M., Peck, J., Lave, R. and Christophers, B. (eds) (2018) *Doreen Massey: Critical dialogues*. Newcastle upon Tyne: Agenda Publishing.

10 Geographies of meaning and experience

Anne Buttimer's lifeworld

Federico Ferretti

Anne Buttimer (1938–2017). Archives of University College Dublin, School of Geography.

Introduction: what is Humanistic Geography?

The French village of Saint-Dié-des-Vosges is now a picturesque tourist destination in the woody Vosges mountain chain laying on the frontier between France and Germany. It was where in 1507 the planisphere of Martin Waldseemüller (1470–1520) was published, the first map that included the name "America" indicating the existence of the New World. From then on the name America was universally used. Today, Saint-Dié-des-Vosges hosts an annual global gathering at the end of September called the International Festival of Geography,[1] where thousands of Geography students and teachers enjoy a rich programme of conferences and events (Figure 10.1). One of the culminating moments of the Festival is the conferring of an international prize named after Waldseemüller's mentor, the geographer Vautrin Lud (1448–1527). It is an international award given to a

DOI: 10.4324/9781003343240-12

Figure 10.1 Banner of the International Festival of Geography.
Photo by the author, 2022.

geographer whose career has been deemed outstanding. Informally, the Lud Prize is considered "the Nobel Prize for Geography" (Alcoforado and Jones 2015).

One evening in September 2014, the Lud Prize was awarded to an woman. Afterwards she gave an interview. Speaking perfect French with a hint of an Irish accent, she said her interest in continental Europe, and especially French-speaking countries, began in the 1960s when she was a student at the University of Washington, Seattle. While there she had travelled to France to study social geography.[2] That student, who became the elderly woman feted in 2014 at Saint-Dié-des-Vosges, was Anne Buttimer (1938–2017). By then she was a leading international geographer, with a long career in the discipline, and best known for her work in, and promotion of, Humanistic Geography. Especially important during the 1970s and 1980s (although still inspiring contemporary geographers), Humanistic Geography is concerned with making the rich texture and meanings of subjective human experience the core of Human Geography.

British geographer Tim Cresswell provides an example of the humanistic approach. He says, imagine that you fancy some fish and chips and you are wondering where to buy it. According to Cresswell, your choice of the fish and chip shop reveals a lot about the importance of Humanistic Geography. Strict 'rationality' suggests that you go to the nearest or cheapest fish and chip place to avoid wasting time and money for an experience that is likely the same across different places. A humanistic approach, however, would say we are not driven by only rationality. Humans are complex beings. Our motivations are multifarious. Rather than going to the closest "chippy", we walk on because we remember that the chips are tastier at a place farther away. Or, we walk on because we want to be served by the cute boy or girl who works behind the counter at a more distant fish and chip shop. Indeed, there are numerous reasons why we might not go to the closest place. As Cresswell concludes: "Humans are complicated. We're supposed to be marked out from other forms of life by our rationality but we also have hopes, dreams, and desires" (Cresswell 2013, 103).

Economic rationality cannot explain everything about our behaviour. Instead, Humanistic Geography places human subjectivity at the centre of geographical research. Our relationships to places are complex. We are not "materialistically motivated robots" as Buttimer (1993, 47) famously argued. The objectivity of statistics and spatial measurements is insufficient to understand the geography of human behaviour, which is often unpredictable because we are irrational (or not completely rational) beings. For the same reason, humanistic geographers distance themselves from positivism, the philosophical doctrine that believes in the progress of scientific, technical and "objective" knowledge and hence increasing social betterment and well-being. For positivists, this occurs because societal behaviours are predictable, conforming to general laws of the kind that exist, for example, in physics or chemistry.

Instead of a naïve belief in linear progress, objectivity and the universal laws that positivism upholds, humanistic geographers consider subjective and unpredictable humans as the proper research object of Geography. This means using qualitative research methods such as interviews and direct observation to investigate, for example, people's wishes, imaginations and real lives.

Why Buttimer's *Values* matters

The definition of Humanistic Geography was first given by Yi-Fu Tuan (1930–2022): "the focus of Humanistic Geography is on people and their condition ... by way of human experience, awareness and knowledge. Humanistic Geography contributes to science by drawing attention to facts hitherto beyond the scientific purview" (Tuan 1976, 266). Humanistic Geography quickly became synonymous with the use of qualitative methods and took seriously people's feelings and experiences as objects of inquiry. Additionally, authors including Buttimer in her *Values in Geography* (1974), strongly contended that Humanistic Geography should be socially relevant and not just theoretically interesting.

In the 1970s, Geography was broadly divided into two fields. The first, which dominated the discipline in the 1960s and was later criticised for its links with positivism, was so-called "quantitative geography" or "spatial science" (see Chapter 4). Quantitative approaches gave geography "scientific" recognition by focusing on statistics, data, mappings and later Geographical Information Systems (GIS) (see Chapter 7). The second was the then rising field of radical geography which resumed forgotten traditions of politically engaged geography by applying ideas such as Marxism and anarchism to geographical inquiry (see Chapter 8). Most radical geographers distanced themselves from quantitative geography because of its pretentions of "neutrality" and "objectivity", focusing instead on the study of social and spatial inequalities that was often linked to social movements in the years that followed the revolts of 1968 (Barnes 2018).

In those years, Buttimer worked at Clark University, a key place for radical geography. It was that movement's earliest base, home to the group who founded the journal *Antipode* (Barnes and Sheppard 2019). Although sympathetic, Buttimer was a Catholic and did not embrace Marxism, rendering her a nonconformist figure in that milieu. Yet, her scholarly talent and strong personality gave her entry, allowing her to engage in public discussions with international leaders of radical geography around her notion of values. *Values in Geography* was published with comments from an array of leading figures in Anglophone Geography, including Edward Soja, Edward Gibson, Yi-Fu Tuan, Jim Blaut and Torsten Hägerstrand. *Values* remains relevant to understanding Buttimer's ideas of engaged geography (it is freely available at the UCD Research Repository).[3]

With Humanistic Geography focused on life and experience, Buttimer brought her own life trajectory to the discussion, declaring what would be called today her "positionality". It was a strategy to make her North American colleagues and students understand "the values conflicts and dilemmas faced by a native Irish girl ... who has become involved in social contexts as varied as Cork, Louvain, Glasgow, Seattle, Paris, Lund and Worcester" (Buttimer 1974, 2). Values were fundamental because for Buttimer being a geographer was not a job, but something like a vocation, an idea that she expressed through the notion of "care". Caring for others was one of the reasons she wanted a geography that was attentive to specific people and living societies. Geography was "not a job to be done; it is one of the regions of my care", Buttimer (1974, 4) wrote.

Although her values were not identical to those of her radical interlocutors, they still referred to social justice and the various forms of impoverishment besetting working classes and marginalised communities. While Buttimer deemed Marxism theoretically incompatible with humanistic views because the former's abstractions erased "real" people, she was not opposed to broad ideas of socialism. She also admired late nineteenth- and early twentieth-century anarchist geographers Peter Kropotkin and Elisée Reclus (Figure 10.2), famous forerunners of later critical and radical geographers (Ferretti 2018). Like them, and like Brazilian educator Paulo Freire whom she equally admired, Buttimer recognised the great importance of education in raising awareness of social justice

Figure 10.2 Example of Anne Buttimer's interest in radical geographical traditions: a folder on anarchist geographer Elisée Reclus from her personal archive.

Photo by the author, 2018.

(Kropotkin 1885). For Buttimer, progressive social changes came through "a change in the consciousness of people" (Buttimer 1974, 31). Even without explicitly embracing radicalism, her social engagement was clear from her ethics.

While Buttimer styled herself an anti-colonialist and anti-capitalist until her last days (Jones and Mennell 2017), in 1974 her critique took the form of rebelling against any institution to which she belonged. Buttimer (1974, 3) wrote: "The only way I can overcome my aversion to the hypocrisy I find in some of these structures – academic, ecclesiastical or national – is by realising that I belong to them only because of a caring commitment to certain persons whom I cherish ... Because I value truth and love, I can, for the present at least, overlook the institutional trappings of each" (Buttimer 1974, 3). Indeed, ideas of caring, loving and cherishing are all aspects of a humanistic approach to Geography, a Humanistic Geography that gives full citizenship to sensations and feelings in explaining human behaviours and approaches to space.

In seeking dialogues between humanistic and radical approaches, Buttimer went beyond the positions of other leading humanistic geographers such as Tuan, David Ley, Edward Relph and others, who were mainly concerned with theoretical matters, specifically, restating the centrality of the human as an alternative to quantitative methods that only captured a "thin" view of human action (Ley and Samuels 1978; Relph 1976). For Buttimer, geography should not only be limited to pure theory but also take an ethical stance and make a social commitment. Humanistic Geography should not only be "human" but also "humane" (Ferretti 2018). It should be an instrument for individual and collective emancipation, challenging dogmatisms and authoritarian powers. As Buttimer noted: "At times when Academy, Church, State, Syndicate, or Proletariat has tried to exercise monopoly power over thought and/or life, a humanist protest has appeared ... an emancipatory cry, one seeking freedom from oppression" (Buttimer 1990, 2).

Some of *Values*' commentators, such as Blaut, criticised the lack of a clear ideology in Buttimer's position. Yet, the very fact she was engaged in scholarly discussion showed that Buttimer, then 36 years old, was gaining attention and being taken seriously by leading geographers, some of whom, like Tuan and Hägerstrand, would later become her friends and collaborators. Certainly part of a scholarly humanistic tradition, Buttimer stressed the importance of holding on to values that inspired social and political engagement.

After *Values*, Buttimer continued to seek dialogue with other geographers. That is evident in her co-edited book with David Seamon, *The human experience of space and place*, whose contributors were a group of Clark scholars and students (Buttimer and Seamon 1980). In that volume Buttimer discussed her fieldwork experience in working-class neighbourhoods of European cities. She argued that planners must consider individual dwellers and their real needs, a quite revolutionary claim at that time. To do that, she contended, meant taking very seriously both how people experienced worlds and the "brute facts" of their material world. This was to be done not only in theory but also in daily life, as exemplified by the notion of lifeworld discussed in the next section.

Place, journey and the dynamism of lifeworld

Lifeworld is a broad concept used in Humanistic Geography to represent the relationship between individuals and their milieus. In her 1976 article, "Grasping the dynamism of lifeworld", Buttimer discussed the origins of the idea of lifeworld, which are in phenomenology. Inspired by philosophers such as Martin Heidegger and Maurice Merleau-Ponty, phenomenology argued for the centrality of the human subject in knowing the world. For phenomenologists, things don't exist in themselves, but only as ideas that we have of them that are part of our knowledge. Yet, Buttimer didn't accept all of phenomenology's principles. She especially argued that it's not only our perception of what happens that counts, but also our constant and relational interaction with the world's elements.

Buttimer defined the "dynamism of lifeworld" as the ceaseless intersubjective relations existing between humans, and between them and their environments. The awareness of these relations gave meaning to our lives. Our lifeworld is "the culturally defined spatio-temporal setting or horizon of everyday life". Conversely, "scientific procedures which separate 'subjects' and 'objects', thought and action, people and environments" block us from grasping the lifeworld, "are inadequate to investigat[ing]" it (Buttimer 1976, 277). Buttimer's idea that humans are both subjects and objects of their own knowledge has important implications for geographical theory. It implies acknowledgement that humans are not all-powerful entities dominating nature as it has been assumed for centuries in Western thought. Instead of assuming, as our "Western intellectual heritage", that there are "barriers ... placed between mind and being", we need to acknowledge their open relationship and "to develop a new respect for space, time, and nature" (Buttimer 1976, 292). As individuals, we are deeply conditioned by and fully inserted in the world, our experience of it ceaselessly modifying our thoughts and behaviours although never annihilating our will for self-determination.

On the one hand, for Buttimer, geographers need to listen to "the liberation plea of lived experience, appealing for more concrete descriptions of space and time, and their meanings in everyday human living". When they do, "intellectual knowledge [will be brought] into closer harmony with lived experience" (Buttimer 1976, 278). On the other, Buttimer did not recognise the absolute centrality of human subjectivity given "the direct [physical] relationships between the human body and its world" (Buttimer 1976, 279). As noted above, subjects are not purely rational beings or immaterial intellectual entities: our bodily experience is daily shaped by our concrete relations with the world. Among all disciplines, Geography is well placed to understand this relationality. As Buttimer (1976, 280) writes, "geographers are aware of the active role of milieu-physical and cultural-in shaping experience" (Buttimer 1976, 280). Grasping the meaning of experience and perception in their relation with the material world becomes a key task for geographers.

Humanistic Geography also deals with concepts such as "dwelling" and "being in the world", associated with the idea of attachment and affection to the native soil. This focus was later criticised for being "static" and "conservative", that places should not be "bounded, and exclusionary" (Seamon and Lundberg 2017, 9). But Buttimer's originality was that she provided a dynamic notion of "dwelling" in which there was a reciprocal and necessary relationship between "home and reach, place and journey" (Buttimer 1980, 16). That is, humans are mobile, and the lifeworld cannot be reduced to a fixed feeling of "home". Reach is the complement to home: we are born somewhere, but we generally aim at going somewhere else to realise our professional or personal ambitions. On that, Buttimer again used examples taken from her own life trajectory. For instance, she identified her family's farm in County Cork as "home", but she engaged in "reach" through her numerous travels, moves and the different houses in which she lived in North America, Scotland, Sweden, Belgium and Ireland (Buttimer 1980, 2016).

Indeed, Buttimer considered human mobility a key feature of a lifeworld. Just as our experience of space is "dynamic" (Buttimer 1976, 285), so is "human subjectivity", that is, understanding the "world as living event" (Buttimer 1976, 287). Because the complex human relationships with places are not static or conservative, we need to grasp the relationalities existing between place, life, movement and experience considering that, as humans, we are mobile (see Chapter 15). Given that the key research objects for Humanistic Geography are humans, Buttimer starts by analysing her own life experiences and concrete lifeworld, the subject of the next section.

Buttimer's own lifeworld and experience

There's a long tradition of Irish peasants and working-class people emigrating to America to seek a better future. But who would have predicted that the daughter of a County Cork farmer, who as a young woman emigrated to Seattle after taking her vows as a nun, would come back to Ireland as an international leading figure in Geography. She was a Full Professor at University College Dublin (UCD) and the first female Chair of the International Geographical Union (IGU) from 2000 to 2004. Buttimer directly recounted her story in several autobiographical writings, recollections and interviews (Buttimer 1979, 1983, 2016; Maddrell 2009). These texts are important as an example of how humanistic geographers such as Buttimer understand geography from an analysis and self-reflection of their own life experience.

In her earliest interview, carried out within the International Dialogue Project (discussed below), Buttimer connected her early "vocation" as a geographer to the fact that she was the daughter of an Irish farmer. That start, she said, influenced "a lot of what I believe, feel and dream also. My biggest dreams come when I am walking barefoot on the grass ... taking care of the home" (Buttimer 1979, 3). Buttimer stressed that what marked her childhood was not carelessness, but the opposite, care (Buttimer 1974). In her recollections, geography came first through experiences in the open air and personal attachment to land that date back to her childhood.

She noted that "we had a large family and weren't very rich, and my parents depended on me to keep the house in order and to have meals on the table" (Buttimer 1979, 4). Caring for the house also led Buttimer to learn about agriculture and soils, inspiring her further interests in Geography. She told of her childhood in very geographical terms: "I was always outside milking cows, feeding calves and helping Dad because my older sister was in the house... I learned about rotation of crops, for example, by seeing how things happened in the fields. I saw erosion, I also saw reclamation. When I was seven we moved to this very big farm where two hundred acres of it had been allowed to become almost 'mountain' land, Dad rotated and reclaimed that two hundred acres himself. ... I saw all that happening. So, agriculture ... was of interest even before I learned physical geography" (Buttimer 1979, 5). For Buttimer, geography understood as learning through experience preceded rationality and was located in early childhood. As I discuss below, this kind of "storytelling" is a key practice in Humanistic Geography, showing how peoples' concrete lives can be considered as a terrain for geographical fieldwork.

Buttimer's later vocation as a scholarly geographer was the decisively shaped by her numerous travel experiences as part of her "reach". Two are particularly important. The first was her move from Cork (Ireland) to Seattle (USA) in the early 1960s, where she obtained her PhD in geography. The second was in the opposite direction, an invitation that Buttimer received in 1964 to go to Belgium for a postdoctoral fellowship. On that occasion, Buttimer also spent some months in France, where she undertook a quite adventurous "Tour de France". She spoke directly to French geographers of various stripes, which she wrote about vividly in her travel diaries (Ferretti 2020a). These lived experiences show how meaning and experience are key to Buttimer's production of knowledge, and how her main contributions to the field of Geography were associated with different phases of her life.

Her early discovery of social geography was associated with her first travels in France, and then reinforced by her postdoctoral research on working-class neighbourhoods in Glasgow (Buttimer 1969). The study of lifeworld and humanistic "values" characterised Buttimer's work at Clark University during the subsequent decade. After 1976, her work in Lund (Sweden) in the Dialogue Project further fostered Humanistic Geography through the analysis of hundreds of scholars' autobiographies. They showed the links between ideas and lived lives (Ferretti 2019; Jones 2019). In 1976, in Sweden, Buttimer relinquished her vows as a Dominican nun, arguably for personal reasons. In 1991, her return to Dublin corresponded to the reach of acknowledged international academic leadership as a world-leading scholar in the fields of Humanistic and Historical Geography. It can be argued that her nonconformist trajectory as an outsider, crossing different countries and intellectual traditions, explains much of her originality and innovation within the Anglo-American academy. Her varied experiences allowed her to bring ideas from one place to another, and then to consolidate her own ideas which she defended tenaciously all her life long.

Buttimer's most important scholarly endeavour, the International Dialogue Project, began in the 1970s. Her starting point was that sociological and geographical theories "are not usually validated through autobiographical or personal accounts" (Buttimer 1976, 285). To fill that lacuna, she argued that life stories should become a source for making sense of experience, that is, producing "meaning". As such, Buttimer conceived autobiography and storytelling as a means to humanise geography by understanding the lifeworlds of geographers. The International Dialogue Project was developed by Buttimer in Lund (Sweden) from 1977 to 1989 in collaboration with Torsten Hägerstrand (1916–2004), ironically a former promoter of quantitative geography (Buttimer 2007).

The practice of geography: storytelling as humanistic geographical practice

One of the main accomplishments of Buttimer's scholarly work, which grounds some of her principal books, such as *The Practice of Geography* (1983) and *Geography and the Human Spirit* (1993), was the dynamics between "home" understood as intimate identity and "reach" understood as travel, discovery and life experience. As she argued in her book *Geography and the Human Spirit*, on the one hand: "*Humanus* literally means earth dweller". On the other, dwelling was constantly related to "journeys and identity, storytelling and interpretation, individual experience and social context" (Buttimer 1993, 3).

For Buttimer, it is through storytelling that meaning emerges from everybody's life trajectory. The Dialogue Project also included Geography's societal relevance, as many of the interviewees were involved in planning, policy elaboration, social/political engagement and "applied" geography in general. As Buttimer and Hägerstrand summarised in their *Invitation to Dialogue*, the Dialogue Project wished to create international and interdisciplinary conversations between geographers and "economists, demographers, technologists, and politicians ... philosophers, sociologists, and psychologists" (Buttimer and Hägerstrand 1980, 4). Collective discussions and individual interviews especially focused on personal experiences and values.

Quite unique in the panorama of social sciences at that time, the Dialogue interviews suggested that there were many possible associations between life and scholarship, between Geography and experience. Key were the ways in which the interviews were conducted. The focus was on personal experience and nobody was requested to represent any specific "school" (Buttimer 1983, 10). Encouraged to speak about themselves, several interviewees provided experiences in which their geographical "vocation" often came through early life experiences inspiring scholarly ideas that they would develop later. As in Buttimer's interview, several participants recounted impressions, travels and walks that took place in their childhood or youth. Together with the interviewees' relations with family figures (typically maternal or paternal), these experiences were associated with geography even before their protagonists knew what scholarly geography looked like.

Some telling recollections are contained in the interview that British geographer Bill Mead gave to Hägerstrand. Mead explicitly said his early geographical inspiration was before he went to school. As he put it, deep geographical sensitivity grows "out of very remote experiences in one's childhood. I certainly think back to all sorts of experiences which happened before the age of six or seven ... one of my earliest recollections is going for walks in the countryside and having every tree and flower, bird and animal pointed out to me and named by my father ... given, as it is natural in the English countryside, Anglo-Saxon names – and I imagine I'm Anglo-Saxon by remote ancestry. Rocks and soils were also brought to my attention early, because I was brought up in a little bit of the English countryside... on the watershed between the Thames drainage and the Ooze drainage ... above all I remember the fossils" (Mead 1979, 3). These picturesque details of Mead's early life experience remained, shaping his later adult experiences as a geographer such as geographical fieldwork.

In certain cases, childhood memories come from intimate and/or idealised relationships with key family figures. Very telling was the interview of North American geographer Fred Kniffen, who reflected on the different roles that his parents played in his early geographical training: while his mother stimulated his interest in their family origins, his father, a lumberman, introduced him "to the traditions of the forest" (Kniffen 1982, 3). In the case of urban geographer Peter Nash, it was the retrospectively reconstructed experience of living his earliest four years on the Main River that he defined as inspirational. Visiting that house many years later, Nash finally got an idea of "what really went on in [his] mind, and what might have planted the seed" (Nash 1984, 3). For Nash, geographical curiosity originated in the balcony of his earliest house on the Main River, by "seeing all types of boats coming and going down the Main and thinking to myself 'where are these boats going?' People were rather imprecise about their answers. 'Oh, they are going to the Ocean', they said. 'How far away is the ocean? Can we go there?' And they said, 'Oh no, the ocean is very far away'. Even at that early age I asked my father to show me a map and I remember he had a big globe in his study too. And he showed me that not far from the river, the Main flows into the Rhein river at Mainz. Later on, on some of our bicycle tours, we went down to see the Main flow into the Rhein ... and then I found out that the river still had to go through Holland to get to its estuary" (Nash 1984, 4).

Also, the deep origins of several geographers' social and political engagement emerged from the Dialogue interviews. Norwegian geographer Aadel Brun-Tschudi, the first woman to obtain a chair of Geography in Norway, discussed at length the importance of being born the daughter of a Protestant missionary who settled in China. Her childhood was completely bicultural and bilingual (Brun-Tschudi 1979). The fact of "belonging to two worlds" had an impact on her subsequent career, as it opened her eyes "to looking for individual people instead of a brash and basically glib bunching together of whole nations" (Buttimer 1983, 37), something that clearly matched some of Buttimer's own "values".

Similarly, the worldviews of Finnish economic geographer, Ilmari Hustic, were strongly informed by his social origins. Born in the working-class slums of Helsinki, he acquired early class consciousness when he had to spend time in a "home for children" of poor people while his mother "slaved in a factory. A background of this kind strongly influences personality" (Buttimer 1983, 103). All this awakened Hustic's life-long interest in politics and social justice. Hustic was also a nonconformist and prickly scholar who refused nationalism, adopting a pacifist and internationalist vision to challenge what he defined as the "prostitution of geopolitics in the service of the power" (Buttimer 1983, 107). On a similar note, at a round table that Buttimer organised with French geographers, anti-colonialist geographer Jean Dresch recollected: "What attracted me to Geography, as well as to History, was a certain taste for political, economic and social problems" (Buttimer 1983, 123). The investigation of scholars' lifeworlds and values through storytelling therefore proved effective not only in placing human life at the centre of Geography's interests, but also in understanding geographers' social and political commitment.

In addition to the printed transcripts of the Dialogue Project conversations, almost 100 of these interviews are available as videos on the IGU YouTube channel.[4] These videos are not only a source for future researchers in the history of Geography: they are also a resource for students at all stages interested in learning about biography, Humanistic Geographies, history and philosophy of Geography, and much more. Containing relatively short dialogues focusing on how life experience shaped professional and academic trajectories, they are arguably an easier and more stimulating means to first approach such matters than most textbooks. At UCD, taught modules such as "Ideas in Geography" included exercises on these videos, which proved very useful for teaching and research at undergraduate level.

Life trajectories and the history of geography: why Humanistic Geography still matters

Although only a few examples can be provided here given the richness of the field, Buttimer's intellectual legacy contributes to the multiple revival of humanistic thought that is currently ongoing in academic Geography. This is first occurring in History of Geography, a field to which Buttimer was personally committed and in which her ideas continue to circulate. It is possible to argue that one of the outcomes of the International Dialogue Project has been to change how we think about the history of our discipline.

Important to this end was the collaboration between Buttimer and Thomas Walter Freeman (1908–1988), a Dialogue Project interviewee and a cofounder of the annual collection, *Geographers, Biobibliographical Studies (GBS)* (Buttimer 2003), which still publishes biographical chapters on geographers. Formerly, the history of geography was presented by most textbooks as a linear succession of disembodied ideas, mainly focusing on "national schools", "paradigms" or celebratory discussions about the thoughts of some "eminent" figures. As the Dialogue interviews showed, though, scholarly lifeworlds of geographers were vital to understanding the history of the discipline. That was true for everybody in Geography, not just those who were well known. Indeed, Buttimer and Freeman strongly contended that the biographies of past geographers should not be limited to the lives of "great" figures, which generally meant straight, white, middle-class academic men.

The same purposes still inspire the *GBS* editorial philosophy, recently defined by its current editors as "not rounding up the usual suspects" (Baigent and Novaes 2020, 2022). This means paying special attention to trajectories of geographers that were neglected in classical disciplinary histories, such as women or scholars from the Global South. More broadly, Buttimer's and Freeman's focus on individuals' concrete life experiences were key in showing the importance of analysing scientific ideas in their places and contexts, a well-established methodology among today's historians of geography (Livingstone 1993; Keighren 2020). That is, historical and local contexts influence the making of Geography, alongside the material circumstances of the concrete individuals who produced these ideas (Ferretti 2020b). Buttimer's work continues to inspire geographers also through her material legacy. Her contribution has become an object of study after she bequeathed her complete archives and library to UCD School of Geography. These materials are nurturing current scholarship by allowing the investigation of Buttimer's work and international scholarly networks (Ferretti and Jones 2018; Ferretti 2019, 2020a).

Another aspect of current histories of geography in which one can perceive Buttimer's influence is in investigating transnational networks based on multilingual sources such as correspondences. These works are inspiring a "multilingual turn" in which scholars studying the history of geography challenge Eurocentric or Anglocentric views by investigating archives of early geographers belonging to other linguistical and cultural areas including the Global South (Ferretti 2022). Coming from a peripheral place such as County Cork, Buttimer was the personification of this kind of multilingual and cosmopolitan scholarly figure, as she used several languages in her scholarship, including English, French, Swedish, Finnish, German and Italian for both the intellectual curiosity of reading texts from different sources and the need to participate in international networks (Ferretti and Jones 2018). In sum, all life experiences matter for grasping the dynamism of lifeworld, and remain an inspiration for several geographers.

Furthermore, the intellectual legacy of Humanistic Geography remained alive in several fields of geographical research even when this label became less fashionable in the 1990s. For instance, it is widely agreed that humanistic ideas about perception and human feelings had an influence in what is called today cultural geography (see Chapter 11), and more recently in fields such as geopoetics (Magrane et al. 2020) and geohumanities (Cresswell et al. 2015). According to Cresswell, Humanistic Geography has been the inspiration for much of the methods that are key today for Human Geography, namely, the "range of qualitative methods including participant observation, textual and visual analysis, and self-reflection" (Cresswell 2013, 119). Even in our world increasingly ridden by technologies and in which anxieties over climate change shatter all former humankind's certitudes including those of our survival on the planet, more and more works call for rediscovering humanistic approaches that address both the roles of material worlds and the continuing relevance of human agency and freedom (Simonsen 2013; Ferretti 2020a; Seamon 2023).

Finally, Buttimer's life and works show how humanistic geography contributed to renovating the field of Geography between the 1970s and the 1980s. Their contributions are behind current claims to consider Geography as a critical discipline with intellectual ambitions beyond the technicalities requested by the neoliberal market for abiding by the dogmas of economic "viability" and students' "employability". On these matters, Buttimer stressed the need for ethics in teaching and research, something that is increasingly relevant in our society dominated by market values—quite the opposite of Buttimer's. As she noted: "There are hazards ... when one assumes the 'Cinderella' cause of human personhood and authenticity in a world where securing bread and butter often demands obliviousness to such matters" (Buttimer 1980, 17). This also has repercussions on campuses: "To be a teacher is an awesome responsibility: inevitably, one's students are subjected to influences which may be intellectually exciting and inspiring but not necessarily those which will guarantee status or career advancement" (Buttimer 1980, 17). This is only one of the provocative insights that Buttimer's nonconformist lifeworld launches. Yet, as she demonstrated, investigating concrete individuals' existences, real lives and feelings beyond abstract formulas, in association with political and social values, contributes to ideas that Geography matters in plural and sometimes unpredictable ways.

Notes

1 See Festival's website: https://www.fig.saint-die-des-vosges.fr/
2 See Buttimer's interview: https://www.youtube.com/watch?v=ovFTmzzCNeI
3 Repository link: https://researchrepository.ucd.ie/home
4 See Dialogue Project's link: https://www.ucd.ie/geography/research/lifeexperienceascatalystforcrossdisciplinarycommunication/

Study task

Variations on this question can be set on any kind of written or oral exam, or as a term-paper assignment, or as a workgroup assignment topic:

With a focus on Anne Buttimer's published works, explore how and why values matter in any attempt to understand the world's human geography.

Starter reading list

Buttimer, A. (1974) *Values in geography*. Washington, DC: AAG Commission on College Geography.

Buttimer, A. (1976) "Grasping the dynamism of lifeworld", *Annals, Association of American Geographers* 66(2): 277–292.

Buttimer, A. (1990) "Geography, humanism, and global concern", *Annals of the Association of American Geographers* 80(1), 5–34.

Buttimer, A. (2001) "Home-reach-journey", in Moss, P. (ed.) *Placing Autobiography in Geography*. Syracuse: Syracuse University Press, pp. 22–41.

Ferretti, F. (2019) "Between radical geography and humanism: Anne Buttimer and the International Dialogue Project", *Antipode, a Radical Journal of Geography* 51(4): 1123–1145.

Harvey, D. (1972) "Revolutionary and counter-revolutionary theory in geography and the problem of ghetto formation", *Antipode* 4(2): 1–13.

Seamon, D. and A. Lundberg (2017) "Humanistic geography", in Richardson, D., Castree, N., Goodchild, M.M. Kobayashi, A., Liu, W. and R.A. Marston (eds) *The International Encyclopedia of Geography*. Hoboken: Wiley, pp. 1–11. doi: 10.1002/9781118786352.wbieg0412

References

Alcoforado, M.J. and M. Jones (2015) "Professor Anne Buttimer, Laureate of the Vautrin Lud International Prize for Geography", *Norsk Geografisk Tidsskrift – Norwegian Journal of Geography*, 69(3), pp. 186–187.

Baigent, E. and Novaes, A.R. (2020) "Not just rounding up the usual suspects: Hugh Clout and Geography's Stories", in Novaes, A.R. and Baigent, E. (eds) *Geographers Biobibliographical Studies* 39. London: Bloomsbury, pp. 1–26.

Baigent, E. and Novaes A.R. (2022) "On Southern Traditions and Geographical trajectories", in Novaes, A.R. and Baigent, E. (eds) *Geographers Biobibliographical Studies* 40. London: Bloomsbury, pp. 1–24.

Barnes, T.J. (2018) "A marginal man and his central contributions: The creative spaces of William (Wild Bill) Bunge and American geography", *Environment and Planning A*, 50(8), pp. 1697–1715. doi: 10.1177/0308518X17707524

Barnes, T.J. and Sheppard, E. (eds.) (2019) *Spatial Histories of Radical Geography*. Oxford: Wiley.

Brun-Tschudi, A. (1979) *Dialogue Project, transcript series G8*. Lund: University of Lund.

Buttimer, A. (1969) "Social space in interdisciplinary perspective", *Geographical Review*, 54(3), pp. 417–426. doi: 10.2307/213485

Buttimer, A. (1974) *Values in Geography*. Washington, DC: AAG Commission on College Geography.

Buttimer, A. (1976) "Grasping the dynamism of lifeworld", *Annals, Association of American Geographers*, 66(2), pp. 277–292. doi: 10.1111/j.1467-8306.1976.tb01090.x

Buttimer, A. (1979) *Dialogue Project, transcript series G10*. Lund: University of Lund.

Buttimer, A. (1980) " Introduction", in Anne Buttimer and David Seamon, (eds) *The human experience of place and space*. London: Croom Helm, pp. 13–18.

Buttimer, A. (1983) *The Practice of Geography*. London and New York: Longman.

Buttimer, A. (1990). "Geography, humanism, and global concern", *Annals of the Association of American Geographers*, 80(1), 5–34. doi: 10.1111/j.1467-8306.1990.tb00001.x

Buttimer, A. (1993) *Geography and the Human Spirit*. Baltimore: Johns Hopkins University Press.

Buttimer, A. (2003) "Thomas Walter Freeman 1908–1988", *Geographers Biobibliographical Studies*, 22, pp. 72–90.

Buttimer, A. (2007) "Torsten Hägerstrand (1916–2004)", *Geographers Biobibliographical Studies*, 26, pp. 119–157.

Buttimer, A. (2016) "Retirement? You must be joking", *The Arab World Geographer*, 19(1–2), pp. 27–37. doi: 10.5555/1480-6800.19.1.27

Buttimer, A. and T. Hägerstrand (1980) *Invitation to Dialogue*. Lund: DIA.

Buttimer, A. and Seamon, D. (eds.) (1980) *The Human Experience of Space and Place*. London: Croom Helm.

Cresswell, T. (2013) *Geographic Thought: A Critical Introduction*. Chichester: Wiley-Blackwell.

Cresswell, T., Dixon, D., Bol, K. and N. Entrikin (2015) "Editorial", *GeoHumanities*, 1, pp. 1–19. doi: 10.1080/2373566x.2015.1074055

Ferretti, F. (2018) *Anarchy and Geography. Reclus and Kropotkin in the UK*. Abingdon: Routledge.

Ferretti, F. (2019) "Between radical geography and humanism: Anne Buttimer and the International Dialogue Project", *Antipode, a Radical Journal of Geography*, 51(4), pp. 1123–1145. doi: 10.1111/anti.12536

Ferretti, F. (2020a) "Travelling in scholarly lifeworlds: New perspectives on (post)humanism, situated subjectivities and agency from a travel diary", *Annals of the American Association of Geographers*, 110(6), pp. 1653–1669. doi: 10.1080/24694452.2020.1715197

Ferretti, F. (2020b) "History and Philosophy of geography I: Decolonising the discipline, diversifying archives and historicising radicalism", *Progress in Human Geography*, 44(6), pp. 1161–1171. doi: 10.1177/0309132519893442

Ferretti, F. (2022) "History and philosophy of geography III: Global histories of geography, statues that must fall and a radical and multilingual turn", *Progress in Human Geography*, 46(2), pp. 716–725. doi: 10.1177/03091325211062170

Ferretti, F. and A. Jones (2018) "Anne Buttimer. 1938–2017", *Geographers Biobibliographical Studies*, 37, pp. 13–40.

Jones, M. (2019) "Anne Buttimer's *The Practice of Geography*: Approaching the history of geography through autobiography", *Geografiska Annaler: Series B, Human Geography* 100(4), pp. 396–405. doi: 10.1080/04353684.2018.1544468

Jones, A. and Mennell, S. (2017) "Anne Buttimer (1938–2017)", *Academia Europaea*, http://www.ae-info.org/ae/Acad_Main/News/Anne%20Buttimer

Keighren, I.M. (2020) "History and philosophy of geography III: The haunted, the reviled, and the plural", *Progress in Human Geography*, 44(1), pp. 160–167. doi: 10.1177/0309132518818725

Kniffen, F. (1982) *Dialogue Project, transcript series G 28*. Lund: University of Lund.

Kropotkin, P. (1885) "What geography ought to be". *The Nineteenth Century*, 18, pp. 940–956.

Ley, D. and M. Samuels (1978) *Humanistic Geography, prospects and problems*. London: Croom Helm.

Livingstone, D.N. (1993) *The Geographical Tradition*. Oxford: Blackwell.

Maddrell A. (2009) "An interview with Anne Buttimer: an autobiographical window on geographical thought and practice 1965–2005", *Gender, Place & Culture*, 16(6), pp. 741–765. doi: 10.1080/09663690903279179

Magrane, E. Russo, L., De Leeuw, S. and C. Santos Perez (eds) (2020) *Geopoetics in Practice*. Abingdon: Routledge.

Mead, W. (1979) *Dialogue Project, transcript series G11*. Lund : University of Lund.

Nash, P. (1984). *Dialogue Project, transcript series G42*. Lund: University of Lund.

Relph, E. (1976) *Place and Placelessness*. London: Pion.

Seamon, D. (2023) *Phenomenological Perspectives on Place, Lifeworlds, and Lived Emplacement: The Selected Writings of David Seamon*. Abingdon: Routledge.

Seamon, D. and A. Lundberg (2017) "Humanistic geography", in Richardson, D., Castree, N., Goodchild, M.M. Kobayashi, A., Liu, W. and R.A. Marston (eds) *The International Encyclopedia of Geography*. Hoboken: Wiley, pp. 1–11. doi: 10.1002/9781118786352.wbieg0412

Simonsen, K. (2013) "In quest of a new humanism: Embodiment, experience and phenomenology as critical geography", *Progress in Human Geography*, 37, pp. 10–26. doi: 10.1177/030913251246757

Tuan, Y.F. (1976) "Humanistic geography", *Annals of the Association of American Geographers*, 66(2), pp. 266–276. doi: 10.1111/j.1467-8306.1976.tb01089.x

11 Landscape as a way of seeing

Denis Cosgrove's symbolic geographies*

Veronica della Dora

Denis Cosgrove (1948–2008) in July 1992. Photo courtesy of Francesco Vallerani.

Introduction

A thin line of dark grey asphalt runs across the cornfields as the morning haze wraps the land in a diaphanous veil. By the side of the road is a narrow ditch rimmed with wild canes and overgrown grass. Amidst the fields lies a small forsaken stone-built structure. Its tiled roof collapsed long ago, under the action of the weather and the weight of time. Brambles have irrupted into its interior and now spread thickly on its derelict walls. Further in the distance, the red-capped campanile of a small-town church breaks the monotony of the flat horizon. On the other side of the road, a row of industrial sheds and grey warehouses blocks the view. The road has two lanes only, but heavy trucks whiz by, careless of the speed limits imposed by its modest carrying capacity. On its side, puzzled drivers suddenly spot a curious, unusual sight: an Englishman is cycling amidst the traffic. He seems unaware, or more probably not caring, that the provincial roads of mainland Veneto are not used for leisure as they normally are in the

* The vignettes in this chapter are based on personal verbal and written accounts by Denis Cosgrove, Stephen Daniels and Francesco Vallerani. I am grateful to all of them for sharing their compelling stories and memories with me, and especially to Steve and Francesco for their helpful comments and suggestions on a previous draft of this chapter.

DOI: 10.4324/9781003343240-13

English countryside. Dodging lorries and cars, the intrepid cyclist bikes kilometre after kilometre as the haze dissipates, and the sun slowly completes its day arc over the plain. In his backpack are a Kodak camera, a couple of sandwiches wrapped in tinfoil, a thickly scribbled notebook and a worn copy of Victorian polymath John Ruskin's *Stones of Venice*.

We are in the mid-1970s on the *terraferma*, the stretch of mainland west of the lagoon of Venice, north-east Italy. The man on the bike is Denis Cosgrove, a young scholar with an ambitious project: tracing the origins and development of the idea of landscape. For the British geographer, the concept of landscape, as we know it today, was not a timeless given. Instead, it emerged, he believed, in a very specific place and moment of early modern Western cultural history: in Renaissance Italy. In his early work, Cosgrove linked the development of the concept to a new, geometrical perception and representation of space, as well as to its practical appropriation by surveyors, hydraulic engineers and architects. The theatrical landscapes of Veneto designed by the Paduan architect Andrea Palladio (1508–1580), he thought, best expressed this new way of seeing and structuring space. Inspired by the architecture of classical Greece and Rome and fully harmonized with the surrounding environment, Palladio's magnificent villas at once overlooked and framed a carefully 'staged' nature. Through his adventurous bike excursions, Cosgrove set out to discover this world of power and beauty.

Cosgrove's 'landscape project' nonetheless ran deeper than historical investigation. It had far more profound an impact than a simple intellectual history of a geographical concept. In fact, it inaugurated a new approach to the study of landscape which, together with the work of Stephen Daniels, Trevor Barnes, James Duncan and others, cast the basis for the 'new cultural geography' of the 1990s. Perhaps even more significantly, it was fundamental in critically repositioning Geography in the humanities. In this chapter, we follow Cosgrove's wanderings through mainland Veneto to explore some of the key aspects of the project: first, the understanding of landscape as a 'way of seeing' emerging under specific historical socio-economic conditions; second, the symbolic and moral dimension of landscape; and third, the interpretation of contemporary cultural landscapes. Before we enter the grand theatre of Veneto, however, some further context is needed. Who was Denis Cosgrove and why did his 'landscape project' end up having such a transformative impact on his chosen discipline (and beyond it)? How were human geographers approaching landscape at the time he was happily biking the Veneto countryside? And more fundamentally, what exactly is 'landscape'?

Denis Cosgrove's path to Geography

Denis Edmund Cosgrove was born in Liverpool on 3 May 1948. The second of six children, he was raised in a lower middle-class Roman Catholic household. His father was a bank clerk deeply devoted to his family and his church. As Cosgrove retrospectively acknowledged, his Catholic upbringing played a crucial role in his formation. Both at home and at his Jesuit school in Liverpool, Rome always loomed larger than London (Freitag and Jons 2005: 210). This was later reflected in the European and cosmopolitan, rather than national, focus of his work and in his longstanding interest in moral questions. It was, however, his hometown that Cosgrove held responsible for triggering his passion for geography at an early age. His family owned neither car nor TV set, and very seldom travelled out of town. Liverpool, however, was one of the major ports in post-war England and what was still the British Empire. Walking along its great docks and watching ships with exotic place names such as Lagos or Montevideo conjured up images of distant, mysterious worlds in the boy's mind.

Much to his dismay, however, at school Cosgrove had to renounce his favourite subject for Greek and Latin in order to remain in the 'A' stream. 'Mrs. Cosgrove, geography is a girls' subject', the headmaster priest responded to his mother's protests. The boy nonetheless continued to read geography textbooks he found by himself. Like the charismatic ships moored in Liverpool's docks, these books were his gateway to the wider world (Freitag and Jons 2005, p. 206).

In the 1960s, Cosgrove went on to study geography at St Catherine's College at Oxford University, under the mentorship of the social geographer Ceri Peach. His fellow students recall him as 'a sharp-talking street-wise Liverpudlian, proud of his Irish ancestry, fresh from selling programmes at [Liverpool's] Goodison Park during the 1966 World Cup' (Driver 2008: 1779–1780). Although he admired his tutor, Cosgrove did not find the regional and topical geography he was taught at Oxford particularly inspiring—it had, he lamented, a dull 'catalogue-like' quality. At the same time, he wasn't drawn to the 'new geography' or 'spatial science' that was then exciting the new generations either (see Chapter 4). Models and numbers had their own elegance, but were too abstract. 'It seemed to me that geography, which I loved, wasn't doing the things that I loved about it. It seemed to have lost the sense of wonder, of the exotic, the rich tapestry of difference across the Earth. This is what had drawn me in', he recalled later in his life (Freitag and Jons 2005: 206).

Cosgrove rediscovered this sense of wonder and excitement during his Master's degree programme. It was 1969 when he was awarded a place at Toronto University. Under the supervision of the late, great Cole Harris, he eagerly turned to historical geography. He also immersed himself in critical landscape literature, though this was mostly written by architects, rather than by geographers. The time in Toronto was an invaluable experience. While 'spatial science' dominated the department, Cosgrove was embedded in a little community of like-minded postgraduate students with a shared interest in experiencing and imaging the world—and moving beyond quantitative geography. Some of them, including Edward Relph, would later become major voices in humanistic geography.

The return to Britain felt less exciting. When Cosgrove went back to Oxford on a government bursary and started to talk about 'things like phenomenology and cultural geography', people stared at him blankly (Freitag and Jons 2005, p. 207). Few knew what he was talking about. So Cosgrove gave up the bursary after a year and worked instead as a research assistant in the Built Environment Research Group at the former Polytechnic of Central London (now Westminster University) on a computer modelling project to locate leisure centres in Westminster. A far cry from cultural or humanistic geography, the job nonetheless had its advantages. For example, Cosgrove shared his office with architects and historians, with whom he exchanged readings and ideas. Most importantly, he had full access to the architecture library. Here he got acquainted with the works of the German Jewish art historian Erwin Panofsky, whose 'iconographic' approach deeply influenced his approach to landscape.

In the following years, Cosgrove's academic life continued to be by necessity multitasked. In Britain there was no such thing as cultural or humanistic geography at that time. While there was a tradition of geographers interested in culture and the land at the University of Wales, where H.J. Fleure and David Bowen had worked, and at the University of Ulster, where Estyn Evans had taught, it had to do with minority cultures and remained a distinctly small tradition—possibly too minor to capture the attention of an ambitious young scholar such as Cosgrove (Freitag and Jons 2005, pp. 209–210). Back in Oxford, Cosgrove thus supported himself by teaching statistics and physical geography at the city's Polytechnic while completing his dissertation on the Palladian townscape of Vicenza and of the Veneto. David Lowenthal, his external examiner and one of the great pioneers of humanistic geography and heritage studies, found Cosgrove's work 'a remarkable synthesis of architectural enterprise, land management and regional history' (Lowenthal 2008). He was so impressed that he boldly upgraded it from a BLitt to a PhD.

One might wonder why a boy from Liverpool ultimately landed in north-east Italy. In the 1970s, Oxford postgraduate students were expected to specialize in a geographical region other than (and possibly culturally distant from) their own. Cosgrove's choice to conduct his research in the Veneto and to learn Italian, however, came less from institutional expectations and more from intellectual curiosity. While writing his Master's thesis on the warehousing district in Toronto, he became interested in John Ruskin and his landscape ideas. Ruskin led him to rethink his own ideas on the English landscape and made him realize that many ideas had come from Italy (Freitag and Jons 2005: 208). Travelling to Veneto, Cosgrove felt, meant journeying to the sources of the 'landscape idea'. His doctoral work and biking adventures thus cast the basis for a 'life-long attachment to an historically-inclined form of cultural geography' (Driver 2008: 1780) and for his 'landscape project'. But what is landscape exactly, and why did it come to 'matter'?

The landscape debate

According to the Webster Dictionary, the English word 'landscape' signifies 'a portion of land or territory which the eye can comprehend in a single view, including all the elements it contains', as well as 'a picture representing a scene by land or sea, actual or fancied, the chief subject being the general aspect of nature, as fields, hills, forests, water, etc.' and 'the pictorial aspect of a country'. Landscape is therefore a polysemic word, a concept designating both a material reality and its representation. Caught somewhere between the lie of the land and the gaze of the viewer, it's a site of encounters and tensions: between proximity and distance; between abstract space and solid matter; between nature and culture. Unlike 'place', 'territory' and 'environment', landscape is an inherently visual concept; it comes into being only when apprehended by an external viewer. It both encompasses framed views of specific places and the scenic character of entire regions (Cosgrove 2006a: 51). Landscape, however, is not only 'looked at', but it is also 'lived with' (Lowenthal 2007). Unlike abstract geometrical 'space', it does have an unquestionably material texture. It is the stage upon and through which we enact our lives, and is, in turn, shaped by our continuously evolving visions and actions.

In late nineteenth- and early twentieth-century geography, the concept of landscape was generally understood and employed as a geographical area imprinted with visible, and thus clearly identifiable, human characteristics. As such, it enjoyed prominence in various national geographic traditions, from German Karl Ratzel's *Kulturlandschaft* to French geographer Paul Vidal de La Blache's *paysage* (see della Dora 2020). The idea of 'cultural landscape' entered twentieth-century Anglophone geography mainly through the work of Carl Sauer from the University of Berkeley in California (see Chapter 6). As with his German and French homologues, Sauer placed special emphasis on humans' transformative activities on the land, especially agriculture. Sauer was deeply influenced by the German *Landschaft* tradition. He conceived landscape as a unit area characterized by specific material features (for example, farmhouses, barns and field patterns). The distribution of such features, he believed, was indicative of the spread of specific cultural groups. The geographer's task was thus, in Sauer's words, to 'chart the distribution over the Earth of the arts and artefacts of man' (Sauer 1969: 1).

While Sauer's work and French regional geography had a long-lasting impact on the discipline, the concept of landscape nonetheless fell from favour in the mainstream Anglophone human geography of the 1950s and 1960s. Economy, cities and infrastructures had been shattered by the war and needed to be reconstructed. Post-war research agendas thus privileged an objective and practical quantitative geography or 'spatial science' over what was perceived as a descriptive discipline (Capel 1987; Livingstone 1992). As a concept, landscape appeared too subjective and fuzzy to post-war Anglophone spatial scientists, preoccupied as they were with the imperatives of economic and social planning. As a 'unit area', landscape lacked the dynamism and scientific feel of their 'functional regions' or 'gravitational models' inspired by physics. Both as a concept and an empirical focus of research, landscape seemed ill-suited to this new type of planning-oriented research (Cosgrove 1985: 45).

In the 1970s, humanistic geographers reacted against this geography, which they perceived as dehumanizing. They made human experience and subjectivity their focus instead. Humanistic geographers thus rehabilitated the concept of landscape for the very reasons spatial scientists had rejected it. In doing so, however, they implicitly, and misleadingly, set landscape in opposition to science. They ignored the fact that, as we shall see below, the 'landscape idea' was originally grounded in *scientific* theory and knowledge. Furthermore, drawing mainly on literary sources, humanistic geographers neglected the visual nature of the concept, and of Geography itself (Cosgrove 1985, p. 46). Finally, by focussing on subjective experience, they implicitly separated the private from the public and thus ignored the politics and oft-contested nature of landscape.

Cosgrove's landscape project took its initial shape between, and largely in response to, these three different streams: 1) a descriptive Sauerian geography criticized for its lack of theoretical reflection and for its seemingly unproblematic, almost deterministic, understanding of culture as a uniform and uncontested agent shaping the landscape; 2) a 'spatial science' ignoring culture and human subjectivity altogether; and not least, 3) an emerging humanistic geography focussed on human experience and subjectivity but lacking critical insight, neglecting Geography's visual heritage, and suffering from 'historical myopia' (Daniels 1985). A self-taught interdisciplinary scholar used to sharing his office with architects and historians, Cosgrove was also aware of the potential impact of his project beyond the discipline. Outside of Geography, Cosgrove believed, tracing the historical and cultural origins of the landscape idea meant extending the concept beyond mere discussions on taste or design (1998[1984]: xiii). It meant broadening out architecture and garden history's focus on site-specific qualities and buildings (1993: 5). More profoundly, it meant uncovering the tension between the insider's personal perception of place and a more critical and socially conscious outsider's perspective: the gaze over the land from on high or afar.

Cosgrove's project took as a starting point the fact that, in its different guises, landscape always linked human geography with culture; that it had always been closely connected with the idea of human agency shaping the land's surface, as stressed by Vidal de la Blache and Sauer. However, more than a collection of material features, Cosgrove believed, landscape was a 'way of seeing', a way of 'composing and harmonizing the external world into a scene, a visual unity' (Cosgrove 1989: 121). The suffix '-scape' conveys precisely this sense of holism, the presence of a unifying principle that makes landscape different from 'region' or 'area'. This unifying principle derives from the mediation of human experience; in other words, from the gazing subject.

Whereas in ancient languages we do have words for spatial concepts such as 'space', 'place' and 'region', there is no such word for landscape. The term emerged only as recently as the sixteenth century. It described a new relationship between humans and their environment, or in Cosgrove's words, 'a way in which some Europeans have represented to themselves, and to others, the world about them' (1998[1984]: 1). How did this relationship come into being?

Geometry and power: landscape as a way of seeing

> Landscape is ... a way of seeing, a composition and structuring of the world so that it may be appropriated by a detached, individual spectator to whom an illusion of order and control is offered through the composition of space according to the certainties of geometry.
>
> (Cosgrove 1985: 55)

Cosgrove linked the emergence of the landscape idea to a new understanding of space in Renaissance Europe. In particular, he connected it to the theorization and development of linear perspective in early fifteenth-century Florence. Linear perspective is a technique that enables the realistic representation of a three-dimensional space on a flat surface. In a painting or drawing employing this technique, parallel lines converge in a single

Figure 11.1 Pietro Perugino, *Christ Giving Saint Peter the Keys to Heaven*, 1481.

Source: Wikimedia.

vanishing point on the horizon (Figure 11.1). As the eye moves from the foreground towards the vanishing point, people, trees, buildings and other objects seem to recede in the composition. Their sizes diminish according to a fixed mathematical proportion. In other words, they appear consistently smaller as their distance from the observer increases. This has the effect of creating an illusion of depth. The painting thus appears to us three-dimensional, even if the canvas is flat. It gives us the impression of being an accurate copy of reality.

Renaissance artists and architectural practitioners did not consider linear perspective simply a pictorial technique or a visual device. For them, it was also an instrument for disclosing truth. Through linear perspective, order was imposed on the external world. The word 'perspective' comes from the Latin verb *perspicere*, which means to 'see through', but also 'to see clearly'. The connection between clear vision, truth and rational understanding still survives in the English language in idioms like 'I see', or 'to put things into perspective'. Linear perspective presupposes a fixed viewer removed from the scene. In this way, the eye is granted absolute mastery over space as an objective entity. For Renaissance artists, linear perspective was therefore a device for controlling the world of objects.

For Cosgrove, however, landscape was not just a graphical representation. It 'mattered' in the literal sense of the word. The landscape idea, he believed, was closely bound up with the practical appropriation of physical space. Like linear perspective, this was achieved through the application of geometrical principles. Geometry literally means 'the measurement of the Earth'. In antiquity and in the Renaissance, it was believed to underpin the structure of the Earth and of the cosmos (ancient Greek philosophers' rationally ordered spherical universe). For this reason, geometry was deemed to be the foundation of knowledge. In the fifth century, the polymath Martianus Capella described it as a distinguished-looking lady wearing shoes worn to shreds from her continuous wanderings around the globe. By the sixteenth century, Geometry had become a detached surveyor floating over the Earth on a cloud (Figure 11.2).

Cosgrove considered the visual appropriation of space by means of linear perspective as part of a broader set of geometrical and mathematical practices. For example, Ptolemaic cartography enabled mapmakers to project the sphere of the Earth on a flat surface and to pinpoint exact locations on a mathematical grid of parallels and meridians. Everyday practical applications of geometry included land survey and 'visual gauging'. The latter was a skill that merchants developed to calculate the volume of barrels, churns and haystacks by eye and the application of simply remembered mathematical formulae (1985: 50). Despite differences in scale and calculation techniques, all these practices presupposed distancing. They presumed an immobile sovereign eye translating the complexity of the world into simple geometrical forms and abstract numbers.

This new, scientific way of thinking and engaging with space, Cosgrove believed, enabled increased human control over nature, and specifically, through the application of linear perspective principles in landscape design. In the book *Social Formation and Symbolic Landscape* (1984), a path-breaking text for cultural geographers of the 1980s and 1990s, Cosgrove connected this attitude to the emergence of a new elite consciousness. This, he argued, had in turn resulted from material and economic changes, namely the shift from feudalism to a capitalist economic system and the consequent commodification of the land and of labour. Landscape was thus a complex cultural product entailing art and vision as much as social and economic relations expressed in terms of patterns of property, commerce and land ownership. These processes were especially visible in the Veneto region, to which Cosgrove later devoted a separate monograph, *The Palladian Landscape* (1993).

Figure 11.2 Allegory of Geometry in E-Series Tarocchi Cards, 1460–1470.

The Elisha Whittelsey Collection, The Elisha Whittelsey Fund, 1992.

Cosgrove characterized the economic development of Venice as 'the greatest achievement of Italian mercantile capitalism in early modern Europe', and the achievements of its painters, architects and urbanists as 'one of the most enduring contributions to the European landscape idea' (1998[1984]: 103). Between the aftermath of the Fourth Christian Crusade (1204) and the fifteenth century, Venice had built its fortune on a maritime empire (*Stato da Mar*) that stretched all the way down the coasts of Dalmatia to the Aegean. In the sixteenth century, it reached as far as Cyprus. As the Ottomans gained control over the eastern Mediterranean and international trade routes increasingly shifted from that part of the world to the Atlantic, however, Venice turned inland, to the *terraferma* (Figure 11.3). Rather than continuing to rely almost exclusively on shipping traffic from the colonies, Venetians decided to boost agriculture in the *Stato da Tera* (Venice's mainland domains) (Figure 11.4).

Agricultural productivity was achieved through extensive land reclamation and large-scale hydraulic interventions. As marshes were converted into profitable cultivations, Venetian patricians took increasing interest in their landed properties and commissioned the construction of opulent mansions on them. Here they retreated during the warmest summer months, when life was quiet and unhealthy in Venice and its lagoon. Together with an army of engineers, surveyors and architects, the landowner thus became the agent of environmental change in the region: a 'land-shaper' overseeing his properties from the heights of his two- or three-story villa.

Palladio, one of the most influential individuals in the history of architecture, knew how to best interpret and realize the agrarian aspirations of the Venetian nobility. Within his

Figure 11.3 NASA satellite image of the lagoon of Venice and the *terraferma*.

Source: Wikimedia.

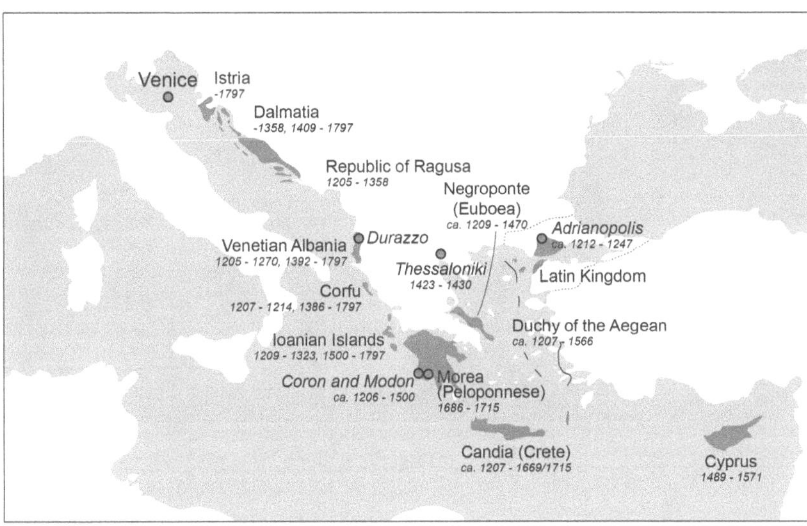

Figure 11.4 Map of Venice's land and sea domains.

Credit: Jen Thornton.

villa designs he combined the needs of an agricultural estate centre with the aesthetic and recreational demands of his powerful clients. Planned at once as belvederes over landed property and as vanishing points of manicured gardens and long promenades, Palladio's villas were always conceived in symbiosis with the surrounding environment and its topography (Figure 11.5). They were normally built at the centre of the estate, so that the master

Figure 11.5 Andrea Palladio's Villa Almerico Capra (also known as La Rotonda) near Vicenza.

Photograph by Mark Beston.

Figure 11.6 Landscape view from the portico of Andrea Palladio's Villa Godi Malinverni, Lugo di Vicenza.

Source: Histouring.com https://www.histouring.com/en/historical-places/villa-godi-malinverni/#

could 'see his possessions, not merely the better to supervise their exploitation but also to enjoy the panorama of nature—the drama of dawn and sunset, the passage of the seasons' (Cosgrove 1993: 105). Merging power and aesthetics, prospect and perspective thus transformed the land into a commodity for the visual consumption of the landowner (Figure 11.6).

In *Social Formation*, Cosgrove unpacked these dynamics and made them 'matter'. He showed how, far from remaining a Venetian *capriccio*, in the following centuries, this

Figure 11.7 Hotel room with a view in Honolulu, Hawaii.
Photograph by Prayitno.

way of seeing and experiencing landscape was exported to Georgian England and to the plantation mansions of Republican America precisely by way of Palladio's designs. As he taught his undergraduate students at UCLA many years later, landscape as a way of seeing had remained part of contemporary Western culture. His house in West Los Angeles, he used to joke, cost ten times less than the same type of house on the top of the hills behind the city, because it 'lacked the view'. Similarly, if we ask for a hotel room 'with a view' today, we are normally charged extra, as if landscape as a way of seeing were a commodity to which is attached a monetary value (Figure 11.7).

Beauty and harmony: landscape as a moral project

> Landscape indicates the significance of human creativity in shaping geography beyond the imperatives of biological and social life. It gestures towards the realms of imagination, creativity and vision.
>
> (Cosgrove 1993: 5)

If Cosgrove's landscape project started on a bike, it culminated on a canoe. The stage for the completion of his book *The Palladian Landscape* was Villa Ramusa, a sober yet imposing early Renaissance rural mansion that had once belonged to the famed Venetian geographer and travel writer Giovanni Battista Ramusio (1485–1557) and was now the home of Cosgrove's friend and colleague Francesco Vallerani and his wife Maria Grazia.

It is August 1992 and we are in San Giorgio in Bosco, a small countryside municipality in the province of Padua. Scattered on the large wooden table in Francesco and Maria Grazia's living room are piles of yellowed handwritten notes and the pages of what will become the final draft of the book manuscript. Cosgrove is working on its final chapters until late. In the mansion, the evening silence is occasionally broken by the raucous bark of a dog from some garden in the neighbourhood, or the popping sound of a motorbike traversing the nearby road. As he navigates his notes, the scholar can hear the rhythmic tolling of the bells carried by the wind from some campanile in the distance, beyond the hedge and line of poplar trees bounding his friend's peaceful retreat. The stress of the impending manuscript deadline is soothed by the peace of the countryside, with its rarefied sounds and silences and its ineffable chromatic shades at dawn and dusk. It is also tempered by Maria Grazia's lovingly cooked dinners; by a good glass of Friuli cabernet sipped by the crackling fireplace; and, not least, by the promise and excitement of a new excursion on the following day.

Francesco is a middle-aged historical geographer based at the University of Venice. With him, Cosgrove shares a passion for Palladian landscapes and for canoeing. Paddling along the minor waterways of Veneto is for the two friends a way to connect with the land, or rather, to develop 'a special fluvial sense of place' (Vallerani 2018). This connection is important for Cosgrove. Landscape is for him not simply a disembodied way of seeing or a representation of itself; it presupposes the land as a realm of practical possibility as much as the viewer looking on: it is a stage to be performed. Paddle stroke after paddle stroke, the two friends slowly make their way along the shady embankments of what once used to be Venice's main inland transport network (Vallerani 2021). They pass small towns and villages, old stone bridges, abandoned traditional boats and charismatic villas reflecting their elegant facades on the iridescent surface of the water.

These canoeing excursions enabled Cosgrove to physically access a distant world of beauty and harmony. Understanding the vision behind this world was yet another fundamental aspect of his 'landscape project' and how it came to 'matter' in the humanities. The Palladian landscape, Cosgrove showed, was shaped by power relations and practical demands, as much as by a vision that transcended the boundaries of the villa estate, and the world itself. Like his contemporaries, Palladio imagined the cosmos as a harmonious and perfectly ordered entity animated by divine love and regulated by mathematical proportions and harmonic ratios. By adopting these proportions in the construction of his own buildings and shaping the land accordingly, the architect brought order into the world. Like the Creator, he fabricated a world out of confused matter. Beauty and harmony in the landscape related to the greater harmony of the Earth and of the cosmos, 'for all things run through correspondence' (Cosgrove 1993: 328). Through the correspondences and proportions that underpinned the fabric of the universe—from the microcosms of the human body, of the villa estate and of the city to the celestial macrocosm—humans could discern divine love as the moving force of creation. They could understand the workings of nature at all scales, and thus enable the soul to seek perfection (Cosgrove 1993, p. 244).

While one might be tempted to read the Palladian landscape as a mere facade of patrician power or as the stage for a languid aristocratic life punctuated with property disputes, Cosgrove showed how at a more profound level it was a project underpinned and guided by a sophisticated cosmological understanding and by spiritual belief. If, after the Marxist art critic John Berger, Cosgrove initially likened landscape to a deceptive curtain behind which 'the struggles, achievements and accidents of its dwellers take place' (1998[1984]: 271), by the time he was drafting *The Palladian Landscape*, his view had broadened. The symbolic landscape was more than a mere aesthetic screen concealing (and normalizing) the labour exploitation and unequal socio-economic power relationships through which the material landscape was built and sustained. In sixteenth-century Veneto, landscape was the visual and material expression of a moral project. This project promoted the belief that humans possessed the means to represent the perfection of the great machine of creation and their place within it. To use a metaphor employed by cultural geographer John Wylie, landscape was for Cosgrove less of a veil to be penetrated so as to uncover power structures than a complex fabric, or 'texture' to be searched 'amidst its folds, along its wave' (2007: 70). But how was this texture navigated?

Decoding and performing the symbolic landscape

> The many-layered meanings of symbolic landscapes await geographical decoding.
>
> (Cosgrove 1989: 126)

Easter 1988. We are travelling on a coach through the Vicentine countryside with a group of British undergraduate students and their two instructors: Cosgrove, by now a Reader at the University of Loughborough in the East Midlands, and his colleague Stephen Daniels from the nearby University of Nottingham. The two are not only good friends, but also close collaborators and intellectual partners: co-authors, co-thinkers, co-teachers. Cosgrove specializes in Renaissance Italy; Daniels in Georgian England. Differences in focus notwithstanding, the two share the same vision and approach to landscape. And this is just one of many successful field trips to the region co-organized during Cosgrove's appointment at Loughborough.

The group arrived the day before on a charter flight operated by Pilgrim Airways, a company, as the name suggests, flying Catholic pilgrims from the heart of England to the shrines of continental Europe—the cheapest and most practical option to reach Veneto in this season. Based in a hotel in the heart of Vicenza, the group is meant to spend the rest of the week undertaking field exercises, working on little projects and visiting the Palladian heritage (Daniels 2018). The students have all taken Cosgrove and Daniels' joint final year Cultural Geography course. For a semester, they have been exposed to slide projections of patrician villas and rural landscapes in the closed space of a darkened classroom. Now, they can immerse themselves in the theatre of the Veneto—and become its protagonists and performers.

It is a rainy day, as it often happens to be in the region this time of the year. The students are returning to their base after a half day-long excursion to the villas in the northern part of the Vicenza province. Induced by an extra glass of wine over lunch, by the grey sky and the rain, or simply by the monotonous hum of the engine, drowsiness is slowly taking over the group. Eyelids become heavy. Some students lie back in their seats and fall asleep. Accompanied by their favourite soundtracks, other students watch the landscape unfold behind the rain-streaked glass. The view is not too dissimilar from what their instructor experienced a decade earlier: cornfields, industrial sheds, two-storey houses, poplar trees, the occasional campanile in the distance. Only, modern workshop and commercial structures seem to have taken over in what is now a key region of post-industrial Italy. The resulting landscape, the two instructors note, is 'more comparable with Orange County, California than either England or the Italy of familiar images' (Cosgrove and Daniels 1989: 177). The Palladian landscape the students had read and heard so much about in their lectures has been swamped by post-industrial urban sprawl, the *città diffusa* (Cosgrove 2006b).

Obscured by a 400-year-old 'palimpsest of erasures and rewritings', the Palladian landscape is hardly detectable to the casual observer. For Cosgrove and Daniels, however, it is precisely such complex texture that makes this fractured and contradictory landscape of cornfields, superconductor factories and furniture hypermarkets a compelling object of study and perfect fieldwork setting for their students. For landscape is seldom uniform and uncontested. In the Veneto, it compresses different, often contrasting or even competing world visions. Excavating an historical landscape of villas and patrician estates out of such a thick palimpsest is fundamentally what the two British geographers call 'an act of visual imagination' (Cosgrove and Daniels 1989: 177).

For Cosgrove and Daniels, reconstructing past landscapes does not simply mean reconstructing land distribution patterns. It means reconstructing worldviews. Whereas for Sauer landscape was an ensemble of material features to be *mapped*, for the two British geographers is was an ensemble of symbols to be *decoded*. Packed with socio-economic, political and cosmological meanings, cultural landscapes, they believe, emerge only when properly set in context. In this sense, the work of the cultural geographer is akin to that of the art historian or the literary critic, for it draws mainly on hermeneutic (or interpretative) methods. The cultural geographer's sources are not simply 'factual documents' enabling reconstruction of bygone material landscapes. They also encompass paintings, poems, novels or music expressing the worldviews and beliefs of those who shaped or experienced those landscapes.

Cosgrove and Daniels insisted on the visuality of landscape and on the centrality of images in both its interpretation and its making. In their influential edited volume *The Iconography of Landscape* (1988), they laid out the principles of their iconographic approach inspired by Panofsky: 'reading what we see'. Through the identification of conventional, consciously inscribed symbols in a painting (for example, the winged lion symbolizing the Venetian republic and its power), iconography enables access to a deeper layer of meaning. It offers, in other words, a key for uncovering the worlds and value systems in which certain images were produced (Cosgrove and Daniels 1988: 2). Landscape is no different: for the Renaissance literate, Palladian villas' design and integration with their natural surroundings symbolized cosmic harmony; to us, a shiny skyscraper housing a bank in central Manhattan speaks of corporate power.

Rather than to a painting, however, landscape is more akin to a theatre (Cosgrove and Daniels 1993). It's not simply looked at and decoded, but also performed. At the same time, far from being a passive backdrop, it also has the power to set certain actors in the foreground and obscure others. More generally, it has the power to influence the actions and behaviours of those occupying it. Today, as during Cosgrove and Daniels' field trips in the 1980s, people enjoy walking up and down the main street or piazza in the historical centre of Vicenza and of many other Italian Renaissance cities originally architected as displays of civic power. In their early evening promenade, 'the citizenry thus parades itself, using the architecture of its city as the stage set for a participatory drama or spectacle' (Cosgrove and Daniels 1989: 174). In England, Victorian parks are likewise conducive of certain habits and behaviours because of their design, which was originally inspired by a specific worldview and iconographic repertoire (Cosgrove 1989: 126). In this sense, landscape is not merely an artefact, but a code of conduct; it communicates and inspires specific sets of values. It is 'not just scenery, but stage for thought and action' (Cosgrove and Daniels 1989: 179). Landscape matters.

Conclusion

As this chapter has shown, the landscape idea is no straightforward concept. It denotes a way of seeing which combines, and in turn mediates, aesthetic, political and moral values. It reminds us that all landscapes—both material and represented—are culturally, socially and historically constructed. Landscapes carry symbolic meaning precisely because they are products of human visions. In this sense, they at once reflect and constitute specific social and cultural environments. Beliefs and ideologies shape the lie of the land. In turn, landscapes shape behaviours and expectations.

Like landscape itself, however, Cosgrove's project was in no way fixed and monolithic, but transformed over time. Emerging primarily in response to different trends in human geography (regional, quantitative and humanistic), it originally adopted a Marxist historical framework to elucidate the origins and workings of the 'landscape idea'. This approach provided the scaffolding for *Social Formation and Symbolic Landscape* and introduced a critical and socially conscious dimension to the study of landscape. Cosgrove, however, subsequently downplayed and even criticized the Marxist narrative for its 'tendency to subordinate both material and imaginative cultural expressions to the imperatives of political economy' (1998[1984]: xv). Pointing at subsequent feminist and postcolonial literature, he also acknowledged its limiting emphasis on social class as the foundation of social action at the expense of other vectors of power and inequality, such as race and gender (see, for example Rose 1993; Mitchell 1996).

Criticisms notwithstanding, the 'landscape idea' opened up new ways of seeing. In Britain it helped create a new subfield in which geographers started to take questions of culture seriously: a new cultural geography attentive to the meanings, representational politics and social relations of power suffusing both the idea and reality of landscape. In the United States, 'the landscape idea' opened cultural and humanistic geography to the study of various representations of the material world (Duncan and Duncan 1988; Duncan 2005). While some criticized *The Iconography of Landscape* and Cosgrove's later work for their dominant focus on art and what they perceived as a 'dematerialization of landscape' (Mitchell 1996; Olwig 1996), the relationship between the land and representation that underpinned this work was far more sophisticated. As Daniels later pointed out, what he and Cosgrove were ultimately attempting in *The Iconography of Landscape* was to describe 'a recurring, historical process in which representation is ingredient to reality, in which surface matters as much as depth, and in which the pleasures and pains of estrangement in landscape apprehension are no less significant than the pieties of dwelling' (2008: 6).

Perhaps even more significantly, the 'landscape idea' helped redefine Geography as a humanistic field of inquiry, for 'the recovery of meaning in ordinary landscapes tells us much about ourselves' (Cosgrove 1989: 134). In this sense, the humanist and moral engagements of Cosgrove's early work are as important as its political concerns. As he declared in an interview at the end of his career and of his prematurely truncated life, for him Geography was primarily about education in the humanist tradition, that is, in the old sense of the study of what make us humans and of our constant dialogue with each other:

> Then out of that dialogue—the fact that we are always social—comes a commonly created world, an architectural world, a world of fields and farms and factories. Where it would be a world of literature and art, it seems to me that we can better come to understand ourselves, which I take it to be the principal task of a human life: to make sense in ourselves. The landscapes that we see around us can serve in the same way as a piece of literature, in the same way as a film, in the same way as a work of art—as expressive aspects of humanity.
>
> (2010: 151)

Study task

Variations on this question can be set on any kind of written or oral exam, or as a term-paper assignment, or as a workgroup assignment topic.

'A landscape is [best understood as] a cultural image, a pictorial way of representing, structuring, or symbolizing surroundings' (Cosgrove and Daniels 1988). Discuss this claim with reference to Cosgrove's work and previous geographers' approaches to landscape.

Starter reading list

Cosgrove, D. (1985) 'Prospect, perspective and the evolution of the landscape idea', *Transactions of the Institute of British Geographers*, 10(1), pp. 45–62.
Cosgrove, D. (1989) 'Geography is everywhere: Culture and symbolism in human landscapes', in Gregory, D. and Walford, R. (eds) *Horizons in Human Geography*. Totowa, NJ: Barnes and Noble, pp. 118–135.
Cosgrove, D. (2006) 'Modernity, community and the landscape idea', *Journal of Material Culture*, 11(1), pp. 49–66.
Cosgrove, D. and Daniels, S. (1993) 'Spectacle and text: Landscape metaphors in cultural geography', in J. Duncan and D. Ley (eds) *Place/Culture/Representation*. London: Routledge, pp. 57–77.
Driver, F. (2008) 'Geography and vision: Denis Cosgrove, 1948–2008', *Environment and Planning A*, 40(8), pp. 1779–1782.
Rose, M. & Wylie, J. (2006) 'Animating landscape', *Environment and Planning D: Society and Space* 24(4), pp. 475–479.

References

Capel, H. (1987) *Filosofia e scienza nella geografia contemporanea*. Milan: Unicopli.
Cosgrove, D. (1985) 'Prospect, perspective and the evolution of the landscape idea', *Transactions of the Institute of British Geographers*, 10(1), pp. 45–62.
Cosgrove, D. (1989) 'Geography is everywhere: Culture and symbolism in human landscapes', in Gregory, D. and Walford, R. (eds) *Horizons in Human Geography*. Totowa, NJ: Barnes and Noble, pp. 118–135.
Cosgrove, D. (1993) *The Palladian Landscape*. Leicester and London: Leicester University Press.
Cosgrove, D. (1998[1984]) *Social Formation and Symbolic Landscape*. Totowa, NJ: Barnes and Noble.
Cosgrove, D. (2006a) 'Modernity, community and the landscape idea', *Journal of Material Culture*, 11(1), pp. 49–66.
Cosgrove, D. (2006b) 'Los Angeles and the Italian città diffusa: Landscapes of the cultural space economy', in Terkenli, T. and D'Hauteserre, A.M. (eds) *Landscapes of a New Economy of Space*. Dordrecht: Springer, pp. 69–91.
Cosgrove, D. (2010) 'The role of geography in the twenty-first century: Interview with Denis Cosgrove by H. Soovali-Sepping', in V. della Dora et al. (eds) *Visual and Historical Geographies: Essays in Honour of Denis E. Cosgrove*, Historical Geography Research Series n. 42. London: Royal Geographical Society – Institute of British Geographers, pp. 147–153.
Cosgrove, D. and Daniels, S. (eds) (1988) *The Iconography of Landscape: Essays on the Symbolic Representation, Design and Use of Past Environments*. Cambridge: Cambridge University Press.
Cosgrove, D. and Daniels, S. (1989) 'Fieldwork as theatre: A week's performance in Venice and its region', *Journal of Geography in Higher Education*, 13(2), pp. 169–182.
Cosgrove, D. and Daniels, S. (1993) 'Spectacle and text: Landscape metaphors in cultural geography', in J. Duncan and D. Ley (eds) *Place/culture/representation*. London: Routledge, pp. 57–77.
Daniels, S. (1985) 'Arguments for a humanistic geography', in Johnston, R. (ed.) *The Future of Geography*. London: Methuen, pp. 143–158.

Daniels, S. (2008) 'Classics in human geography revisited: Denis Cosgrove and Stephen Daniels (eds) (1988), *The Iconography of Landscape: Essays on the Symbolic Representation, Design of Past Environments*. Cambridge: Cambridge University Press', *Progress in Human Geography*, 35(2), pp. 268–270.

Daniels, S. (2018) 'Fieldwork, landscape and performance', in Vallerani, F. (ed.) *Everyday Geographies and Hidden Memories: Remembering Denis Cosgrove*. Venice and Egham, Surrey: Ca' Foscari University of Venice and RHUL Centre for the GeoHumanities, pp. 46–57.

della Dora, V. (2020) 'Landscape and history', in Heffernan, M., Withers, C. and Domosh, M. (eds) *Sage Handbook of Historical Geography*, vol. 1. London: SAGE, pp. 121–143.

Driver, F. (2008) 'Geography and vision: Denis Cosgrove, 1948–2008', *Environment and Planning A*, 40(8), pp. 1779–1782.

Duncan, J. (2005) 'Classics in human geography revisited: Cosgrove, D. 1985: *Social Formation and Symbolic Landscape*. Totawa, NJ: Barnes and Noble', *Progress in Human Geography*, 29(4), pp. 477–479.

Duncan, J. and Duncan, N. (1988) '(Re)reading the landscape', *Environment and Planning D: Society and Space*, 6(2), pp. 117–126.

Freitag, T. and H. Jons (2005) 'Vision and the cultural in geography: A biographical interview with Denis Cosgrove', *Die Erde – Journal of the Geographical Society of Berlin*, 136(3), pp. 205–216.

Livingstone, D. (1992) *The Geographical Tradition*. Cambridge: Cambridge University Press.

Lowenthal, D. (2007) 'Living with and looking at landscape', *Landscape Research*, 32(5), pp. 635–656.

Lowenthal, D. (2008) 'Professor Denis Cosgrove, cultural and historical geographer', *The Independent*, 8 April, p. 35.

Mitchell, D. (1996) *The Lie of the Land: Migrant Workers and the California Landscape*. Minneapolis: University of Minnesota Press.

Olwig, K. (1996) 'Recovering the substantive nature of landscape', *Annals of the Association of American Geographers*, 86(4), pp. 630–653.

Rose, G. (1993) *Feminism and Geography*. Cambridge: Polity Press.

Sauer, C. (1969) *Agricultural Origins and Dispersals*. Cambridge, MA: MIT Press.

Vallerani, F. (2018) 'Shared waterscapes: meandering along a sentimental waterway', in Vallerani, F. (ed.) *Everyday Geographies and Hidden Memories: Remembering Denis Cosgrove*. Venice and Egham, Surrey: Ca' Foscari University of Venice and RHUL Centre for the GeoHumanities, pp. 95–120.

Vallerani, F. (2021) *I piaceri della villa: vivere e raccontare la campagna tra abbandoni e ritorni*. Milan: Le Monnier.

Wylie, J. (2007) *Landscape*. London: Routledge.

12 Boundaries and borders matter

Ron Johnston's electoral geography

Charles J. Pattie

Ron Johnston (1941–2020). Photo credit Chrystal Cherniwchan, with permission.

Introduction

Tip O'Neill, Democrat Speaker of the US House of Representatives between 1977 and 1987, famously opined that "All politics is local". Citizens' political opinions, outlooks and actions are influenced by economic, social and political conditions in the local communities where they live (Agnew, 1987; Enos, 2017). Threats to those local communities generate political protests as citizens seek to protect their area. Politicians build reputations as defenders of the places they represent. During elections, parties structure their campaign activities to best exploit local variations in support to maximise their eventual representation. And so on.

But how might we define a politically consequential place? How local is local – and local for what? Where does one local community end and another begin? No simple consensus exists. Different purposes imply different definitions of locality, at different scales, and the boundaries of these areas are liable to be fuzzy and contested. Each of us lives in a complex, interlocking and imprecisely delineated web of places that matter to and affect us for different reasons: what I consider as my local community when I think of my friends and acquaintances is rather different to the locality within which I work, or

DOI: 10.4324/9781003343240-14

which defines my sense of local or regional identity. What I see as local for any one of these purposes is unlikely to overlap precisely (or perhaps even remotely) with what my next-door neighbour might consider as her local area.

Despite the difficulties, however, some means of delineating boundaries between places is necessary in most political systems. And it matters. Which side of a political boundary, whether between countries or subnational administrative districts, one falls on can affect one's citizenship, the laws under which one lives (and how fairly they are drawn up and enforced), one's taxes, the services one accesses, the nature of one's political representation and so on. Changing boundaries can have far-reaching consequences for those who suddenly find themselves on a different side. So where and how should these boundary lines be drawn? That puzzle has for many years concerned political geographers.

This chapter examines how one influential geographer, Ron Johnston (1941–2020), tackled it through his research on electoral redistricting and how his work informed actual practice. A detailed account of his work is beyond this chapter's remit (though see Pattie et al., 2021). But a brief overview gives a sense of its remarkable range and scale. Over a long career (from the early 1960s to his death in 2020), he was astonishingly prolific, producing more than 800 refereed papers, over 40 single- or co-authored books (and more edited collections), and hundreds of book chapters. His interests encompassed urban social geography (e.g., Johnston, 1971, 1980), quantitative methods (e.g., Johnston, 1978), political geography (Johnston, 1979a, 1982), and the history of human geography (Johnston, 1979b).

This chapter will focus on one aspect of his activity, his work on electoral geography (Taylor and Johnston, 1979; Johnston and Pattie, 2006) – in particular on electoral systems and redistricting (Johnston et al., 2001a, 2021).

We begin by looking at Johnston's development as a geographer, before focusing on his research on electoral systems and redistricting. We'll explore a key tension in drawing up political districts: should the map of districts reflect 'organic' considerations such as the geography of distinct local communities, or should it reflect 'arithmetic' considerations around administrative efficiency and equal representation? In truth, both factors are in play – but (as Johnston's work showed) they are not amenable to easy or neat solutions. As the political scientist E.E. Schattschneider commented over 60 years ago, "organisation is the mobilization of bias" (Schattschneider, 1960, 71; Taylor, 1985). And organising space into discrete defined political territories inevitably creates political biases, either deliberately or accidentally. As we'll see, understanding why that might be the case, and finding ways of uncovering the nature and extent of such biases lay at the core of Johnston's work in this area.

Placing the person

One of Johnston's longstanding academic interests was how geographical context – for instance, the place one is born and raised – influences people. Appropriately, his own outlook was influenced in important ways by where he lived and worked. Born in 1941, he grew up outside the Wiltshire railway town of Swindon. Throughout his life, he retained both his Wiltshire accent and a strong attachment to the county and to Swindon itself – an identity he first consciously embraced when, shortly after he went to university, his family moved to Yorkshire and away from what he saw as his local roots.

He entered Manchester University as a Geography undergraduate in 1959. At the time, human geography was in the grip of the 'quantitative revolution' (see Chapter 4). But the new approaches had not yet reached Manchester. Johnston's early encounters with them came largely from his own work on his Bachelor's and Master's dissertations on central place theory (Johnston, 1984a). A move to Monash University, Australia, in 1964 as a teaching fellow and PhD candidate researching urban social geography (his main focus until the late 1970s) extended the encounter with the 'new geography'.

Taking up a Geography lectureship at the University of Canterbury, New Zealand in 1967, he became increasingly frustrated by the heavily empiricist nature of early quantitative geography (Johnston, 1984a, 44). The more critical approaches to human geography emerging in the late 1960s and early 1970s helped identify ways to make data interpretation more meaningful. Moving back to the UK in 1974 to a Chair in Geography at the University of Sheffield (where he remained for almost 20 years), his teaching included courses on David Harvey's (1973) new politically radical urban geography (see Chapter 8) and on approaches to human geography (the beginnings of his interest in the history of geography: Johnston, 1979b). Increasingly, too, his research focus shifted from urban to electoral geography (Taylor and Johnston, 1979; Johnston and Pattie, 2006). While his research remained quantitative, a theoretical focus on the socially constitutive effects of place on social and political life helped him move beyond 'spatial science' (Johnston, 1986, 1991; Gregory, 1978; Thrift, 1983; Giddens, 1984).

Following a term as Vice Chancellor of the University of Essex between 1992 and 1995, he returned to full-time research and teaching, taking a post as Professor of Geography at the University of Bristol, where he remained for the rest of his life. This brought him back to his roots in the south-west of England and also enabled a return to Wiltshire to live in Salisbury, the county town.

Research and writing were central to Johnston's sense of self. But he wasn't simply an 'ivory tower' academic. Much of his work, particularly on electoral geography, entailed active engagement with the policy world. He served in a variety of roles including as a trustee for the McDougall Trust (a UK-based electoral reform charity), as a Deputy Electoral Commissioner for the Boundary committee for England (the body charged with revising local government boundaries) and as a member of the British Law Commission's advisory group on electoral reform, and often appeared as an expert witness before parliamentary committees, political parties and public bodies, both in the UK and abroad.

Placing the voter

Johnston's interest in elections was first piqued by reading English-American geographer Kevin Cox (1969) on the 'neighbourhood effect' in voting – the tendency for people to adopt the dominant political views in their neighbourhood. He was sceptical: surely people voted based on their personal political principles, not on what others thought? So he set out to disprove Cox – and was unable to do so (Johnston, 1972, 1974; Forrest and Johnston, 1973).

From this beginning, his work on elections grew. One strand continued to explore the neighbourhood effect, linking it to socio-economic influences on political party support. Traditionally, working-class voters leaned to parties of the Left, and middle-class voters to parties of the Right (Lipset and Rokkan, 1967). This implies the geography of party support should follow that of major social and economic patterns, an expectation confirmed by studies combining election results and socio-economic data for electoral districts (Crewe and Payne, 1976). But, Johnston found, things were more complicated. In British general elections, the class cleavage in party support was not nationally uniform, but varied from place to place (Johnston, 1985; Johnston et al., 1988). Where a party's 'core class' was particularly numerous, it tended to pick up more support in all classes. This was linked to a variety of factors, including voters' conversations with their friends and families (Pattie and Johnston, 2001; Johnston, Jones et al., 2005) and local economic conditions (the more buoyant the local economy, the more likely voters there were to support the incumbent government and the less likely to support the main opposition, irrespective of the political hue of either party: Johnston and Pattie, 2006; Johnston, Sarker et al., 2005).

Another strand focussed on the geography of election campaigning. Overturning the received academic wisdom that local campaigning was redundant in a television (and internet) age, he demonstrated that the harder parties campaigned locally, the better they did, and by enough to affect the result in close contests (e.g., Johnston, 1987; Johnston and Pattie, 2014). Major parties clearly recognised this. In Britain's first-past-the-post electoral system, for instance, they concentrate on marginal seats, where a few votes either way could alter who wins, while putting much less effort into seats where they either have little chance of winning, or their victory is all but guaranteed.

Defining political places

But our main focus here is on a third aspect of his electoral work, on electoral redistricting (dividing a country up into smaller electoral districts from which MPs are elected). Most countries are divided internally into smaller political areas – local government areas, parliamentary constituencies and so on. But how are those areas to be defined? In practice, it is not easy.

One solution might be to draw boundaries so as to include all members of a defined community in the same area, with each community in its own demarcated zone. This was one of the criteria used after World War I, when Central and Eastern European multinational empires were broken up into smaller nation-states according to a principle of national self-determination. But, as the American geographer Isaiah Bowman (1928) recognised, this was easier said than done. Since ethnic and national groups were intermingled on the ground, 'national minorities' were inevitable in all new states. And what of people who didn't see themselves as members of the national community to which they had been assigned? Far from providing a once-and-for-all solution for Europe's political map, the post-World War I settlement was just one among many possible new maps which could have been produced – and it contained the seeds of future conflict and change.

Alternatively, boundaries might be drawn on the basis of functional, rather than community, criteria. The major 1974 reorganisation of English local government provides an example (Dearlove, 1979). New Metropolitan County authorities were created to govern England's major city-regions, enabling greater co-ordination and economies of scale. But their boundaries often ignored local sensibilities, lumping disparate communities into the same overarching authority. Just over a decade after they were first established, the 'Met Counties' were abolished, partly because they proved unpopular with the Conservative government of the 1980s. But they also failed to generate significant local community support: people didn't identify with them.

Nor is the problem simply one of competing (and unclear) criteria. Any attempt to draw new boundaries between political places inevitably runs into the modifiable areal unit problem (or MAUP: Openshaw, 1984). Each definition of new boundaries creates a new geography, even if the underlying patterns do not change. To see why, imagine a region divided into 36 small areas, each with the same number of people in it (Figure 12.1a). It is inhabited by two equally-sized groups, the Uppies and the Downies. The Uppies all live in the areas shaded grey, while the Downies all live in the unshaded areas. If we were given the task of dividing the region into three districts of equal population

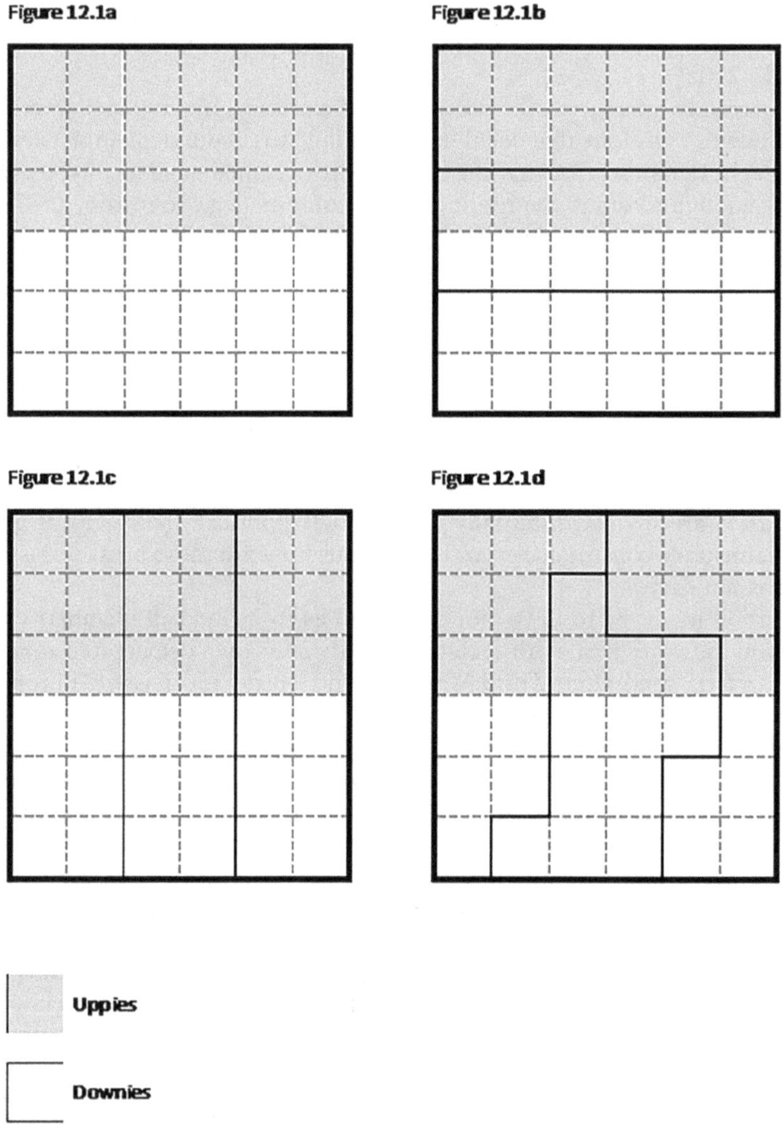

Figure 12.1 Change the boundaries, change the geography: the MUAP. a) The region; b) Solution 1: 1 Uppy, 1 Downie, 1 mixed; c) Solution 2: All mixed d) Solution 3: 2 Uppy, 1 Downy.

size (i.e., each containing 12 of the small areas), how might we do it? Figures 12.1b–d show three different possibilities (there are others: how many can you find?). Although the Uppies and Downies haven't changed locations, the three different sets of districts produce rather different apparent geographies. The solution displayed in Figure 12.1b would create one district which was 100% Uppy, one which was 100% Downy and one which was split 50:50 between the two groups. The map proposed in Figure 12.1c, meanwhile, would create three 50:50 districts – neither the Uppies nor the Downies would have a majority in any district. And the third solution (Figure 12.1d) would give the Uppies a majority over the Downies in two of the three new districts.

And the new geographies so created can be politically consequential. If the Uppies and Downies were opposing political groups, and the districts were local government areas, the alternative maps produced by the different district proposals could have quite far-reaching consequences for the political control of each new local authority, and hence for the sorts of policies enacted there (and so for locals' quality of life).

Johnston's research on political boundaries explored these issues. He worked on the politics of residential segregation in US cities, where fragmented residential geographies were reflected in fragmented patterns of local government (Johnston, 1984b). Boundaries were often drawn (and redrawn) to exclude some groups from the benefits that others living within a local government jurisdiction might receive. Richer communities tried to draw local government boundaries around themselves, maximising the local government tax base while minimising the numbers of (poorer) residents who might make potentially expensive demands on local services. Predominantly white communities attempted to design school districts to minimise the number of non-white students attending local schools. Lines on the map had profound political and social consequences.

Redrawing the electoral map

But Johnston's most substantial work was on electoral boundaries. An early step came in New Zealand, when he encountered Ralph Brookes' work on electoral bias in single-member plurality elections (Brookes, 1960). In elections employing such first-past-the-post (FPTP) rules, results tend to be disproportional. Winning parties tend to gain a higher share of seats in Parliament than they do of votes in the country, and smaller parties, especially those whose support is not geographically concentrated, tend to be seriously under-represented. Brookes' insight was that FPTP could also be structurally biased – one party might consistently receive a better return of seats relative to its vote than a rival (even if it lost the actual election). He proposed a measure for the relative bias between two parties: how many more seats would one party win than the other if both had obtained the same national share of the vote? In addition, his method showed how any bias was created, and how important each source was. Johnston extended Brookes' method to examine electoral bias in New Zealand (Johnston, 1976), the USA (Johnston, Pattie, and Rossiter, 2001b; Tamas et al., 2022), Australia (Johnston and Forrest, 2009) and the UK (Johnston et al., 1994, 2001a, 2021; Johnston, 2002).

Electoral bias occurs when one party is more efficient than another in converting its votes into seats. This happens because of two different effects: malapportionment and gerrymandering. Malapportionment occurs when one party tends to be popular in seats with below average numbers of voters, and another party tends to be popular in seats with above average numbers. Under those circumstances, the former party gains an advantage over the latter, as it needs fewer votes to elect each MP. Gerrymanders, meanwhile, are created when electoral districts' boundaries are drawn so as to favour one party over another (even if all seats have the same number of voters). For instance, in a packed gerrymander, one party might have its supporters disproportionately packed into a few seats that it's then bound to win by large majorities, while its rival's voters are more evenly spread across a larger number of seats. Under these circumstances, the 'packed' party is disadvantaged, as many of its votes are 'wasted' in super-majorities that do not elect it extra MPs.

Malapportionment and gerrymandering take us into the realms of electoral redistricting, the periodic redrawing of electoral district boundaries to take account of the effects of changing population geographies. Failure to redraw electoral districts regularly can produce a creeping malapportionment if some parties are over-represented in constituencies with declining electorates, and others are over-represented in constituencies with growing electorates. The malapportionment so produced gives the former parties an advantage over the latter, as fewer votes are required to elect each of their MPs. But redrawing the boundaries to return to equal-sized electorates can produce new gerrymander-like effects, due to the MAUP: change the boundaries and potentially change the outcomes! Understanding how the process was conducted in Britain, and with what effect, was a major strand of Johnston's work (see, e.g., Rossiter et al., 1999; Johnston et al., 2021).

Electoral redistricting in the United Kingdom

On the face of things, the British redistricting process might seem uncontroversial. In contrast to countries (notably the USA: McGann et al., 2016) where electoral redistricting is largely in the hands of politicians (and hence is open to overt political manipulation and gerrymandering), in the UK the process has, since the 1940s, been conducted every few years by strictly non-partisan and impartial Boundary Commissions (one for each of the four constituent countries of the UK). By law, they can consider only two factors: reducing inequalities in electorates between seats (an 'arithmetic' goal) and representing local communities (an 'organic' goal). In each review, they produce initial proposals for the new constituency map. These are put out for public consultation, and are then revised in the light of the consultations. During the consultation phase, the political parties (and others) can suggest possible changes to the Commissions' proposals. The parties' suggestions usually serve their partisan interests, though they can only be argued on the basis that they better represent communities or equalise electorates. At no point do the Commissions give consideration to political arguments and they are not obliged to make any changes. But they will consider options which seem to better achieve their goals. They then produce final recommendations for the new seats: politicians cannot tinker with them (though they can delay their implementation).

Even non-partisan redistricting processes can have political consequences, however (Gudgin and Taylor, 1979). Because of the MAUP, any given redistricting exercise could produce many alternative constituency maps, all compatible with creating more equal electorates. As we have seen, different solutions may have different partisan consequences. But how many different solutions are there, and how much variation might there be in political outcomes? To get an idea, Johnston and his collaborator David Rossiter wrote, in the early 1980s, a computer programme that identified all possible ways Sheffield City Council's 27 local government wards could be combined to create six parliamentary constituencies, each with an electorate within ±12% of the city average (Johnston and Rossiter, 1982).

Table 12.1 Redistricting Sheffield in 1979: the political effects of alternative sets of constituencies (from Johnston and Rossiter, 1982)

Number of unique solutions	*Seats won by Labour*	*Seats won by Conservatives*
12,327	5	1
2,913	4	2
697	6	0

They found 15,937 unique solutions (Table 12.1). In most, Labour would have taken five of the seats and the Conservatives one (based on voting in the 1979 local elections) – but some solutions would have given the Conservatives two of the six seats, and some none. Across an entire country (the UK currently has 650 constituencies), there are vastly more alternative ways of drawing the map – with a wide range of partisan consequences.

When the Boundary Commission for England produced its preliminary proposals for new seats in Sheffield during the Third Periodic Review in the early 1980s, Johnston and Rossiter's programme identified alternative constituency maps for the city which proved better at achieving equal electorates than the Commission's own initial efforts. So they submitted written evidence to the Sheffield public inquiry. Their alternative recommendations were accepted by the Commission.

As a consequence of their evidence to the Sheffield inquiry, they were approached by a local Labour councillor (and prospective MP) on behalf of a group of senior party figures who were planning a legal challenge against the Commission's final recommendations for England (the case was launched in 1982). They wanted to delay the adoption of new seats until after the upcoming general election, as they feared Labour would lose out as a result of the new map. But, as political considerations have no legal force in British boundary reviews, this couldn't be the formal basis of their case. Instead, they argued the Commission had wrongly applied its own rules, by setting too much store on representing communities, and not enough on equalising electorates. Johnston and Rossiter acted as expert witnesses, demonstrating that it was possible to produce seats that achieved greater equality in electorates than those proposed by the Commission. The case was not successful, the judge ruling that the law then in place gave the Commissions considerable leeway in interpreting the relative merits of achieving equal electorates, representing communities and minimising disruption to existing arrangements. But it helped establish Johnston's credentials as an academic expert on redistricting. Through his research into the operation of the Third and subsequent periodic reviews of Britain's constituency map, he interviewed (and was respected by) the key figures in all the main parties with responsibility for electoral policy.

Estimating the effect of changing (and not changing) electoral boundaries

Worry over the MAUP was not why the Labour Party figures who engaged Johnston and Rossiter for the 1982 case hoped to delay the adoption of new constituencies, however. Rather, the reasons lay in two trends that, for most of the last 80 years, typified Britain's population geography: suburbanisation, as people moved out of city centres; and a general drift from the North and Midlands to the South-East (Gudgin and Taylor, 1979; Rossiter et al., 1999; Johnston et al., 2021). Both meant areas where the British Labour party was traditionally strong tended to lose population relative to areas where the

Conservatives had more support. The longer a particular constituency map had been in effect, therefore, the smaller average electorates became in Labour seats, and the larger in Conservative ones. That created a malapportionment bias that increasingly advantaged Labour relative to the Conservatives. By reducing the number of constituencies in areas experiencing relative population decline and increasing the number in areas experiencing relative growth, boundary reviews reduced this pro-Labour bias (only for it to re-emerge as the new seats aged in their turn).

In addition, the legislation which governed the UK's review process for much of the post-war period guaranteed minimum numbers of seats in Westminster for Scotland and Wales (Rossiter et al., 1997). As both countries were losing population relative to England for much of the period, this meant both became increasingly over-represented in Westminster (an effect which was not substantially changed by recurring reviews, as the minimum number of seats for each country were fixed). And as Labour was more popular than the Conservatives in Scotland and Wales, this also resulted in an unintentional pro-Labour bias.

Not surprisingly, therefore, Conservatives looked forward to regular reviews, while Labour did not (other, smaller, parties generally had more to gain from shifting the whole basis of UK elections to a more proportional system). Adapting Brookes' bias measure, Johnston estimated just how large an advantage Labour gained over the Conservatives due to ageing seats and the over-representation of Scotland and Wales (Johnston et al., 2021). Between 1950 and 2019, the average advantage Labour gained over the Conservatives from variations in constituency electorates was in the order of 21 more seats if the parties tied on votes. This dropped to just a nine-seat advantage for Labour over the Conservatives in the first election after each review (an advantage remained as the reviews did not achieve complete equality in electorates) and then increased again as the seats aged. The effect was noticeable, but not large relative to the overall size of the UK parliament. Scottish and Welsh over-representation, meanwhile, on average gave Labour a seven-seat advantage over the Conservatives – and was not reduced by reviews as both countries' representation was protected.

But Johnston's work on bias at all post-war British general elections also revealed a bigger picture. Variation in electorates between seats is not the only source of bias. It can also arise from variations in turnout (if turnout tends to be lower in seats won by one party than in those won by another, the former party gains an advantage), and by the relative efficiency with which each party's vote is spread relative to the geography of seats (parties which tend to win seats with very large majorities gain less of an advantage than parties which win by narrow margins). Put all that together, and the overall bias at any given election can tell a somewhat different story to the bias resulting just from varying electorates. Through the 1950s, for instance, Britain's redistricting rules allowed smaller seats in rural than in urban areas. As a result, overall net bias tended to favour the Conservatives rather more than Labour! The removal of protection for rural areas removed that bias, and through the 1960s, 1970s and 1980s, the overall net bias between Labour and Conservatives was very small, and Labour's 'gains' from ageing seats were largely cancelled out by the Conservatives' more efficient distribution of votes.

But, from the 1992 election, Britain's elections became increasingly and substantially biased in Labour's favour. This was partly because of lower turnout in Labour seats, but especially because of Labour's increasingly efficient vote distribution – a result of the party campaigning more efficiently by focusing on marginal seats, which links back to

Johnston's research on campaigning. By the 2001 election, Labour would have gained 111 more seats than the Conservatives, had they tied on votes. What is more, after 1992, boundary reviews ameliorated but did not remove Labour's substantial advantage. As a result, the Conservatives began to look seriously at the rules for redistricting (though, ironically, the major sources of the pro-Labour bias – turnout and efficiency – had nothing to do with variations in constituency electorates).

From analysing redistricting to reform

But what changes were needed? Johnston identified three interconnected problems in the rules governing UK boundary reviews between the 1950s and the 2000s (Rossiter et al., 1999; Johnston et al., 2001a). First, the arithmetic goal was vaguely defined: the Commissions were required to propose seats which were 'as near the (average electorate) as practicable', but gave no exact guidance on how close electorates should be to each other. Second, the rules were ambiguous (and potentially contradictory) on whether the Commissions should prioritise the arithmetic (equalising electorates) or the organic (community representation) criterion (Johnston, Rossiter et al., 1993). Third, the review process was painfully slow. The period between reviews could be long (from the late 1950s to the 1990s, reviews only had to be held every 10–15 years) and could be altered by governments arbitrarily accelerating or slowing down the process. And reviews themselves, once under way, could take several years to complete (the Third Periodic Review began in 1976 but the new constituencies were ready only for the 1983 election; the Fourth, beginning in 1991, didn't come into effect until 1997).

The first two issues gave the Commissions, and third parties making representations through the public consultation process, considerable room for manoeuvre, as no exact yardsticks were available. How equal should seat electorates be? What was a local community? No-one knew. Johnston's close examination of public inquiries throughout the Fourth Periodic Review showed that even in the same areas, very different and sometimes mutually contradictory claims about community ties could be put to the Commission, which had no means of adjudicating between them. And because reviews use data on local electorates as they were at the start of the review process to plan seats, the slow progress of reviews meant that the new seats were already substantially out of date when they were first used, because of population change during the review process (with inevitable consequences for malapportionment effects).

Johnston and his colleagues suggested new rules to deal with these issues (Rossiter et al., 1996; Johnston et al., 2001a). Their main proposals involved: fixing the size of the House of Commons at 650 MPs (the previous rules had been ambiguous on this, creating a tendency for Parliament to increase in size over time); no over-representation of Scotland, Wales or Northern Ireland; removing the requirement to represent 'local community ties' (as their research showed that these were almost impossible to evaluate definitively, and were in practice used by parties in their representations to public inquiries as a means of justifying proposals which were really intended to forward their political interests); ensuring that the achievement of greater electoral equality was the top priority for the Commissions; and requiring them to ensure that all seats proposed would have electorates within a defined range (±10%) of the average. They also proposed that the Commissions should be required to complete each review no later than 10 years after the end of their previous review (to reduce the problems of over-long gaps between reviews).

Although not produced with any particular party in mind, these proposals influenced Conservative thinking, while in opposition in the 2000s, on how to reform the boundary review system: the party consulted Johnston and his team on the issue (along with other academic experts). Once in coalition government with the Liberal Democrats after the 2010 election, the Conservatives published a Bill outlining their proposed new rules. While not exactly the same as Johnston's, these were recognisably similar. Like Johnston, they proposed: faster and more frequent reviews; a fixed size for the House of Commons; a single electoral quota for all seats in the UK (so no over-representation for the Celtic nations); clear priority for the goal of establishing electoral equality; and a clearly defined target range within which (almost) all new seats' electorates should sit.

But the government's proposals also differed in important ways from Johnston's. The most headline-catching was a proposal to cut the size of the Commons by 50 to 600 MPs. But the plans also suggested a much tighter range (±5% of the quota) within which seats' electorates should lie. They retained a role (albeit now clearly subordinate to achieving more equal electorates, which was to be the Commissions' primary goal) for local ties. And they envisioned much more frequent reviews, every five years (a further proposal, to remove the public consultation phase, was in part intended to reduce the time reviews would take). Combined with another of the coalition government's new laws, the Fixed-term Parliaments Act 2011, this implied that every future election would be fought in a new constituency map.

While welcoming the greater fairness and consistency of the proposed new rules, Johnston pointed out, in both academic papers and in evidence to various parliamentary committees, that they also contained potential problems (e.g., Johnston et al., 2012a; Balinski et al., 2010). In particular, the tight range for permissible electorates would produce much more disruption to the constituency map than had been the case in past reviews, especially if the Commissions held to their practice of using local government wards as the basic building blocks of new seats (Johnston et al., 2021). Importantly, this would be a feature not just of the first review under the new rules (where many expected it due to the proposed reduction in the number of MPs), but of all subsequent reviews, with adverse implications for continuity in MPs' representation of their constituents. The problem was that under the proposed new rules, even constituencies with electorates already within the required range at the start of a review (and which therefore might be thought to be safe from change) would be likely to have their boundaries altered in order to accommodate the knock-on consequences of necessary changes elsewhere in a region.

An example from the first – uncompleted – review to be conducted after the passage of the new legalisation illustrates the point. The Boundary Commission for England's initial proposals for the Yorkshire and the Humber region made substantial changes to seats in urban West Yorkshire, where the electorates of many old constituencies were outside the required range. But to make the arithmetic work, they had to include North Yorkshire wards in some new West Yorkshire seats. This created local ties issues, placing some rural communities in largely urban constituencies. But it also created pressures for change throughout North Yorkshire, where most existing constituencies' electorates were within the required margins (and so might have been thought safe from redistricting). Those that lost wards to solve problems in West Yorkshire now had electorates which were too small. To solve that problem, they had to be redrawn, including wards from neighbouring North Yorkshire seats, which then fell below quota, necessitating further change ... and so on, through the county. Although initially it had seemed largely safe from revision, North Yorkshire's constituency map would actually have been substantially altered – and much more so than under the old rules.

During the Bill's progress through Parliament, Johnston's warnings about the potential for serious disruption at each review were raised by critics (and critical supporters) of the measure. However, when enacted in 2011, the key clauses creating the risk were untouched (the Act did, however, reinstate a public consultation phase, in a revised and faster form). The Boundary Commissions promptly set to work. As the review rolled out, it did indeed prove highly disruptive in much the way predicted (Johnston et al., 2012b). It did not result in a new constituency map, however: a dispute between the coalition partners prevented that. In 2018, the Commissions began a new review, also under the 2011 rules: the review was completed, but was not implemented.

Several attempts were made to amend the 2011 Act. Johnston and his colleagues once again acted as expert witnesses. In a report written in part to guide possible revisions to the law, they showed that the disruption to the existing constituency map consequent on a review could be considerably reduced simply by adopting a slightly wider target range for constituency electorates after a review, of ±8% around the national average electorate (Johnston et al., 2014). A greater willingness on the part of the Commissions to split wards between seats would also help.

In 2019, the Conservative government introduced a Bill reforming the 2011 Act. Its main effects were to fix the size of the House of Commons at 650 MPs (the target of 600 MPs had never been achieved), to increase the time between reviews from five to eight years, and to remove the requirement that Parliament should approve the Commissions' recommendations (they were to take effect automatically). But in other respects, including, crucially, the rules on the permitted variation in electorates, the 2011 rules remained unchanged. Johnston and his colleagues once more made the case for a wider tolerance around the average constituency electorate (a view shared by many opposition MPs). When he died in May 2020, he was preparing for an upcoming Parliamentary Committee hearing on the new Bill. It was subsequently enacted, largely unchanged.

Conclusions

Political boundaries are inescapable features of all existing political systems. They are one part of an elaborate structure of material and symbolic distinctions that all societies create in different ways. They define the territorial limits of a state's authority, the area over which citizens might be eligible for some services or subject to some rules and taxes, and – in their electoral manifestation – who is represented by whom. They are also human constructs. There are no 'natural' political communities to fall back on when drawing up boundaries. Nor are there uncontested and unambiguous criteria which can be readily applied to achieve 'optimum' results. Each and every attempt to carve up political space has political consequences, and, thanks to the MAUP, has many different possible solutions, none of which is incontestably 'correct'.

As a result, in many political systems, drawing electoral units is a task which politicians manipulate to maximise their own partisan benefits. But, as Johnston's work on redistricting in the UK demonstrates, even where the task is given to a scrupulously non-partisan body, tensions remain, and redistricting is politically consequential (as is not redistricting, it bears saying). Those tensions partly reflect competing principles – the arithmetic goal of achieving near-equal electorates in all constituencies and the organic goal of representing (inevitably fuzzy and contested) communities of interest. As Johnston showed, Schattschneider was right – all organisation is indeed bias, not least in electoral systems. But, as his work also illustrated, we can at least understand some of its sources, its size and its consequences. And we can use that understanding in efforts to achieve reform.

Study task

This task could form the basis of a term paper, a seminar discussion or a practical assignment.

Compare the arrangements for designing electoral districts in two countries of your choice. How does each system work, what roles (if any) do politicians and the public play in the process, and to what extent is each vulnerable to partisan meddling?

Starter reading list

Johnston, R.J. (2002) 'Manipulating maps and winning elections: measuring the impact of malapportionment and gerrymandering'. *Political Geography* 21: 2–31.
Johnston, R.J. (2015) 'Which Map? Which Government? Malapportionment and gerrymandering, UK-style'. *Government and Opposition* 50: 1–23.
Johnston, R.J. (2018) 'America gerrymanders on – for the moment?' *The Political Quarterly* 89(4): 667–678.
Johnston, R.J. and Pattie, C.J. (2008) 'Place and vote'. In Cox, K.R., Low, M. and Robinson, J. (eds), *The Sage Handbook of Political Geography*. London: Sage Publications.
Johnston, R.J., Pattie, C.J. and Rossiter, D.J. (2013) 'Manipulating territories: British political parties and new parliamentary constituencies'. *Territory, Politics, Governance* 1: 223–245.
Pattie, C.J., Taylor, P.J. and Jones, K. (2021) 'Johnston, Ron, 1941–2020'. *Biographical Memoirs of Fellows of the British Academy* 20: 89–109.

References

Agnew, J. (1987) *Place and politics: The geographical mediation of state and society*. Boston: Allen and Unwin.
Balinski, M., Johnston, R., McLean, I. and Young, P. (2010) *Drawing a new constituency map for the United Kingdom: The Parliamentary Voting System and Constituencies Bill 2010*. London: British Academy.
Bowman, I. (1928) *The new world: Problems in political geography* (4th edn). London: George Harrap and Co.
Brookes, R.H. (1960) 'The analysis of distorted representation in two-party single-member elections', *Political Science*, 12, 158–167.
Cox, K. (1969) 'The voting decision in a spatial context', *Progress in Geography*, 1, 81–118.
Crewe, I. and Payne, C. (1976) 'Another game with nature: an ecological regression model of the British two-party vote ratio in 1970', *British Journal of Political Science*, 6, 43–81.
Dearlove, J. (1979) *The reorganisation of British local government: Old orthodoxies and a political perspective*. Cambridge: Cambridge University Press.
Enos, J. (2017) *The space between us: Social geography and politics*. Cambridge: Cambridge University Press.
Forrest, J. and Johnston, R.J. (1973) 'Spatial aspects of voting in the Dunedin City Council elections of 1971', *New Zealand Geographer*, 29, 166–181.
Giddens, A. (1984) *The constitution of society*. Cambridge: Polity Press.
Gregory, D. (1978) *Ideology, science and human geography*. London: Hutchinson.
Gudgin, G. and Taylor, P. (1979) *Seats, votes and the spatial organisation of elections*. London: Pion.
Harvey, D. (1973) *Social justice and the city*. London: Edward Arnold.
Johnston, R.J. (1971) *Urban Residential patterns: An introductory review*. London: G. Bell and Sons Ltd.
Johnston, R.J. (1972) 'Spatial elements in voting patterns at the 1968 Christchurch City Council elections', *Political Science*, 14, 49–61.
Johnston, R.J. (1974) 'Local effects in voting at a local election', *Annals, Association of American Geographers*, 64, 418–429.

Johnston, R.J. (1976) 'Spatial structure, plurality systems and electoral bias', *The Canadian Geographer/Le Géographe Canadien*, 20, 259–352.
Johnston, R.J. (1978) *Multivariate statistical analysis in geography: A primer on the general linear model*. London: Longman.
Johnston, R.J. (1979a) *Political, electoral and spatial systems*. Oxford: Oxford University Press.
Johnston, R.J. (1979b) *Geography and geographers: Anglo-American human geography since 1945*. London: Edward Arnold.
Johnston, R.J. (1980) *City and society: An outline for urban geography*. Harmondsworth: Penguin.
Johnston, R.J. (1982) *Geography and the state*. London: Macmillan.
Johnston, R.J. (1984a) 'A foundling floundering in world three', In Billinge, M., Gregory, D. and Martin, R. (eds) *Recollections of a revolution: Geography as spatial science*. London: Macmillan.
Johnston, R.J. (1984b) *Residential segregation, the state and constitutional conflict in American urban areas*. London: Academic Press.
Johnston, R.J. (1985) *The geography of English politics: The 1983 general election*. London: Croom Helm.
Johnston, R.J. (1986) 'The neighbourhood effect revisited: Spatial science or political regionalism?' *Environment and Planning D: Society and Space*, 4(1), 41–55.
Johnston, R.J. (1987) *Money and votes: Constituency campaign spending and election results*. London: Croom Helm.
Johnston, R.J. (1991) *A question of place: Exploring the practice of human geography*. Oxford: Basil Blackwell.
Johnston R.J. (2002) 'Manipulating maps and winning elections: Measuring the impact of malapportionment and gerrymandering', *Political Geography*, 21, 1–31.
Johnston, R.J. and Forrest, J. (2009) 'Geography and election results: Disproportionality and bias at the 1993–2004 elections to the Australian House of Representatives', *Geographical Research*, 47, 95–108.
Johnston, R.J., Jones, K., Propper, C. Sarker, R., Burgess, S. and Bolser, A. (2005) 'A missing level in the analysis of British voting behaviour: The household as context as shown by analyses of a 1992–1997 longitudinal survey', *Electoral Studies*, 24, 201–225.
Johnston, R.J. and Pattie, C.J. (2006) *Putting voters in their place: Geography and elections in Great Britain*. Oxford: Oxford University Press.
Johnston, R.J. and Pattie, C.J. (2014) *Money and electoral politics: Local parties and funding in general elections*. Bristol: Policy Press.
Johnston, R.J., Pattie, C.J. and Allsopp, J.G. (1988) *A nation dividing? The electoral map of Great Britain 1979–1987*. Harlow: Longmans.
Johnston, R.J., Pattie, C.J., Dorling, D. and Rossiter, S. (2001a) *From votes to seats: The operation of the UK electoral system since 1945*. Manchester: Manchester University Press.
Johnston, R.J., Pattie, C.J. and Rossiter, D.J. (1994) 'The geography of voting and representation: Regions and the declining importance of the cube law'. In Heath, A., Jowell, R., Curtice, J. and Taylor, B. (eds). *Labour's last chance? The 1992 election and beyond*. Aldershot: Dartmouth.
Johnston, R.J., Pattie, C.J. and Rossiter, D. (2001b) 'He lost... but he won! Electoral bias and George W Bush's victory in the US presidential election, 2000', *Representation*, 38, 150–158.
Johnston, R.J., Pattie, C.J. and Rossiter, D.J. (2012a) 'The principles and processes of redistribution: Issues raised by recent UK legislation', *Commonwealth and Comparative Politics*, 50: 3–26.
Johnston, R.J., Pattie, C.J. and Rossiter, D.J. (2012b) '"Somewhat more disruptive than we had in mind": The Boundary Commission for England's 2011 proposed redistribution of Parliamentary constituencies', *The Political Quarterly*, 83, 44–59.
Johnston, R.J., Pattie, C.J. and Rossiter, D. (2021) *Representative democracy? Geography and the British electoral system*. Manchester: Manchester University Press.
Johnston, R.J. and Rossiter, D. (1982) 'Constituency building, political representation and electoral bias in urban England'. In Herbert, D.T. and Johnston, R.J. (eds) *Geography and the urban environment Volume V*. Chichester: John Wiley.
Johnston, R.J., Rossiter, D and Pattie, C.J. (2014) *Equality, community and continuity: Reviewing the UK rules for constituency redistributions*. London: McDougall Trust.
Johnston, R.J., Rossiter, D.J., Russell, A.T. and Pattie, C.J. (1993) 'The changing interpretation of ambiguous rules. the Boundary Commission for England', *Urban Geography*, 14, 507–515.

Johnston, R.J., Sarker, R.E., Jones, K., Bolster, A., Propper, C. and Burgess, S. (2005) 'Egocentric economic voting and changes in party choice: Great Britain, 1992–2001', *Journal of Elections, Public Opinion and Parties*, 15, 129–144.
Lipset, S.M. and Rokkan, S. (1967) 'Cleavage structures, party systems and voter alignments: an introduction'. In Lipset, S.M. and Rokkan. S. (eds) *Party systems and voter alignments*. New York: The Free Press.
McGann, A.J., Smith, C.A., Latner, M. and Keena, A. (2016) *Gerrymandering in America: The House of Representatives, the Supreme Court and the future of poplar sovereignty*. Cambridge: Cambridge University Press.
Openshaw, S. (1984) *The modifiable areal unit problem: Concepts and techniques in modern geography 34*. Norwich: Geo Books.
Pattie, C.J. and Johnston, R.J. (2001) 'Talk as a political context: conversation and electoral change in British elections, 1992–1997', *Electoral Studies*, 20, 17–40.
Pattie, C.J., Taylor, P.J. and Jones, K. (2021) 'Johnston, Ron, 1941–2020', *Biographical Memoirs of Fellows of the British Academy*, 20, 89–109.
Rossiter, D.J., Johnston, R.J. and Pattie, C.J. (1996) 'Reforming the Act and the procedures'. In McLean, I. and Butler, D. (eds) *Fixing the boundaries: Defining and redefining single -member electoral districts*. Aldershot: Dartmouth.
Rossiter, D.J., Johnston, R.J. and Pattie C.J. (1997) 'The evolution and partisan impact of Scottish and Welsh over-representation in the redrawing of British Parliamentary constituencies', *Regional and Federal Studies*, 7, 49–65.
Rossiter, D., Johnston, R. J. and Pattie, C.J. (1999) *The boundary commissions: Redrawing the UK's map of parliamentary constituencies*. Manchester: Manchester University Press.
Schattschneider, E.E. (1960) *The Semisovereign People: A realist's view of democracy in America*. New York: Holt, Reinhart and Winston.
Tamas, B., Johnston, R. and Pattie, C. (2022) 'The impact of turnout on partisan bias in US House Elections, 1972–2018', *Social Science Quarterly*, 103, 181–192.
Taylor, P.J. (1985) 'All organization is bias: The political geography of electoral reform', *The Geographical Journal*, 151(3), 339–343.
Taylor, P.J. and Johnston, R. J. (1979) *Geography of elections*. Harmondsworth: Penguin.
Thrift, N.J. (1983) 'On the determination of social action in space and time', *Environment and Planning A*, 1, 23–57.

13 Mobility matters

Movement, meaning, and practice in the context of power

Tim Cresswell

Tim Cresswell (1965–)

Introduction

During the recent COVID-19 lockdowns across the Global North it became very clear how important various forms of mobility – geographical movement – are to everyday life. Moving to or from work, international travel, daily walks, the movements of goods to and from shops, and to and from our homes were all dramatically altered. How our mobilities were altered depended a lot on whether we lived in the Global North or Global South, in affluent areas or impoverished ones. Covid's impacts on our mobilities were socially and geographically variable. It was also quickly apparent how important mobility was to the virus itself. It needed humans to move in order to spread, and may well have been caused, in part, by humans moving into areas of the 'natural' world

DOI: 10.4324/9781003343240-15

which encouraged inter-species transmission and mutation. Along with all its other impacts, COVID-19 has made clear the importance of mobility to everyday life across the globe.

Why mobility matters

'Mobility' is clearly a central concept for the discipline of Geography. It sits alongside other obviously geographical concepts like space, place, territory, region, scale, and landscape. And yet it's different from these other concepts in that, at first glance at least, all of the others are based on some idea of fixity and stasis while mobility is about flow and dynamism. In this sense, mobility is the opposite of what some have conventionally thought of as Geography. Indeed, in many instances, the pervasiveness of mobility has been equated with the end, or 'death', of Geography. Perhaps this is why, despite its clearly geographical character, mobility hasn't been conceptualized within Geography until relatively recently. So, what is mobility? When I use this word in non-academic contexts, people often think first of social mobility – the ability (or more often lack of opportunity) for people to move between positions in a social hierarchy. What I mean by it is the movement of people, things, images, data, or ideas across, within, or between locations. The discipline of Geography has long considered the movement of people, things, and ideas within different sub-disciplines. Two kinds of movement stand out – circulation and migration. Circulation has traditionally referred to temporary and repetitive movement such as commuting or the delivery of goods. It's primarily been dealt with by the subdiscipline of transport geography. Migration has most often referred to more permanent and less frequent movements where the intent is to leave one place and arrive in another. Migration has traditionally (before the 1980s) been dealt with by population geographers. Clearly, these are not discrete categories and are quickly broken down by, for instance, the recognition of circular migration. Movement was central to the quantitative geography that arose in the late 1950s and continues to this day. Advocates of this 'spatial science' placed spatial interaction (interactions of one kind or another between places) and gravity models at the heart of their endeavours (Abler, Adams, and Gould 1971). Questions of why people, things, or information move, how often they move, how far they move, how fast they move, and by what routes they move have long fascinated geographers as well as other social scientists.

The fact that movements of one kind or another have been central to Geography should not come as a surprise. Moving is, after all, central to our experience of the world. Our days are filled with movement as we make our way through time and space – going to and from work, school, shops, or leisure spaces. We go for walks or bike rides, drive cars or take buses. Things come to us too. Packages get delivered, energy and water flow in and out of our houses. Shelves get restocked. The geography of our world is as much one of movement as it is one of boundaries, regions, places, or scales.

Despite the observations that forms of movement have been central to both the discipline of Geography and our experience of the world, it remained the case that the concept of mobility had been largely taken for granted by geographers and others until around the year 2000. A lot of conceptual work has been done on other concepts in Geography since the 1970s. Space, place, territory, region, landscape, and scale had all been subjected to increasing theoretical scrutiny since at least the 1970s and the advent of both humanistic and Marxist conceptualizations within Geography. To take just one example here, place wasn't really thought of as a concept in Geography before humanistic geographers in the 1970s began to ask what exactly place was and what role it played in human experience of the world (Tuan 1977; Buttimer and Seamon 1980; Relph 1976). Before then place was taken for granted. We broadly knew what places were and the idea of place did not seem worthy of conceptualization. A place was a dot on a map. Humanists asked, for the first time, what place is and why it is important to humans. They began to show how place provided a focus of identity, how places gave meaning to human life. Critical geographers took this in different directions, showing how place is important in the creation of social hierarchies, forms of exclusion, and the production of power (Massey 1997; Sibley 1995). By 2000, there was a well-developed set of conceptual arguments around the importance of place in geography. The same could not be said for mobility, despite the fact that it was clearly just as important to human meaning making and politics as place.

It was in this context that, twenty years ago, I wrote the book *On the Move: Mobility in the Modern Western World* (Cresswell 2006). In that book I defined mobility as a combination of the fact of movement, meanings associated with movement, and movement as practice. All this in the context of power. By movement, I simply meant motion across motion between locations – the line that links A to B on a map. Movement is amenable to mapping and quantification and is often used by transport planners and technocrats in towns and cities. Movement was central to 'gravity models' and spatial interaction theory in the work of spatial scientists. Movement was the line on the map in the same way that location was the dot on the map. As a cultural geographer I was keen to show how movement was *meaningful* for people. Just as the dot of location became place when humanists like Anne Buttimer started to consider the role of meaning, so movement became mobility when we recognized that the movements of people, things, and ideas, often in combination, were deeply meaningful parts of human life. Journeys, after all, are at the heart of many foundational stories in Western and other cultures, from the Odyssey to Norse sagas to early novels such as *Don Quixote*. We often describe our life metaphorically as a journey. Forms of movement from everyday commutes to immigration have been layered with meaning by those undertaking the journeys as well as by media, politicians, and others seeking to represent them. Mobility is also practised. What I mean by that is that we enact mobility in various ways – we walk, dance, drive, or fly. We might be a driver or a passenger. Our walk might be a one-hour routine during a COVID-19 lockdown or it might be a hike in the mountains. It might be a five-mile walk to collect water in Mali. Clearly how mobility is practised matters too.

Questions of movement, meaning, and practice are always related to power. Mobilities are both a reflection of arrangements of power in a given society and important tools in the production, maintenance, and possible transformation of those arrangements of power. You can see this on the dance-floor, where ballroom dance codes urge women to follow men, in journeys to work, where men travel further but have simpler journeys thanks to the multi-tasking of work, school, and shopping often experienced by women, and in immigration policy where Ukrainian refugees are given reasonably easy access to the United Kingdom and Syrian refugees, for instance, are not. We can see the politics of mobilities in the way cities are arranged to facilitate travel by automobile at the expense of safe and easy routes for cyclists, and in airport terminals where people who are able to pay for business-class tickets get allocated to fast lanes and move more swiftly through security. All these examples of mobilities are both the outcome of systematically asymmetrical arrangements of power, and tools in producing these arrangements (Figures 13.1 and 13.2).

Figure 13.1 Syrian and Iraqi refugees arrive at Skala Sykamias, Lesvos, Greece. (Source, Ggia) Creative Commons Attribution-Share Alike 4.0 International Wikimedia Commons https://commons.wikimedia.org/wiki/File:20151030_Syrians_and_Iraq_refugees_arrive_at_Skala_Sykamias_Lesvos_Greece_2.jpg

Figure 13.2 Driver-free car (Source, BP63Vincent) Creative Commons Attribution-Share Alike 3.0 Unported Wikimedia Commons https://commons.wikimedia.org/wiki/File:Driver_free_car.jpg

In *On the Move* I illustrate these ideas through case studies that form an unlikely array of seemingly disparate stories. The case studies include ballroom dancing, mobilities in the workplace and kitchen, US legal cases surrounding the right to move, international migration, and the mobilities of global hub airports. I chose this strategy to show how mobility as a concept can inform research across a range of scales when it is taken seriously and not taken for granted. Previously in Geography, any of these individual case studies might have existed – but confined to a sub-disciplinary silo like transport geography, or population geography. Mobility, I argued, is a fundamental concept for the discipline. Every bit as important as space or place, territory or region. It had either been ignored or taken for granted. It hadn't been theorized as a key concept for the discipline. That was what I wanted to achieve in the book.

Autobiography and the mobilities project

In the rest of this chapter I explain how I came to the 'mobilities project'. The first point to note is the mobility of my own life. I'm the son of a Royal Air Force ground engineer and a paediatric nurse – both of whom left school at 16 and were pretty quickly on the move. As a child I lived in Norfolk, Berlin, Oxfordshire, Singapore, and Suffolk. As an adult I have never lived anywhere longer than seven years, moving between universities in the UK and USA as a student and as an academic. In more recent years this has been part of a family team, including my wife who is a US citizen, which partly explains the trans-Atlantic moves. Having a trans-Atlantic family necessitates moving, just as being part of an Air Force family did when I was a child. In short, I have been constantly on the move, and never really lived anywhere long enough to feel local to a place.

Most people born to parents from modest backgrounds with no tertiary education would not get to see the world in the 1960s and 1970s. The armed forces is one place where this actually happens. I remember flying to Singapore at the age of eight, stopping along the way in Cyprus and the island of Gan in the Maldives. I remember the feel of getting off a plane into the warm close heat of the tropics. I recall being struck by immediately sensed difference. Singapore and RAF Tengah could not have been more different from Brize Norton, Oxfordshire. I learned to eat different food, became fascinated with the 'chit-chats' (gheckos) that lived on our walls, and fished for catfish in monsoon ditches behind our colonial bungalow. I didn't know it, but I was experiencing the tail-end of empire with all the privileges that brings even for a relatively lowly member of a foreign military. In no other circumstance, for instance, would my parents have had a servant to do laundry. It must have been very strange for them. I was equally struck when I returned to Brize Norton at the age of ten. I distinctly recall arriving back on a chilly, wet, grey day and drinking full fat milk for the first time in years. I remember how the fat coated my mouth. I recount these experiences in order to note an early embrace of the experience of difference that travel brings – an unarticulated and inchoate sense of the possibilities of multiplicity that come with geography – possibilities that are primarily experienced through mobility.

The mobile trajectory of my autobiography may well have informed my more academic interest in mobility, but this was more clearly derived from my connections with particular sites of geographical knowledge production, both as a student and, later, as an academic geographer. My geographical imagination was sparked into life by the classes I took at University College London as an undergraduate. I was lucky enough to be taught by Jacquie Burgess, Peter Jackson, and Denis Cosgrove, who were all busy blending the insights of humanistic geography with those of the Marxist-tinged scholars of the Centre for Contemporary Cultural Studies in Birmingham as well as their forebears. So-called "new cultural geography" (Cosgrove and Jackson 1987; Cosgrove 1983) was being born out of this mélange and to be a student in 1985 at UCL was to witness this live. There was a very real sense that something important was happening and that I was at its epicentre (as a spectator at least). At the heart of the "new cultural geography" was a combination of a recognition, derived from broadly humanistic geography, that our geographical worlds are meaningful to us, and an equally important recognition that these meaningful geographies (places, space, mobilities) are contested. There are, new cultural geographers insisted, both dominant geographies that help to form our common-sense views of the world as well as practices in the world, and geographies of resistance where these seemingly common-sense views and practices are contested. My life changed at UCL and, as part of this life-changing experience I encountered books such as Yi-Fu Tuan's *Space and Place* (Tuan 1977) and Cosgrove's *Social Formation and Symbolic Landscape* (Cosgrove 1984), as well as the notes for Peter Jackson's *Maps of Meaning* (Jackson 1989) as they emerged.

It was in this context that I started to think about mobility. In Jacquie Burgess's final-year undergraduate class we had to read a piece of travel writing and analyze it using some of the conceptual tools we had been introduced to. I was assigned William Least Heat Moon's *Blue Highways* (Heat Moon 1982). I was enchanted by its tales of diner stops on America's back-roads. Somewhere along the line, I also started reading J.B. Jackson's influential essays on landscape and, particularly, moving *through* landscape (Jackson 1957). This led to my undergraduate dissertation. It was on "Metaphors of Mobility in Bob Dylan". New cultural geography was still so new that I had to justify it as properly geographic in front of a sceptical geomorphologist, the head of the department, Ron Cooke. No-one would have to justify that now. The central argument of the dissertation, following the insights gained from my exposure to Peter Jackson's teaching in particular, was that mobility was both made meaningful in Bob Dylan's music and lyrics, and that it represented a form of resistance to a dominant culture at the time. At the end of my time at UCL I asked Peter Jackson how to go about post-graduate education. He suggested writing to scholars who had inspired me. I wrote to Yi-Fu Tuan and received a lovely letter back. I decided to go to Madison, Wisconsin, to undertake an MA and PhD.

Clearly a 'mobility project' had begun to emerge at UCL, and it took further shape at Madison. For my Master's thesis I expanded on the topic of my undergraduate dissertation. I referred to my work as 'go-egraphy' and it explored the role of mobility in the popular culture of the United States, including literature (Kerouac and Ginsberg) and film (*Easy Rider* and *Bonnie and Clyde*). Informed by the Gramscian work of the Centre for Contemporary Cultural Studies (Hall 1980) and Peter Jackson's call for a fully developed 'geography of resistance', I was keen to show how mobility, as a geographical theme, was both at the centre of constructions of American identity and at the centre of attempts to question or resist it. It was, in other words, a contested 'map of meaning'. As this developed, I gradually realized that I was on to something bigger – the observation that mobility itself was under-theorized within Geography. While Tuan was my supervisor, Robert Sack was on my thesis advisory panel and Tuan and Sack worked as a team. Both of them existed outside of the main currents of disciplinary theory and both insisted on geographical conceptualization that did not play second fiddle to other kinds of theory – particularly the social and cultural theory that was at the heart of the new cultural geography.

Tuan was a central figure in the rise of humanistic geography in the 1970s. His work insisted on the role of space and place in human experience and how our geographies give meaning to our lives (Tuan 1977). I had read his work at UCL and it had immediately entranced me as being both wonderfully clear, and, at the same time, profound. The possibility of thinking and writing like Tuan had been a very significant factor in my commitment to becoming a geographer. Sack's work also focused on concepts at the heart of the discipline including space, territory, and place, and he spent his academic life building up a theory of place that positioned geography at the heart of understanding what it is to be human in the world (Sack 1997). By the time I arrived in Madison in 1986, neither of them was at the heart of the geographical firmament. In many ways, they stood in opposition to the excited uptake of cultural and social theory that the work of Peter Jackson at UCL had represented. Madison seemed to be a place that stood proudly against whatever new intellectual currents of the time were being championed by geographers elsewhere. Some of Sack's insistence on properly geographical theory rubbed off on me even as I resisted it at the time. Space, place, territory, and landscape were all clearly 'geographical primitives' that had their associated key texts from within the discipline. All were being discussed fervently in the conferences I attended. Mobility, as far as I could tell, had no obvious key text and it was not something that was being explored in the meeting rooms and corridors of conferences. It increasingly seemed to me that mobility had not been given the same scrutiny as the seemingly more fixed spatial preoccupations of the discipline. Often, this went even further, with commentators and theorists describing the rise of a mobile world as the 'death of geography', as though geography could only ever refer to things that are fixed. In other words, geography was part of the problem I was becoming aware of – the designation of mobility and all things mobile as secondary (at best) to all that is fixed.

Following my Master's degree thesis I returned to Peter Jackson's call for a "geography of resistance" (Jackson 1989) and turned to the theme of place and 'transgression' for my doctoral work. While mobility was not the primary focus of this work it was nevertheless a necessary part of it. The thesis, which became my book *In Place/Out of Place: Geography, Ideology, and Transgression* (Cresswell 1996), explored the role of place in the constitution of hegemonic expectations about who could do what – about judgements of appropriateness. As transgression meant crossing a boundary, this often meant moving. This was particularly true of graffiti in New York City as it moved on trains around the landscape, and the activities of so-called 'new-age travellers' in 1980s Britain. Both examples of what Stuart Hall and colleagues were calling "moral panics" (Hall 1978). Mobility, in other words, was rarely far from my mind. It was at that time (1988–1993) that versions of mobility began to appear as prominent themes in the humanities and social sciences. I am thinking particularly of work that asked fundamental questions of the pervasive, seemingly organic, link between forms of identity and bounded ideas of place. Anthropologist James Clifford's essay "Travelling Cultures" was an influential example (Clifford 1992), as was Liisa Malkki's *National Geographic* paper which defined the problem of a "sedentarist metaphysics" (Malkki 1992). The idea of a "sedentarist metaphyics" describe pervasive patterns of thought that see a fixed and bounded nation-state as a moral container of identity, culture, and politics. Anything that does not conform to this – particularly mobile or itinerant people – is seen as immoral or deviant and treated as such. Malkki's work focused on refugees but her idea can be extended to think of all the ways morality, politics, culture, and society are attached to seemingly fixed and bounded places and how this sense of moral geography makes forms of mobility suspicious at best. At around the same time as Malkki and others were exploring the moral geographies of fixity, boundedness, and stasis, others, particularly continental philosophers, were putting metaphors of mobility such as nomads and rhizomes at the centre of their work (Deleuze and Guattari 1986). There was, in other words, a 'nomadic metaphysics' in the air.

By the time I arrived at the University of Wales, Lampeter in 1993, for my first permanent academic position, mobility was at the front of my mind. If "new cultural geography" had been fermenting while I was at UCL, it had become well established and expanded at Lampeter. Lampeter, a very small town in rural West Wales, had become what the French sociologist Bruno Latour called a "center of calculation" – a particularly productive node in a network of ideas and the people that carry them (Latour 1987). The human geographers at Lampeter were predominantly around the same age as I was (or younger) and had been supervised by leading figures in critical cultural geography. To take just a few examples, Catherine Nash had been supervised by Stephen Daniels at Nottingham – a leading figure in the theorization and history of the idea of landscape in cultural geography (Daniels 1990). Ian Cook had been in classes with me at UCL, done a Master's at the University of Kentucky – a hotbed of post-structural theory – and returned to Bristol where he was supervised by the leading critical geographer Nigel Thrift. Other geographers who'd arrived before the three of us had worked with Derek Gregory who was, at the time, the leading translator of social theory into human geography. Each of us, to a greater or lesser degree, were carriers of the contexts we had arrived from and Lampeter became a particularly creative assemblage of theoretically informed cultural, historical, political, and economic geography.

On the Move and the new mobilities paradigm

By 1999 I had moved to nearby Aberystwyth as the extraordinary Geography department at Lampeter gradually folded thanks to the eagerness of colleagues to find somewhere more central in the UK's academic firmament and the reluctance of potential students to move to a small town of several thousand people in the west of Wales. It was during my time at Aberystwyth (1999–2006) that I wrote *On the Move: Mobility in the Modern Western World* (Cresswell 2006). I also started to become part of a network of scholars working in different disciplines on projects with mobility at the core. For the first time, I had a sense that my work was part of something much bigger – an academic movement across the social sciences and humanities that John Urry and Mimi Sheller would later label the "new mobilities paradigm" (Sheller and Urry 2006). My first sense of this was when I co-organized a conference held in the University's conference centre at Gregynog, in the countryside near Newtown in Wales, with the theme of Mobilities. We invited people from a range of disciplines who were working on mobile themes including the cultural theorist Iain Chambers, whose book *Migrancy, Culture, Identity* had inspired me (Chambers 1994), Don Mitchell, whose work on the labouring bodies of migrant workers was important (Mitchell 2001), and the literary scholar Georges Van Den Abbeele who had written *Travel as Metaphor: From Montaigne to Rousseau* (Van den Abbeele 1992). The conference was full of scholars from a range of disciplines across the humanities and social sciences including Peter Merriman, a geographer who went on to be a leading figure in the mobilities field who would later co-instigate the interdisciplinary project of "mobility humanities" (Merriman and Pearce 2017) and become, briefly, my colleague at Aberystwyth. There were scholars from Lancaster who would later form the field-defining Centre for Mobilities Research in 2003 and launch the journal *Mobilities* in 2006. I met Ginette Verstreate whom I had invited on the recommendation of Caren Kaplan, a trailblazer in thinking about links between travel, displacement, feminism and literary theory (Kaplan 1996). Following the conference we worked on a multidisciplinary edited collection around the intersections of place and mobility (Verstraete and

Cresswell 2002). I sensed a palpable buzz about an emergent set of ideas that were beginning to cohere. Three papers from this meeting – my own, Verstraete's and Mitchell's – appeared in Volume 43 of *New Formations* in 2001 – an interdisciplinary journal that had been a favourite of mine since my days working in a radical bookshop in Madison (Cresswell 2001; Verstraete 2001; Mitchell 2001).

By the time of the Gregynog conference geographers and others were fully engaged with the diffuse literatures that placed differently labelled forms of mobility at the heart of theory and philosophy. In addition to Kaplan's book *Questions of Travel*, James Clifford's book *Routes* was being widely read (Clifford 1997), and Bauman's diagnosis of liquid modernity read through the twin mobile figures of the tourist and the vagabond was well established (Bauman 1996). Work on hybridity, diaspora, and transnationalism was everywhere from literary theory to anthropology and political science. I would read all of these and see them through the lens of the idea of mobility which, I believed, underpinned all of them. Mobility was not undoing geography, as some still insisted, rather it was producing new geographies and enabling new ways of thinking geographically. Other geographers had begun to embrace mobility thinking in ways that were different from transport or migration geography. Peter Merriman joined the staff at Aberystwyth from Reading and we were briefly colleagues before I moved on to Royal Holloway in 2006. We frequently discussed mobilities in our offices, the staff room, and along the short corridors of Llandinam Tower. The book that we later edited together, *Geographies of Mobilities* (Cresswell and Merriman 2010) brought together geographers to think about mobilities collectively and in print for the first time. The underpinning of everything from identity to power to governance by a fixed spatial imagination had lost its monopolistic hold on theorization, so something new had to emerge from the inchoate soup of mobile ontologies and epistemologies that surrounded us.

Gradually, mobility had emerged as a key term well beyond geography. In 2001 I read John Urry's book *Sociology Beyond Societies: Mobilities for the Twenty-First Century* (Urry 2000). In this book Urry argued for a transformation of the discipline of sociology to one that had mobilities as its core focus. It did for sociology what Clifford and Malkki had argued for anthropology – questioning the metaphysical, ontological, epistemological, and methodological fixation on fixity. Here it was the fact and idea of mobility that was given centre stage for the first time as opposed to travel, tourism, migration, transport, or any of its variants. From within Geography, Doreen Massey's concept of "power-geometries" had also become influential as a way of thinking about how different mobilities exist in relation to each other which both reflected and reproduced power relations (Massey 1998). Small conferences began to spring up all over the place that explicitly placed mobility at their heart. One I remember very fondly, in which Urry invited me to take part, was held in the Normandy villa of Cerisy le Salle in 2003. The chateau had been used for such meetings since at least the 1950s and the walls were covered in black-and-white photographs of previous participants in meetings there including Sartre, Barthes, and Derrida. The meeting was called *Le Sens du Mouvement* and was organized by, among others, the French geographer, Jacques Lévy. In addition to John Urry, Ed Soja and Nigel Thrift were also there. While I note the overwhelming masculinity of the names listed above, I also confess to this being one of the moments where I felt I had a place in the intellectual landscape as I was surrounded by the men whose books and papers had inspired me when I was a graduate student – and now I was one of them. Despite my uncertain beginnings as a first-generation student at UCL, I was now part of what Susan Christopherson called "the project" (Christopherson 1989).

While I'd been developing my thoughts about mobility, I'd only gradually become aware of this broader community of scholars moving in the same direction. I was never planning a "new paradigm" and would never have dreamed that my work would become a part of one. It was largely luck that my own predilections were mirrored by those of others and that my book and associated papers going back at least to the 2001 edition of *New Formations* would become part of something much bigger. Sheller and Urry's 2006 paper 'The new mobilities paradigm' is one of the most highly cited papers in social science and the most cited paper ever from *Environment and Planning A* (Sheller and Urry 2006). In 2010 my paper 'Towards a Politics of Mobility' was published in *Environment and Planning D* and is now the most cited paper in that journal's history (Cresswell 2010). Since 2006 work in the new mobilities paradigm or mobilities turn has its own journals including *Mobilities*, *Applied Mobilities*, and *Transfers*. More recently the field of "mobility humanities" has emerged (Merriman and Pearce 2017) with Mobility Humanities Centres in Padua, Italy, and Seoul, South Korea. There is a *Mobility Humanities* journal. Every year there are multiple mobilities-based workshops and conferences across the world including the regular meetings of the Cosmobilities network and the T2M group. The words "mobility" and "mobilities" appear regularly in the titles of papers within Geography journals and at presentations at conferences. Before 2006 this was far less common. There was no way any of this would have seemed likely when I was writing my undergraduate dissertation or Master's thesis. It did not even seem likely when I was planning the 1999 Gregynog meeting or writing my application for an ESRC fellowship on "The Politics and Poetics of Mobility". I am fortunate that my work has found a hospitable ecosystem to circulate within. I am grateful that others were planning a new paradigm, even if I was not.

Conclusions

I've been fascinated with the rise of biographical accounts of the development of geographic thought in our discipline. The work of Trevor Barnes and various co-authors to situate the development of ideas with people in places has always been enlightening and convincing (Peck and Barnes 2019; Barnes 2014). The collection on the *Spatial Histories of Radical Geographies* was particularly fascinating, with its warts-and-all coverage of the behaviour of various radical geography titans in Worcester, Berkeley, Michigan and elsewhere (Barnes and Sheppard 2019). In all these accounts there is a combination of individuals making choices, places that provide ecosystems for even the most solitary thinkers, and the wider structural elements of history and society that frame the interactions of people in places. While it is easy to discern these layers of explanation in the development of ideas in biographical accounts, it is slightly less clear how to do this when thinking about myself.

Reading what I've written above, it seems as though I was a precocious twenty-year-old who wanted to change the discipline of Geography by taking mobility seriously and then made a series of choices to do exactly that. That would not be accurate. While my biography includes a particular form of interaction with mobility and place from childhood on, this is not the only element of my biography that informed the "mobilities project". My interactions with others have been instrumental in informing my project. To start with, I recognize that my partner and children – my family – have both allowed and enabled various moves that helped me in my career. We've been collectively ready and able to move in ways that other scholars, particularly women, have not. Our children are now proficient and eager travellers themselves. We have tried to make these moves advantageous for everyone but I would be wrong if I failed to recognize that a critical account of the gendering of mobility can be applied to my own moves which have benefitted me in ways that surpass any benefits that have accrued to my family.

Second, I was lucky enough to have transformative and caring teachers and mentors at UCL and Wisconsin. I didn't know, while at UCL, that a revolution in critical cultural geography was happening in front of my eyes – a revolution that I know inspired at least three other classmates to become academic geographers. I have always wanted to teach as though the same thing was happening wherever I happened to be. While I struggled with the maverick geographies of Tuan and Sack I recognize, in hindsight, that they provided invaluable lessons in keeping clear objectives and working towards them while centering the very specific things that geography has to offer. Third, I recognize the impact of the smart and supportive colleagues in all the places I have been. Lampeter, Aberystwyth, Royal Holloway and now Edinburgh have all been good places to think in, where I've been encouraged to follow the sometimes idiosyncratic paths I have chosen. I was never made to believe I had to apply for a big grant I did not need, or make a change in research direction I did not want. It amazes me to list the colleagues whose insights I have benefitted from. If I had moved from a PhD in Wisconsin to one of the over 50 places in North America where I applied for jobs, I would probably have been the only critical cultural geographer within 100 miles. By moving to Lampeter, a small town in the wilds of West Wales, I was surrounded by scholars who broadly spoke the same language.

A fourth way I have been fortunate is with the doctoral students and post-doctoral scholars I have supervised and mentored who have taken mobility into new areas. I taught Peter Adey as an undergraduate and then supervised him as a PhD student at Aberystwyth: while he was exploring the worlds of Liverpool Airport, I was doing the same at Schiphol (Adey 2010). We have since worked together on a project on mobility transitions to socially just low-carbon futures (Adey et al. 2021). While at Royal Holloway (2006–2013) I learned from Kimberley Peters and Weiqiang Lin, who between them developed work on mobilities on the sea and by air (Peters, Steinberg and Stratford 2018; Lin 2015). I was fortunate to mentor Jenny Middleton while she developed her work on walking (Middleton 2009). All of these have taken mobilities across land, air and sea into places I would never have imagined. I hope I have passed something on to them in the same way my mentors did to me. Mostly, however, these scholars have expanded my imagination and encouraged new ways to think about mobility that go way beyond anything I might have come up with.

Finally, I acknowledge that the mobilities project arose within wider social, cultural, economic, and political contexts that, no doubt, influenced the turns the project has taken. The world of globalization, one marked by new forms of mobility that are more frequent and faster, has led others to formulate ideas of network societies (Castells 1996) or liquid modernity (Bauman 2000). Mobility was and is in the air. It still seems odd to me that mobility as a theme had not been critically addressed in 1986 when I began to seriously think about these things. Its centrality seems obvious now. I also have to recognize that the particular way that academic recognition works has benefitted me. Work that speaks to wide themes or concepts, such as mobility, is often recognized in ways that equally valuable, but more empirically specific, work is not. Papers and books on these themes, if done reasonably well, are likely to attract citations which then feeds into promotions and academic headhunting – all of which have benefitted me as I have developed my career across multiple sites and though my own mobilities.

Study task

Variations on this question can be set in an unseen examination, an open-book examination or as a term-paper assignment

Mobility can be defined as a combination of movement, meaning, and practice in the context of power. Illustrate this with reference to walking.
(NB: *walking* could be replaced with cycling, driving, or flying)

Starter reading list

Cresswell, Tim (2006) *On the Move: Mobility in the Modern Western World* (Routledge). Chapter One – Introduction.

Cresswell, Tim (2010) Towards a Politics of Mobility. *Environment and Planning D: Society and Space* 28(1): 17–31.

Cresswell, Tim (2011) Mobilities I: Catching Up. *Progress in Human Geography* 35(4): 550–558.

Cresswell, Tim (2015) Black Moves: Moments in the History of African-American Masculine Mobilities. *Transfers* 6(1): 12–25.

Lorimer, Hayden (2011) Walking: New Forms and Spaces for Studies of Pedestrianism. In Tim Cresswell and Peter Merriman (eds) *Geographies of Mobilities: Practices, Spaces, Subjects* (Ashgate/Routledge), pp. 19–34.

Middleton, Jennie (2009) 'Stepping in Time': Walking, Time, and Space in the City. *Environment and Planning A* 41(8): 1943–1961.

Middleton, Jennie (2018) The Socialities of Everyday Urban Walking and the 'right to the city'. *Urban Studies* 55(2): 296–315.

Munt, Sallie (2000) *The lesbian flâneur*. In Borden, I., Kerr, J., Pivaro, A., and Rendell, J. (eds) *The Unknown City: Contesting Architecture and Social Space* (MIT Press), pp. 247–262.

Oliver, Michael (1996) 'What's So Wonderful about Walking? In *Understanding Disability* (MacMillan). Chapter 7.

Sheller, Mimi (2019) *Mobility Justice: The Politics of Movement in the Age of Extremes* (Verso). Chapter 2 – 'Bodily Moves and Racial Justice'.

References

Abler, R., J. Adams, and P. Gould. 1971. *Spatial Organization: The Geographer's View of the World*. Englewood Cliffs, NJ: Prentice Hall.

Adey, P. 2010. *Aerial Life: Spaces, Mobilities, Affects*. Malden, MA: Wiley-Blackwell.

Adey, P., T. Cresswell, A. Nikolaeva, A. Novoa, and C. Temenos. 2021. Moving Towards Transition: Commoning Mobility for a Low-Carbon Future. In *Just Sustainabilities*, edited by Julian Agyeman. London: Zed Books.

Barnes, T.J. 2014. "What's old is new, and new is old: History and geography's quantitative revolutions." *Dialogues in Human Geography* 4 (1):50–53. doi: 10.1177/2043820614525707

Barnes, T.J., and E.S. Sheppard. 2019. *Spatial Histories of Radical Geography: North America and beyond, Antipode book series*. Hoboken, NJ: Wiley.

Bauman, Z. 1996. "*Tourists and Vagabonds: Heroes and Victims of Postmodernity", Reihe Politikwissenschaft/Institut für Höhere Studien, Abt. Politikwissenschaft*. Wien: Institut für Höhere Studien (IHS), Wien.

Bauman, Z. 2000. *Liquid Modernity*. Cambridge: Polity Press.

Buttimer, A., and D. Seamon. 1980. *The Human Experience of Space and Place*. New York: St. Martin's Press.

Castells, M. 1996. *The Rise of the Network Society*. Edited by Manuel Castells, *Information age; v. 1*. Cambridge, MA: Blackwell Publishers.

Chambers, I. 1994. *Migrancy, Culture, Identity*. London: Routledge.

Christopherson, S. 1989. "On Being Outside 'the Project'." *Antipode* 21 (2):83–89. doi: https://doi.org/10.1111/j.1467-8330.1989.tb00181.x

Clifford, J. 1992. "Traveling cultures." In *Cultural Studies*, edited by Lawrence Grossberg, Cary Nelson and Paula Treichler, 96–111. London: Routledge.
Clifford, J. 1997. *Routes: Travel and Translation in the Later Twentieth Century*. Cambridge, MA: Harvard University Press.
Cosgrove, D.E. 1983. "Towards a radical cultural geography: Problems of theory." *Antipode* 15 (1):1–11.
Cosgrove, D.E. 1984. *Social Formation and Symbolic Landscape*. London: Croom Helm.
Cosgrove, D.E., and P. Jackson. 1987. "New directions in cultural geography." *Area* 19 (2):95–101.
Cresswell, T. 1996. *In Place/Out of Place: Geography, Ideology and Transgression*. Minneapolis: University of Minnesota Press.
Cresswell, T. 2001. "The production of mobilities." *New Formations* 43 (Spring):3–25.
Cresswell, T. 2006. *On the Move: Mobility in the Modern Western World*. New York: Routledge.
Cresswell, T. 2010. "Towards a politics of mobility." *Environment and Planning D: Society and Space* 28 (1):17–31.
Cresswell, T., and P. Merriman. eds 2010. *Geographies of Mobilities: Practices, Spaces, Subjects*. London: Ashgate.
Daniels, S. 1990. "Marxism, Culture and the Duplicity of Landscape." In *New Models in Geography*, edited by N. Thrift and R. Peet, 177–220. Boston: Allen and Unwin.
Deleuze, G., and F. Guattari. 1986. *Nomadology: The War Machine*. New York: Semiotext(e).
Hall, S. 1978. *Policing the Crisis: Mugging, the State, and Law and Order, Critical Social Studies*. London: Macmillan.
Hall, S. 1980. *Culture, Media, Language: Working Papers in Cultural Studies, 1972–79*. London: Hutchinson in association with the Centre for Contemporary Cultural Studies, University of Birmingham.
Heat Moon, W.L. 1982. *Blue Highways: A Journey into America*. 1st ed. Boston: Little, Brown.
Jackson, J.B. 1957. "The Abstract World of the Hot-Rodder." *Landscape: Magazine of Human Geography* 7:22–27.
Jackson, P. 1989. *Maps of Meaning*. London: Unwin Hyman.
Kaplan, C. 1996. *Questions of Travel: Postmodern Discourses of Displacement*. Durham, NC: Duke University Press.
Latour, B. 1987. *Science in Action: How to Follow Scientists and Engineers through Society*. Cambridge, MA: Harvard University Press.
Lin, W. 2015. "'Cabin pressure': Designing affective atmospheres in airline travel." *Transactions of the Institute of British Geographers* 40:287–299. doi: 10.11.11/tran.12079
Malkki, L. 1992. "National geographic: The rooting of peoples and the territorialization of national identity among scholars and refugees." *Cultural Anthropology* 7 (1):24–44.
Massey, D. 1997. "A global sense of place." In *Reading Human Geography*, edited by Trevor Barnes and Derek Gregory, 315–323. London: Arnold.
Massey, D. 1998. *Power-geometries and the Politics of Space-time*. Heidelberg: Hettner-Lecture, Department of Geography, University of Heidelberg.
Merriman, P., and L. Pearce. 2017. "Mobility and the humanities." *Mobilities* 12 (4):493–508. doi: 10.1080/17450101.2017.1330853
Middleton, J. 2009. "'Stepping in time': Walking, time and space in the city." *Environment and Planning A* 41 (8):1943–1961.
Mitchell, D. 2001. 'The Devil's Arm: Points of passage, networks of violence and the political economy of landscape." *New Formations* 43:44–60.
Peck, J., and T.J. Barnes. 2019. "Berkeley In-Between." In *Spatial Histories of Radical Geography*, 211–246.
Peters, K.A., P.E. Steinberg, and E. Stratford. 2018. *Territory beyond terra, Geopolitical Bodies, Material Worlds*. London; New York: Rowman & Littlefield International.
Relph, E. 1976. *Place and Placelessness*. London: Pion.
Sack, R.D. 1997. *Homo Geographicus*. Baltimore: Johns Hopkins University Press.
Sheller, M., and J. Urry. 2006. "The new mobilities paradigm." *Environment and Planning A* 38 (2):207–226.
Sibley, D. 1995. *Geographies of Exclusion: Society and Difference in the West*. London: Routledge.

Tuan, Y.-F. 1977. *Space and Place: The Perspective of Experience*. Minneapolis: University of Minnesota Press.

Urry, J. 2000. *Sociology beyond Societies: Mobilities for the Twenty-first Century, International Library of Sociology*. London; New York: Routledge.

Van den Abbeele, G. 1992. *Travel as Metaphor: From Montaigne to Rousseau*. Minneapolis: University of Minnesota Press.

Verstraete, G. 2001. "Technological frontiers and the politics of mobility in the European Union." *New Formations* 43 (Spring):26–43.

Verstraete, G., and T. Cresswell. 2002. *Mobilizing Place, Placing Mobility: The Politics of Representation in a Globalized World*. Amsterdam: Rodopi.

14 Scale matters

The case of workers and their geographies*

Andrew Herod

Andrew Herod (1964–)

Why scale is an important concept for geographers and others

In 1968, 46,000 dockers from Maine to Texas went on strike for 116 days in a fight over geographical scale. This was not the first time they had done so. It would not be the last. Of course, they did not talk about their actions in such academic terms. But their strike, one of the longest in US history, was nevertheless centrally about scale. Specifically, it was about the scale at which their labour contracts should be negotiated – locally, as

* I dedicate this chapter to Robert Q. Hanham (1947–2023), who passed away during its writing.

DOI: 10.4324/9781003343240-16

employers preferred, or nationally, as dockers wished. The fact that this strike was so long and so geographically widespread is testament to the important roles played by political struggles over the spatial scales at which social life is organised.

'Scale' is one of Geography's weightiest concepts. This chapter explores my own and others' research into how what has been called the 'production of scale' has been analysed and why understanding something of this production is important for shedding light upon many real-world matters. Before proceeding, though, it's important to recognise that the term scale has two distinct meanings. The first relates to map-making and how we represent the world cartographically. The second concerns the material processes and practices that shape how landscapes are made and bounded. Although my own research has principally been focused upon this second, to set the scene for what follows I begin with an overview of both ways of comprehending geographical scale.

What is scale?

In the cartographic realm, scale refers to the ratio between something's size on a map and what it represents on the Earth's surface – a centimetre on a map being equivalent to 10 km on the ground, for instance. Whilst this view is often presented as a neutral, 'scientific' understanding of scale, in truth it is a culturally specific concept. Although many cartographers rarely acknowledge this fact publicly, the different ways in which scale has been thought about historically do clearly illustrate this point. Hence, in European map-making scale as a representation of spatial ratios has long been important, given maps' centrality in navigation and recording landownership (miscalculating the size of one's property has significant economic implications!). For example, Roman cartographers spent much time making cadastral (property) maps, meaning that concepts of scale and geographical precision were deeply embedded in the Roman world's economics and politics. However, these mapping practices largely fell out of use with the collapse of the Roman Empire and alternative means of delineating property boundaries came into favour, ones less dependent upon meticulous visual representations of territory. Hence, landownership increasingly came to be demarcated via written descriptions (called 'Metes and bounds' in Britain) – "Sir Thomas owns property stretching northward from the riverbank to where the old oak tree stood in King Richard's time and thence 150 rods westward to land owned by Lord de Warenne and thence 172 rods southwards to land owned by the Bishopric and thence eastward back to the riverbank", for instance. However, this was hardly a precise way in which to describe landownership, especially as, over time, collective memories could fade – 'Now, where exactly did that old oak tree stand?' – and things like riverbanks could change their configurations.[1]

The Roman Empire's collapse, however, also provided a power vacuum in which new sources of cartographic authority could develop. One institution that wielded much influence over people's lives in this regard was the Catholic Church, which imposed new systems of meaning and interpretation in many aspects of people's lives. With regard to mapping, many in the Church promulgated cartographic representations in which the use of scale to represent distance and size largely disappeared. This was because a key desire of such mapmakers was not to produce maps for navigational or tax collection purposes but, rather, to use cartography to evidence piety. This was reflected in the dominance of *mappa mundi* ('maps of the world'), which often put Jerusalem at the centre of the world map, and so-called 'T and O' maps (see Figure 14.1). However, as the Church's power was challenged by the Reformation and the scientific revolution propagated by people

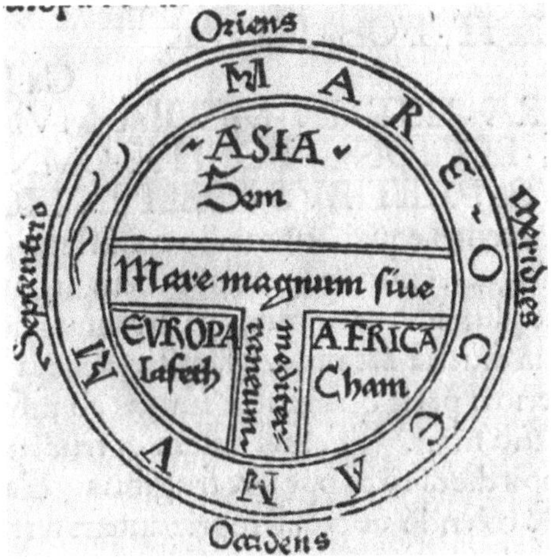

Figure 14.1 A typical mediæval T&O map. The T represents both the Mediterranean Sea and the cross of the crucifixion whilst the O represents both an encircling ocean and infinity – an O is something without beginning or end and served as a powerful Biblical allusion (in Revelation 22:13 Christ states "I am the Alpha and the Omega, the first and the last, the beginning and the end"). Such maps typically placed East ('Oriens') at their top, associating this direction with Heaven – the Sun's daily 'resurrection' was seen to parallel Christ's resurrection. Although European cartographers only knew of three continents at the time, fortuitously these three seemed to reflect Biblical teachings – they were often associated with the lands to which Noah's sons (Shem ['Sem'], Ham ['Cham'], and Japheth ['Iafeth']) went after the Flood in the Old Testament and with the Trinity in the New Testament.

Source: https://en.wikipedia.org/wiki/T_and_O_map

like Galileo, forms of cartographic representation that are more familiar to us today and in which scale as a tool for measuring size and distance came to (once again) dominate – as illustrated by Flemish cartographer Abraham Ortelius's 1570 *Theatrum Orbis Terrarum* ('Theatre of the World') (see Figure 14.2).

With the subsequent coming of the Renaissance and the rise of capitalism, concerns about having precise surveys of who owned what rose ever more to the fore and so cadastral mapping – and scale – made a resurgence, if in a geographically uneven manner (Kain and Baigent 1992). An early place of re-emergence was the Netherlands in the early 16th century, where it was associated with mapping (and taxing) lands newly created from drained estuarine and marine areas. In Sweden, King Gustavus Adolphus began a campaign in 1628 to establish a national cadastre. In what would become Germany and in the Austrian Hapsburg lands, though, major cadastral mapping activities did not begin until the 18th century. For its part, in France accurate land mapping to ensure equitable taxation was central to the ideals of the 1789 revolution. However, in Norway such practices really did not come along until the 20th century, given how much longer pre-capitalist forms of communal land tenure endured. Equally, in the Austrian-controlled parts of northern Italy and south-eastern Europe longstanding struggles between central royal authority and provincial lords and clergy who recognised that accurate mapping would provide the Austrian Emperor with better land ownership data, thereby reducing

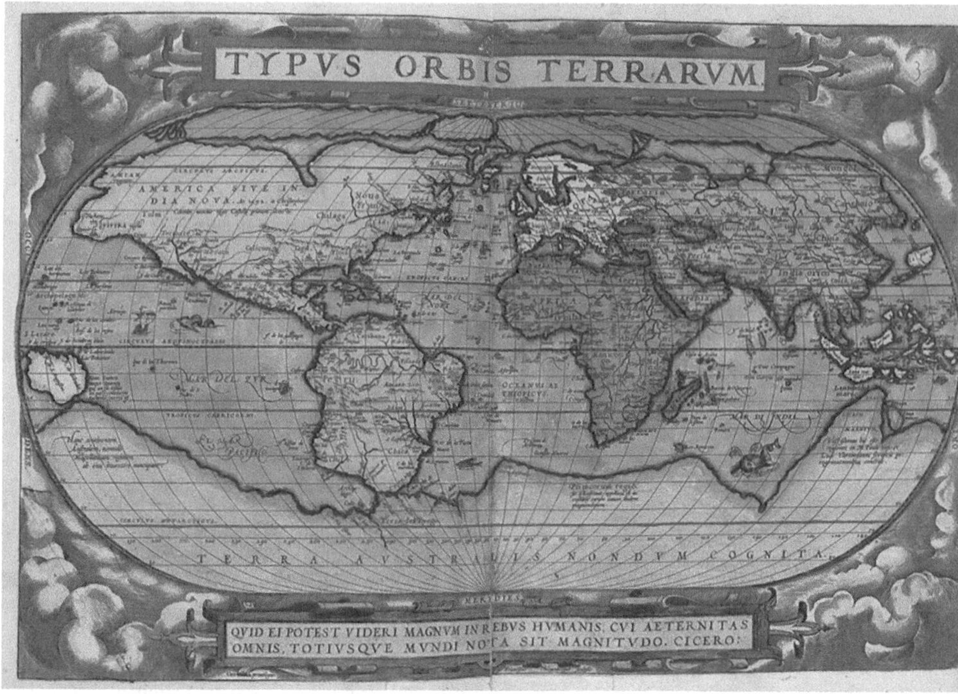

Figure 14.2 Abraham Ortelius's 1570 *Theatrum Orbis Terrarum* ('Theatre of the World'). In this typical Renaissance-style map, accurately representing landmasses' sizes and locations was crucial to European explorers' abilities to 'discover' new lands and claim them. Scale as a measurement of dimension and distance was integral in this. These cartographic developments were thus central to the spread of capitalism. Nevertheless, many such maps still contained elements of religious conviction. Trusting that God would not have created a world in disequilibrium, Europeans had long believed that there 'must' be a large landmass in the southern hemisphere to balance out the landmasses in the northern hemisphere. Consequently, their maps frequently recorded a *Terra Australis Nondum Cognita* (the 'southern land not yet known') centuries before any European would actually set eyes upon it.

landowners' abilities to avoid taxation, fuelled their resistance to a national rural cadastre. Associated with the rise of powerful rulers, cadastral mapping was thus a central – if contested – element in statecraft. Scale as a means of ensuring accurate measurement was central to this. Meanwhile, maps themselves became prized as decorative objects and were often displayed on walls as symbols of landowners' dominions. In the process, they shaped public perceptions of how maps 'should' look and thus the subsequent development of European cartography – they were, in other words, central in developing various practices and politics of representing the world. The power of these maps to shape public perceptions should not be underestimated. The Mercator projection created in 1569, for instance, famously misrepresents land masses' sizes and shapes, making Europe appear proportionately larger than it actually is in relation to continents like Africa. Significantly, because of how it distorts things the projection requires a different scale for each latitude, as a centimetre at the Equator does not represent the same distance as a centimetre at, say, 60°N. Consequently, to overcome misrepresentations of Europe's size, in the 1970s the Peters Projection was presented as an alternative to Mercator's projection. Although the Peters Projection represents land masses in their correct proportions, it, too, misrepresents scale because it must distort landmasses' shapes in order to represent their relative sizes correctly. These two maps, then, both distort cartographic scales, thereby illustrating the broader point that scales are not neutral metrics for measuring distances on maps.

The fact that political struggles shaped the spread of cadastral mapping and its central concept of scale as a means to measure size and distance mean that scale as a 'social construct' within cartography has an important historical geography to it. Although I'm not much of a cartographer, I do find this fact thought-provoking and examine it in one of the large undergraduate classes I teach, challenging students to think of maps not as somehow 'objective, scientific' things but as texts that we create to represent the world. In this regard, I have been greatly influenced by geographer Brian Harley (see, for instance, Harley 1989) and am particularly interested in questions of how political power has shaped systems of representation not just in mapping but also in art and architecture.[2]

If cartographic scale is a central element in geographical representations of the world, one that reflects various social, economic, and political contexts within which its use has evolved, then the second way in which scale has been central to geographical comprehension is as a framework for ordering the world. Often, this latter approach has presented scales in terms of a spatial hierarchy, with 'the global scale' at the top and other scales – 'the national' or 'the regional', for instance – nested below it. Indeed, this language even shaped a whole approach to geographical scholarship – Regional Geography – in the early to mid-20th century. This approach was endorsed by geographers like the American Richard Hartshorne (1899–1992), who saw 'the region' as the central concept that would allow Geography to claim its place as the "Mother of sciences" (Hartshorne 1939, p. 373), French geographer Paul Vidal de la Blache (1845–1918), often considered the "Father of French geography" (Buttimer 1971), who focused (1903 [1994]) upon how France's various regions (*pays*) were shaped by the local cultures which had moulded their natural environments over time, and British geographer Andrew Herbertson (1865–1915), who argued for studying what he called (1905) "natural regions". For much of my career it is this second understanding of scale – scale as an instrument of imposing order and structuring organisation – with which I've been centrally involved.

Engaging with scale

If the above provides a brief overview of why scale is important, below I want to talk about my own engagement with it. Specifically, I've investigated what's come to be called the 'production of scale'. It's perhaps no surprise that this is not something I spent a great deal of time thinking about as a proto-geographer in secondary school nor in my undergraduate degree programme at Bristol University in the 1980s. At that time I viewed scales rather uncritically as things that just seemed to exist. They represented, it seemed to me (and many others), just a natural or logical way in which to carve up the world's spaces – certain things were *inherently* seen to occur/operate at particular scales, like 'the urban' or 'the national'. Indeed, for much of Geography's history as an academic discipline most geographers never really gave much thought to why these particular scales seem to be important for shaping how we understand the world and how it's organised spatially. However, as it turns out, this view of scale is problematic. Nonetheless, it wasn't really until I began my PhD research in 1988 at Rutgers University, working with the late Marxist geographer Neil Smith (1954–2012), that I started to consider scale from a more measured point of view.

I had first met Neil when he came to West Virginia University (WVU) to give a talk on gentrification, the area of research for which he is perhaps most famous (e.g., see Smith 1979). Like so many things in life, my own journey to WVU had been fortuitous. My father had grown up in extremely impoverished circumstances in the English town of Peterborough during the Depression of the 1930s (his father had worked in the shipyards of the River Tyne but moved the family south in search of work). Consequently, my dad always argued for the need for economic security in life and suggested that I look at a profession like accountancy upon graduation. But I didn't find the thought of a future in accountancy particularly appealing (no offence to aspiring accountants or to anyone with one in the family!). Just at that moment, fate stepped in and a month after graduating I was off to the USA, to study for a Master of Arts in Geography at WVU. This opportunity arose because my tutor at Bristol (Richard Dunn) had been contacted by WVU geographer Robert Hanham. Robert was seeking students to work on a project about unemployment patterns in US cities (Professor Dunn had conducted similar research on British cities). It was truly a matter of being in the right place at the right time. Anyway, I and another student applied. As it happened, Robert was also the Department Head and so secured some extra money to bring us both over (the other student later completed a PhD in City and Regional Planning at Cornell University and now works for the Library of Congress in Washington, DC). After WVU I went off to Rutgers in New Jersey to complete my PhD in Geography and to work with Neil, having been inspired by his WVU presentation. Finishing in 1992, I took a job at the University of Georgia (UGA), a large university that's the oldest public institution of higher education in the USA (all of the older US universities are private). I've been here for my whole professional working life, conducting research and teaching – much of it geared towards questions of scale and its politics.

Smithian understandings of scale

Smith's conception of scale would be fundamental to developing my own understandings of how the geography of capitalism is constructed. He'd laid out a theory of scale production in his 1984 book *Uneven Development: Nature, Capital, and the Production of Space*. This book had been his doctoral dissertation, written as Professor David Harvey's student at Johns Hopkins University in Baltimore, Maryland (see Chapter 8). For Smith, scale is central to understanding the uneven development that characterises landscapes made under capitalism. In particular, he argued that there are two contradictory tendencies shaping how capitalism develops. On the one hand, investment must be fixed in place – accumulation must occur somewhere and this requires that capitalists invest in particular locations, sometimes for longer and sometimes for shorter periods of time. On the other hand, capitalists are always looking for better investment opportunities elsewhere and so seek to retain mobility. According to Smith, negotiating this geographical tension between fixity and mobility shapes firms' behaviour. Significantly, he argued, capital – by which he meant capitalists collectively – seeks to manage this tension at a number of spatial resolutions. Initially Smith identified four. At the lower end of his scalar hierarchy is the urban scale, which he suggested is defined by the geographical extent of daily labour markets. Hence, a city's 'travel-to-work area' represents the general spatial limit to workers' day-to-day migrations between home and places of paid employment, with its boundary enclosing areas within which there's a certain coherence to the city's labour market but beyond which it fragments. The regional scale, he averred, is defined by capital's geographical concentration to form particular territorial divisions of labour – steel-making or fruit-growing regions, for instance. Whilst such regions might initially have been the product of Nature's bounty – plentiful iron ore here or good soil there – over time, Smith maintained, capitalism's regionalisation has increasingly been produced by the needs of accumulation. Hence, previously barren agricultural lands can be made productive by applying chemical fertilisers and constructing irrigation systems. A region, then, is usually a collection of localities (towns and cities) with elements of economic complementarity amongst them. Smith contended that the national scale emerges from different capitals' needs to compete in a world market. Initially they might come together on the basis of things like cultural similarities – a common language, say – or because they are enclosed by physical boundaries like mountain ranges or wide rivers. But, over time, they develop a scale of political organisation – that of the nation-state – which facilitates their economic interests, like establishing legal mechanisms to protect private property and defend against working-class agitation. This scale typically results from the historical amalgamation of smaller territorial entities – duchies, city-states, various minor kingdoms, and so forth – into larger spatial units. Whilst they may compete with one another domestically, these capitals often work together to defend their collective interests against foreign capitals, both at home and overseas. Finally, the global scale, Smith continued, is not defined by the Earth's geologically given limits but results from capital's quest to spread across the planet waged labour, which lies at the heart of the capitalist mode of production. Consequently, over time capital has transformed the pre-capitalist world of international trade that it inherited in the 18th century into "a world economy based on production and the universality of wage[d] labour" (Smith, 1984 [1990], p. 139). The global scale is therefore made; it is not simply discovered.

Smith's schema fundamentally challenged long-held views that had seen scales simply as convenient and/or logical mental contrivances for carving up the world into different absolute spaces. Instead, Smith argued, scales are social products that are made and remade over time by the actions of capital. I found this argument that scales are 'social products' that exist in the material world very stimulating as it opened the door to thinking about how the scales at which human life is organised are the result of political contests rather than simply the outcome of how we superimpose on the world pre-conceived frameworks for understanding it. It highlighted, in other words, a fundamental philosophical division between idealist ways of understanding (which see any perceived order in the world as imposed upon it by the intellectual frameworks employed by those analysts who investigate it) and materialist ways of understanding (which see any order as resulting from material processes and not from the researcher's brain). Two things shaped my early interaction with this Smithian understanding. First, for part of my time as a PhD student I worked for the City of New York's Department of Housing Preservation and Development, the largest such municipal agency in the United States. The city government had a programme wherein it would take buildings from landlords who hadn't paid their city taxes and would then refurbish them to house low income people, using federal, state, and city funds to do so. Given that it was cheaper for landlords to walk away from their buildings than it was for them to bring them back up to a state sufficient for human habitation, these buildings were mostly in pretty poor shape, even as many had good bones. Whilst our main office was around the corner from Wall Street, a centre of global finance, the redevelopment projects upon which I worked were all in Harlem, Spanish Harlem, and the Bronx, with many of these areas at that time (1990) looking like photos I had seen of German cities bombed by the Allies. My job was to interview residents in buildings being managed by companies that wanted to bid for contracts to manage the newly renovated buildings in these areas. Some parts of these neighbourhoods were considered sufficiently dangerous by our office manager that she would only let us travel to them in pairs, and sometimes only if accompanied by a New York City police officer. This experience spoke to me of the politics of uneven development that are at the heart of Smith's concept of scale as I pondered how there could be such poverty and dystopia existing alongside areas of tremendous affluence. Moreover, it also raised questions of where, exactly, the economic boundaries differentiating wealthy from poor neighbourhoods were spatially demarcated across the city's built environment and how these shifted over time as processes like gentrification changed things – that is to say, at what spatial scales such neighbourhoods were constituted.

Second, I was becoming increasingly interested in the political geography of labour and working-class movements. I had written my MA thesis at WVU on the spatial politics surrounding the closure of a glass manufacturing facility (see Herod 1991b, 1994). I knew that exploring labour's geographical practices was an area of research that I wished to pursue further. I also took a class on what's called 'social theory' with Neil, writing a paper on the geography of union contract negotiating and how various national agreements in sectors like steelmaking and trucking had been dismantled in the Reagan era. This seemed to me implicitly to be a story of struggles over the scales at which workers can exert influence with their employers – as commentator Mike Davis (1986, p. 143) observed at the time, breaking up such national contracts was about "shrinking [the] perimeter of unionism", a particularly powerful geographical image. A revised version of this paper ended up being my first publication (Herod 1991a). A key matter that I explored in it was how working people play roles in creating scales. This is because, despite the innovativeness of Smith's approach, I was always a little uneasy with what I perceived to be its somewhat 'capital-centric' approach to questions of scale, as viewing scale as produced out of capital's internal tensions seemed to me to write living people – and especially workers – out of the narrative.

Towards less capital-centric understandings of scale

Whilst my 1991 article marked an initial foray into thinking about how workers' struggles play roles in constructing geographical scales of political and economic organisation and thereby shape how capitalism's unevenly developed landscapes evolve, my work for my doctoral dissertation really marked my much deeper consideration of such matters. I had long had an interest in the work process and how that intersects with the making of economic landscapes. In particular, my Welsh trade unionist grandfather had been centrally involved in negotiating labour contracts in the British steel industry in the 1950s and 1960s and had always remarked to me as I was growing up that it was important to understand how the improvements workers secure at any particular steel mill intersect with conditions elsewhere – something I now recognise as being centrally about the 'politics of scale' and the relationship of the local to the regional and national, even if I did not at the time.[3] Consequently, when it came to choosing a dissertation topic I decided to explore how the introduction of containerisation into the shipping industry in the 1950s had affected work in ports and how the East Coast dockers' union, the International Longshoremen's Association (ILA), had responded to the challenges new labour processes initiated in the industry. However, when I first began reading some of this history

I really had little appreciation that scalar politics would play such an important role in the story. In particular, not only did containerisation destroy thousands of port jobs, as many fewer workers were required on waterfronts than was previously the case, but it also unleashed powerful new geographical forces in the industry. Two were particularly significant.

First, whereas historically the most labour-intensive aspects of loading/unloading ships involved handling cargo in ports, containerisation now allowed much work – packing freight into containers and removing it from them, called 'stuffing' and 'stripping' in the industry – to occur inland. Now at ports only the containers themselves needed to be handled, which required many fewer dockers. Second, containerisation's introduction spawned what could have been intense conflicts between dockers in different ports. This is because the industry had historically employed a system of labour relations that was decidedly localist in orientation, with negotiations between dockers and employers taking place on a port-by-port basis – dockers in New York negotiated with New York employers, those in Boston with Boston employers, and so forth. The fact that containerisation was introduced initially into New York, though, meant that dockers there were the first to feel its effects but also the first to organise to protect their jobs in the face of new cargo-handling modes. This threatened to create a situation in which New York dockers enjoyed work protections that others did not, with the result that shippers might decide to divert their cargoes to other ports for handling.

What was interesting to me about these developments was that the ILA developed two explicitly geographical strategies to try to save dockers' jobs at the waterfront. Both involved creating new material scales of operating. With regard to employment moving off waterfronts, the union negotiated a series of agreements whereby some work that could now be done at inland warehouses would nevertheless be retained at the waterfront, under certain conditions. These rules – known as the Rules on Containers – established a 50-mile perimeter around each East Coast port. Such perimeters effectively constituted new scales of economic activity around these ports, creating spaces within which particular types of cargo-handling activities would be prohibited. With regard to containerisation's impacts upon the geographical scale of contract bargaining, union officials quickly realised that the traditional port-by-port negotiating system left dockers vulnerable to employer strategies of playing different ports against each other. This was especially so because containerisation and the spread of the interstate highway system meant that consumers and/or manufacturers in, say, New York could increasingly be served by importing/exporting goods not just through New York but also other nearby ports like Philadelphia or Boston. In response, union leaders determined to remake the scale of bargaining in their industry – from a port-by-port to a coastwise, multi-port system – as a way to prevent shippers from utilising this strategy. Although it took them two decades and many strikes to achieve their goal, in 1977 they eventually managed to negotiate a contract covering the 34 ports from Maine to Texas in which the ILA represented dockers (this story is covered in detail in Herod 1997a).

For me, this research on the ILA's scale-making efforts had two noteworthy aspects to it. The first was that it represented a much more political and less capital-centric understanding of spatial scales' production than had Smith's formulation. I think that, at some point, Neil himself also recognised that his initial conceptualisation perhaps needed re-envisioning a little, especially as he worked on anti-gentrification struggles in Manhattan's Lower East Side where he explored efforts by homeless and fair housing advocates to expand struggles from specific neighbourhoods into a citywide campaign – as he put it (1992a, p. 60), by so "jumping scales ... evictees [could] dissolve spatial boundaries that [we]re largely imposed from above and that contain[ed] rather than facilitate[d] their production and reproduction of everyday life" (see also Smith 1992b). The second aspect of my thinking about scale at that point was that I largely thought of scales in topographical/areal ways, seeing them as what French theoretician Henri Lefebvre (1974 [1991], p. 351) termed "space envelopes". In such a view, scales are understood to enclose particular absolute spaces – those of the urban or the regional or the national or even the global – that have a certain coherence to them and to differentiate these spaces from one another, so that it is clear when one has jumped from one scale of activity to another. This language of scales as space envelopes has some concordance with Michel Foucault's (1975 [1977], pp. 141–143) work on how enclosing and partitioning space is central to controlling social actors. Hence, employers can enclose spaces, like those of factory grounds, so as to control who can enter them (keeping out union organisers, for instance) whilst also internally partitioning them so as to separate individuals from one another (think office cubicle!) and so reduce the likelihood they will develop any collective consciousness. Equally, enclosing workplaces represents a simultaneous partitioning of them from the broader economic landscape that exists beyond them in much the same way that scalar boundaries separate different absolute spaces.

The 1990s and 2000s, then, were a productive time within Geography for theorising about scale. Several other scholars interested in labour also contributed to the debate. For instance, Don Mitchell (1998) detailed how commercial farmers in California manipulated scalar narratives to present union organisers as non-local outsiders as a way to delegitimise them, whilst Jane Wills (2002) explored how hotel workers in different countries went about developing complex, multi-scalar responses to challenges raised by globalisation. Meanwhile, scholars like urban theorist/sociologist/geographer Neil Brenner (1997, 1998, 2000) theorised how questions of scale production intersected with investigations of globalisation's impacts upon urban governance. However, not everyone was on board. Marston et al. (2005), for instance, argued for what they called a "flat ontology" that would eliminate the concept of scale because, they suggested, it presented too hierarchical a way of understanding the world.[4] Nevertheless, the period since the 1990s has been one in which research into scale's production has been a key driver of geographical theorising in a variety of arenas.

From a topographical to a topological view of scale

In my research on the ILA's response to containerisation I largely saw scales as areal units, with the union seeking to both carve out 50-mile scales of operation around ports and to expand its scale of bargaining from a focus upon what happened within each individual port to operating as if the entire East Coast were a single bargaining space. However, as I later began to work upon issues of transnational labour organising, I started to think of scales in a slightly different way, one which saw them not just in *areal* terms but also in *networked* terms. This shift in my thinking was driven by exploring new research questions as part of my efforts to develop a 'Labour Geography' (Herod 2001) that sees working people as active shapers of economic landscapes (rather than as simply passive inhabitants of landscapes created by collective capital), and by reading some of the post-structuralist Economic Geography being produced in the late 1990s. In particular, as I started to think about globalisation and its impacts upon workers I began to question how the relationship between the local and the global is represented. This was a time when there was much talk of local-global relationships, especially in terms of considering how each of these scales is implicated in the other – concepts of 'the glocal' and 'glocalisation' dominated much analysis (e.g., Swyngedouw 1997). My own thought processes were influenced in this regard by the work of post-structuralist geographers like J.K. Gibson-Graham (the pseudonym of two geographers, Julie Graham and Kathy Gibson). In 1999 both came to UGA to give the keynote address to a mini-conference I organised with two colleagues (Melissa Wright, now at Pennsylvania State University, and Paul Plummer, now at the University of Western Australia) on 'Theorizing Space and Time at the End of the Millennium'. In this keynote lecture, which was published in a book based upon the conference (Herod and Wright 2002), they talked about how 'the global' is often presented discursively as inherently more powerful than is 'the local' and that this narrative plays an important role in political struggles (Gibson-Graham 2002 – see Chapter 21 by Kathy Gibson). This argument spoke to me because I was interested at the time in how corporations often use references to their 'global' reach to try to kowtow workers into submission by making them believe that they can never hope to similarly organise themselves 'globally'. However, describing such corporations not as 'globally organised' but instead as, say, 'multi-locational' entities, even if these locations are indeed in different parts of the globe, changes how we conceptualise such corporations and thus what we imagine the possibilities to be for challenging them – challenging a 'globally organised' corporation may seem insurmountable, whereas we may see challenging a 'multi-locationally organised' corporation as simply requiring a well-coordinated transnational campaign focused upon particular local places where the firm has operations. I found this an interesting gestalt shift.[5]

My work on labour and globalisation, though, took two related, if somewhat different, approaches to trying to understand workers' activities and scalar politics. The first involved challenging the then-dominant narrative that globalisation is largely a product of capital. In some ways, this narrative had connections with Smith's argument that the global scale is created out of capital's urge to spread across the planet, though more generally it reflected a focus in myriad disciplines upon transnational corporations as today's dominant global actors. However, I wanted to show how workers and their organisations have also been active in bringing about globalisation and that they are not just its victims. For instance, US unions played important transnational roles in helping both to rebuild the economies of Eastern Europe after communism's collapse (Herod 1998) and facilitating US capital's expansion into Latin America during the 20th century (Herod 1997b). Workers and their organisations, in other words, have been active producers of global scales of economic organisation. However, another area of my research at this time involved thinking about how, through organising locally, workers can sometimes challenge employers who are organised trans-locally. In particular, I examined how a dispute by autoworkers at just two plants in Michigan in 1998 led General Motors to shutter most of its assembly and components plants across North America (and even in Singapore), in the process showing that locally focused actions can, in fact, be effective against a 'globally organised' corporation, at least under particular circumstances – in this case, GM relied heavily upon just-in-time systems of component delivery and so any disruptions to its supply chain had momentous operational impacts (Herod 2000) (see Figure 14.3a and b).

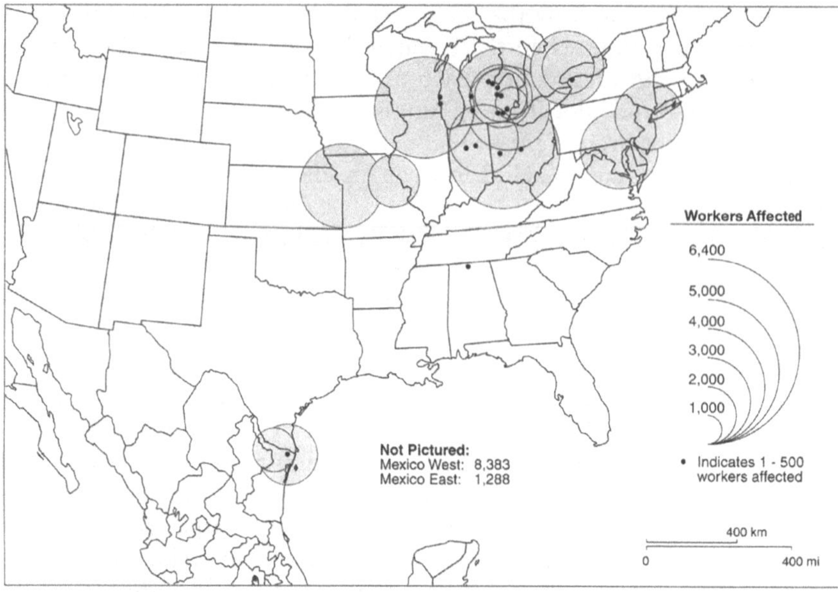

(a)

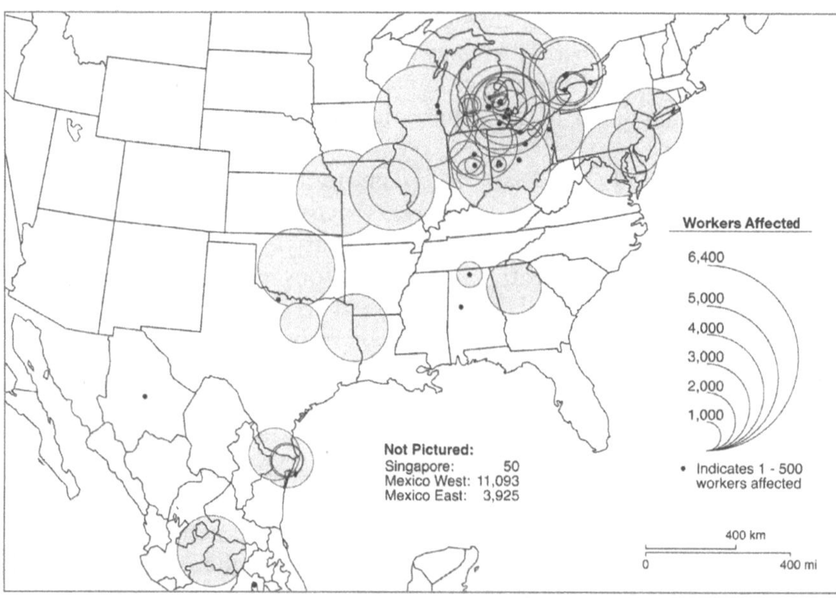

(b)

Figure 14.3 (a) Number and location of workers affected by the UAW strike against General Motors, as of 4 p.m. on 12 June 1998. (b) Number and location of workers affected by the UAW strike against General Motors, as of 4 p.m. on 19 June 1998. These two maps show how strikes at two GM components-manufacturing factories in Flint, Michigan, spread across North America and beyond. The speed with which workers were affected reflected GM's use of just-in- time inventory control systems, whilst the geography of the dispute's spread reflects where various plants were located organisationally within the company's structure. Thus, plants that were geographically proximate were not necessarily impacted in close succession. Rather, it was being proximate logistically that shaped the geography of the strike's spread. The geography of plant closure, in other words, reveals something about the spatial organisation of GM's supply chain.

Through my research and reading of various post-structuralist writers, then, I came to see that scale could be viewed not just *topographically* but also in *topological*/networked terms. In particular, work by the late Bruno Latour (1947–2022) was instrumental in shaping my thinking. Latour (1996, p. 370) had argued that the world's complexity cannot be encapsulated by "notions of levels, layers, territories, [and] spheres" – that is, the kinds of approaches to thinking about scale exemplified through Lefebvrian space envelopes. Instead, he averred, we should think of it as "fibrous, thread-like, wiry, stringy, ropy, [and] capillary". This means that scales might more appropriately be conceptualised in 'rhizomic' than in territorial terms. In such an approach, rather than seeing, say, the regional scale as contained within/operating below the national scale or the national scale as contained within/operating below the global scale, scales are theorised as connected rather like the strands of a spider's web, in which everything is connected to everything else but some parts are closer than are others, or the roots of a tree, with some roots going deeper into the Earth and some being shallower but all being connected to the tree's trunk (see Herod 2010, pp. 49–52 for illustrations). Consequently, for Latour (1993, p. 122) the global and the local are not part of a fixed hierarchy of separate arenas but instead "offer points of view on networks that are by nature neither local or global, but are more or less long and more or less connected". They are, in other words, not opposite ends of a scalar spectrum but, rather, are part of a language for juxtaposing shorter and less-connected networks with longer and more-connected ones, with the global and local (and other scales) simply different 'takes' on the same universe of networks, connections, abstractness, and concreteness.

Final thoughts

I've argued above that scale is a central geographical concept. In mapping, the concept of scale has a significant historical geography to it. This recognition moves us beyond comprehending cartographic scales as neutral, scientific elements of spatial representation. In terms of political and economic practices, recent theorising has challenged longstanding views that saw scales as simply natural/logical ways of dividing up the world to understand it. However, whereas this latter work has made it clear that scales of economic or political organisation are socially produced and contested things, it has promulgated at least two different ways of thinking about them – scales as Lefebvrian 'space envelopes' and scales as Latourian networks.

Throughout the above discussion I've illustrated how geographical knowledge – about scale or anything else – develops both relationally and contextually. Theorists have been influenced by each other's work, whether through agreement, disagreement and debate, or even indifference. Equally, ideas develop in particular places, with there often being a degree of serendipity in this – in my own case, being in the audience when Smith came to talk at WVU. There is, then, an important geography to knowledge creation, as ideas are advanced in specific places but then become geographically dispersed through various mechanisms – publishing books/academic articles, presenting papers at conferences, discussing research in the classroom, emails between academics working on particular topics, and so forth. In this way, certain places became central to scale theorising at particular times – Johns Hopkins University in Baltimore, where Smith developed his initial ideas, Rutgers University in New Jersey, where he honed some of them and where I first really interacted with them, the University of Massachusetts Amherst and the Australian National University in Canberra, where Julie Graham and Kathy Gibson were based, the

University of Georgia, where I am presently located, together with many other places where contributors to debates over how we understand scale are or have been situated. There's also, though, clearly a scalar politics to scalar theorising, as some contributions to the debate have remained rather localised, whilst others have achieved considerable national and/or global exposure.

Notes

1 This system was transferred to British colonies – many land deeds in the USA laid out prior to the 1785 Land Ordinance contain such descriptions.
2 I regularly teach a course in Paris where students learn about the politics of urban planning and architecture from the mid-18th century to the early 20th (Herod 2014).
3 Harvey (1995) makes similar observations when recounting his participation in researching a dispute involving workers at the Cowley car plant in East Oxford, especially how he felt that what happened at that particular plant needed to be understood within a much broader geographical context.
4 'Ontology' is a philosophical term that refers to the nature of what exists.
5 A 'gestalt shift' refers to how an interpretation of something changes depending upon how one views it. A famous example was first published in 1892 in a German humour magazine, *Fliegende Blätter* (Nr. 2465, p. 147), which asked '*Welche Thiere gleichen einander am meisten*?' ('Which animals resemble each other the most?'), responding '*Kaninchen und Ente*' ('Rabbit and duck') – from one perspective the viewer sees a duck's face; from a different perspective a rabbit appears.

Study task

Variations on this question can be set on an unseen examination, an open-book examination or as a term-paper assignment:

Geographical scale matters, but how and to whom? Answer this question with a focus on the research of a labour geographer such as Andrew Herod.

Sample reading list

Brown, J.C., and M. Purcell (2005) 'There's nothing inherent about scale: Political ecology, the local trap, and the politics of development in the Brazilian Amazon'. *Geoforum* 36 (5): 607–624.

Chapura, M. (2009) 'Scale, causality, complexity and emergence: Rethinking scale's ontological significance'. *Transactions of the Institute of British Geographers*, New Series, 34 (4): 462–474.

Cox, K. (1998) 'Spaces of dependence, spaces of engagement and the politics of scale, or: Looking for local politics'. *Political Geography* 17 (1): 1–23.

Crump, J. and C. Merrett (1998) 'Scales of struggle: Economic restructuring in the US Midwest'. *Annals of the Association of American Geographers* 88 (3): 496–515.

Jonas, A. (2006) 'Pro scale: Further reflections on the "scale debate" in human geography'. *Transactions of the Institute of British Geographers*, New Series, 31 (3): 399–406.

Marston, S. (2000) 'The social construction of scale'. *Progress in Human Geography* 24 (2): 219–242.

McMaster, R., and E. Sheppard (eds) (2004). *Scale and Geographic Inquiry: Nature, Society and Method.* Oxford: Blackwell.

Moore, A. (2008) 'Rethinking scale as a geographical category: From analysis to practice'. *Progress in Human Geography* 32 (2): 203–225.

Smith, N. and W. Dennis (1987) 'The restructuring of geographical scale: Coalescence and fragmentation of the northern core region'. *Economic Geography* 63 (2): 160–182.

Taylor, P. (1982) 'A materialist framework for political geography'. *Transactions of the Institute of British Geographers*, New Series, 7 (1): 15–34.

References

Brenner, N. (1997) 'State territorial restructuring and the production of spatial scale: Urban and regional planning in the Federal Republic of Germany, 1960–1990', *Political Geography* 16 (4): 273–306.

Brenner, N. (1998) 'Between fixity and motion: Accumulation, territorial organization and the historical geography of spatial scales', *Environment and Planning D: Society and Space* 16 (5): 459–481.

Brenner, N. (2000) 'The urban question as a scale question: Reflections on Henri Lefebvre, urban theory and the politics of scale', *International Journal of Urban and Regional Research* 24 (2): 361–378.

Buttimer, A. (1971) *Society and Milieu in the French Geographic Tradition*. Chicago: A.A.G. Monograph No. 6, Rand McNally.

Davis, M. (1986) *Prisoners of the American Dream: Politics and Economy in the History of the US Working Class*. New York: Verso.

Foucault, M. (1975 [1977]) *Discipline and Punish: The Birth of the Prison*. New York: Pantheon Books.

Gibson-Graham, J.K. (2002) 'Beyond global vs. local: Economic politics outside the binary frame', in A. Herod and M. Wright (eds), *Geographies of Power: Placing Scale*, pp. 25–60. Oxford: Blackwell.

Harley, J. (1989) 'Maps, knowledge, and power', in D. Cosgrove and S. Daniels (eds), *The Iconography of Landscape*, pp. 277–312. Cambridge: Cambridge University Press.

Hartshorne, R. (1939) *The Nature of Geography: A Critical Survey of Current Thought in the Light of the Past*, Lancaster, PA: Association of American Geographers (Fourth printing, 1951).

Harvey, D. (1995) 'Militant particularism and global ambition: The conceptual politics of place, space, and environment in the work of Raymond Williams'. *Social Text* 42 (Spring): 69–98.

Herbertson, A. (1905) 'The major natural regions: An essay in systematic geography'. *The Geographical Journal* 25(3): 300–310.

Herod, A. (1991a) 'The production of scale in United States labour relations'. *Area* 23 (1): 82–88.

Herod, A. (1991b) 'Local political practice in response to a manufacturing plant closure: How geography complicates class analysis'. *Antipode* 23 (4): 385–402.

Herod, A. (1994) 'Further reflections on organized labor and deindustrialization in the United States'. *Antipode* 26 (1): 77–95.

Herod, A. (1997a) 'Labor's spatial praxis and the geography of contract bargaining in the US east coast longshore industry, 1953–89'. *Political Geography* 16 (2): 145–169.

Herod, A. (1997b) 'Labor as an agent of globalization and as a global agent', in K. Cox (ed.), *Spaces of Globalization: Reasserting the Power of the Local*, pp. 167–200. New York: Guilford Press.

Herod, A. (1998) 'The geostrategics of labor in post-Cold War Eastern Europe: An examination of the activities of the International Metalworkers' Federation', in A. Herod (ed.), *Organizing the Landscape: Geographical Perspectives on Labor Unionism*, pp. 45–74. Minneapolis: University of Minnesota Press.

Herod, A. (2000) 'Implications of Just-in-Time production for union strategy: Lessons from the 1998 General Motors-United Auto Workers dispute'. *Annals of the Association of American Geographers* 90 (3): 521–547.

Herod, A. (2001) *Labor Geographies: Workers and the Landscapes of Capitalism*. New York: Guilford Press.

Herod, A. (2010) *Scale*. London: Routledge.

Herod, A. (2014) 'The rational city: Paris', in P. Knox (ed.), *Atlas of Cities*, pp. 88–105. Princeton: Princeton University Press.

Herod, A. and M. Wright (eds) (2002) *Geographies of Power: Placing Scale*. Oxford: Blackwell.

Kain, R. and E. Baigent (1992) *The Cadastral Map in the Service of the State: A History of Property Mapping*. Chicago: University of Chicago Press.

Latour, B. (1993) *We Have Never Been Modern* (translated by C. Porter). Cambridge, MA: Harvard University Press.

Latour, B. (1996) 'On actor-network theory: A few clarifications'. *Soziale Welt* 47: 369–381.

Lefebvre, H. (1974 [1991]) *The Production of Space*. Oxford: Basil Blackwell.

Marston, S., J. Jones III, and K. Woodward (2005) 'Human geography without scale'. *Transactions of the Institute of British Geographers*, New Series, 30 (4): 416–432.

Mitchell, D. (1998) 'The scales of justice: Localist ideology, large-scale production, and agricultural labor's geography of resistance in 1930s California', in A. Herod (ed.), *Organizing the Landscape: Geographical Perspectives on Labor Unionism*, pp. 159–194. Minneapolis: University of Minnesota Press.

Smith, N. (1979) 'Toward a theory of gentrification: A back to the city movement by capital, not people'. *Journal of the American Planning Association* 45 (4): 538–548.

Smith, N. (1984 [1990]) *Uneven Development: Nature, Capital, and the Production of Space*. Oxford: Basil Blackwell.

Smith, N. (1992a) 'Contours of a spatialized politics: Homeless vehicles and the production of geographical space'. *Social Text* 33: 54–81.

Smith, N. (1992b) 'Geography, difference and the politics of scale', in J. Doherty, E. Graham, and M. Malek (eds), *Postmodernism and the Social Sciences*, pp. 57–79. London: Palgrave Macmillan.

Swyngedouw, E. (1997) 'Neither global nor local: "Glocalization" and the politics of scale', in K. Cox (ed.), *Spaces of Globalization: Reasserting the Power of the Local*, pp. 137–166. New York: Guilford Press.

Vidal de la Blache, P. (1903 [1994]) *Tableau de la géographie de la France* [*Picture of the Geography of France*], Paris: Table Ronde.

Wills, J. (2002) 'Bargaining for the space to organise in the global economy: A review of the Accor-IUF trade union rights agreement'. *Review of International Political Economy* 9 (4): 675–700.

15 Proximity, distance, and difference

The global and the intimate

Geraldine Pratt

Geraldine Pratt

Introduction

I recall that it was spring 2005 when I had my first meeting with Victoria Rosner, a telephone call between College Station, Texas and Vancouver, arranged through two of our closest friends, Cindi Katz (mine), Nancy Miller (Victoria's). Cindi and Nancy were co-editing *Women's Studies Quarterly* and they thought it would be terrific if Victoria and I guest edited an issue on a theme of our choice. Victoria was then Associate Professor of English at Texas A&M University. She had just finished a book entitled *Modernism and the Architecture of Private Life*. She was slightly intimidating to me, as English professors can be. We talked around our research interests in those pre-zoom days, two voices on a phone. I remember standing in my kitchen looking out of the window as we talked through possible threads of connection. My memory is that we came to our theme quickly; the intimate (Victoria's terrain) and the global (mine). We put out our call for papers. Over the next year or so we spoke to each other often – sometimes by phone, often by email, sometimes many times a week, because of the journal issue but more because each of our lives took such an awful turn. I was diagnosed with breast cancer. When she was seven months pregnant, Victoria's husband decided he didn't want to be

DOI: 10.4324/9781003343240-17

married any more (Rosner, 2010). She moved back home to New York City. Life was more eventful than either of us wanted and I found such comfort in Victoria's long-distance company. We stayed together as co-editors long enough to turn the journal issue into a book (Pratt and Rosner, 2012). This is a feminist friendship story. But it is also a feminist research story. In many ways a very ordinary feminist research story.

The Worcester project: gender, work and 'space'

My first academic job was in the School of Geography at Clark University in Worcester Massachusetts. In my first year there, Professor Susan Hanson took seriously her role as mentor and brought me into a research project she'd conceived to explore the ways women's labour market participation is structured by the places in which we live and the work we do in the home. We were interested in learning whether the paid work that women did and the incomes that they earned were dependent on where they lived in Worcester and the unpaid work they did at home and in their communities.[1] Susan's mentorship is another important facet of this research story. Although they are often buried in accounts of research, friendships – relationships, proximities, intimacies, and interdependencies – often lie at the heart of the research process.

Susan and I carried out a standard social science survey. In the first phase, we surveyed households. In the summer of 1987 through to early 1988 we sent a team of graduate and senior undergraduate students into the field (the Worcester metropolitan area). Their instructions were to knock on the doors of particular houses on particular blocks of the city in order to produce a randomly selected representative sample of the population of the metropolitan area. We asked these research assistants to approach the identified household at least three times on different days of the week and at different times of the day before giving up on that household's participation. The survey was 53 pages long, and consisted of many closed questions (that is, there were defined answers for respondents to choose from), with spaces to record additional comments (which we later coded for quantitative analyses). It included questions about family structure, housing history, who did what kinds of domestic chores, and histories of paid employment. Ultimately, we surveyed 492 women and 206 men in 613 different households. With this information we documented the importance of geography (home–work journey and local labour markets) and domestic responsibilities for women's paid employment options and decisions. Particularly working-class women found employment close to home, typically part-time employment, once they had children. Whether they worked in a factory or office was to a significant extent an accident of their residential location and whether they had access to a car. Claims that the sample was representative of the urban population were essential to the impact of the research. I still find disturbing our finding that only one woman in the entire sample of almost 500 had an annual income of at least $35,000 and a school-aged child at home, while one in three men in the sample earned this much (and of course a good proportion of these lived in households with school-aged children). It spoke volumes to the gendered relationship between work at home and in paid employment.

It's now easy to see what we didn't see until after the survey was completed. Other studies by feminist geographers began to show that our insight that women find paid employment close to home did not apply to Latinx and Black women. Given racialised housing and labour markets in the New York metropolitan area, Sara McLafferty and Valerie Preston (1991), for instance, found that Black women in particular, who were reliant on public transit that was ill-conceived in relation to their transportation needs, travelled exceptionally long times to reach their paid employment. This has serious implications for their capacity to care for themselves and their own families. The whiteness of our gender story was also brought into focus through the vibrant feminist, Black and postcolonial scholarship of the time. The work of Crenshaw (1989), hooks (1990), Moraga and Anzaldua (1981), and Spivak (1987), for instance, made it abundantly clear that gender is lived at the intersections, through relations of race and racism, sexuality, imperialism, colonialism, and class.

As we wrote some years after the survey was completed (in the book that brought the research together in one place: Hanson and Pratt, 1995), to some extent the whiteness of our study reflected the place in which we were doing our research (Worcester Massachusetts), and the methodology we deployed. Sampling for the population, with a designated proportion of households from each area, failed to bring the relatively small but significant Latinx population in Worcester into visibility. Attempting to sample five households on 140 blocks distributed throughout the metropolitan area (as was required by our sampling methodology) meant we could miss this population entirely. As it turned out, only 10% of the sample were non-Anglo, non-white persons, and their experiences were thus lost in our analysis.

The critique of our middle-class white feminism deepened in a personal way. When writing the book on the Worcester project, we asked graduate and undergraduate students who were sent into the field as research assistants to reflect on that experience. A good number offered observations that gently but forcefully revealed how we'd built our positionality as white middle-class women into the closed-question format of the survey 'instrument'. We'd made assumptions about the division of labour within the households being surveyed. For instance, we'd not imagined that older children would regularly prepare dinner. And so, we failed to ask about children's domestic work. We also began to see how we'd inadvertently built patriarchal (and capitalism-centric) assumptions into the survey. Recording an individual's work history, we didn't ask women what else they were doing (beyond domestic work) when not employed in the formal labour market. We thus replayed in our survey the invisibility and under-valuing of all kinds of community-based work outside paid employment. One graduate student research assistant wrote, reflecting on the power dynamics of the research more generally: "[Most] disconcerting was the exclusion of survey participants from the survey design. The survey was designed to answer questions of interest to the academic literature. The unidirectional flow of information from the community to the university rarely benefits the community."

Addressing our request to reflect on their experience as research assistants, one wrote: "Ironically, this exchange continues to promote many of the pitfalls discussed above; in this case students have now replaced community respondents as the source of information" (Hanson and Pratt 1995, 84, 239, ft 24). These observations about power dynamics within the research process were sharpened when, at a research presentation, an audience member advised us to stop looking at exploitation by employers and look more closely at our own (as researchers). The student research assistants had affectionately dubbed our survey The Worcester Expedition. These critiques of the power dynamics within our survey methodology brought new meaning to this naming.

I have come back to Donna Haraway's influential essay, 'Situated Knowledges: The Science Question in Feminism and the Privilege of Partial Perspective' (1988), many times over the years. It has resonated because I've lived it. Haraway argued that we can get closer to something we might call objectivity through conversations across differently situated knowledges. What she means by situatedness is expansive. It includes one's social location (in my case, white, heterosexual, middle class, settler Canadian, among other categories of lived experience). And – crucial for geographers – it also includes place (Cindi Katz [2001] reminds feminists of this in her writing on 'counter-topographies'), along with the technologies used for 'seeing' evidence (for instance, a survey instrument administered to a randomly selected sample meant to be representative of a population). We learnt some of the lessons of situated knowledge through the Worcester project, from our students and colleagues, in conversation across those situated differently in relation to it. The Worcester project was not inherently flawed. Indeed, it has stood the test of time and is still widely cited 30 years on. But like all research, it bears the imprint of the place in which it was carried out, the researchers who conceived it, and the methodology we used. It offers significant but partial insight into the way that some relations of gender are lived through geographies of place and spatial relations.

Working with the Philippine Women's Centre: the labour of care

Absorbing the criticisms of the Worcester project, I put race and class inequality *between* women – and relative white privilege – at the centre of the next project. From the early 1990s I was living and working in Vancouver, Canada and I began a project focused on domestic workers and their employers. This is an employment arrangement in which difference among women could not be ignored. I began by interviewing Canadians who were advertising in a community newspaper for nannies to take care of their children in their home, as well as Canadian nanny agents who listed themselves in the telephone directory. This was an easy point of entry because parents were keen to talk. Indeed, I've never since encountered such an eager group of interviewees. Whereas a 35% positive response to this kind of (cold-call) research invitation is normal, even quite good, about 90% of the parents contacted immediately agreed to an interview. They were in a crisis: finding affordable, reliable, reasonable quality childcare was urgent and they were eager to talk, to share their stories and to hear during their interview what other parents had shared with me. In other words, they positioned me as other than a one-way recipient of their information.

Because this research project was originally framed within a US-based academic literature (which for the most part contexualises domestic work within the aftermath of slavery), I had not properly understood that a study of private childcare in Canada was also a study of contemporary transnational labour migration. Place matters, including for the research we do. Canada has a long history of importing domestic workers, and by the early 1990s – when this research began – the majority of a relatively small number of live-in care workers were coming from the Philippines on temporary work visas. Soon after the research began, the Canadian government significantly expanded temporary foreign worker programmes. These programmes are in essence a geographical or spatial strategy of opening a national labour market to workers outside national boundaries, but without conferring the rights and benefits of citizenship. They often place these workers in conditions of precarity, both on the job, and in many other facets of their lives. The Live-in Caregiver Program (LCP) was instituted in 1992, and is one of the oldest and largest of such programmes in Canada. It grew massively over the course of the research: only 2,684 people were admitted through it in 2000, rising to 13,775 by 2007 (Kelly, Park, de Leon and Priest, 2011). So, research into an important but relatively small phenomenon turned into something bigger.

If interviewing Canadian employers and nanny agents was easy, by the early 1990s the moment had passed in feminist theory and practice for a white researcher to solicit the life stories of racialised women, perhaps especially those from the global South (Mohanty, 1988). The questions were and remain: what right did I have as a researcher to request these stories, particularly when the expected storyline was a painful one of labour exploitation and family separation? Is it not deeply problematic for a feminist scholar to advance her career by telling the painful stories of marginalised women? I stalled before contacting domestic workers. I stalled until my research funds were about to be rescinded.

The Center for Research in Women's Studies and Gender Relations was founded at the University of British Columbia in 1991. In the initial years we instituted a community scholars' programme that enabled community activists and other scholars to spend a concentrated period of time at the university. I met Cecilia Diocson, the director of the Philippine Women's Centre (PWC) in Vancouver through this programme. Having met her on my terrain at the university, I arranged to meet Cecilia and two other organisers from the PWC on theirs. We met in a basement office in a small house on a residential street on the other side of the city, to discuss whether the PWC might be interested in a research collaboration. It helped, I believe, that with the assistance of our savvy department administrator I would be able to transfer research monies to the PWC in advance of the research. This may seem like a small thing but it switched up the power relations within our research relationship. The PWC did not have to itemise and justify payment for services already rendered, and our relationships of accountability were allowed to grow along different, more personal lines. This point about funding and material resources is an important one. As university researchers we have access to significant resources that can be shared with community researchers. But it goes beyond this. On the basis of their track record as community researchers working in collaboration with a number of university researchers, the PWC later secured their own funding from Status of Women Canada and Heritage Canada for subsequent research projects of their own.

What emerged from the first meeting at the PWC was an agreement to work together on a project documenting the lives of Filipina domestic workers in Vancouver, around themes that we would identify together. This step in the process was crucial. Some years later, Cecilia spoke of her many refusals to work with researchers who approached the Centre simply to access Filipino immigrants' experiences. "Collaboration," she said "means that we need to sit down and formulate what kind of project we really need to do: what kind of research? That is what we call collaboration" (Pratt, in collaboration with the Philippine Women Centre of BC and Filipino-Canadian Youth Alliance, 2007). On six different Saturdays, we met in that same small basement in East Vancouver, typically with around a dozen domestic workers who were members of the PWC, often from noon until 8pm, talking, laughing, sometimes crying and sharing meals.

At the first meeting we decided on the topics to be documented and methods of doing so (Figure 15.1). The methodologies came from the PWC as much as from me. As community organisers they were familiar with a much wider repertoire of methodologies, including role-playing and various art practices such as sculpture and montage (the latter discussed but not in the end used). In our next meetings, we met to work in small groups to record narratives and role-play around the agreed upon themes. In later weeks we met to read through the transcripts from previous sessions together. Literally. The transcripts were passed around the group sitting in a circle and read aloud like a script, commented up, verified, and embellished throughout the reading. This sharing of life

Figure 15.1 Collaborating with the Philippine Women's Centre in 1995.

Photo courtesy of the Philippine Women's Centre.

stories and employment circumstances documented the lives of Filipina live-in care workers in Vancouver. It was a means of telling a collective community story. It was also a form of organising for the PWC. Domestic workers learned from hearing about each other's lives. I recall one woman discovering that, while she thought that she had health care insurance, it seemed from what others were saying that she likely did not. She left that session knowing that she needed to pursue this with her employer. This was research certainly *and* it was part of a process of building a community of engaged activists. Learning about others' experiences, that their issues were not theirs alone or the result of a random 'bad employer', we emerged with a more structural analysis of temporary labour migration. The analysis situated the LCP in a fundamentally geographical frame: within histories of US imperialism in the Philippines, white supremacy in Canada, and global uneven economic development that compels some women to migrate for economic reasons.

We carried out more research over the next fifteen years or so when the PWC felt the urgency to document other aspects of Filipino migrant experience in Vancouver. I also worked with a related organisation, the Filipino-Canadian Youth Alliance, to tell stories of Filipino youths. In all of these projects there was the understanding that the data was common property and that any academic papers should be distributed and commented upon by the organisation before distribution. One of my next research projects looked at the lives of the women who participated in the first study who by then had become permanent residents of Canada. It documented their deskilling through the LCP. In another study, we documented the experiences of family reunification after women were able to secure permanent resident status and sponsor their families to join them in Vancouver (Pratt, 2012).[2] We found that a median of eight years of family separation had taken its toll. There was conflict within many reunifying families and male Filipino youths were dropping out of high school at disturbing rates, amongst the highest of any group of immigrant male youths in Vancouver. These were difficult stories to tell and hear, and the risks of victimisation and voyeuristic witnessing of suffering were ever present. Community ownership of the research was essential.

There's no doubt that the research was strengthened immeasurably through the collaboration with the PWC, by the fact that it was done in a combination of Tagalog, Taglish, and English, among women who knew and trusted each other, and because the community researchers had a depth of knowledge about the Philippines, migration processes, policies and structures in the two countries, and the circumstances within their community in Vancouver. It was strengthened by the fact that the PWC was accountable to the Filipino community. I was a beneficiary of all of this. From this perspective, my role could be framed within what cultural critic Sara Ahmed has termed the "phenomenology of whiteness". She conceives this as the freedom of white bodies to move through the world with relative ease and to "inhabit the world as if it were a home" (2007: 153). This is an orientation that informs a great deal of Western knowledge production and it's one that my collaborators gently and persistently pushed against. The first time I worked with the Filipino-Canadian Youth Alliance, for instance, and we met to review a draft of a paper based on our focus groups, they told me to work further to "deepen" my analysis. This was an important moment of unsettling my authority as an older white university researcher and theorist.

But still, one might ask, why did the PWC invite me into their community with such patience and generosity over so many years? What further lessons can be learnt about collaborating across difference? Conversations with my collaborators offer some answers (Pratt, in collaboration with the Philippine Women Centre of BC and Filipino-Canadian Youth Alliance, 2007). Prolonged engagement was essential. From Charlene Sayo, of the Filipino-Canadian Youth Alliance, "When we first did the project with Filipino youth, in particular, it wasn't like you just left and that was it. There was more that you wanted to explore, and that was really important" (2007, 99). The PWC and related organisations could see the utility of working with an independent academic researcher: "It's just a statement if it's not backed up by a university researcher" (2007, 102). University researchers have resources that our privilege often prevents us from seeing. Access to research grants is one obvious resource but there is much beyond this that we typically take for granted, such as access to a university library. One of the data sets that revealed the crucial finding that a high proportion of Filipino male youths in Vancouver were not completing high school was not publicly available. It seems probable that the PWC would not have had access to it. Our collaboration involved a traffic of ideas between the academy and community. Some contemporary academic theory resonated: for instance, concepts of 'social death', 'slow death', and 'bare life'. A possibly idealised rendering of our collaboration is that we sought reciprocity and built collective strength through our differences. We shared a common project: to document and fight against the persistent devaluation of women's care labour, and to hold the Canadian government accountable to its liberal ideals of justice and equity. As improbable as these liberal ideals often seem in practice, they set the course for holding governments accountable. Having a clear common project, it seems to me, is essential for meaningful collaboration across difference.

A travelling play

In 2007 the possibility emerged for another collaboration that we hoped would work on public opinion and government policy from a different direction. I first met Caleb Johnston in 2003 when he was an undergraduate Geography student. I supervised a directed studies course in which he researched as participant-observer a theatre project aimed at addressing poverty in Vancouver (Pratt and Johnston, 2007). The opportunity to supervise this directed studies course is just one instance of what most academics know well: the deep and sometimes unexpected connections between teaching and research.

Drawing on his long personal history in Vancouver's theatre community, Caleb asked in 2007 whether the PWC and I might be interested in collaborating with him and a professional theatre friend, Alex Ferguson, to create a testimonial play. They were interested in working with our now extensive archive of testimonial interviews, as well as my cache of employer interviews, within this theatrical form. Testimonial, verbatim or documentary theatre is well suited to research because the script for the most part is created by quoting research or other archival materials verbatim, essentially word for word (though edited). It brings research to a public audience in an immediate way, and can gather separate communities into a shared space.

We (Caleb, Alex, PWC and I) created a play through a two-year development process. *Nanay: A Testimonial Play* has a flexible format (for script see Pratt and Johnston, in collaboration with the Philippine Women Center of British Columbia, 2014). Small audiences (of 50 people in total, broken into four smaller groups of about 13 for much of the performance) are invited to move through nine scenes staged in different rooms to hear the testimony of individual domestic workers, their children, a nanny agent, and Canadian employers.

There is no linear narrative and different groups are exposed to the testimonial narratives in different orders. The entire audience of 50 then meets for an extended discussion at the end of the play.

This flexible format has allowed the play to travel, to absorb what we have learned through criticism of previous versions, and to add new scenes from research done in each new context in which it is being performed (Johnston and Pratt, 2019). Academic research travels into the world in ways that we typically don't fully know. In the case of *Nanay* we could follow its travels, nudge it on its way and revise it in light of what we learned through each new iteration.

We took the play to the PETA Theatre in Manila in 2013 (Figure 15.2). One criticism of the earlier productions is that they represented Filipino domestic workers as sad victims of their circumstances. This is now a common criticism of representations of Black and Indigenous women and women from the global South (Hartmann, 1997; Mohanty,

Figure 15.2 Mirrored movements, *Nanay* performed at the PETA Theatre Center, Manilla 2013.

Photo courtesy of Caleb Johnston.

2002). And so, we built into one scene a video of domestic workers talking about their lives in Vancouver, laughing and making fun of their employers. The intent was to represent domestic workers in Vancouver in another emotional register, with friendships and lots of humour, and subversive energy and organising.

New criticisms nonetheless emerged in Manila. We were reminded of our privilege by the organisation Migrante International. Migrante noted that, as a professional theatre, PETA Theatre was inaccessible to migrants. Graduate students in a dramaturgy class at the University of the Philippines (Diliman) told us that the play was Western-centric. They wrote: "White man's play for white man's problems" and "Where is the Filipino dramaturg? Sensibility of the director is not attuned to the sensibilities of the audience."

And so we returned a year later to partner with Migrante International and a Filipino director to create a community performance of *Nanay* in the poor migrant-sending community of Bagong Barrio (Figure 15.3).

Working with Migrante International, we did research in Bagong Barrio to build the experiences of local residents into the revised play. We integrated scenes from Canada with stories of overseas Filipino workers (OFWs) working in the various countries in the Middle East to which residents in Bagong Barrio are more likely to go. Children left behind in Bagong Barrio testified to their experience. We absorbed the director's criticism of the realist genre of the original monologues (which he viewed as Western) and wrote a chorus from what we had learned in Bagong Barrio (here in English translation):

We are the people of Bagong Barrio
Do you know your history? Do you know what you are capable of?
In the 1970s Bagong Barrio was known for resistance to land issues, to water issues, to housing issues.
We could mobilise 7,000 people in an afternoon
We had jobs and could feed our families
And then our jobs were contractualised and our factories moved to export processing zones
We work and we work and we work and we work and we work
We make paper bags for National Books
We sew scraps of fabric for cleaning cloths
We peel garlic all day for 70 pesos a bag
We use our ingenuity to invent jobs out of air
But there are no more secure jobs and people have to leave in order to survive
We need to analyse and organise
We will create our own security together
Here at home
Not in some foreign land

A long way from academic authoring, to be sure, but I can still feel the pleasure of writing this chorus, and seeing a short video of it being performed by the young community performers on the street in Bagong Barrio. This chorus is based on traditional interview-based and archival research in Bagong Barrio, and was written to communicate this research directly to residents there. With reference to the short street performances, a Migrante organiser had this to say: "They had another purpose. They led to breakthroughs and new revelations. They weren't just to advertise the play. They helped to build confidence and shed fear [among the performers]. They are a way of talking to people" (Johnston and

Figure 15.3 Calling the audience to attend community performance in Bagong Barrio, 2014.

Photo courtesy of Vanessa Banta.

Pratt, 2019: 109). They were a way of talking to and bringing a forgotten history of the local community directly to the people of Bagong Barrio. This then is another shift in my research practice over the years, a shift that has occurred across the social sciences and humanities. We can (and many of us feel we should) produce our research so that it is accessible to different communities, both within and beyond the academy.

What have we learned from the travels of *Nanay*? We learned in the first instance that creating and staging a play can be a mode of doing (and not just disseminating) geographical research. We took up the Manila-based director's insistence that the experiences of local residents be represented in the Bagong Barrio production by doing a new round of research there. Building in narratives from Bagong Barrio allowed us to address a common criticism of migration research, that migrants' lives are typically narrated from the perspective of the global North. Migration scholars have called this abbreviated geographical imagination "methodological nationalism" (Wimmer and Schiller, 2002). In feminist postcolonial studies, Gayatri Spivak (2000: 335) notes that this tendency effectively strips migrants of their history and geography, their agency, and their complex identifications and dis-identifications with the places they leave and the places to which they migrate. Our dramaturg for the first production of *Nanay* in Vancouver put it more simply: "It's as if Filipina domestic workers come to life half way across the Pacific." For Filipino domestic workers in Canada what lies on the other side of the Pacific very often includes a university education, personal ambitions as well as family responsibilities, layers of social relationships, and the active negotiation of networks of information, including rules and regulations of a complex migration infrastructure. This complexity is part of their story. What we learned through the travels of *Nanay* is that we need not reduce individuals to caricatures of themselves to mount a strong critique of state policies, histories of imperialism, and the uneven economic development that shape their lives. We learned that migrant life under the LCP in Canada cannot be fully understood without seeing the rest of migrants' ongoing transnational lives.

Taking *Nanay* to the Philippines and collaborating with (astonishingly generous) partners has been a means not just of extending our knowledge claims but, possibly more importantly, of understanding how deeply situated these claims are in our own geography, and hence the limits of them. Perhaps, as Dylan Rodriguez has suggested in his critique of the "transparent I" of the white European subject, destabilising our knowledge claims' "presumed dominion over the lived meanings of race/place/body" is one of the most important lessons of all (2010: 199). That is, the claims and categories developed within 'Western' knowledge are also partial and situated. This is a place from which to start more worldly and world-affirming conversations across social and geographical difference.

Final reflections

I came to Geography half-way through my master's degree, with no undergraduate training in the discipline. (My bachelor's major was Psychology, with a minor in Art History.) I've found the discipline of Geography to be a capacious interdisciplinary home that invites and fosters all kinds of experimentation. In my case, it's accommodated a methodological trajectory from social survey to theatre. The devaluing of the work of social reproduction – of the essential care of families and communities, and the profound unacknowledged linkages between this undervalued gendered work in the home and formal employment in the labour market – is a preoccupation that has run through all of my research since the Worcester Expedition. Criticisms of that early research (along with the changing *zeitgeist* of feminist theory and the on-the-ground empirics of the LCP) led me to focus more explicitly on race, specifically how race and colonial histories maintain the devaluation of the work of social reproduction even after it is commodified as paid employment. Geography is a discipline that has a deeply complicit history within various colonial and imperial ventures but it also has been a location from which to critically examine the ways global colonial and imperial histories endure into the present, including in the intimate spaces of our lives: the care of our children and elders. In all of my work since the Worcester Expedition, this has been a connective thread.

Our geographies shape and limit what we see. Binaries between private–public, home–workplace, global North–global South are critical to the devaluation and relative invisibility of the work of social reproduction. Geographies also shape what we see as researchers and scholars, and this is both a limitation and an opportunity. It's both an invitation and injunction to open ourselves to conversations across difference in order to reach towards the kinds of world-making to which we aspire.

This orientation to conversation across difference merges with a basic feminist principle: that the notion of the 'autonomous self' is a destructive fiction and we are strengthened rather than weakened through our dependencies and interdependencies. Autobiography is a strange genre from the perspective of such a self. I hope that this autobiographical account has made clear that my learning and unlearning and relearning about the world has been an ongoing collaborative venture, with valuable debts gathered all along the way, only a fraction of which have been named here.[3]

Notes

1 We didn't frame this in these terms at the time, but we were in effect empirically documenting the impacts of women's unpaid work of social reproduction on their paid employment and earning capacity. Making visible the significance of this work of social reproduction to the maintenance of both capitalist economies and the dynamics of gender and race oppression has been a center point of generations of feminist political economists. See Fraser 2022.

2 I still feel shame about sole authorship of this book. An eleventh-hour conflict with the publisher over co-authorship led to our joint (PWC and me) decision to proceed with sole authorship. Thus, the book's dedication to the PWC of BC ('This book is yours') is literal.

3 To this point, I thank my colleagues and friends, Rafi Arefin, Trevor Barnes, Noel Castree and Jess Dempsey for reading and commenting on a draft of this chapter.

Study task

This task can be set as a term paper or as part of a seen or unseen timed examination:

Focussing on the research of Geraldine Pratt and collaborators, explain how the relative distance between the workplace and the homeplace affects many women's attempts to combine paid employment with family life. In your response, consider how differences between women can influence the way they combine waged work with domestic commitments.

Starter reading list

Hanson, Susan, and Geraldine Pratt (1988) 'Reconceptualizing the links between home and work in urban geography', *Economic Geography* 64, 4: 299–321.

Hanson, Susan, and Geraldine Pratt (1991) 'Job search and the occupational segregation of women', *Annals of the Association of American Geographers* 81, 2: 229–253.

Pratt, Geraldine, and Susan Hanson (1994) 'Geography and the construction of difference', *Gender, Place and Culture* 1, 1: 5–29.

Pratt, Geraldine (2009) 'Circulating sadness: Witnessing Filipina mothers' stories of family separation', *Gender, Place & Culture* 16, 1: 3–22.

Pratt, Geraldine, Caleb Johnston and Vanessa Banta (2017) 'Lifetimes of Disposability and Surplus Entrepreneurs in Bagong Barrio, Manila' *Antipode* 49, 1:169–192.

References

Ahmed, Sara (2007) 'A phenomenology of whiteness', *Feminist Theory* 8:2, 149–168.

Crenshaw, Kimberlé (1989) 'Demarginalizing the intersection of race and sex: A black feminist critique of antidiscrimination doctrine, feminist theory and antiracist politics', *University of Chicago Legal Forum*. Article 8: 139–167.

Fraser, Nancy (2022) *Cannibal Capitalism: How Our System Is Devouring Democracy, Care, and the Planet – and What We Can Do About It*. London and New York: Verso.

Hanson, Susan and Pratt, Geraldine (1995) *Gender, Work and Space*. London and New York: Routledge.

Haraway, Donna (1988) 'Situated knowledges: The science question in feminism and the privilege of partial perspective', *Feminist Studies* 14: 579–599.

hooks, bell (1990) *Yearning: Race, Gender, and Cultural Politics*. Boston: South End Press.

Hartmann, Saidiya (1997) *Scenes of Subjection: Terror, Slavery, and Self-Making in Nineteenth-Century America*. New York and Oxford: Oxford University Press.

Johnston, Caleb and Pratt, Geraldine (2019) *Migration in Performance: Crossing the Colonial Present*. London and New York: Routledge.

Katz, Cindi (2001) 'On the grounds of globalization: A topography for feminist engagement', *Signs* 26:4, 1213–1234.

Kelly, Philip, Park, Stella, de Leon, Conely, and Jeff Priest (2011) Profile of Live-in Caregiver Immigrants to Canada, 1993–2009 TIEDL (Toronto Immigrant Employment Data Initiative) Analytical Report 19, Toronto.

McLafferty, Sarah and Preston, Valerie (1991) 'Gender, race, and commuting among service sector workers', *The Professional Geographer* 43: 1–14.

Mohanty, Chandra Talpade (1988) 'Under Western Eyes: Feminist scholarship and colonial discourses', *Feminist Review* 30: 61–88.

Mohanty, Chandra Talpade (2002) '"Under Western Eyes" revisited: Feminist solidarity through anticapitalist struggles', *Signs* 28:2, 499–535.
Moraga, Cherrie and Anzaldua Gloria E. (eds.) (1981) *The Bridge Called My Back: Writings by Radical Women of Color*. Persephone Press.
Pratt, Geraldine (2012) *Families Apart: Migrant Mothers and the Conflicts of Labor and Love*. Minneapolis: University of Minnesota Press.
Pratt in collaboration with the Philippine Women Centre of BC and Ugnayanng Kabataang Pilipino sa Canada/Filipino-Canadian Youth Alliance 2007 Working with migrant communities: Collaborating with the Kalayaan Centre in Vancouver, Canada 95–103.
Pratt, Geraldine and Johnston, Caleb (2007) 'Turning theatre into law, and other spaces of politics', *Cultural Geographies*, 14: 92–113.
Pratt and Johnston in collaboration with the Philippine Women Centre of British Columbia 2014 'Nanay (Mother): A Testimonial Play', in *Once More, With Feeling: Five Affecting Plays*, edited by Erin Hurley, 49–90. Toronto: University of Toronto Press.
Pratt, Geraldine and Rosner, Victoria (eds) (2012) *The Global and the Intimate: Feminism in Our Time*. New York: Columbia University Press.
Rodriguez, Dylan (2010) *Suspended Apocalypse: White Supremacy, Genocide, and the Filipino Condition*. Minneapolis: University of Minnesota Press.
Rosner, Victoria (2005) *Modernism and the Architecture of Private Life*. New York: Columbia University Press.
Rosner, Victoria (2010) Could I Forgive Him One Last Time? *New York Times*, January 3, Section ST, 5.
Spivak, Gayatri Chakravorty (1987) *In Other Worlds: Essays in Cultural Politics*. London and New York: Routledge.
Spivak, Gayatri Chakravorty (2000) 'Thinking cultural questions in "Pure" literary terms', in Paul Gilroy, Lawrence Grossberg and Angela McRobbie, eds *Without Guarantees: In Honour of Stuart Hall*, 335–357. London and New York: Verso.
Wimmer, Andreas and Schiller Nina Glick (2002) 'Methodological nationalism and beyond: Nation-state building, migration and the social sciences', *Global Networks* 2:4, 301–334.

16 Which realities are we trying to understand?

The workings of a physical geographer in the quest to respect river diversity

Gary Brierley

Gary Brierley (R) with friend and colleague Kirstie (L) at a lookout overseeing the Upper Yellow River at Dari, Qinghai province, Western China.

Introduction

Respecting diversity is a foundational premise of my life. I cherish prospects that people and rivers can express themselves as they wish, so long as their actions do not harm others. To me, such thinking underpins notions of a just sustainability in the face of global inequity, biodiversity loss, land use development and climate change (Agyeman et al., 2003). My commitment to create and maintain a safe and just environmental space for humanity deliberates between idealism and pragmatism, recognising humans as part of nature and living with variability and emergent traits in the quest to address concerns for environmental quality and human equality.

In this chapter I recount elements of my personal journey in ways of being, knowing and making rivers at different stages of my long career in Geography (Table 16.1). In simplistic terms, high-school, undergraduate and postgraduate training in the United Kingdom and Canada provided state-of-the-art 'textbook' understanding of geomorphic river science, and controls upon river diversity. A move to Australia, where I completed post-doctoral research and began my career as a lecturer, quickly instilled a strong sense that a different approach to river science and management was required to look after the

DOI: 10.4324/9781003343240-18

Table 16.1 Ways of knowing and making rivers at differing stages in my career

Career stage	*Ways of knowing and making rivers*
1973–1981. School and undergraduate days. Bury Grammar School and Durham University, UK	Cherished childhood experiences on the family farm at the confluence of the Roch and the Irwell Rivers in northwest England firmly grounded my sense of place, identity and belonging. Textbook training of a generalist Geographer sought to make sense of differences and diversity of people, places, landscapes, cultures, etc., scoping ways to address moral concerns for societal justice, care and wellbeing (e.g., Bunge, 1973; Harvey, 1973). A third-year course on "Principles of Geography" empowered students to think and question independently: "Geography is what Geographers do". Travel experiences and field trips including research on semi-arid rivers in Tunisia and the West Bank complemented early encounters with river diversity in various parts of the United Kingdom and Europe.
1981–1989. Postgraduate experiences. Geography Department, Simon Fraser University, Canada	Positivist-realist, empirical case study work in reductionist discipline-bound science (geomorphology, sedimentology). Specialist postgraduate and post-doctoral training provided the passport for career progression, as I learnt how to play the academic game. My supervisor, Ted Hickin, somehow managed to impose the required level of application to an appropriately substantive but narrowly defined topic. Field experiences in a wide range of settings, including majestic fiord rivers of western Canada and playful sojourns to classic geomorphic and cultural landscapes of the American southwest, supported my professional development, recurrently recontextualising earlier understandings of river diversity.
1989–1991. Post-doctoral fellowship, Biogeography and Geomorphology, Research School of Pacific Studies, Australian National University	Liberating and exhilarating opportunities to pursue Geography as I wanted. Inspired by my boss, John Chappell, to make sense of river diversity as part of a big-picture project on geographies of global and environmental change (climate change, deforestation, land use change), comparing and contrasting ancient, low-slope, tectonically stable landscapes in central and northern Australia with other types of rivers in different parts of the Pacific region – New Zealand, Papua New Guinea, Fiji.
1991–2004. Lecturer, Senior Lecturer, Associate Professor, Macquarie University, Sydney, Australia	Emergence of a Physical Geographer. My first Head of School, Russell Blong, adopted a hands-off approach to guidance and mentoring, empowering me to approach professional life as I wished. A focus on interactive field-based teaching and learning helped a group of talented postgraduate students to generate foundation research on river responses to human disturbance following European settlement. Recognising similar management problems in dry Gondwanan settings, attendees at a cross-disciplinary collaboration between river scientists and managers in Australia and South Africa in 1994 were tasked to craft an integrative classification procedure to support water management. The quest proved to be elusive (Uys, 1994). Co-development of the River Styles Framework (Brierley and Fryirs, 2005) emerged from these deliberations.

(*Continued*)

Table 16.1 (Continued)

Career stage	*Ways of knowing and making rivers*
2004–present. Professor, Chair of Physical Geography, University of Auckland, New Zealand	International research collaborations support the development and use of emerging technologies to facilitate the next generation of River Styles applications. Professional short courses embed international uptake (Fryirs et al., 2019a). A student-led group at the University of Auckland promoted approaches to Critical Physical Geography that challenged conventional framings in the knowledge and governance of rivers in Aotearoa (e.g., Blue et al., 2012; Tadaki et al., 2012, 2015). Contemplations on Indigenous relations to rivers prompted my re-emergence as a generalist Geographer. Expressions of societal connections to place expressed through poetry, artistic and theatrical representations (act/dance me a river) accompany scientific presentations on riverbanks during field trips. An Arts-Science-Dance collaboration supported a choreographed expression of changing socio-cultural relations to water as part of the Auckland Arts Festival (Longley et al., 2013). Adoption of a more-than-human lens promotes uptake of a plural knowledges approach to Find the Voice of the River (Brierley, 2020) and Let the Rivers Speak (Salmond et al., 2019).

distinctive values of Australian rivers. This frame of reference underpinned the co-development and application of the River Styles Framework with Dr Kirstie Fryirs. My appointment as a professor at the University of Auckland, Waipapa Taumata Rau, in Aotearoa New Zealand brought about a shift in thinking beyond a conventional scientific lens, reconceptualising my appreciation of socio-cultural relations to rivers and approaches to protecting river diversity through a more-than-human perspective. I've ended up as a somewhat unconventional physical geographer, using my academic freedom to push for new ways of knowing the world and acting in it.

My efforts to respect river diversity now seek to enact and embed conceptualisations of the material world that extend beyond understandings of Earth surface processes and landforms, acknowledging how values, power and political choices are intrinsic to the ways rivers are known, used and managed.

This chapter is structured as follows. First, I provide an overview of the scientific principles that underpin geomorphic understandings of river systems. I then present an overview of two distinct approaches to knowing and making rivers that I developed and applied at different stages of my career. In the discussion I reflect upon personal agency in the design and conduct of my work, showing how the disciplinary lens of Geography shaped and enabled my understandings of and actions in the world.

Contextual scientific considerations that shape geomorphic understandings of rivers

The rules of the academic game require that particular standards must be met, adhering to certain procedures to the satisfaction of peers. In scientific terms, this entails a commitment to rigour, reliability and replicability in the design, conduct and reporting of work, appropriately rationalising and contextualising the questions asked (hypothesis testing), selection of methods (e.g., approach to experimentation and measurement) and the ways data are analysed and results are documented and communicated. Science is a systematic, objective and ostensibly neutral (apolitical) quest to build and organise knowledge about the world (and the universe). Meaningful description and approaches to explanation (multiple working hypotheses) provide a basis to predict and test understandings. Sometimes profound transformations in understandings and practices bring about paradigm shifts – new scientific ways of knowing and doing (Kuhn, 2012).

Geomorphology is the scientific study of landscapes. Approaches to geomorphic enquiry differ markedly in space and time. Notable differences are evident, for example, between concerns for social and environmental justice expressed through the Geography disciplinary home in British and Commonwealth universities relative to the technocentric focus of enquiry in the Geology (Earth Science) home in the United States (Church, 2010). Over time, geomorphic practices have advanced from regionally framed descriptions of landscape form through phases dominated by quantitative, theoretical, process-based measurements and understandings to increasingly remotely sensed worlds and virtual realities expressed through contextualised place-based understandings of a Digital Earth (e.g., Preston et al., 2011). In all instances, field-derived observations, measurements and understandings provide a critical basis to inspire, ground and test geomorphic analyses (Brierley et al., 2021; Fryirs et al., 2019b).

As a branch of environmental science concerned with Earth and life history, geomorphology is primarily an abductive science that tasks practitioners to make the best interpretation of patchy, incomplete information to craft a logical account of the formative processes and sequences of events that created a particular landscape. Much must be inferred from the selective record of what's left behind in developing a body of evidence that supports a logical, well-reasoned argument (Brierley et al., 2021). Abductive reasoning differs markedly from deductive (linear, cause-and-effect) analyses conducted through experimental procedures under controlled (readily repeatable) conditions. Empirical analyses of 'real' geomorphology embrace understandings of complexity, non-linear dynamics, chaotic behaviour and contingent and emergent traits (Richards et al., 1997). Sometimes inductive reasoning provides flashes of insight from a certain instance that may provide generalisable insights to guide analyses of other situations. Geographers are ideally situated to relate the local to the general, guiding appropriate transfer of understanding from one situation to others.

Key challenges are faced in relating theoretical principles that fashion the complexity of geomorphic diversity to place-based understandings of the character, behaviour and evolutionary traits of a given river system. Here I outline two approaches to knowing rivers that I developed and applied at different stages of my career, reflecting upon the differing realities that are envisaged and enacted through these different framings.

Development and application of the River Styles Framework as a scientific approach to analyse and manage geomorphic river diversity

Slope, discharge, bed material size and bank strength are the key determinants of geomorphic river diversity (Figure 16.1; Brierley and Fryirs, 2022). However, each river is 'perfect' in its own right (Phillips, 2007), and there's no magic number of river types. Accordingly, open-ended procedures are required to identify, respect and categorise 'new' types of river, appropriately situating each type on the spectrum that ranges from bedrock-forced morphologies to fully alluvial, self-forming rivers (Figure 16.1A–E). As

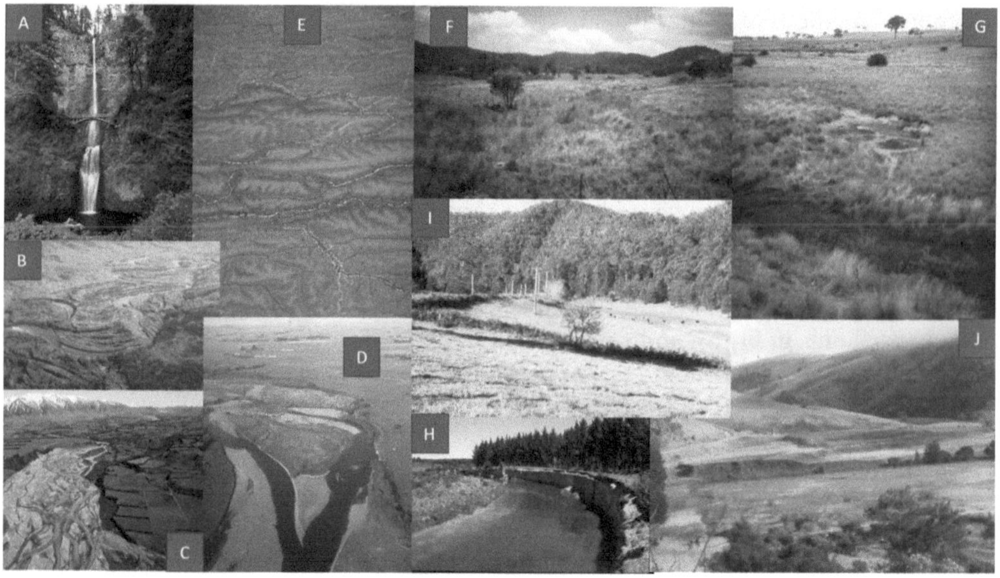

Figure 16.1 Examples of the wondrous range of geomorphic river diversity. See text for details.

aquatic biodiversity is innately adapted to the geodiversity and unique attributes of river systems, careful attention must be given to exceptional instances, recognising how riparian vegetation, wood and ecosystem engineers influence river structure and function (Figure 16.1F–J).

Steep headwater streams characteristically have a bedrock-imposed morphology (**A**, Waitakere Ranges, West Auckland, New Zealand). Alluvial rivers with single or multiple channels have variable sinuosity and adjust their form and position in different ways. Active meandering channels laterally migrate (**B**, Northern Canada). Braided channels and bars recurrently shift their position on the valley floor (**C**, Rakaia River, Canterbury Plains, New Zealand). Wandering gravel-bed rivers have attributes of both active meandering and braided rivers, with some channels prone to avulsion, whereby they jump from one position to another on the valley floor (**D**, Waiau River, Southland, New Zealand). Elsewhere, fine-grained deposits of low gradient multichanneled anastomosing rivers create stable bed and bank conditions such that these rivers passively adjust (incremental geomorphic adjustment over decades; **E**, Channel Country, southwestern Queensland, Australia). Some alluvial rivers do not have a continuous channel, as flood flows deposit fine-grained suspended load sediments across valley floors (an upland swamp) in the 'fill' phase of 'cut-and-fill' landscapes (**F**, Wolumla Creek, Bega Catchment, New South Wales, Australia). Particular sets of historical conditions create opportunities for 'chain-of-ponds' systems to develop as a distinct variant of 'cut-and-fill' river (**G**, Murrumbateman Creek, New South Wales, Australia). Pumice-rich streams create distinctive river morphologies comprised of low-density gravel-sized particles (**H**, Rangitaiki River, Bay of Plenty, New Zealand). Although by definition floodplains are typically relatively flat, some floodplains have a pronounced levee-floodchannel complex (**I**, Macdonald River, New South Wales, Australia). In some settings sequences of floods may catastrophically strip pockets of floodplain deposits (**J**, Manning River at Bretti, New South Wales, Australia).

Biophysical attributes of rivers, among many factors, influence societal relations to rivers. Prospects for agricultural use, water supply for domestic and industrial purposes, sand and gravel resource extraction, navigability and recreational activities (fishing, swimming, boating, etc.), among many other considerations, vary from river to river (Figure 16.1). Both direct/purposeful (Figure 16.1A–D) and indirect/inadvertent (Figure 16.1E) human activities modify river structure and function. Dams alter flow and sediment regimes, silencing a river (Figure 16.1A; McCully, 1996). Artificial levees (stop-banks) strangle a river, disconnecting the channel from its floodplain (Figure 16.1B; Brierley et al., 2023). Many urban streams are channelised within concrete-lined channels or 'buried alive' in subterranean pipes (Figure 16.1C and D). Land use changes have induced profound changes to the flow and sediment regimes of rivers across the world (Figure 16.1E).

Research in Geography provides key insights into the influence of geology, topography, climate and human activities upon geomorphic river diversity (Figures 16.1 and 16.2). A fascination with exceptional circumstances and end-member situations is amply nourished by riverscapes in Australia where ancient, tectonically stable, low-relief landscapes, alongside profound climate variability in the land of droughts and flooding rains, impacts of intensive fires, and distinctive flora and fauna on the driest inhabited continent on Earth, create a diverse suite of geomorphologically distinctive river types (Figure 16.1E–J). Platypus are simply fascinating creatures, while burrowing frogs that can stay underground for years until the next flood flows occur are simply weird.

Figure 16.2 Examples of human impacts on geomorphic river diversity. **A.** Benmore Dam disrupts the longitudinal connectivity of the Waitaki River, South Island, New Zealand, silencing the river (McCully, 1996). **B.** Stop-banks (artificial levees) along the Ishikari River in Hokkaido, Japan, laterally disconnect the channel from its floodplain, strangling the river (Brierley et al., 2023). **C, D.** Urban streams are often channelised or buried alive in culverts (**C** is the engineered channel of Wairau stream on Auckland's North Shore, New Zealand; **D** is an artificial channel in western Sydney, Australia). **E.** Across much of the world, systematic deforestation has indirectly altered flow and sediment regimes of rivers (East Coast, New Zealand, early C20th).

As the longest continuously inhabited continent on Earth, aboriginal peoples learnt to adapt and live with the diversity and variability of rivers across the Australian continent. Such understandings were largely disrespected and ignored by British colonisers who applied a command-and-control ethos that conceptualised land and water as resources in the service of a settler society (Beattie and Morgan, 2017). Mal-adapted land use and management practices that imposed Old World realities upon New World rivers induced profound biophysical and socio-cultural impacts (e.g., Fuller and Rutherfurd, 2022). Eventually, recognition of the severity of anthropogenic impacts alongside the emergence of a global era of environmentalism brought about moves to redress concerns for river health. Unfortunately, however, conservation and restoration programmes in Australia perpetuated the history of use of inappropriate externally derived knowledges and practices. In scientific terms, this was characterised by two major limitations. First, discipline-bound approaches to river repair pitched concerns for hydrology, ecology, water quality and geomorphology against each other, failing to embrace the generative coherence of a 'riverscape' approach (Brierley and Fryirs, 2008). Second, understandings of geomorphic river diversity failed to respect the distinctive biophysical attributes of Australian rivers. Co-development and use of the River Styles Framework with Kirstie Fryirs set out to address this oversight.

The River Styles Framework generates knowledges-in-place that characterise and explain geomorphic river diversity (Brierley and Fryirs, 2005; Fryirs et al., 2019a). Building on longstanding geomorphic principles, a cross-scalar, nested hierarchical approach relates local attributes of rivers at the site/reach scale to catchment-specific relationships and broader considerations at the ecoregion scale. A coherent package of biophysical information documents understandings of what the river looks like, how it behaves and functions, why it is in the condition it is today, its evolutionary trajectory and its associated conservation or recovery potential. This coherent place-based package of information provides baseline guidance to help determine which attributes to protect for any given river system (e.g., assets, rare and endangered/threatened species). This supports proactive and precautionary management plans that outline how to conserve or rehabilitate key attributes/values at local, catchment, and (inter)national scales. Working directly with practitioners, managers and policy makers, field days, workshops, community hall meetings and professional short courses helped to engender notable transformations in river management practice in New South Wales and elsewhere (Table 16.2; Brierley et al., 2002, 2011; Fryirs et al., 2013, 2019a, 2021). Although developed in Australia, this distinctively Geographic approach to river science and management was conceived, designed and enacted through generic, non-prescriptive and open-ended procedures that can be adapted to different circumstances.

Principles adopted in the River Styles Framework resonate with contemporary approaches to scientific endeavour – appreciation of complexity, non-linear dynamics, chaotic, emergent and contingent traits, no analogue states, tipping points, inherent uncertainty and unpredictability, etc. Place-based, catchment-specific applications recognise that managing rivers as particular types – predictable entities that operate as linear, cause-and-effect systems – is doomed to fail (Brierley and Fryirs, 2022). Fighting nature does not work. Nature fights back, and eventually it wins! The emergence of fluvial geomorphology as an increasingly predictive data-rich science in recent years now generates datasets and modelling applications that feed information directly into applications of the River Styles Framework (Fryirs et al., 2019b). This supports the generation of living databases that recurrently update place-based understandings of river diversity, adjustment and evolutionary trajectory (e.g., Boothroyd et al., 2021, 2023; Fryirs et al., 2021).

Table 16.2 Geomorphologically informed transformations in river management practices in Australia since the 1990s (Brierley and Fryirs, 2022; Fryirs et al., 2013, 2019a)

• Proactive, precautionary and pre-emptive plans and actions respect diversity, working with the river in ways that appropriately frame catchment-specific applications in their regional and broader context.
• Realistic goals work with the realities of rivers as they are today. Coherent cross-disciplinary science informs decision-making with a clear rationale and evidence base, avoiding the limitations of compromise solutions (there is no such thing as half a habitat; life cycles either work or they don't).
• Approaches to treatment response minimise negative off-site impacts and legacy effects (opportunity costs set by negative path dependencies), ensuring that problems are not simply transferred elsewhere.
• Cost-effective programmes apply fit-for-purpose process-based measures that manage at source and at scale, addressing the causes not the symptoms of environmental degradation. Realistic forecasts of river futures use geomorphic principles to prioritise effectively. A conservation ethos protects and/or enhances local values (things that matter), looking after good-condition remnant reaches before they become a problem (prevention is cheaper and more effective than cure). Targeted measures in strategic reaches protect conservation goals, addressing threatening processes that imperil sustainability programmes and biodiversity goals (a stitch in time).
• Whenever possible, passive rehabilitation practices work with recovery, allowing the system to self-heal.

Perhaps inevitably, contestations abound at the interface of science and management (policy, practices, governance). Management demands for classification procedures that make 'science' more accessible and practicable presented serious challenges in the development and applications of programmes that respect diversity. Inevitably, an inherent politics underpins classification schemes and the outcomes engendered through their application (Tadaki et al., 2014). Programmes that strive to make rivers the same are the antithesis of the underlying goals of the River Styles Framework. By design, the framework does not provide a cookbook of simplified, consistent and systematic procedures that tells practitioners who have no training in geomorphology how to 'manage' rivers by type (cf., Lave, 2012). Rather, a scaffolded framework of questions encourages critical thinking, not a prescriptive suite of answers. Informed adaptation and tweaking are required to ensure that practices and outcomes are appropriate for a given situation. Such reflections prompted a personal reappraisal and recontextualisation of approaches to the derivation and use of knowledge about rivers, framed as part of emerging approaches to Critical Physical Geography.

Envisaging and enacting rivers through a more than human lens grounded in Critical Physical Geography

> The river has taught me to listen; you will learn from it, too. The river knows everything; one can learn … from it.
>
> Herman Hesse, 1951, *Siddhartha, p. 105*

My appointment as Professor and Chair of Physical Geography in the Faculty of Science at the University of Auckland (UoA) brought about two major reconceptualisations in my

Figure 16.3 Multiple meanings of landscapes. *This painting by Nigel Brown sits on my living room wall. Each day that I'm home it nudges me to reflect upon the ways landscapes are known and made. A scientific perspective is just one of many lenses, as differing mindsets, value systems and worldviews inscribe landscapes with meanings, connections and memories. The painting shows Charles Cotton's geomorphology textbook laid out on a table with biblical overtures, surrounded by representations of the Earth's elements. 3-D block diagrams in the foreground exemplify Cotton's insight into landscapes as patterns of physical landforms. An image of Colin McCahon in the background is surrounded by his 2-D re-representations of landscapes inspired by Cotton's work. As expressed by the Australian writer, Murray Bail, McCahon's work imbued the New Zealand landscape with meaning, recreating the 'Land of the Long White Cloud' as the 'Land of the Long Black Shadow' (Bail, 2003; see Paton, 2019). The artist (Nigel Brown) was a former student of Colin McCahon's.*

approach to protecting river diversity (see Figure 16.3). First, I realised that my previous work seldom stepped outside a scientific lens, and my awareness and appreciation of Indigenous understandings was paltry. I was a novice in the world of epistemic cultural pluralism. Being based in Aotearoa New Zealand progressively created greater familiarity with Indigenous (Māori) worldviews and associated meanings of a more-than-human lens. Being 'of' the land rather than separate from it underpins a mindset that knows and makes rivers in quite different ways (e.g., Kikuroa et al., 2019; Stewart-Harawira, 2020; Te Aho, 2019). My eyes have now been opened wide to the realities and possibilities of the pluriverse (multiple ways of knowing, acting and being). Second, related to this, I experienced a dawning realisation that science is not value-free. There's no single science. Rather, science is socially constructed, and practices are far from the 'pure' and objective pursuit that they purport to be. My reformation echoed the sentiment expressed in earlier work by Levins and Lewontin (1985, p. 4):

> The denial of the interpenetration of the scientific and the social is itself a political act, giving support to social structures that hide behind scientific objectivity to perpetuate dependency, exploitation, racism, elitism, colonialism.

In a similar vein, Erik Swyngedouw (1999) asserted that Geography unravels and reconstructs contested processes in the "production of nature", wherein historical manifestations in the intertwining of the social and the natural, and perpetual changes in relations between them, generate a particular, but inherently dynamic, geographical configuration that has profound consequences for those that follow. Inspired by such conceptualisations, I explored new ways to know and make the world through a changing sense of my relationship to rivers.

In part, my re-evaluation of professional practice reflected greater awareness of critiques of the objectivity and independence of science – the ways findings are derived and used. One of the greatest frustrations in my earlier work was the realisation that pitching discipline-bound scientific endeavours against each other (e.g., water quality versus ecology versus engineering versus geomorphology) will never promote, facilitate and enact the whole-system (life-cycle, Mountains to the Sea) understandings and applications that are required to support effective conservation and rehabilitation interventions. Alongside this, it became self-evident that unless coherent approaches to river science work alongside mindsets and practices that engage generatively with socio-cultural relations to rivers, the transformative changes that are required to engender healthier rivers will not be achieved.

A student-led Freshwater Geographies research group at the University of Auckland played a significant role in the emergence of an Antipodean perspective on Critical Physical Geography (Blue et al., 2012; Tadaki et al., 2012, 2015). These ventures put people back into science, highlighting and questioning the role of the practitioner in the design and conduct of research – what is done (aims, motivations), how it is done (what is measured where, when, how and why; Blue and Brierley, 2016), the ways information/data are analysed, how findings (results) are represented and communicated, and the outcomes (consequences, implications) that ensue. Work in Critical Physical Geography recognises that choices must be made in selecting from a range of theoretical approaches and practices. Such framings juxtapose the notional objectivity of facts alongside values, methodological considerations and associated ethical concerns (Lave et al., 2019). The conduct and use of research has material impacts and potentially political consequences. Accordingly, practitioners have an ethical obligation to select and enact progressive approaches in the quest for social and environmental transformation, explicitly justifying and documenting the values, aims and methods which underpin a body of work. Accordingly, a statement of positionality and situatedness should accompany the conduct and reporting of research, never speaking on behalf of others and taking care to ensure that approaches do not dilute and hence disrespect the integrity of coherent understandings that may already be in place (Hoskins and Jones, 2022). Recognising that landscapes are co-constituted (people shape landscapes and vice versa), and that human values are embedded in accounts, has underpinned the emergence of approaches to landscape science such as ethno-geomorphology (Wilcock et al., 2013) and socio-geomorphology (Ashmore, 2015). Such framings embrace and enact a different philosophical lens to conventional scientific practices in geomorphology (see Rhoads, 2006).

Embracing principles of Critical Physical Geography, efforts to respect river diversity in my recent work have moved beyond the perspective of a specialist scientist and now reflect practices of a generalist Geographer. Enhanced appreciation of Māori ways of being gained through ongoing collaborations with Professor Anne Salmond and Associate Professor Dan Hikuroa, colleagues in the School of Māori Studies at UoA, opened my eyes to a frame of reference that has turned the science–management nexus on its head. A Māori lens conceptualises rivers as living, indivisible entities, with no separation from the people who live within (as part of) the river system (Stewart-Harawira, 2020; Te Aho, 2019). Such framings reject the ontological dualism that separates humans and nature (Levins and Lewontin, 1985; Wesselink et al., 2017). Focus on mutual interdependence, reciprocity and co-evolutionary traits highlights inherent indeterminacy in ways of knowing, doing and being. Working with nature, working with the river, is a given (e.g., Brierley et al., 2019; Koppes and King, 2020; Lent, 2021; Wilkinson et al., 2020). My work now emphasises approaches to 'living generatively with living rivers' rather than seeking to 'manage' rivers in ways that assert human authority over them (Brierley, 2020). A project entitled 'Let the Rivers Speak' (with Anne, Dan and Billie Lythberg) embraces a more-than-human lens in the conduct of river ethnographies (Hikuroa et al., 2021; Salmond et al., 2019, 2022). Cross-boundary explorations of landscape wisdom (Trudgill, 2003a) package together insights into Digital Rivers, field-based conversations and scientific analyses, and river stories. A place-based eco(river)centric lens develops and applies a duty of care in which ethical concerns for social and environmental justice look after things that matter. Concerns for sentience, the community of life and the agency of the river no longer draw a line between the living and non-living: "Oh, that tree, same as me" (Neidjie, 1989; see Farrell, 2013). Herein lies the prospect that geographies of hope can embrace and enact an environmental consciousness that engenders equitable approaches to collective wellbeing as part of a just sustainability.

Poetic assertions expressed through personal endeavours in 'Finding the Voice of the River' (Brierley, 2020) were not well received by publishers, as the book title failed to serve the functional requirements of search engines. Hence I added the subtitle: 'Beyond Restoration and Management'. Such framings seek to know and make rivers in quite different ways to earlier endeavours in my career (Table 16.1).

Reflections on the personal agency that underpinned key elements in my career

> ... we, and the subject we profess, are well served by mavericks and debate rather than by consensus .... Even for a maverick, the need for 'approval' is very strong. so it reads: we, and the subject we profess, are well served by mavericks and debate rather than by consensus ... Even for a maverick, the need for 'approval' is very strong.
>
> Stephen Trudgill, 2003b, p. 275, p. 273

The academic literature contains few accounts of reflections upon career developments in Geography, as intimated by Sir Ron Cooke (2003, p. xi): "Alas, the motivation, vision, enthusiasm and inspiration of most scientists is not a matter of public record." This personal account of changing research practices highlights how independent mission-led research was facilitated and enacted outside conventional power relations and the institutional arrangements that support them. By design and intent, my passion and professional commitment to respect river diversity was not a targeted project with a specific goal and a deadline/endpoint. Rather, it represents an emergent 'way of being', never wanting to get 'locked in' to a specific way of doing things, never forgetting that learning forever adapts to changing circumstances.

Inevitably, personality and personal perspectives play a critical role in the aims, design, conduct, communication and embedding of research. While personal circumstances shape opportunities, personality traits shape what we make of them. All too often, personal perspectives and practices clash with institutional realities. Fortunately, my mentality and confidence have allowed me to respond positively to opportunities presented or created in particular situations and moments. Patience, persistence and support come to the fore in the face of adversity. Experience surely helps. On the one hand I can now draw on lessons learnt from previous situations. On the other, as outcomes of past work are increasingly acknowledged, growing credibility creates further opportunities, including invitations to engage in activities and decision-making rooms never previously envisaged. The River Styles book (Brierley and Fryirs, 2005) provided the platform for international engagement, enactment and embedding of geomorphic principles to inform place-based management applications that respect diversity (Fryirs et al., 2019a). Impact and network development accompanied delivery of professional short courses both nationally and internationally. Many personal friendships that emerged from these experiences provided critical support at later stages of my career.

As indicated in this summary account of the ways I've known and sought to make rivers at different stages in my career, high-school and undergraduate experiences provided remarkable foundations that opened my eyes to the wider world and what was possible (Table 16.1). Key advisers and supervisors provided superb guidance in my postgraduate and post-doctoral life, always making themselves available, occasionally hinting at corrections or redirections, but critically never obstructing aspirations and approaches proposed by my highly opinionated self. They helped to provide a safe space for a soft landing when I fell over, while allowing me to develop and express myself.

In terms of research development and profile, I consider standing back and supporting student-led groups to achieve their individual and collective potential to have engendered my greatest career successes. I'm especially proud of the number of researchers that I've helped to write their first academic paper. In pedagogic terms, I've always endeavoured to give students opportunities to express their own perspectives in their own way, allowing space for imagination and experimentation. Recognising that academic life in general and research in particular are worlds of intense criticism, I've spent much time and effort striving to create armour and defence mechanisms to help students and colleagues protect themselves in the face of adversity. Much time and energy are spent negotiating difficult circumstances and harmful relationships in efforts to work around or push aside impediments and roadblocks. It pays to balance opportunities for spontaneity with moments of critical reflection – finding solace and meaning through calm, meditative relations (Lane, 2017).

As an opinionated individual with pronounced anti-institutional tendencies, these student-led groups have been my primary collaborators, though international partnerships have become increasingly prominent in recent years. Working independently and outside (or at the margins of) the corridors of power and influence, we've collectively co-developed and co-enacted practices and approaches to river research. Ongoing collaborations with Kirstie Fryirs have provided the key basis to operationalise activities. Our complementary personalities balance big-picture perspectives and attention to detail alongside diverse approaches to learning and communication.

The sociology of academia is very strong as regards both the communities and networks within which we choose (or are allowed) to interact, and the pressure to conform, play it safe and observe local 'rules of engagement' (Trudgill, 2003b). However, as stepping outside comfort zones is key to collaborative learning, concerns for psychological wellbeing come to the fore in learning to cope with power games. To stay the same is to lag behind in the research process, but fighting convention is hard, as this challenges prevailing dogmas and the authorities that underpin them.

Discussion: Geography as an enabler

> Where is the wisdom we have lost in knowledge?
> Where is the knowledge we have lost in information?
>
> T.S. Eliot, 1934, The Rock

Understanding river diversity is a critical step in learning to live generatively with it. As each and every river reflects a particular entanglement of values and perspectives – social, cultural, political, ecological, spiritual – which conception of a river determines how we seek to live with it as power-laden relationships shape the generation and (ab)use of knowledge (see Figure 16.4)?

Figure 16.4 The land the law forgot (Williams, 2023). Marked differences in river morphology are evident for the braided Waimakariri River in Aotearoa New Zealand (A) relative to textbook understandings of an active meandering river derived from the northern hemisphere (B). Legal procedures in Aotearoa New Zealand are struggling to come to terms with how to define a river. What differentiates the bed from the banks? How can meaningful account be made in drawing lines on maps that determine where water ends and land begins, as channels adjust and evolve? Pressure to develop 'land' that makes up part of the active channel bed of braided river systems (the dry bits of bars, islands and adjacent floodplains on (A) reflects inappropriate legal definitions of rivers that are imported from the mother country (B). Failure to respect and protect diversity has profound implications in biophysical, socio-economic and cultural terms (Brierley et al., 2023). Quite different priorities and outcomes come to light when opportunities for land development on valley floors are related to scientific assertions of river character and behaviour, or Māori conceptualisations of rivers as living and indivisible entities (e.g., Stewart-Harawira, 2020; Te Aho, 2019).

Notionally, at least, environmental decision-making is scientifically informed, with a clear aim, rationale, justification and evidence base. However, particular agendas often shape the design, conduct and use of research – the questions asked, approaches to enquiry and the ways findings are used. Practices and the outcomes they engender cannot be understood separately from the values and practices that underlie them. Most commissioned research works within prevailing paradigms that selectively and purposefully seek to address certain interests. Implicitly, framings that support existing power relationships strive to maintain the status quo. Rather than claiming (and hiding behind) assertions of notional objectivity in science, making values explicit is required to facilitate meaningful comparison of multiple means and ends.

Geography provides a disciplinary vehicle to envisage, enable and enact transformative change in ways that address concerns of social responsibility, morality and stewardship. A relational lens embraces respect for place (situated perspectives upon the local) and fairness (relations to each other) that underpin assertions of a better world. Geoethical practices apply a duty of care that works with the river, respecting its diversity and its capacity to adjust and evolve as a living entity (Brierley et al., 2019; Sharp et al., 2022). Inclusive and participatory approaches strive to prevent bad situations from occurring, proactively mitigating ethical dilemmas in ways that address concerns for environmental, generational and distributional justice.

Embracing diversity in the forms and workings of the world is a very geographical project. By design and intent, pedagogic experiences in Geography strive to prepare students to be ethically informed citizens who are able to make defensible decisions in difficult situations, rather than unthinking rule-followers who push aside ethical responsibility to the institutions and bureaucracies within which they are ensconced. Geography is innately sensory and experiential. Enactive pedagogies and research practices support critical thinking skills through transboundary work in fuzzy spaces. Flexible, open-ended practices embrace a plurality of framings and discourses, actively and generatively engaging in the fertile ground of cross-over spaces (see Figure 16.5). Adaptability, not certainty, is a core virtue of geographical reasoning. Being affective to be effective empowers disruptive behaviours that question existing realities, seeking to enact better, more hopeful alternatives – more equitable ways of living and working in a fast-changing world. Embodied and enactive pedagogies engage the whole person (spirit, body and mind) in teaching and learning, embracing difference as a source of inventiveness in creative and experimental practices (Couper, 2023). To me, fieldwork keeps it real. Dirty-boots Geography comes alive through active place-based encounters with rivers. Hands-on learning appropriately grounds and contextualises place-based knowledges, striving to craft and tell each river's story. Such assertions resonate with the well-known quote from Confucius (c. 551 – c. 479 BCE): "Tell me and I'll forget, Show me and I'll remember, Engage me and I will understand."

Concluding comment

This autobiographical account of a career dedicated to respecting and looking after river diversity reflects upon a personal transition from a conventional foundation in reductionist river science co-developed and enacted through the River Styles Framework – a geomorphic platform to support the use of integrative river science – to practices in socially situated science conducted under the auspices of Critical Physical Geography. I now apply a more-than-human lens that works with plural knowledges in efforts to Let the River Speak. This lens contemplates and seeks to enact the rights of the river, striving to

Figure 16.5 The Upper Yellow River at Dari, Qinghai Province, Western China (3800 m asl). I love this photograph. The river didn't listen to the textbook. In geomorphic terms, it cannot decide if it wants to be braided, anabranching or something else. I use this image in a postgraduate course asking students to determine measures of river health. Most responses emphasise biophysical measures such as geomorphological, ecological, hydrological or water-quality attributes. However, the key to this exercise is to think 'outside the box', reflecting upon socio-cultural values and transitional land uses, the agency of the river itself, etc. (see Blue, 2018; Blue and Brierley, 2016).

live generatively with living rivers rather than managing rivers in ways that assert human authority over them. Why would we want to classify and manage rivers by type, divorced from their local setting and 'meaning'? Why wouldn't we want to respect the place-based values of each and every river?

This account shows how Geography provided a supportive disciplinary vehicle in the personal quest to respect river diversity. This disciplinary lens facilitated self-discovery and self-reflection while allowing me to work out what mattered and what I wanted to do, where, with whom and how. Resulting perspectives and practices now respect the agency of rivers in ways that reflect and express ethical concerns for collective wellbeing, for ourselves and the planet of which we are a part.

Acknowledgements

I dedicate this chapter to my bike. Large parts were composed while trundling along back roads of Waiheke Island en route to an inspirational bench. I thank Jenny Salmond and Noel Castree for this invitation to humanise my approach to work and practice in Geography. Sincere thanks to high-school Geography teachers, amazing lecturers at Durham University, formative supervisory experiences with Ted Hickin at Simon Fraser University, liberating and supportive mentoring by John Chappell at ANU and Russell Blong at Macquarie University, inspiring collaborations with members of the 'Let the Rivers Speak' team, and countless colleagues along the way. Special thanks to Kirstie Fryirs for longstanding friendship and a shared commitment to looking after rivers.

Study task

Variations on this question can be set on an unseen examination, an open-book examination or as a term-paper assignment:

A scientific approach to understand the physical environment offers not so much 'objectivity' as a particular angle of vision on reality. Discuss, with reference to Gary Brierley and colleagues' research into river restoration and management in Australia and New Zealand.

Starter reading list

Brierley, Gary, et al. "A geomorphic perspective on the rights of the river in Aotearoa New Zealand." *River Research and Applications* 35.10 (2019): 1640–1651.

Brierley, Gary J. *Finding the voice of the river: Beyond restoration and management.* Springer Nature, 2019.

Church, Michael. "Observations and experiments." *The SAGE handbook of Geomorphology.* London, SAGE Publications (2011): 121–141.

Hikuroa, Dan, et al. "Restoring sociocultural relationships with rivers: Experiments in fluvial pluralism." *River restoration: Political, social, and economic perspectives* (2021): 66–88.

Lane, Stuart N. "Slow science, the geographical expedition, and critical physical geography." *The Canadian Geographer/Le Géographe Canadien* 61.1 (2017): 84–101.

Salmond, Anne, Gary Brierley, and Dan Hikuroa. "Let the rivers speak: Thinking about waterways in Aotearoa New Zealand." *Policy Quarterly* 15.3 (2019).

Tadaki, Marc, et al. "Cultivating critical practices in physical geography." *The Geographical Journal* 181.2 (2015): 160–171.

Tadaki, Marc. "Rethinking the role of critique in physical geography." *The Canadian Geographer/Le Géographe Canadien* 61.1 (2017): 73–83.

References

Agyeman, J., Bullard, R. D. and Evans, B. (Eds). 2003. *Just sustainabilities: Development in an unequal world.* MIT Press.

Ashmore, P. 2015. Towards a sociogeomorphology of rivers. *Geomorphology, 251*, pp. 149–156.

Bail, M. 2003. Out of darkness. *Financial Review*, June 4, 2003.

Beattie, J. and Morgan, R. 2017. Engineering Edens on this 'Rivered Earth'? A review article on water management and hydro-resilience in the British Empire, 1860–1940s. *Environment and History, 23*(1), pp. 39–63.

Blue, B. 2018. What's wrong with healthy rivers? Promise and practice in the search for a guiding ideal for freshwater management. *Progress in Physical Geography: Earth and Environment, 42*(4), pp. 462–477.

Blue, B. and Brierley, G. 2016. 'But what do you measure?' Prospects for a constructive critical physical geography. *Area, 48*(2), pp. 190–197.

Blue, B., Gregory, C., McFarlane, K., Tadaki, M., van Limburg-Meijer, P. and Lewis, N. 2012. Freshwater geographies: Experimenting with knowing and doing geography differently. *New Zealand Geographer, 68*(1), pp. 62–66.

Boothroyd, R.J., Williams, R.D., Hoey, T.B., Barrett, B. and Prasojo, O.A. 2021. Applications of Google Earth Engine in fluvial geomorphology for detecting river channel change. *Wiley Interdisciplinary Reviews: Water, 8*(1), p. e21496.

Boothroyd, R.J., Williams, R.D., Hoey, T.B., MacDonell, C., Tolentino, P.L., Quick, L., Guardian, E.L., Reyes, J.C., Sabillo, C.J., Perez, J.E. and David, C.P., 2023. National-scale geodatabase of catchment characteristics in the Philippines for river management applications. *Plos One, 18*(3), p. e0281933.

Brierley, G.J. 2020. *Finding the voice of the river: Beyond restoration and management.* Springer Nature.

Brierley, G.J. and Fryirs, K.A. 2005. *Geomorphology and river management: Applications of the river styles framework*. John Wiley & Sons.

Brierley, G.J. and Fryirs, K.A. (Eds). 2008. *River futures: An integrative scientific approach to river repair*. Island Press.

Brierley, G.J. and Fryirs, K.A. 2022. Truths of the Riverscape: Moving beyond command-and-control to geomorphologically informed nature-based river management. *Geoscience Letters*, *9*(1), pp. 1–26.

Brierley, G.J., Fryirs, K., Outhet, D. and Massey, C. 2002. Application of the River Styles framework as a basis for river management in New South Wales, Australia. *Applied Geography*, *22*(1), pp. 91–122.

Brierley, G.J., Fryirs, K., Cook, N., Outhet, D., Raine, A., Parsons, L. and Healey, M. 2011. Geomorphology in action: Linking policy with on-the-ground actions through applications of the River Styles framework. *Applied Geography*, *31*(3), pp. 1132–1143.

Brierley, G. J., Tadaki, M., Hikuroa, D., Blue, B., Šunde, C., Tunnicliffe, J. and Salmond, A. 2019. A geomorphic perspective on the rights of the river in Aotearoa New Zealand. *River Research and Applications*, *35*(10), pp. 1640–1651.

Brierley, G. J., Fryirs, K. A., Reid, H. and Williams, R. 2021. The dark art of interpretation in geomorphology. *Geomorphology*, *390*, p. 107870.

Brierley, G. J., Hikuroa, D., Fuller, I. C., Tunnicliffe, J., Allen, K., Brasington, J., Friedrich, H., Hoyle, J. and Measures, R. 2023. Reanimating the strangled rivers of Aotearoa New Zealand. *Wiley Interdisciplinary Reviews: Water*, p. e1624.

Brower, A., Hoyle, J., Gray, D., Buelow, F., Calkin, A., Fuller, I., … & Vosloo, R. (2024). New Zealand's braided rivers: The land the law forgot. *Earth Surface Processes and Landforms*, 49(1), pp. 10–14.

Bunge, W. 1973. Ethics and logic in geography. In: Chorley, R.J. (Ed.), *Directions in geography* (pp. 317–331). Routledge.

Church, M. 2010. The trajectory of geomorphology. *Progress in Physical Geography*, *34*(3), pp. 265–286.

Cooke, R. 2003. Preface. In: Roy, A. and Trudgill, S. (eds), *Contemporary meanings in physical geography: From what to why?* (pp. xi–xii). Arnold.

Couper, P.R. 2023. Interpretive field geomorphology as cognitive, social, embodied and affective epistemic practice. *The Canadian Geographer/Le Géographe canadien* 67(3), 430–441.

Eliot, T.S. 1934. *The Rock*. Faber & Faber Limited.

Farrell, M. 2013. The geopoetics of affect: Bill Neidjie's story about feeling. *Journal of the Association for the Study of Australian Literature*, *13*(2).

Fryirs, K., Chessman, B., & Rutherfurd, I., 2013. Progress, problems and prospects in Australian river repair. *Marine and Freshwater Research*, 64(7), pp. 642–654.

Fryirs, K. and Brierley, G. 2021. How far have management practices come in 'working with the river'? *Earth Surface Processes and Landforms*, *46*(15), pp. 3004–3010.

Fryirs, K., Hancock, F., Healey, M., Mould, S., Dobbs, L., Riches, M., … & Brierley, G. (2021). Things we can do now that we could not do before: Developing and using a cross-scalar, state-wide database to support geomorphologically-informed river management. *Plos One*, 16(1), p. e0244719.

Fryirs, K.A., Brierley, G.J. and Dixon, T., 2019a. Engaging with research impact assessment for an environmental science case study. *Nature Communications*, *10*(1), pp. 1–10.

Fryirs, K.A., Wheaton, J.M., Bizzi, S., Williams, R. and Brierley, G.J. 2019b. To plug-in or not to plug-in? Geomorphic analysis of rivers using the River Styles Framework in an era of big data acquisition and automation. *Wiley Interdisciplinary Reviews: Water*, *6*(5), p. e1372.

Fryirs, K., Hancock, F., Healey, M., Mould, S., Dobbs, L., Riches, M., Raine, A. and Brierley, G. 2021. Things we can do now that we could not do before: Developing and using a cross-scalar, state-wide database to support geomorphologically-informed river management. *PloS One*, *16*(1), p. e0244719.

Fuller, I.C. and Rutherfurd, I.D., 2022. Geomorphic responses to anthropogenic land-cover change in Australia and New Zealand. *Reference Module in Earth Systems and Environmental Sciences*. Elsevier. https://doi.org/10.1016/B978-0-12-818234-5.00104-8

Harvey, D. 1973. *Social justice and the city*. University of Georgia Press.

Hesse, H. 1951. *Siddhartha*. New Directions Publishing Corporation (originally published in Germany, 1922).

Hikuroa, D., Brierley, G., Tadaki, M., Blue, B. and Salmond, A. 2021. Restoring sociocultural relationships with rivers: Experiments in fluvial pluralism. In: Morandi, B., Cottet, M. and Piégay, H. (eds), *River restoration: Political, social, and economic perspectives*. Wiley, pp. 66–88.

Hoskins, T.K. and Jones, A., 2022. Indigenous inclusion and indigenising the University. *New Zealand Journal of Educational Studies*, *57*, pp. 305–320.

Koppes, M.N. and King, L. 2020. Beyond x, y, z (t); Navigating new landscapes of science in the science of landscapes. *Journal of Geophysical Research: Earth Surface*, *125*(9), p. e2020JF005588.

Kuhn, T.S. 2012. *The structure of scientific revolutions*. University of Chicago Press.

Lane, S.N. 2017. Slow science, the geographical expedition, and critical physical geography. *The Canadian Geographer/Le Géographe Canadien*, *61*(1), pp. 84–101.

Lave, R. 2012. Bridging political ecology and STS: A field analysis of the Rosgen Wars. *Annals of the Association of American Geographers*, *102*(2), pp. 366–382.

Lave, R., Biermann, C., & Lane, S. N. 2019. *The Palgrave handbook of critical physical geography*. Palgrave.

Lent, J. 2021. *The web of meaning: Integrating science and traditional wisdom to find our place in the universe*. New Society Publishers.

Levins, R. and Lewontin, R. 1985. *The dialectical biologist*. Harvard University Press.

Longley, A., Fitzpatrick, K., Šunde, C., Ehlers, C., Martin, R., Brown, C., Brierley, G. and Waghorn, K. 2013. Streams of writing from a fluid city. *Qualitative Inquiry*, *19*(9), pp. 736–740.

McCully, P. 1996. *Silenced rivers: The ecology and politics of large dams*. Zed Books.

Neidjie, B. 1989. Story about feeling. (Edited by Keith Taylor). Magabala Books, Broome.

Paton, J. 2019. *McCahon Country*. Penguin Books.

Phillips, J.D. 2007. The perfect landscape. *Geomorphology*, *84*(3–4), pp. 159–169.

Preston, N., Brierley, G. and Fryirs, K. 2011. The geographic basis of geomorphic enquiry. *Geography Compass*, *5*(1), pp. 21–34.

Richards, K.S., Brooks, S.M., Clifford, N.J., Harris, T. and Lane, S. 1997. Theory, measurement and testing in 'real' geomorphology and physical geography. In: Stoddart, D. R. (ed.), *Process and form in geomorphology*. Routledge.

Rhoads, B. L. (2006). The dynamic basis of geomorphology reenvisioned. *Annals of the Association of American Geographers*, 96(1), pp. 14–30.

Salmond, A., Brierley, G. and Hikuroa, D., 2019. Let the rivers speak: Thinking about waterways in Aotearoa New Zealand. *Policy Quarterly*, *15*(3).

Salmond, A., Brierley, G., Hikuroa, D. and Lythberg, B., 2022. Tai Timu, Tai Pari, the ebb and flow of the tides: working with the Waimatā from the Mountains to the Sea. *New Zealand Journal of Marine and Freshwater Research*, *56*(3), pp. 430–446.

Sharp, E. L., Brierley, G. J., Salmond, J., & Lewis, N. (2022). Geoethical futures: A call for more-than-human physical geography. *Environment and Planning F*, 1(1), pp. 66–81.

Stewart-Harawira, M.W. 2020. Troubled waters: Maori values and ethics for freshwater management and New Zealand's fresh water crisis. *Wiley Interdisciplinary Reviews: Water*, *7*(5), p. e1464.

Swyngedouw, E. 1999. Modernity and hybridity: Nature, regeneracionismo, and the production of the Spanish waterscape, 1890–1930. *Annals of the Association of American Geographers*, *89*(3), pp. 443–465.

Tadaki, M., Salmond, J., Le Heron, R. and Brierley, G. 2012. Nature, culture, and the work of physical geography. *Transactions of the Institute of British Geographers*, *37*(4), pp. 547–562.

Tadaki, M., Brierley, G. and Cullum, C. 2014. River classification: Theory, practice, politics. *Wiley Interdisciplinary Reviews: Water*, *1*(4), pp. 349–367.

Tadaki, M., Brierley, G., Dickson, M., Le Heron, R. and Salmond, J. 2015. Cultivating critical practices in physical geography. *The Geographical Journal*, *181*(2), pp. 160–171.

Te Aho, L. 2019. Te Mana o te Wai: An indigenous perspective on rivers and river management. *River Research and Applications*, *35*(10), pp. 1615–1621.

Trudgill, S. 2003a. Meaning, knowledge, constructs and fieldwork in physical geography. In: Roy, A. and Trudgill, S. (eds), *Contemporary Meanings in Physical Geography: From What to Why?*. Arnold, pp. 25–46.

Trudgill, S. 2003b. Conclusion: Contemporary meanings in physical geography. In: Roy, A. and Trudgill, S. (eds), *Contemporary meanings in physical geography: From what to why?*. Arnold, pp. 269–277.
Uys, M.C. (ed). 1994. Classification of rivers and environmental health indicators. Proceedings of a joint South African/Australian workshop, February 7–14, South Africa. Water Research Commission Report No TT 63/94.
Wesselink, A., Kooy, M. and Warner, J. 2017. Socio-hydrology and hydrosocial analysis: toward dialogues across disciplines. *Wiley Interdisciplinary Reviews: Water*, *4*(2), p. e1196.
Wilcock, D., Brierley, G. and Howitt, R. 2013. Ethnogeomorphology. *Progress in Physical Geography*, *37*(5), pp. 573–600.
Wilkinson, C., Hikuroa, D.C., Macfarlane, A.H. and Hughes, M.W., 2020. Mātauranga Māori in geomorphology: existing frameworks, case studies, and recommendations for incorporating Indigenous knowledge in Earth science. *Earth Surface Dynamics*, *8*(3), pp. 595–618.
Williams, D. 2023. Defining braided rivers will help avoid future disasters. Newsroom, February 7 2023. Available at: Defining Braided Rivers Will Help Avoid Future Disasters | Newsroom.

17 Beyond science

Climate change in a 'wicked world'

Mike Hulme

Mike Hulme (1960–)

Introduction

For over 40 years I have sought to understand climate change – initially as a physical phenomenon, but in the last 20 years as also an idea that has a certain life of its own in society. In my career, I've operationalised the perspectives of both a physical and human geographer, starting with heavily quantitative research into climate dynamics, later studying how people think and talk about climate. Climate and its changes is not something that geographers can ignore. And nor have they. For geographers pursuing their task of 'writing and drawing the Earth', the geographical and historical variations in the world's

DOI: 10.4324/9781003343240-19

weather offer a tableau – a vivid, lively scene – which can inform our understanding of many other physical and cultural facets of planetary and human life.

Seeking an understanding of climate has been integral to a geographer's mission, from the earliest shaping of Geography as an academic discipline in the late nineteenth century. But how exactly should geographers conceive of climate? Is climate a self-evident property of the physical world, "a largely taken for granted setting" (Stehr, 1997: 167) for human life, much like soil, vegetation and rivers? This was how the influential Scottish and Oxford geographer Andrew Herbertson (1865–1915) regarded climate, as one of the four "fundamental distributions" that defined his reading of the Earth in terms of 'natural regions'. Herbertson's natural regions were really climatic in basis, and "climate ... lay at the root of his thinking about the natural world. The vaults of the Oxford School of Geography are packed with the library of climatological literature that he accumulated" (Hare, 1966: 99).

Or should climate rather be understood as an *idea* of the human imagination, an idea which co-constitutes people and places in a multitude of different ways (Wright & Tofa, 2021) and which is bound up with the collective symbolism of human cultural life, belonging and memory? In this reading, it is climate's "near infinite plasticity" (Hulme, 2009: 28) that gives climate such imaginative power and consequently makes it such a rich object of geographic study. The idea of climate is also used to stand in for the more general notion of 'environment', a description of the unity and inter-dependency of all things which offers the pre-condition for all emplaced life. This is how the German naturalist Alexander von Humboldt imagined climate two hundred years ago, climate as a description of the "unity of landscape" emerging from a confluence of nature and mind.

And then what of the slippery difference between weather and climate? The aphorism 'climate is what you expect, weather is what you get' is too trite, but it does seem to capture something important. It distinguishes between the embodied human experience of 'weather in place', on the one hand, and the mediating role of 'climate in the mind', on the other (Simonetti, 2019). Amidst the ceaseless volatility of the atmosphere's physical behaviour, the idea of climate – as distinct from weather – seems to offer the human body-mind something important: both the prospect of semi-predictable patterns and regularities in physical phenomena *and* some means of psychological orientation in a lively world of constant change and variability.

However one approaches the idea of climate, it's clearly something that is bound up with the ways in which people live, think and act in the world. It's an intimate idea, almost (but not quite) intuitive at one level, but also an idea difficult to fully grasp in formal terms. Climate is everywhere: wherever one is in the world one lives 'in a climate'. But it's also nowhere: it escapes rigid boundaries or fixed manifestations. Making sense of climate and its change is therefore a task warranting the serious efforts of geographers, a task that has occupied my entire professional life. This autobiographical essay reflects on how I have conducted this inquiry, sometimes outside the discipline of Geography but always with the training and instincts of a geographer. My journey from undertaking quantitative research about climate trends and dynamics to research and thought leadership in how societies think about climate change has been a relatively unusual, but highly rewarding one.

Climate as 'climatology'

Geography degree courses have traditionally offered compulsory or elective courses in 'climatology', literally the study of climate. Recognising spatial and temporal variations in climate, climatology divides into distinct sub-fields depending on analytical method or theme. For example, English geographer A. Austin Miller's classic textbook *Climatology* went through nine editions over the 34 years of its publishing life (1931–1965): it offered students "a systematic geographical treatment of the subject in which the climatic type is adopted as the unit, while the important regional peculiarities are dealt with separately. Particular attention is paid to the climatic control of vegetation" (Miller, 1931: title page). These sub-fields of climatology, familiar to many past geographers, later expanded to include, among others, statistical, synoptic, mountain, urban, tropical, agricultural and applied climatologies, each spawning distinct literatures of their own.

It was therefore as 'climatology' that I first encountered the formal study of climate as an undergraduate geographer in 1979 at the University of Durham. This was in an optional second-year course taught by climatologists Ray Harris and Joan Kenworthy. Their synopsis stated that the course would "examine the ways in which we observe the atmosphere and the models developed to explain the behaviour of climate", including some "regional and local climates". The course concluded with "an examination of the changes in our climate which have occurred and which continue to occur". My climatological lectures 45 years ago consisted of topics such as 'climatological methods', 'climatic classification', 'synoptic climatology', 'urban, monsoon and mountain climates', 'climate modification' and 'climatic change'.

As Julie Winkler observes in her entry 'Climatology' in *The International Encyclopedia of Geography* (Richardson et al., 2017), interest in climatology "has increased dramatically, partly in response to concerns about natural and anthropogenic climate change. The terminology 'climate science' has grown in popularity, and appears to be replacing 'climatology' as the umbrella term for the study of climate" (Winkler, 2017). But even in 1979, geographer climatologists such as Harris and Kenworthy could alert their 19-year-old students to "the flood of literature relating to current anxieties about climatic change" and would preface the introduction to their course by stating that "In recent years, our climate has been brought to the attention of many scientists".

Having taken this second-year course, and with my own interest in the subject matter probably piqued by an especially cold British winter in 1978/1979, I titled my undergraduate dissertation 'A Synoptic Study of Recent Winters in the British Isles'. This got me working with historical instrumental meteorological data to construct and analyse statistical indexes of winter weather and to place this analysis in the context of longer-term trends or cycles in British climate. I was somewhat astonished to find later that what became known as the "Hulme winter severity index" had been promoted by geographer John Thornes at the University of Birmingham and used by a number of local authorities in the development and application of salting regimes for motorway winter maintenance (Gustavsson, 1996). It demonstrated to me very early in my career how statistical representations of climatic variability could find direct practical applications.

But even as my Geography peers and I were being introduced to the sub-discipline of 'climatology' around 1980, a new science of climate was already emerging internationally. This brought together of a range of disciplines – for example, meteorology, geochemistry, oceanography, Quaternary science, glaciology – which, by the mid-1980s, had crystallised as a new inter-discipline called Earth System science (NRC, 1986). Since the International Geophysical Year of 1957/1958, new global infrastructure of planetary-scale monitoring and computerised calculation, sustained by huge public investment in a Cold War world, had facilitated the emergence of this more exclusively scientific notion of a global climate system (Heymann & Dalmedico, 2019). The goal of Earth System science (ESS) was to make climate amenable to integrated scientific enquiry, to advanced numerical simulation, visualisation and prediction, and hence, presumptively, amenable to control.

Geographers of the later twentieth century were not central to this development of ESS, nor even for the most part peripheral. The new climate science developed with little regard to geographers' earlier understanding of climate as rooted in places and connected to distinctive local environments and cultures. It was this latter framing that had guided my own PhD studies in the period between 1981 and 1985. My PhD followed a fairly traditional (geographical) path of statistical climatology combined with infusions of thinking about human–environment interactions, inspired by Hewitt and Hare's (1973) conceptual framework for environmental geography. My PhD thesis on semi-arid Sudanese rainfall variations barely acknowledged the global climatic context of these changes, and was completely uninformed by the rapidly developing science of climate modelling (Hulme, 2022).

Towards the end of my PhD studies, I gained my first academic job at the University of Salford in 1984, as lecturer in physical geography. As well as teaching some introductory physical geography courses, and geographical statistics, I designed my own specialist third-year course which I titled, 'Contemporary climatic change'. The aims of this course were to introduce students to:

- The sensitivity of the atmospheric system to 'man-made' and natural changes in the environment;
- The techniques and methods of studying climate changes and variations which result from the above sensitivity;
- The impacts of an unstable climate on human society and the natural environment.

I taught this course for three years between 1985 and 1988, but my approach remained quite distinctively climatological and geographical, rather than embracing the new computation and systems approach to climate science. I delivered lectures on the historical context of present-day climate, on long-term trends in global climate, on the causes of drought in North Africa and on climate research programmes; the four recommended core books were: Ausubel and Biswas (1980), Houghton (1984), Lamb (1982) and Tickell (1977).

Climate as Earth System science

By the later 1980s, public concerns about anthropogenic climate change – in particular the 'enhanced greenhouse effect' – had surged on the rising tide of the new climate science and modelling. It was not geographers or Geography journals that the media and policy advisers looked to. It was to the new breed of globally oriented climate scientists and their connections to powerful and well-funded national and international science programmes emerging under the auspices of the International Union for the Conservation of Nature (IUCN) and the UN. I was acutely aware of this at the time. If I wanted my enthusiasm for researching climate and its changes to gain public funding and visibility, and if I wanted to make my own contribution to this rising public policy issue, then maybe I'd be better off leaving the discipline of Geography.

The Climatic Research Unit (CRU) is located in the School of Environmental Sciences at the University of East Anglia (UEA) and by the late 1980s had been aligning itself with these new international climate science networks for at least a decade. It was been founded in 1971 by the geographer-meteorologist Hubert Lamb, but since 1978 had been led by the Australian physicist Tom Wigley. Wigley was well tuned to the 'epistemic

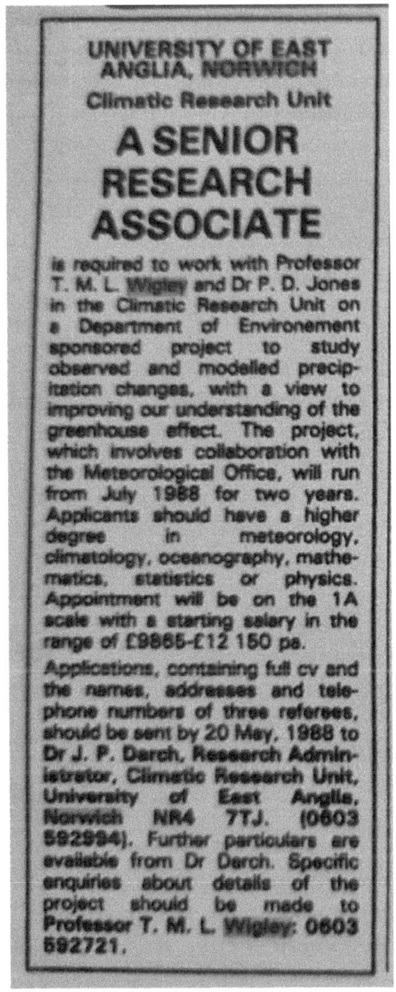

UNIVERSITY OF EAST ANGLIA, NORWICH
Climatic Research Unit

A SENIOR RESEARCH ASSOCIATE

is required to work with Professor T. M. L. Wigley and Dr P. D. Jones in the Climatic Research Unit on a Department of Environement sponsored project to study observed and modelled precipitation changes, with a view to improving our understanding of the greenhouse effect. The project, which involves collaboration with the Meteorological Office, will run from July 1988 for two years. Applicants should have a higher degree in meteorology, climatology, oceanography, mathematics, statistics or physics. Appointment will be on the 1A scale with a starting salary in the range of £9865-£12 150 pa.

Applications, containing full cv and the names, addresses and telephone numbers of three referees, should be sent by 20 May, 1988 to Dr J. P. Darch, Research Administrator, Climatic Research Unit, University of East Anglia, Norwich NR4 7TJ. (0603 592994). Further particulars are available from Dr Darch. Specific enquiries about details of the project should be made to Professor T. M. L. Wigley: 0603 592721.

Figure 17.1 Advertisement for a research position in the Climatic Research Unit, for which I applied and was successful.

From *New Scientist* magazine, 12 May 1988. Copyright University of East Anglia, with permission.

power' and public and political kudos of the new generation of climate models and had astute awareness of sources of public funding for work in this area, often from the USA.

When a fixed-term research position was advertised at CRU in May 1988 to work on a project validating global climate models, funded by the UK Department of Environment, it seemed the perfect exit route for me out of Geography (see Figure 17.1). Working within this internationally recognised Unit, connected to most of the world's leading climate research and modelling centres, I would have a much bigger platform from which to pursue my research into climatic change. I therefore relinquished my lectureship in physical geography at Salford and in September 1988 embarked on what became a series of fixed-term post-doctoral positions at UEA. At the time I gave little thought to the significance of embracing this new paradigm of climate science. To an enterprising 28-year-old there seemed little alternative.

Not only did I leave the discipline of Geography behind, I also left teaching behind. CRU was a research-oriented unit, with neither undergraduate nor (until rather later) Master's level teaching. I pursued various lines of investigation, working with quantitative data and models, creating and promoting what I later called "global kinds of knowledge" (Hulme, 2010a). These were the forms of knowledge that the newly established Intergovernmental Panel on Climate Change (IPCC) was called on to assess and package for policy audiences. Not only was this knowledge 'global', it was exclusively numerical and often quite detached from the social-cultural world (ironically the world that was both responsible for anthropogenic warming and being impacted by it). Geographers and their semi-quantitative, eclectic and somewhat disjointed forms of climatological knowledge, appeared to me a backwater for climate research.

Around this time, c.1990, quite a lot of the more environmentally oriented geographers – at least a lot of those that I knew in the UK – seemed to agree. The international networks, and the governmental and inter-governmental funding bodies that kept them well oiled, began to absorb a fair number of geographers, among them Ann Henderson-Sellers, Martin Parry, Tim Carter, Jean Palutikof, Diana Liverman, Nigel Arnell and Paula Harrison. Climatology courses began to disappear from undergraduate Geography programmes and at some stage during the 1990s, the joint specialist study group on climate – the Association of British Climatologists -- run jointly by the Institute of British Geographers (IBG) and Royal Meteorological Society, disbanded. When the IBG and the Royal Geographical Society merged in 1995, I let my 11-year long membership of the IBG lapse, a symbolic severing from my intellectual roots. Working in a School of Environmental Sciences, where there was no Geography taught, I was now a long way from my Geography homeland.

The limits of Earth System science

By the late 1990s, my work was deeply embedded within the international networks of global climate science. I was committed to making 'global kinds of knowledge' and to assessing this knowledge and communicating it through various roles I had within the IPCC, notably in the Third Assessment Report (1996–2001). I was regarded as 'a climate scientist' both within my epistemic community, but also by myself. In my passport and on visa application forms I entered my occupation as 'climate scientist'. I had very little connection with the academic discipline of Geography.

In 1998, as an outcome from New Labour's first spending review under Chancellor Gordon Brown, several major new investments in UK science were announced. One of these was the calling of an open competition to establish a national climate change centre, a new inter-university research network to be based at a British university and with a vision to design and deliver integrated research for sustainable solutions to climate change. Very substantial funding of £10m (c. £22m in today's value) was available for the successful bid.

The University of East Anglia saw itself at the time as probably the leading university in the country for climate change research. Discussions between my colleagues and I in CRU, and with senior management, quickly confirmed the institution's desire to lead one of the bids for this centre. A number of circumstances led to my name being put forward to lead the bid development and, if successful, to become the founding executive director. Not least of these reasons, I now believe, was my background as a geographer. This was as someone who could be perceived by researchers affiliated to the new centre, and by the three relevant funding research councils, as holding credibility across different disciplinary traditions: natural science, social sciences, economics, ecology and engineering. My Geography background, unexpectedly, had now become a trump card within the academic communities that we sought to bring together around the £10m honeypot. Following success in the competition process, the newly named Tyndall Centre for Climate Change Research was formally launched in November 2000, boldly promising a range of stakeholders – public, private, governmental, non-governmental – solution-oriented integrated science, one of the first bespoke research centres worldwide to do this with regard to climate change.

Although the Centre attracted stakeholders and secured research impact, my experience of running the Centre started to trouble me. I had direct high-level exposure to the worlds of government, business, NGOs, media and academia, and I saw at much closer quarters how the framing, communication and search for solutions of the growing public policy issue of climate change was heavily shaped by the knowledge emerging from research. I began to question the knowledge framework/epistemic paradigm that we had adopted in the Centre. I was clear that Earth System models were a limited technology/methodology for studying the dynamics and politics of climate change. That much we were clear about in our original bid and proposed research programme. But I now saw that our favoured paradigm of integrated assessment models (IAMs), to which the Tyndall Centre would make a pioneering contribution jointly with the Potsdam Institute for Climate Impact Research (PIK) through the mediation of my research director John Schellnuber, was also deficient. IAMs were one way of trying to integrate the human world with the physical world to provide decision support tools for policy advisers and decision makers. But IAMs provided a very impoverished representation of human agency, of how people individually or collectively form beliefs, make decisions and act in the world.

There was something else that troubled me. Working closely with up to 100 researchers from different disciplines spread across nine institutions seeking jointly to design and deliver an integrated programme of research, I increasingly observed that Tyndall Centre researchers were not all working from the same starting assumptions. We might all be studying climate change – and making our careers out of doing so. But when it came to choices about practical interventions, or proposed policy measures to research, and how to evaluate such measures, we were guided by different worldviews, values and preferences. Examples of such measures at the time included the idea of domestic tradeable carbon quotas, the role of nuclear energy, preferences for carbon taxes or carbon trading, how to capture non-market valuations, and whether the energy transition was hampered more by inadequate innovation or by weak-willed implementation.

And yet we rarely made these value differences explicit in our work. I observed that we seemed to hold the implicit belief that integrated interdisciplinary knowledge within a (predominantly) scientific and computational paradigm would inevitably yield the universal insights and tools that decision makers were wanting. Between the years 2003 and 2006, this observation formed the seeds from which my later book, *Why We Disagree About Climate Change*, emerged.

At the same time, I began to realise that generating more 'integrated scientific knowledge' may be as much a part of the problem of climate change as it was part of a solution. The Tyndall Centre straddled the ESS and IAM paradigms, making some limited space for critical social science, with the ambition for this knowledge to be useful for public policy. But reflecting on this ambition, I found myself re-visiting a dormant element of my earlier geographical education. This concerned the philosophy of knowledge or what, as a Durham undergraduate geographer, we had encountered in our courses on 'Principles of Geography'. From around 2004 onwards I began to read literature from science and technology studies, including Jerry Ravetz's and Silvio Funtowicz's idea of "post-normal science" (Funtowicz & Ravetz, 1993) and the notion of "wicked problems" introduced by Horst Rittel and Melvin Webber (Rittel & Webber, 1973). More generally, I was reading the work of people such as Thomas Kuhn, David Bloor, Dan Sarewitz, Sheila Jasanoff and Steve Rayner. Not much of this literature was written by geographers. But these writings afforded me new windows into how knowledge is made and how it gains its social status, connecting me with half-forgotten questions we had discussed as undergraduate geographers about positivism, humanism, models and phenomenology through reading the work of geographers such as Derek Gregory, Ron Johnston, David Harvey and Yi-Fu Tuan.

These two emerging lines of discontent now crystallised in my thinking: the masking or suppression of the plural values that underlie our approach to framing and solving climate change; and the ignorance or naivety of how researchers think science gains and retains trust in (different) society(ies) (see Figure 17.2). These concerns were pressing in on me. They appeared far more important, and exciting, subjects of research than the questions that had motivated my earlier research, such as 'What were the observed trends

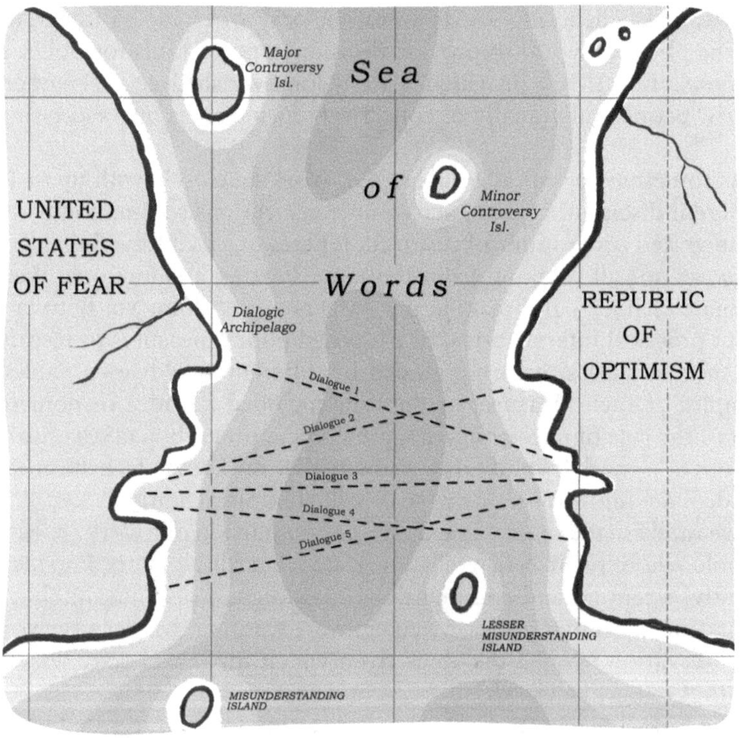

Figure 17.2 Cartoon depicting the need for dialogue and discussion between those holding different viewpoints on climate change.

Source: Hulme (2013) *Exploring Climate Change Through Science and in Society* (Routledge).

in precipitation?', 'Were climate models getting better?' or 'How can future climate scenarios be designed?'

Put simply, climate change was not an environmental problem (that had been scientised), but best seen as a conundrum of the human condition (linking culture, society, knowledge and politics). The relationship between facts and values, and how that relationship was perceived differently within and among different societies and polities, now seemed to me of central importance. Climate change was indeed a wicked problem. These were not questions that we were openly pursuing within the Tyndall Centre. Nor were they the primary concerns of the wider international research networks of IGBP, WCRP, IPCC or even the IHDP, within which my work and professional life had previously been embedded.

During the years 2005 and 2006 I therefore experienced a major intellectual crisis (together with a minor personal one, which I won't elaborate here). My need to escape the suffocating way in which climate change was being framed, studied and acted on became overwhelming. Initially, I did this by re-entering the classroom, enrolling in an evening class on memoir writing and registering for a postgraduate diploma in history, both undertaken at the UEA. A vastly different reading agenda also formed part of my escape route, with Julie Cruikshank's *Do Glaciers Listen?* (2005), Thomas Dunlap's *Faith in Nature: Environmentalism as Religious Quest* (2004) and David Livingstone's *Putting Science in its Place* (2003) just three examples of books that offered signature moments of mind-shaping enlightenment. I also found myself connecting my thinking about climate change more explicitly with my own religious faith, traditional evangelical Christianity, which is quite comfortable with discussions about faith, truth and myth in the search for the grounds for just and moral action in the world.

It was the limits to science, the embeddedness of climate in culture and the sheer diversity of ways that peoples around the world made sense of their climatic environments and its vicissitudes that now excited me. I began to re-discover what had animated my interest in climate in the first place as an undergraduate student and then a PhD student in Sudan. It was an excitement that demanded my re-connection with Geography.

'Geographical work at the boundaries of climate change'

Much of my thinking around this time became expressed in a shortish essay I published in *Transactions of the Institute of British Geographers* as one of their series of commissioned essays called 'boundary crossings' (Hulme, 2008). This became one of the journal's top-ten cited articles of the last 15 years. My choice of this journal to publish in was deliberate. It signalled the growing importance of Geography (again) in my thinking about climate change – a discipline I had left behind 20 years earlier. I wanted to pose my recent (re-)thinking about climate change as a challenge specifically to geographers: "We need new ways of thinking about and understanding the hybrid phenomenon of climate change. Geographers have a unique role to play in this task" (p.6). I articulated this role in terms of "repairing the damage done" – not the damage done to the physical functioning of the climate system through our actions, but *damage done to our way of thinking about climate* in narrowly scientific and reductionist terms.

> Geographers … understand that meaning is lost once the physical and the cultural are cleaved. By purifying climate [of cultural meaning] we de-stabilise the idea of climate, allowing it to be appropriated in support of one-eyed agendas. We cannot allow this to happen … By stripping climate of its flowing cultural and psychological symbolism,

> by ignoring the multiple meanings of climate, we are in danger of letting the idea of climate change get out of control. And why does any of this matter for geography and geographers? For too long geographers have forgotten that climate is an idea that emerges from the heart of the discipline – or rather that Geography lies at the heart of the idea of climate.
>
> (p. 10) – see also Hulme (2015b)

I went on to argue that the naïve climate determinists of the first half of the twentieth century had inadvertently prompted geographers to distance themselves from the idea of climate as something that brought the physical and social worlds together. Geographers had "become increasingly happy to leave climate well alone. Distancing themselves from the idea, climate was left first to the physical geographers, who in turn handed it over to the meteorologists who most recently have been usurped by the Earth system scientists. This distancing has not been helpful" (p. 10).

At the same time, I was writing what was to be published as my first monograph, *Why We Disagree About Climate Change* [*WWDACC*] (Hulme, 2009). This was both the culmination of the factors that had shaped my (re-)thinking about how to study climate change which had surfaced during the years I had been running the Tyndall Centre and my attempt to make sense of climate change's 'human dimensions'. These factors were a growing conviction about the limits and reductionism of science; frustration with the inadequacy of how scientific knowledge connected to people's lives; an intellectual fatigue and staleness; and a personal mid-life soul-searching: did I really want to spend the next 15–20 years doing the same thing?

Among the various reviews of *WWDACC*, I noticed how often readers attached significance to me being a climate "scientist". Thus Silke Beck in the journal *Futures*, "Drawing upon 25 years of professional work as an international climate change scientist and public commentator"; Chris Caseldine in *The Geographical Journal*, "Essentially this is a book about ideas, written by a scientist who not only recognises the importance of his science but also .... wants to understand what it is about that science that needs to be communicated widely and how best to achieve that objective"; Rupert Read in *Green World*, "It's great virtue is that it is a book by a climate scientist that, unlike many previous pronouncements by climate scientists, does not seek to pretend that being a climate scientist gives one any special political, ethical or philosophical privilege"; and David Demeritt in *Progress in Human Geography*, "Steeped in this tradition of predictive environmental science, Hulme offers a powerful critique of its limitations and of the problems caused by 'disconnecting climate from its cultural forms [and] framing climate as overtly physical and global' (p. 28). Hulme began his career as a climate scientist and continues ... to make important contributions to precipitation climatology and climate scenario analysis." People received the book, and gauged its significance, partly on the grounds it was written by 'a scientist' as opposed to 'a geographer'.

The wider reception to *WWDACC* was astonishing. As I wrote a few years later in an essay in the journal *Dialogues in Human Geography* (Hulme, 2015a: 323–324):

> The subsequent reaction to the book surprised me. Unknowingly, I had written a climate change text which seemed of not much interest to climate scientists and geoscientists, but which was of considerable interest to disciplines beyond environmental science. The book offered a framework for scholars, analysts and other professionals to engage with climate change on their own terms, rather than on the terms being

offered by global climate science or the IPCC. The central argument of the book was that it is necessary to reveal our (many) reasons for disagreement – our underlying beliefs and motivations – before we can act effectively in the world. The intellectual bandwidth which Geography occupies, and from which it launches its engagement with those outside geography, now offered a natural home for me. After writing *Why We Disagree...*, an eventual return to a geography department was perhaps inevitable.

With my conviction strengthened by the success of *WWDACC*, I now widened the range of academic journals in which I published. Thus in one of the earliest issues of the new climate change journal from the Nature-Springer stable, *Nature Climate Change*, I published an essay titled 'Meet the humanities' (Hulme, 2011a), Here, I argued that, "To shed new light on the multiple meanings of climate change in diverse cultures, and to create new entry points for policy innovation, the interpretative social sciences, arts and humanities need new spaces for meeting as equals with the positivist sciences" (p.179). In the same year, in the American history of science journal *Osiris*, I published a warning about reducing the future to climate (Hulme, 2011b). Geographers' older dalliance with environmental determinism (see Chapter 3) was worryingly being resurrected under the guise of what I called "climate reductionism". This essay has been the most highly cited in the journal's 85-year history. And then in 2016, writing a chapter entry for the *Routledge Handbook of Religion and Ecology*, I argued that climate change would not be understood in today's world without engaging with the thought and practice of the world's major religions. "Different regions and diverse groups of stakeholders understand the threat of climate change according to particular and often distinct religious frames of reference. These religious narratives and rituals shape the nature and credibility of different knowledge claims about climate – what is happening to it and why – as well as shaping individual and communal ethical and social behaviors" (Hulme, 2016: 243).

These moves were all part of a deliberate expansion of my analytical purchase on the idea of climate change, beyond the quantitative paradigm of natural science or the study of social and institutional behaviours in the social sciences. I gained inspiration from the challenge laid down in Jerome Kagan's book *The Three Cultures* (2009): "... every society needs a cohort of intellectuals to check the dominance of a single perspective when its ideological hand becomes too heavy" (pp. 266–267). I was arguing that to fully understand the mutating idea of climate change in contemporary life, the imaginative disciplines of the humanities were essential for exploring and establishing meaning and purpose, and for reaching people in new ways, whether this be through literature, art, poetry or religion (see Figure 17.3).

Whether or not "a return to a Geography department" (cf. above) was inevitable, after writing *WWDACC* I did indeed return to academic Geography to continue studying climate change on my own terms. By 2013, I was finding myself increasingly isolated in a School of Environmental Sciences which, ironically, in 1988 I had been very happy to leave Geography in order to join. A catalysing example of how this came about concerned a new MA in Environmental Science and Humanities that I had designed, along with some colleagues in the Faculty of Humanities at UEA, and that included environmental history, philosophy and literature. Establishing this taught programme was a struggle, largely because the School of Environmental Sciences had no vision for what I was seeking to do. Very little financial or institutional support was forthcoming. The MA admitted its first students for academic year 2012/2013, but by then I was already looking for opportunities to move back into Geography and in 2013 an opportunity arose at King's College London (KCL).

Figure 17.3 Cover from my book, *Exploring Climate Change Through Science and in Society* (Routledge, 2013), which captures a range of my writings about climate change over the previous 25 years.

After 25 years outside the discipline of Geography, I was now back. Two years after arriving at KCL, I delivered my inaugural lecture which I titled 'Studying Climate and its Changes: In Places, With Numbers, Through Myths'. This laid out my new manifesto for how climate change should be studied through science, social sciences and humanities, and a department of Geography seemed the perfect home for such a programme of work. In 2015, with the full support of senior management in the Faculty of Social Science at KCL, a contrast to the lukewarm support I had received at UEA, I launched a new MA in Climate Change: History, Culture, Science. The advertising blurb was now explicit about how I believed climate change should be studied. Students, it said, would "Explore the ways that science, policy, beliefs, values and the imagination shape human responses to climate change". The MA was suitable for students "with either a science or humanities background who wish to develop their understanding of how climate change is historically and culturally situated". This was a more expansive taught programme on climate change than existed elsewhere at the time. And Geography provided a welcoming intellectual home.

Making Geography matter for climate change

Having read the foregoing pages, I can imagine my readers asking whether 'the project' narrated here is really about Geography rather than being one that has been conducted by someone who self-identifies as a geographer. My project certainly isn't the way in which all practising academic geographers today would approach the study of climate change; it also far from exhausts what geographers have to offer to our understanding of climate change. On the other hand, this project has given me the vantage point to contribute a recent book-length version of my framing of climate change in the Routledge 'Key Ideas in Geography' series (Hulme, 2021). This was my attempt to show why Geography matters for climate change.

Moving between and across these different cultures of knowing – sciences, social sciences and humanities – has not always been smooth or painless. I have been criticised in my later work for not giving sufficient weight to scientific claims about climate change, been charged with being a climate 'delayer' if not a 'denier', and been accused of not doing enough to galvanise climate activism. At other times I have found myself on the wrong side of the peer review process. Peer reviewers 'police' what should or should not be published and at times my work hasn't fitted the standard scientised narrative of climate change, or else as a 'mere' geographer or as 'a scientist' my work appears to violate the theories and tacit epistemic commitments of a particular sub-discipline. For example, when publishing my 2010 article in the journal *Science as Culture* (Hulme, 2010b), one of the reviewers sniffily pronounced that,

> Having said all that, I can now see why I am undecided. This has all the hallmarks of an essay by a highly intelligent and committed scientist, *who (in my opinion) shares the limitation of many scientists, in believing that writing on broader issues does not require the same intellectual discipline as doing real research.* This lack of awareness of the 'tacit knowledge' (to quote Polanyi) of social studies of science may not be too damaging in the case of a reflective piece for a newspaper. But since this piece is intended for a scholarly journal, its deficiencies obtrude. The editors ... should try a second opinion.
>
> [emphasis added]

Moving into different epistemic territories is not always welcome by the incumbent dwellers.

But if Geography is what is done by geographers, then my 'project' of pursuing climate change beyond science, described here, has achieved certain things for Geography by drawing upon sensibilities I first cultivated as a Geography undergraduate in the 1970s. It has opened new approaches for how a new, younger generation of geographers might engage with the idea of climate change. For example, it has helped inspire the 'cultural turn' in climate change studies of the last 15 years, as signalled by the work in the early 2010s of the cultural geographer Hilary Geoghagen (Geoghagen & Leyshon, 2012). An element of this turn has been to attach new emphasis to religious myth, symbolism and practice for understanding climate change sense-making, as is well illustrated in the work of the young geographer Hannah Fair (Fair, 2018). My project of representing climate as much more than a physical phenomenon has also been picked up and developed in significant new ways by more experienced geographers, notably Harriet Bulkeley as captured in her 2019 essay 'Navigating climate's human geographies: Exploring the whereabouts of climate politics' (Bulkeley, 2019). Bulkeley's signature work was recognised in her receiving the Chair for the 2023 RGS-IBG Annual Conference in London, on her own theme, 'Climate changed geographies'.

And this is why making Geography matter for climate change matters. Geography provides the intellectual space to explore the world in its holistic integrity, without the necessity of narrow disciplinary or reductionist thinking. In 1953, the philosopher Isaiah Berlin published a short book called *The Hedgehog and the Fox* (Berlin, 1953). Berlin used this metaphor to divide thinkers into two categories. Hedgehogs view the world and write about it through the lens of a single defining idea. They dig deep along a single line of sight, but are closed off to seeing things differently. Foxes, on the other hand, scavenge widely. They draw on a wide variety of experiences to make sense of the world they encounter in all its diversity. Both hedgehogs and foxes are survivors, but their survival strategies are different. My instincts when it comes to studying and understanding climate change are those of a fox: as per my KCL inaugural lecture, 'Studying Climate and its Changes: In Places, With Numbers, Through Myths'.

But they are also the instincts of a geographer. When I took on the task in 2008 of designing, and then executing for 15 years as Editor-in-Chief, the interdisciplinary review journal *WIREs Climate Change*, I had a definite vision in mind. It was that of a geographer. As I wrote in one of my *WIREs* editorials (Hulme, 2018):

> You will know that I make no apologies for my belief in the primacy of "geography" as a disciplinary space in which the many dimensions of climate change as a socio-natural phenomenon can best be studied and taught. Geography, at least in the UK, knows no intellectual boundaries and this is reflected in my approach to curating the content of *WIREs Climate Change*. The scope of climate change reviews published here very deliberately covers humanities disciplines and interpretative social sciences, as well as the social sciences more broadly, and environmental geography, the natural sciences and economics.
>
> (p. 1)

Study task

This task can be set as a term paper or an exam question:

On a BBC radio programme in April 2019, Greta Thunberg was asked 'What do you want people to do, what do you want governments to do?'. Thunberg replied ' ... listen to the science, listen to the scientists ...' Drawing upon Mike Hulme's work, explain why Thunberg's reply is inadequate?

Alternative:

Drawing upon Mike Hulme's work, why are "global kinds of climate change knowledge" (Hulme, 2010a) unlikely to bring about necessary change?

Starter reading list

Hulme, M. (2020) One Earth, many futures, no destination. *One Earth* 2(4): 309–311.

Hulme, M. (2018) "Gaps" in climate change knowledge: Do they exist? Can they be filled? *Environmental Humanities* 10(1), 330–337.

Hulme, M. (2015a) (Still) Disagreeing about climate change: Which way forward?. *Zygon: A Journal of Religion and Science* 50(4), 893–905.

Hulme, M. (2015b) Climate and its changes: A cultural appraisal. *GEO: Geography and Environment* 2(1), 1–11.
Hulme, M. (2011a) Meet the humanities. *Nature Climate Change* 1(4), 177–179.
Hulme, M. (2011b) Reducing the future to climate: A story of climate determinism and reductionism. *Osiris* 26(1), 245–266.
Hulme, M. (2010) Problems with making and governing global kinds of knowledge. *Global Environmental Change* 20(4), 558–564.

References

Ausubel, J. and Biswas, A.K. (eds) (1980) *Climatic Constraints and Human Activities*. Oxford: Pergamon Press. 205 pp.
Berlin, I. (1953) *The Hedgehog and the Fox: On Tolstoy's View of History*. London: Weidenfeld & Nichsolon.
Bulkeley, H. (2019) Navigating climate's human geographies: Exploring the whereabouts of climate politics. *Dialogues in Human Geography* 9(1): 3–17.
Cruikshank, J. (2005) *Do Glaciers Listen? Local Knowledge, Colonial Encounters and Social Imagination*. Vancouver: UBC Press. 312 pp.
Dunlap, T.R. (2004) *Faith in Nature: Environmentalism as Religious Quest*. Seattle: University of Washington Press. 206 pp.
Fair, H. (2018) Three stories of Noah: Navigating religious climate change narratives in the Pacific Island region. *Geo: Geography and Environment* 5(2): e00068.
Funtowicz, S.O. and Ravetz, J.R. (1993) Science for the post-normal age. *Futures* 25: 739–755.
Geoghegan, H. and Leyson, C. (2012) On climate change and cultural geography: Farming on the Lizard Peninsula, Cornwall, UK. *Climatic Change* 113(1): 55–66.
Gustavsson, T. (1996) Test of indices for classification of winter climate. *Meteorological Applications* 3: 215–222.
Hare, F.K. (1966) The concept of climate. *Geography* 51: 99–110.
Hewitt, K. and Hare, F. K. (1973) *Man and Environment: Conceptual Frameworks*. Washington, DC: Association of American Geographers.
Heymann, M. and Dahan Dalmedico, A. (2019) Epistemology and politics in Earth system modeling: Historical perspectives. *Journal of Advances in Modeling Earth Systems* 11(5): 1139–1152.
Houghton, J.T. (ed.) (1984) *The Global Climate*. Cambridge: Cambridge University Press.
Hulme, M. (2008) Geographical work at the boundaries of climate change. *Transactions of the Institute of British Geographers* 33(1): 5–11.
Hulme, M. (2009) *Why We Disagree About Climate Change: Understanding Controversy, Inaction and Opportunity*. Cambridge: Cambridge University Press. 393 pp.
Hulme, M. (2010a) Problems with making and governing global kinds of knowledge. *Global Environmental Change* 20(4): 558–564.
Hulme, M. (2010b) Claiming and adjudicating on Kilimanjaro's shrinking glaciers: Guy Callendar, Al Gore and extended peer communities. *Science as Culture* 19(3): 303–326.
Hulme, M. (2011a) Meet the humanities. *Nature Climate Change* 1(4): 177–179.
Hulme, M. (2011b) Reducing the future to climate: A story of climate determinism and reductionism. *Osiris* 26(1): 245–266.
Hulme, M. (2013) *Exploring Climate Change Through Science and in Society: An Anthology of Mike Hulme's Essays*. Abingdon, UK: Interviews and Speeches Routledge, 330 p.
Hulme, M. (2015a) Changing what exactly, and from where? A response to Castree. *Dialogues in Human Geography* 5(3): 322–326.
Hulme, M. (2015b) Climate and its changes: A cultural appraisal. *GEO: Geography and Environment* 2(1): 1–11.
Hulme, M. (2016) Climate change: Varieties of religious engagement. Chapter 25 (pp. 239–248) in: *Routledge Handbook on Religion and Ecology*. (eds) Jenkins, W., Tucker, M.E. and Grim, J. Abingdon: Routledge. 448 pp.
Hulme, M. (2018) *WIREs Climate Change* 2018: an editorial essay. *WIREs Climate Change*. 9(1): e503, 6 pp.

Hulme, M. (2021) *Climate Change: Key Ideas in Geography*. Abingdon: Routledge. 288 pp.
Hulme, M. (2022) Reflections on the afterlives of a geography PhD thesis. *Area* 54(2): 280–289.
Kagan, J. (2009) *The Three Cultures: Natural Sciences, Social Sciences and the Humanities in the 21st Century*. Cambridge: Cambridge University Press. 310 pp.
Lamb, H.H. (1982) *Climate, History and the Modern World*. London: Methuen.
Livingstone, D.N. (2003) *Putting Science in its Place: Geographies of Scientific Knowledge*. Chicago: University of Chicago Press. 234 pp.
Miller, A.A. (1931) *Climatology* (1st ed.) London: Methuen & Co. Ltd.
NRC (1986) *Earth System Science. Overview: A Program for Global Change*. Washington, DC: National Academies Press.
Richardson, D., Castree, N., Goodchild, M.F., Kobayashi, A.L., Liu, W. and Marston, R. (eds.) (2017) *The International Encyclopedia of Geography: People, the Earth, Environment, and Technology*. Malden, Oxford: John Wiley & Sons, 9,120 p.
Rittel, H.W.J. and Webber, M.M. (1973) Dilemmas in a general theory of planning. *Policy Sciences* 4: 155–169.
Simonetti, C. (2019) Weathering climate: Telescoping change. *Journal of the Royal Anthropological Institute* 25: 241–264.
Stehr, N. (1997) Trust and climate. *Climate Research* 8: 163–169.
Tickell, C. (1977) *Climate Change and World Affairs*. Cambridge, MA: Harvard University Press.
Winkler, J. (2017) Climatology. Entry in: *The International Encyclopedia of Geography: People, the Earth, Environment, and Technology*. (eds) Richardson, D., Castree, N., Goodchild, M.F., Kobayashi, A.L., Liu, W. and Marston, R.. Malden, Oxford: John Wiley & Sons, 9, 120 pp.
Wright, S. and Tofa, M. (2021) Weather geographies: Talking about the weather, considering diverse sovereignties. *Progress in Human Geography* 5(5): 1126–1146.

18 'Other' geographies

Engaging with different ways of knowing, valuing, and acting in post-colonial Australia

Sue Jackson

Sue Jackson (1965–)

Introduction

Australia, like many settler-colonial countries, built its wealth and independence on Indigenous dispossession and displacement. Nation-building dramatically transformed the territories occupied by Indigenous peoples for millennia. Today, the political projects of reconciliation and constitutional recognition have gained traction in efforts to redress the historical legacies and injustices of exclusion, racism, and other forms of discrimination, and to enhance the freedom of Indigenous minorities. At the same time there is growing acknowledgement that Australia needs to reckon with the colonial legacies of settler forms of environmental management: the systematic marginalisation of the knowledge traditions of Indigenous peoples and dismissal of Indigenous ways of managing and governing 'natural resources'. As we collectively grapple with how to embrace diverse

DOI: 10.4324/9781003343240-20

knowledges and ontologies,[1] while respecting the inherent rights of Indigenous peoples, the disciplines and organisations that generate environmental knowledge will also need to be decolonised.

With Geography's interest in human–environment relations, place dynamics, and community mobilisation (Coombes et al. 2012) it offers much to the pursuit of sustainability and equitable co-existence of diverse cultures. However, it's only in the past 20 years or so that Australian geographers have demonstrated a concerted effort to reflect on and transcend the discipline's colonial heritage (Howitt and Jackson 1998). Here, and in other parts of the world, human geographers are engaging more meaningfully with Indigenous environmentalism[2] (Coombes et al. 2012), and Indigenous geographies is a topic of growing academic interest, but it's a relatively recent subject area that wasn't defined or practised when I started my career in the 1990s.

The project which I've pursued for the past three decades has entailed learning about the complexity of Indigenous environmental relationships, philosophies and knowledges in the context of contestation over development projects and conservation initiatives. In my research I question whose knowledge and authority counts, so too the ontological assumptions that underpin dominant forms of environmental governance. The most satisfying aspects involve trying to solve problems that are identified by Indigenous communities, including Indigenous experts, and applying our collective expertise to the development of new forms of dialogue and collaboration. Advocating for wider institutional change – to laws, policies, organisations – has been integral to making this research relevant and rewarding. Such a project calls for continual and sometimes uncomfortable reflection on my role as a researcher of settler (English) heritage and presumptions about appropriate ways to engage and effect change. Yet my research has subjected government and non-Indigenous environmental institutions[3] to scrutiny and critique as much as it has responded to the urgent need to understand and elevate the regard with which Australian society holds Indigenous knowledges and customary institutions.

More specifically, my research trajectory has tracked the field of water governance, exposing major gaps in disciplinary environmental knowledge of water–society relations. Water is widely studied and regulated as a resource to be stored and allocated. Indigenous conceptions and modes of governing water are not formally recognised and there is a stark inequity in water rights distribution. Water rights, or ownership of entitlements, are woefully skewed against Indigenous peoples (who have less than 1% of Australian water holdings), in marked contrast to the degree to which Indigenous people control land (over 30% of the continent).[4] Water governance is thus a field suffused with settler-colonial power relations. It is also rapidly changing in response to new policy imperatives like the commodification of water and acute water scarcity. Indigenous claims for equitable distribution of water entitlements come with an expectation of meaningful participation in the institutions of water governance which will need to be reshaped to recognise customary law and Indigenous environmental philosophies. This is the context in which my career as a geographer has unfolded over the past three decades.

A final point needs to be made about the contextual drivers of my career choices, which have also influenced the turns the project has taken. A life-long career in Geography was unexpected. I started out as an environmental activist and serendipitously ended up a researcher. While captivated by what's sometimes termed the 'geographical imagination' and always driven by my own political commitments, in truth my career has unfolded by seeking out supportive institutions, creative alliances and funds to sustain field research activities. While I've never worked in a Geography department, I believe that being a geographer has enabled me to explore and contribute to real-world geographies of difference and co-existence: to places, processes and agendas of intellectual interest and of political importance.

Environmental activism in the Northern Territory

Drawn to environmental campaigning, I moved to the Northern Territory to work as the coordinator of its peak environmental NGO in 1990, soon after graduating with an Honours degree in Geography. The Territory was a fascinating place, with its sweltering tropical heat, raw provincial politics and its capital, Darwin, closely entwined with Asia. I was taken by the atmospheric intensity and drama of the monsoonal climate, the huge open, scrubby savanna landscapes that drain some of Australia's largest rivers, and its red, arid centre. But I found its socio-geographical distinctiveness equally compelling. It had been more recently colonised than southern Australia, had been developed according to a different economic model, and had the country's strongest land rights legislation. Traditional owners held recognised rights to an Indigenous estate that comprised almost half of the jurisdiction and 80% of its coastline. Indigenous peoples had a prominent stake in its politics of place. This was not true in much of the rest of Australia.

All the significant political contests of that era centred on Indigenous defence of place (Escobar 2001) and the pursuit of political autonomy to make decisions about the management of Country, according to inherited systems of governance and local priorities (Palmer 2004a, 2004b). The provincial government strongly resisted Indigenous territorial claims and opposed nearly every land claim in a relentless colonial agenda to secure white possession.

Attuned to geographic ideas of place, I pored over maps that hinted at the immense complexity of the statutory legal regime and the "equally complex, often incommensurate, maze of indigenous social relations" (Povinelli 2006: 155). They showed huge tracts of inalienable freehold title (Figure 18.1). This was land that had successfully been reclaimed and was controlled by local descent groups, much of it considered marginal to white settler society. Within these "Indigenous territories of difference" (Escobar 2008) were small homelands that had first been established as missions or government settlements. In many places Indigenous lands were surrounded by pastoral leases or encroached on by transnational mining corporations. Some of these maps would show Indigenous place names referencing the marvels of creator beings who had created the landscape and its social institutions. These were customary estates "shaped by *a priori* spiritual forces and imbued with spiritual power" and agency (Langton 2002: 94).

Immersed in cartographic exploration of these Indigenous geographies, I thought about the implications of the co-existence and interaction of plural laws – ancient customary forms and more recent statutory ones – which governed relations between people, land, water, fire, vegetation, minerals, and other non-human life. This was by no means a symmetrical co-existence: competing claims and overlapping jurisdictions combined with colonial scales of governance and administrative neglect to constrain the ability of Indigenous peoples to pursue their lifeways (Howitt 2006). Shifting governmental and public attitudes to Indigenous place-based orientations required constant adjustment.

In the 1990s, scientists were coming to realise how much Indigenous peoples had shaped the physical landscape of Australia and its biota over 99% of human history (Head 1993). Australian science was discovering Indigenous fire regimes, following archaeological research and several studies that had revealed the way in which Indigenous

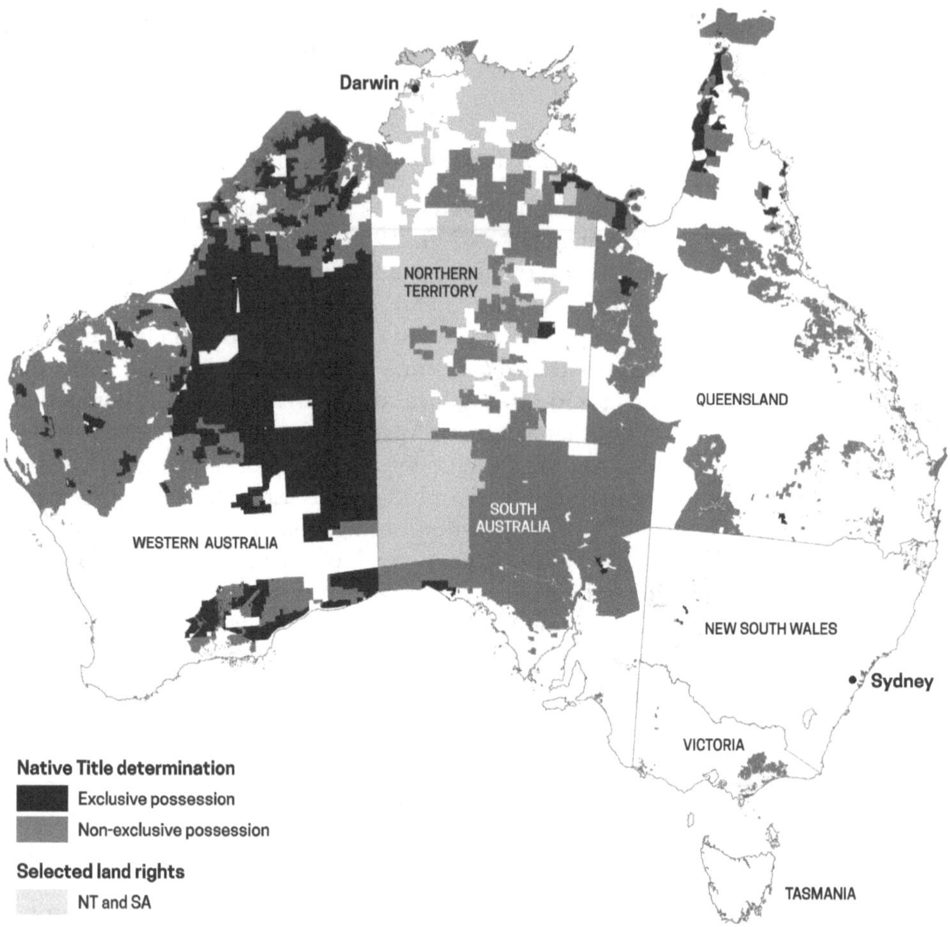

Figure 18.1 Map of Indigenous tenures, Australia, 2020.

Source: Jon Altman and Francis Markham, ANU.

peoples "through purposeful means, actively changed aspects of their environment" (Langton 1998). A re-evaluation of Indigenous peoples as managers, and not merely resource users, had profound implications for environmental management (Head 2000). With Indigenous voices growing ever more prominent it was clear that their influence over the settler-colonial institutions, and the issues that preoccupied the environment movement (land tenure, cultural heritage, competing development plans), would similarly grow.

Every environmental protection campaign would inevitably touch on matters of importance to Indigenous peoples: for their claims were fundamentally about the control of Country (a term I will discuss more later). Their struggles challenged me, as they did the small, community-based organisation and larger national environmental groups of which I was also part.[5] We were obliged to consider our environmental moralities and strategic positions, to reflect on Australia's colonial histories of settlement and nationhood, and the heritage of the ideas about the 'nature' we were trying to conserve.

Conservation in Australia had emerged from the bush-walking cultures of white Australia (Palmer 2004a) and settler notions of wilderness, or "external, primordial

Nature", empty of people (Jacobs 1996: 135). This ideal was then under direct fire from Indigenous people. Through their theorising of the cultural landscape idea, Australian geographers Lisa Palmer, Marcia Langton, Elspeth Young, Jane Jacobs, Lesley Head, Richard Baker, and Jocelyn Davies did much through the 1990s to unsettle and counter the chauvinistic and exclusionary cultural politics of this era (see Howitt and Jackson 1998 for a fuller survey). It didn't seem possible, however, to develop or articulate alternatives from within the NGO (non-governmental organisation) circles I then occupied. A more open space was needed to reflect deeply on how environmentalism was to respond – in a just way – to Indigenous claims to control Country and exercise jurisdiction. I returned to an academic setting to find it. I felt this might give me some critical distance.

Post-doctoral research: Macquarie University and the North Australia Research Unit

I enrolled in a PhD programme at Macquarie University (in Sydney) under the supervision of Richie Howitt. Richie was more interested in settler-colonial power relations than the geographers of my undergraduate education. He had put his progressive politics to work in 'action research' around the social impacts of resource development on Indigenous peoples of remote Australia (Howitt 1991, 1993). Working at the intersection of environment and development, Richie wanted to disrupt and re-shape relationships between Indigenous and non-Indigenous peoples. He was also thinking about scale and how local and translocal links could be made to further Indigenous rights agendas.

It was an exciting time to embark on a PhD. In 1992 the Australian High Court's *Mabo* judgement had recognised the fact that Indigenous people owned the land at the time of British conquest. This judicial revolution promised an expansion of restitution processes. The existence of two tenure systems (colonial and Indigenous) also raised hopes for a rebalancing of the power relations that determined how land should be used and managed in Australia and whose knowledge should count in those decisions.

Richie introduced me to Marcia Langton, a descendant of the Yiman and Bidjara nations of Queensland, when she took up a professorship in Indigenous Studies at Charles Darwin University. Marcia was an anthropologist[6] who had worked for decades for Aboriginal organisations and government agencies while maintaining an active research and creative life (she later completed a PhD in Geography). Marcia was then working on her critique of the colonial foundations of some forms of environmentalist thought and politics, especially the wilderness ideal. She published influential correctives to misconceptions about the Australian environment and the impact of Indigenous occupation of the continent held by many Australians and actively promulgated by some scientists[7] (Langton 1998, 2002). Langton argued that conservationists were perpetrating 'science fictions' by applying the term 'wilderness' to large tracts of Aboriginal land and denying the long and continuous presence and actions of Aboriginal peoples. In this critique of the Australian version of the 'noble savage', Marcia identified a tendency in much conservation practice to attempt to suppress Indigenous economic activity.

Environmental thought and practice of this era was further influenced by Indigenous geographical concepts that were translated and explicated by researchers affiliated with the North Australia Research Unit in Darwin, where I was based. Anthropologist Deborah Bird Rose brought the concept of Country to the Australian public in her essay, *Nourishing Terrains: Australian Aboriginal Views of Landscape and Wilderness* (1996). Country, she explained, is

> not a thing, it is a subject. Country is multi-dimensional, it consists of people, animals, plants, Dreamings, the dead, and the yet to be born, the underground – earth, soils, minerals, and waters .... There is sea country and land country, and sky country, too. Country has origins and a future... Country exists in and through time ... Country has its own story; its people are part of it and the story isn't over. (https://www.youtube.com/watch?v=suSbvoAw0g4)

With an understanding that was also gained from involvement in land claims processes, the geographer Elspeth Young publicly championed the Indigenous practice of 'caring for Country'. Elspeth wrote of the "universal belief" of all Aboriginal people "that all Country must be cared for. And caring for country was first and foremost a spiritual affair, through which its ecological and environmental future would be assured" (1987: 170). In 1990, with Helen Ross, she shared leadership of the team that produced a benchmark report for the Australian National Parks and Wildlife Service. It reviewed support for Indigenous land management and described the challenges of moving from the era of claiming ancestral lands (1970s–1990s) to the contemporary one of reoccupying and managing very substantial tracts of claimed land.

Caring for Country gradually moved into mainstream English. It was institutionalised as part of the policy lexicon (Jackson and Palmer 2015). That Country needs people and relations of care[8] assisted the wider intercultural project of environmental management and stewardship, such that by 2021 Indigenous landowners were the single largest contributor of land to Australia's protected area system (Altman and Kerins 2012). Interest in the concept of Country later took hold in Geography departments with the pluralistic ontological basis of research on Indigenous environmental action inspiring numerous studies around the 2010s (see Suchet-Pearson et al. 2013; Howitt et al. 2013; Jackson and Palmer 2015; Weir 2009).

It was assumed by many at the time I was embarking on my research career that 'caring for Country' was exclusively about *land* management. I had started my PhD research with the intention of studying sea space and customary marine tenures because I was intrigued by the relationships Indigenous peoples maintained with *sea Country*. Conflicts over desecration of sacred sites in the sea and commercial and recreational fishing were ever present in Darwin (Palmer 2004b). These flashpoints had prompted traditional owners to 'close the seas' and restrict access to others under NT law. Indigenous peoples made no proprietorial distinction between land and sea estates, as Matthew Dhulumburrk of the Ubulkarra clan in Arnhem Land eloquently explained in a public sea closure hearing:

> The reason why we are closing the seas and the land, off-shore two kilometres ... is that the earth and the sea, the water is not empty. It's not like a man building a house, leaving a house without furniture ... the earth contains things that you don't understand or Europeans don't understand. The reason we are closing the seas — we got something in it, we always have it and we'll be having it all the time ... What belongs to Ubulkarra belongs to Ubulkarra ... The land and the sea are not empty sheds that man has built. There's something in it.
>
> (cited in Jackson 1995)

The terrestrial bias in environmental governance, as well as in Geography, was the focus of my published paper titled 'The water is not empty: Cross-cultural issues in conceptualising sea space' (Jackson 1995). As my first contribution to cultural geography, it highlighted "what the neglect of the sea reveals about Western geographies and their imperial historical foundations". The paper confirmed that economic logics work in and through cultural processes, giving everyday sense to the uneven outcomes of capitalist, settler-colonial society. The importance given to property as a system of rights that gives people legal control over things and the meaning imbued in landed environments explained the terrestrial bias. It went back at least to imperial European laws of the freedom of the seas and played out in hostile responses from white Territorians seeking to protect their lifestyle (Palmer 2004b) – many of whom argued that the sea could not be owned, nor could navigational passage be impeded, or other profitable or pleasurable uses excluded. These cultural politics rendered Indigenous maritime relationships and sea tenures invisible and naturalised the claims of settlers to regulate the seas and the coast for their unfettered use and enjoyment. When subsequently exploring freshwater issues, the insights drawn from sea claims and accounts of juridical responsibility to saltwater estates were invaluable.

The maritime legal strategies of the Northern Land Council (representing traditional owners of the Top End) did not align with the timing of the PhD programme, so I went west to the town of Broome in Western Australia at the invitation of the Kimberley Land Council. Over three years, I researched the efforts of traditional owners of Broome and Darwin to reclaim their Country, exposing environmental planning's complicity in the assertion of settler control and the dynamics of settler-colonial urbanism. Conducted under a formal agreement with the Land Council and in collaboration with lawyers and anthropologists as well as the Yawuru traditional owners of Broome, the PhD was used to support claims to customary estates in a national heritage dispute, the native title claim and legal settlement over Broome, and by traditional owners to establish coastal management plans. Several published papers revealing the profound force and material effects of spatial technologies of mapping and zoning came from that period of ethnographic immersion in Broome, and similar observations from Darwin.

Working as a scientist in Australia's national science organisation

Following a few years of teaching, I took up a post-doctoral fellowship at CSIRO. The Commonwealth Scientific and Industrial Research Organisation is Australia's national science organisation and there were only a few social scientists amongst its 6,000 staff. Over the 11 years I was with CSIRO, the pool of social scientists grew significantly. Most researchers in the environmental area were geographers. At the Darwin lab, I was responsible for research on Indigenous land and water management and the integration of social and environmental sciences.

Charting a course of research on Indigenous water rights and governance had acquired a special urgency. Australia's system of water governance was being restructured, with the most significant changes conforming to neoliberal doctrine. Yet no one in government was thinking about the implications for Indigenous Australians and very few Indigenous organisations were ready to address what might happen when land and water titles were separated, or when water markets were rolled out. There had been such great strides in the 'caring for Country' agenda, but what part were Indigenous peoples to take in addressing the country's water crisis, and how would new institutions like water markets affect Indigenous rights and interests?

While at CSIRO my water governance research expanded to become the largest Indigenous project in the organisation. National interest in the prospects of 'developing the north' through irrigated agriculture brought funding for water research and saw the establishment of multi-disciplinary environmental research programmes, such as the Tropical Rivers and Coastal Knowledge Hub (TRaCK). It was from this institutional base that my project became a distinctly interdisciplinary one and one that more deliberately sought to transfer research capacity to Indigenous people. For many years I served as a member of the research executives of these large multii-organisational consortia and led the development of their social science research themes, as well as collaborating widely with ecologists, hydrologists, economists and lawyers. Indigenous participation across all organisational scales was crucial to the success of these initiatives, and it was consistently and effectively championed by their research leader, ecologist Michael Douglas.

From 2004–2010, I undertook research in regions beyond the 'frontier' of water regulation and planning (Fitzroy, (WA), Cape York (Qld)), to the relatively intensively developed aquifers of the Katherine-Daly (NT), Darwin (NT), and Pilbara (WA) regions (see Figure 18.2). They were experiencing their first water plans designed to underpin greater use of water. Several papers and reports from the Daly River region, including my 2006 paper, 'Compartmentalising culture: the articulation and consideration of Indigenous values in water resource management', were produced during this time. That paper analysed theoretical weaknesses in the treatment of 'culture' within the field of water management, which is a subject revisited in other publications written for scientists and policy makers (see Jackson 2017). My intention was to invite and enable government

Figure 18.2 Fish River Gorge, Daly River, country of the Wagiman, Malak Malak, Kamu and Labarganyan peoples.

Credit: Author.

environmental managers and scientists to reflect on the socio-cultural context of their endeavours – their 'cultures of nature' (Head 2017) – and the power asymmetries Indigenous people encounter in cross-cultural research.

In this and other publications, I was also trying to challenge the dominant policy narrative by elucidating the richness of Indigenous water cultures: the meaning of water, its symbolic significance and material value. Australians were unaware of the nation's history of water colonialism (Hartwig et al. 2022) and what was at stake. Land rights struggles focused attention on property rights regimes, but how had historical transformations in state systems of water governance and management affected Indigenous peoples, and what kinds of exclusions were being enacted as water laws and policies were being remade to align with global 'best practice' in water governance? There was an urgent need to formulate institutional responses to the denial of water rights in partnership with Indigenous organisations (see Taylor et al. 2017). With Indigenous experts such as Joe Morrison, Marcia Langton and later Brad Moggridge, as well as other scholars with a keen interest in resource rights, such as economic anthropologist Jon Altman and water lawyer Poh Ling Tan, we generated many critiques of the substantial legal and structural impediments to Indigenous access to water. This work showed that water governance systems were failing to provide sufficient recognition, respect and autonomy for Indigenous laws, values and norms.

A highlight was the establishment of a model of academic activism that took the form of an Indigenous water think-tank called the Indigenous Water Policy Group. Together with Joe Morrison, then CEO of the influential North Australia Indigenous Land & Sea Management Alliance, we obtained funds to improve the capacity of Indigenous people to develop their own policy responses to the water reforms that were sweeping across Australia. The Group set about improving Indigenous allocation outcomes in advance of the establishment of water trading and a significant increase in water use in north Australia. Indigenous leaders from the Land Councils of tropical northern Australia, who shared common interests in the region's waterscapes, would meet, discuss research results and work up policy options, with input from government water managers (see Jackson and Altman 2009).

The other orientation in the research of this period was socio-ecological, with many papers directed to science and environmental management journals. In projects undertaken with ecologists and hydrologists, conducted in partnership with several Indigenous groups, such as the Wagiman and Wardaman (see Figure 18.3), the aim was to contribute to river conservation via the new concept of environmental flows. 'The environment' was in the process of being personified; it was considered an embodied stakeholder and scientists were being tasked with determining its 'water requirements' (Jackson 2022). With this authority it was extremely difficult for Indigenous custodians to speak for Country. Instead, their rights and interests were treated as 'cultural values' somehow projected onto the asocial 'realities' of water (Jackson 2006). Outputs from this effort included publications co-authored with Indigenous traditional owners on fish and the economic value of aquatic species obtained from rivers and wetlands. All this co-produced knowledge was designed to inform water allocation plans, the programmes of Indigenous Ranger groups, and the strategies of larger Indigenous organisations.

Working from within a science organisation, this socio-ecological research was framed around the multifaceted dependence of Indigenous peoples on river ecosystems and the need for models of co-management. This framing carried some weight in policy circles, but it remained wedded to the Western separation between subject and object, nature and

Figure 18.3 Working on the 'fish and flows' project in the Daly River, here with Wagiman traditional owners and aquatic ecologists. Credit: Michael Douglas. I'm second left at the back.

culture. The scientists with whom I worked were concerned with the state or status of discrete environmental elements – for example, fish or food webs (which tended not to include people) – and they sought Indigenous knowledge to contribute to wider understanding of those elements.[9] The Aboriginal peoples with whom we worked, on the other hand, held that water is a source of life with its own agency and is part of social identity. As Biddy Lindsay and Paddy Huddleston from the Daly River recounted (Jackson 2005):

> Living water – like animal – I think of it like that – is living thing. White man way – don't care for that water – how you going to live?
>
> (Biddy Lindsay, Malak Malak traditional owner)

> That river and human being are connected. They're alive. It's killing Aboriginal life (all that water that's going to get used for farms). The water and the trees are connected to Aboriginal life.
>
> (Paddy Huddleston, Wagiman traditional owner)

Biddy, Paddy and so many others who were struggling for epistemic and legal recognition in the water contests of the time stressed the importance of the *relationship between water and society*. Relationality was central to the project, but I hadn't then found the conceptual tools to fully embrace or the means to activate it, which is perhaps not surprising given where I worked. The 2006 paper, 'Compartmentalising culture', provided a pointer to a more productive direction I was to later take, in collaboration with cultural geographer Lisa Palmer, and anthropologist and hydrologist, Sarah Laborde. For the people I was working with along the Daly River, it was their relationship with a living river system that was so highly valued, alongside their ability to reproduce those relationships.

Research with Indigenous communities was directed towards increasing their representation in state systems of decision making and advancing local knowledge projects. Communities were concerned about the inter-generational loss of knowledge, few people had engaged in catchment or water management processes, and the bureaucratic terminology, such as 'consumptive water', 'entitlements', 'recharge and discharge', was becoming increasingly technical. Journal articles were not going to deliver, so we produced community education materials, such as seasonal calendars for use in schools (see https://www.csiro.au/en/research/indigenous-science/Indigenous-knowledge/Calendars; Woodward et al. 2012). Information was translated into comprehensible formats, including a film narrated by Indigenous knowledge holders, and assistance with management skills like monitoring was offered.

Rejoining the university sector: the Australian Rivers Institute

In 2013 I left CSIRO and Darwin. The naturalism[10] of sustainability science, as I encountered it, and the assumption that 'integration' would bridge the gaps between disciplines and knowledge systems, limited the ability of critical geographers like me and other social researchers to frame productive research questions from within environmental scientific programmes. Rejoining the University sector allowed collaboration with a wider circle of scholars (especially geographers) and the opportunity to extend the geographic reach of my water research to the Murray Darling Basin (MDB), Australia's most intensively developed agricultural region.

Water in the MDB is over-allocated, severely degraded, and Indigenous peoples are locked out of the power structures that determine its use. These characteristics are all paradigmatic features of modern water management. The region was always held up as the model that north Australia must avoid. As in the north, the First Nations[11] of the MDB engage in and with the water sector to defend their relationships with Country against the most overwhelming efforts to eradicate them. Indigenous peoples control less than 2% of the land and 0.2% of the water of the Basin (Hartwig et al. 2022). Water struggles draw in political issues of nation-building, institutional development, self-determination, and knowledge contestation (Hemming et al. 2019). For example, organisations like the Murray Darling Rivers Indigenous Nations and the Northern Basin Aboriginal Nations have advanced their idea for 'cultural flows' to leverage water allocations off the success of the environmental flow concept, but in a manner that stakes out a distinct and vital role for First Nations in water governance (Weir 2009; Hemming et al. 2019).

I also tried to confront the poor appreciation within the water management sector, and amongst wider society, of the importance of historical events and structural processes in explaining Australia's pattern of inequitable access to water and Indigenous exclusion. Governments have been working on settler-colonial institutions for over 150 years, building hydraulic infrastructure of all kinds, including the rules and norms designed to privilege the 'white water citizen' and settle the inland. This was a term that US geographer Kate Berry and I coined in our paper on the racialised socio-spatial processes that bound whiteness to water in Australia and the Western US (Berry and Jackson 2018). The publication in the *Annals of American Geographers* required a detailed understanding of key moments in the colonial regulation of water and the evolution of Basin governance arrangements, which are critical to understanding dispossession and its enduring effects (Hartwig et al. 2022).

My recent research has benefited from heavily drawing on other geographical concepts and the ideas used by political ecologists in Geography to understand hydro-social relations and modern Western conceptions of water (Linton and Budds 2014). The hydro-social cycle, which construes water and society as co-constituted, provided the conceptual basis for several papers (two with Lesley Head) that showed how altering water flows affects social relations and structure, and vice versa (Jackson and Head 2020). Much of this work is concerned with the basic question of what water is – sometimes what a river is – and what it would mean for environmental governance to take seriously many ways of being with water. It challenges the limiting idea often encountered in collaborations with scientists: that water has a singular material existence (Yates et al. 2017). Earlier multi-disciplinary exercises had been unable to embrace ontological differences between Western scientific and Indigenous ways of knowing because science needed to adhere to modern constructions of nature with its strict separation between the biophysical, human, and supernatural worlds (Escobar 2001). The hydro-social cycle, however, provided a framework for revealing the existence of multiple waters and for exploring the emergence of many river realities under different configurations of social power, material properties of water, and technologies/infrastructures.

The comparison between different waters was made most graphically through a recent project with Sarah Laborde and Indigenous communities in the Kimberley region. It involved four years of work with language groups holding native title rights to much of the Martuwarra/Fitzroy River catchment. The Martuwarra is a sentient ancestral being and traditional owners have a custodial responsibility to maintain reciprocal relationships with it. In collaboration with Nyikina elders (one of many language groups of the catchment), we argued that a relational ethos, by which people interact with non-humans in their environment as relations or kin, rather than an external nature or objectified resource, is a core feature of Indigenous life (Milgin et al. 2020).

In the first paper from this project, 'Sustainability Crises are Crises of Relationships' (Milgin et al. 2020), we showed how a Nyikina relational ethics shapes socio-ecological sustainability within Nyikina Country. This paper also explained our use of a hydrograph[12] to connect seasonal Nyikina hydro-ecological knowledge with flow regimes in the Martuwarra/Fitzroy River. This visual representation (see Figure 18.4) allowed us to 'bridge' knowledge systems in a tangible way, and to facilitate interpretation of Nyikina hydro-ecologies by water planners and managers for whom the hydrograph is a familiar heuristic. Then, in 'Living Waters or Resource? Ontological differences and the governance of waters and rivers' (Laborde and Jackson 2022), we drew a picture that we hoped would be accessible to all beyond the academy. We used conceptual models to compare the 'living waters' that are in relationship with Aboriginal custodians to the 'modern water' that underpins much of aquatic science and policy. This work stressed the relations between things (the river, waterholes, plants, animals, people and ancestors), and showed how things came into being through those relationships. In that paper, and in a story map (https://arcg.is/0OLSeb), we urged water managers and scientists working with the Martuwarra/Fitzroy River to acknowledge ontological diversity and Indigenous jurisdiction and embed them in state environmental policy.

Figure 18.4 Traditional owners from the Martuwarra/Fitzroy River holding the Nyikina seasonal calendar during their protest over Western Australian heritage protection laws, Perth, WA, 2021.

Source: Walalakoo Aboriginal Corporation.

Conclusion

Much of this chapter has been concerned with institutions, and so it's fitting to conclude with a few comments on academic Geography and what kinds of geography matter there. I have never practised Geography from within a Geography department (except while doing my PhD). Operating largely from within institutes of applied science, Geography has been grounding for me, but also capacious. It has enabled me to cross boundaries and pursue decades of interdisciplinary research across a large continent. Geography, Kay Anderson argues, is expansive: it opens an intellectual space that is "across and beyond the established boundaries of disciplinary discussion in the human sciences" (1996: 123). To Anderson "the human geographer is not so much the reckless poacher smuggling fragments from here and there, but the creative disobedient who trespasses the arbitrarily erected borders of the academy and confronts the indivisibility of knowledge in their research practice".

I wonder how different it all might have been had I located myself within an Australian Geography department for my whole career. It's likely that an academic environment would not have encouraged as much applied research, particularly with community-oriented policy research being so undervalued in universities until recently. Would the nature–culture divide have been any easier to cross? It's hard to imagine so, given how Geography's version of this binary continues to define the practice of physical and human geographers in university departments. Few physical geographers are willing to embrace social theory, as I have found to be the case with most ecologists and other 'natural' scientists. While there are some, such as Gary Brierley and Rebecca Lave, we need more. Indigenous palaeoecologist Michael-Shawn Fletcher stands out in Australian Geography, not only for engaging in contemporary policy debates about Country and colonisation's environmental legacies. Michael participates in the cultural geography sessions of the Institute of Australian Geographers annual meetings and his recent accounts of environmental change build on the intellectual contributions of some of the human geographers cited in this chapter (Fletcher et al. 2021).

In 1998 Richie Howitt and I discussed the relationship between Australian Geography and Indigenous peoples (Howitt and Jackson 1998). We argued that although there'd been significant change, there was a pressing need to examine imperialistic and paternalistic legacies of Geography in Australia and to do more to decolonise the discipline. Twenty-five years later there are now many more Indigenous geographers in universities and Indigenous researchers in cognate disciplines (Head 2022), all of whom are heavily involved in important discussions over land, water, fire, and environmental change (Fletcher et al., 2021; Freeman et al., 2021; Martuwarra RiverOfLife 2021). Similarly, Indigenous 'environmental' organisations continue to insist on a fundamental reconception of research partnerships, including the adoption of Indigenous methodologies. While Geography provides some of the tools with which to analyse the issues central to the project of co-existence, other epistemological boundaries remain. What room is being made from within Geography to listen to, accredit and centre the knowledges of Indigenous people who remain outside the academy? The work of negotiating institutional spaces of co-existence and differences across knowledge traditions has only just begun.

Acknowledgements

I am grateful to Lisa Palmer, Lesley Head and Sangeetha Chandrashekeran of the University of Melbourne for their comments on this chapter, and to the editors for the opportunity to contribute to the collection. I also wish to acknowledge the many Aboriginal people with whom I have worked over my research career and to thank them for their generosity. Thanks also to Jon Altman and Francis Markham for the use of their map of Indigenous tenures and to Michael Douglas and Walalakoo Aboriginal Corporation for permission to reproduce their photos.

Notes

1 Ontology is concerned with the nature of existence and the structure of reality.
2 See the three-part report on this topic by Coombes et al. in *Progress in Human Geography* (2012).
3 Institution is meant in the sociological sense here, "as a rule, norm, or custom simultaneously enabling and constraining human agency" (Bakker 2002: 767) as well as the political institutions such as state and federal parliaments, government bureaucracies and scientific organisations.
4 Altman and Markham (2015) calculated that either statutory land rights or 'native title' are recognised over more than 30% of the Australian continent.
5 During the 1990s I served as a Council member of the Australian Conservation Foundation and a Vice President.
6 Marcia was the first Indigenous honours graduate in Anthropology in Australia.
7 Colonial geographers were also responsible for facilitating and justifying imperial occupation and expansion (Howitt and Jackson 1998).
8 Caring for Country is now the name of a network of Indigenous land managers and supporters.
9 This was not the only motivation or foundation for the research collaborations. Other motivations included a sincere commitment to partnerships that could support Indigenous forms of land management in the future and an awareness that to conduct research without Indigenous consent and participation was unethical, even if it had been the norm for decades. One project,

on fish and flows, continued for 16 years of "talking, listening and working together", demonstrating how the scientific practice of partnering was in this case implicitly relational even if the object of analysis could not be recognised as such (https://nesplandscapes.edu.au/impact/wiltma-nargun-lahan-walking-on-wagiman-land/).

10 Latour defines naturalism as the view that there is "a single natural world, comprehensible through Science" (cited in Gandy 2008).

11 First Nations is a term that is preferred by many Indigenous peoples of southern Australia.

12 A hydrograph is a way of displaying water level information over time.

Study task

Variations on this question can be set on an unseen examination, an open-book examination or as a term-paper assignment:

With a focus on Sue Jackson's research into water use in Australia, show how and why it's essential to make space for cultural difference and cross-cultural dialogue in the governance of 'natural resources'.

Starter reading list

Anderson, Elizabeth P. et al. "Understanding rivers and their social relations: A critical step to advance environmental water management." *Wiley Interdisciplinary Reviews: Water* 6.6 (2019): e1381.

Burdon, Peter et al. "Decolonising Indigenous water 'rights' in Australia: Flow, difference, and the limits of law." *Settler Colonial Studies* 5.4 (2015): 334–349.

Finn, Marcus, and Sue Jackson. "Protecting indigenous values in water management: A challenge to conventional environmental flow assessments." *Ecosystems* 14 (2011): 1232–1248.

Hartwig, Lana D. et al. "Water colonialism and indigenous water justice in south-eastern Australia." *International Journal of Water Resources Development* 38.1 (2022): 30–63.

Jackson, Sue. "Compartmentalising culture: The articulation and consideration of indigenous values in water resource management." *Australian Geographer* 37.1 (2006): 19–31.

Leonard, Kelsey et al. "Water Back: A review centering rematriation and indigenous water research sovereignty." *Water Alternatives* 16.2 (2023): 374–428.

McGregor, Deborah, et al. "Towards meaningful research and engagement: Indigenous knowledge systems and Great Lakes governance." *Journal of Great Lakes Research* 49.1 (2023): S22–S31.

Salmond, Anne et al. "Let the rivers speak: Thinking about waterways in Aotearoa New Zealand." *Policy Quarterly* 15.3 (2019): 45–54, https://doi.org/10.26686/pq.v15i3.5687

Searle, Tania, and Samantha Muller. "'Whiteness' and natural resource management: Let's talk about race baby, let's talk about sovereignty!" *Geographical Research* 57.4 (2019): 411–424.

Woodward, Emma et al. "Utilising Indigenous seasonal knowledge to understand aquatic resource use and inform water resource management in northern Australia." *Ecological Management and Restoration* (2012): 58–64.

References

Altman, J. and S. Kerins (eds), 2012. *People on Country, Vital Landscapes, Indigenous Futures.* Sydney: The Federation Press.

Altman, J. and F. Markham 2015. Burgeoning indigenous land ownership: diverse values and strategic potentialities. In: Brennan, S., Davis, M., Edgeworth, B., Terrill, L. (eds), *Native Title from Mabo to Akiba: A Vehicle for Change and Empowerment?* Sydney, Australia: The Federation Press, pp. 126–142.

Anderson, K. 1996. In the place of trans-disciplinary space. *Australian Geographical Studies* 34(1): 121–125.

Bakker, K. 2002. From state to market? Water mercantilización in Spain. *Environment and Planning A* 34: 767–790.

Berry, K. and S. Jackson 2018. The making of white water citizens in Australia and the US: Racialization as a transnational project of irrigation governance. *Annals of the Association of American Geographers* 108(5): 1354–1369.

Coombes, B., Johnson, J. and Howitt, R. 2012. Indigenous geographies I: Mere resource conflicts? The complexities in Indigenous land and environmental claims. *Progress in Human Geography* 36: 810–821.

Escobar, A. 2001. Culture sits in places. *Political Geography* 20(2), 139–174.

Escobar, A. 2008. *Territories of Difference: Place, Movements, Life, Redes*. Durham and London: Duke University Press.

Fletcher, M-S, R. Hamilton, W. Dressler et al. 2021. Indigenous knowledge and the shackles of wilderness. *Proceedings of the National Academy of Sciences of the United States of America* 118: e2022218118.

Freeman, D., B. Williamson and J. Weir. 2021. Cultural burning and public sector practice in the Australian Capital Territory. *Australian Geographer* 52(2): 111–129.

Gandy, M. 2008. Above the treetops: Nature, history and the limits to philosophical naturalism. *Geoforum* 39: 561–569.

Hartwig, L., S. Jackson, F. Markham and N. Osborne. 2022. Water colonialism and Indigenous water justice in south-eastern Australia. *International Journal of Water Resources Development* 38(1), 30–63.

Head, L. 1993. Unearthing prehistoric cultural landscapes: A view from Australia. *Transactions of the Institute of British Geographers* 18: 481–499.

Head, L. 2000. *Second Nature: The History and Implications of Australia as an Aboriginal Landscape*. Syracuse, NY: Syracuse University Press.

Head, L. 2017. Cultures of nature. In *The International Encyclopedia of Geography: People, the Earth, Environment and Technology*. eds: Castree et al. New York, NY: Wiley-Blackwell. https://onlinelibrary.wiley.com/doi/10.1002/9781118786352.wbieg0342

Head, L. 2022. Beyond Green: Human–environment geographies for the 'new' century. *Environment and Planning F* 1: 93–103.

Hemming, S., D. Rigney, S. Bignall, S. Berg and G. Rigney. 2019. Indigenous nation building for environmental futures: *Murrundi* flows through Ngarrindjeri country. *Australasian Journal of Environmental Management* 26(3), 216–235.

Howitt, R. 1991. Aborigines and restructuring in the mining sector: Vested and representative interests. *Australian Geographer* 22(2): 117–119.

Howitt, R. and S. Jackson. 1998. Some things *do* change – Indigenous rights, geographers and geography in Australia. *Australian Geographer* 29(2): 155–173.

Howitt, R. 1993. Social impact assessment as 'applied peoples' geography'. *Australian Geographical Studies* 31(2): 127–140.

Howitt, R. 2006. Scales of coexistence: Tackling the tension between legal and cultural landscapes in post-Mabo Australia. *Macquarie Law Journal* 6: 49–64.

Howitt, R., G. Lunkapis, S. Sandie Suchet-Pearson and F. Miller. 2013. New geographies of coexistence: Reconsidering cultural interfaces in resource and environmental governance. *Asia Pacific Viewpoint* 54: 123–125.

Jackson, S. 1995. The water is not empty: Cross-cultural issues in conceptualising sea space. *Australian Geographer* 26(1): 87–96.

Jackson, S. 2005. Indigenous values and water resource management: A case study from the Northern Territory. *Australasian Journal of Environmental Management* 12(3): 136–146.

Jackson, S. 2006. Compartmentalising culture: The articulation and consideration of indigenous values in water resource management. *Australian Geographer* 37(1): 19–32.

Jackson, S. 2022. Enacting multiple river realities in the performance of an environmental flow in Australia's Murray-Darling Basin. *Geographical Research* 60: 463–479.

Jackson, S. and J. Altman. 2009. Indigenous rights and water policy: Perspectives from tropical northern Australia. *Australian Indigenous Law Review* 13(1): 27–48.

Jackson, S. and T. Head. 2020. Australia's mass fish kills as a crisis of modern water: Understanding hydro-social change in the Murray Darling Basin. *Geoforum* 109: 44–56.

Jackson, S. and L. Palmer. 2015. Reconceptualising ecosystems services: Possibilities for cultivating and valuing the ethics and practices of care. *Progress in Human Geography* 39(2): 122–145.

Jackson, S. 2017. How much water does a culture need? Environmental water management's cultural challenge and Indigenous responses, in Horne, A., Webb, A., Stewardson, M., Richter, B. and M. Acreman, (eds) *Water for the Environment*. London: Elsevier, 173–188.
Jacobs, J. 1996. *Edge of Empire: Postcolonialism and the City*. London: Routledge.
Laborde, S. and S. Jackson. 2022. Living waters or resource? Ontological differences and the governance of waters and rivers. *Local Environment* 27(3), 357–374.
Langton, M. 1998. Burning Questions: Emerging Environmental Issues for Indigenous peoples in Northern Australia, Centre for Indigenous Natural and Cultural Resource Management, Northern Territory University, Darwin.
Langton, M. 2002. The 'wild', the market and the native: Indigenous people face new forms of global colonization, In W. Adams and M. Mulligan (eds) *Decolonizing Nature*. Routledge: Strategies for conservation in a post-colonial era.
Linton, J. and J. Budds. 2014. The hydrosocial cycle: Defining and mobilizing a relational-dialectical approach to water. *Geoforum* 57: 170–180.
Martuwarra RiverOfLife, Alessandro Pelizzon, Anne Poelina, Afshin Akhtar-Khavari, Cristy Clark, Sarah Laborde, Elizabeth Macpherson, Katie O'Bryan, Erin O'Donnell and John Page. 2021. Yoongoorrookoo, *Griffith Law Review*. 10.1080/10383441.2021.1996882
Milgin, A., L. Nardea, H. Grey, S. Laborde and S. Jackson. 2020. Sustainability crises are crises of relationships: Learning from Nyikina ecology and ethics. *People and Nature* 2: 1210–1222.
Palmer, L. 2004a. Bushwalking in Kakadu: A study of cultural borderlands. *Social & Cultural Geography* 5(1): 109–127.
Palmer, L. 2004b. Fishing lifestyles: 'Territorians', traditional owners and the management of recreational fishing in Kakadu National Park. *Australian Geographical Studies* 42(1): 60–76.
Povinelli, E. 2006. Finding Bwudjut: Common land, private profit, divergent objects, In T. Lea, E. Kowal, G. Cowlishaw (eds) Moving Anthrolpology: Critical Indigenous Studies, Charles Darwin University, Darwin, pp. 148–165.
Rose, D. B. 1996. *Nourishing Terrains: Australian Aboriginal Views of Landscape and Wilderness*. Canberra: Australian Heritage Commission.
Suchet-Pearson, S., S. Wright, L. Lloyd, L. Burarrwanga and Bawaka Country. 2013. Caring as country: Towards an ontology of co-becoming in natural resource management. *Asia Pacific Viewpoint* 54: 185–197.
Taylor, K., B. Moggridge and A. Poelina. 2017. Australian indigenous water policy and the impacts of the ever-changing political cycle. *Australasian Journal of Water Resources* 20(2): 132–147.
Weir, J. 2009. *Murray River Country*. Canberra: Aboriginal Studies Press.
Woodward, E., S. Jackson, M. Finn and P. Marfurra McTaggart. 2012. Utilising indigenous seasonal knowledge to understand aquatic resource use and inform water resource management in northern Australia. *Ecological Management and Restoration* 13(1): 58–64.
Yates, J., L. Harris and N. Wilson. 2017. Multiple ontologies of water: Politics, conflict and implications for governance. *Environmental Planning D: Society Space* 35: 797–815.
Young, E. 1987. Resettlement and caring for the country: The Anmatyerre experience. *Aboriginal History* 11(2): 156–170.

Part III

Making geographical knowledge matter beyond Geography

19 Geographers and the national state

Dudley Stamp plans Britain's towns and countryside

Trevor Barnes

Sir L. Dudley Stamp (1898–1968) Copyright National Portrait Gallery, London. (by Elliot and Fry) https://www.npg.org.uk/collections/search/person/mp78655/sir-laurence-dudley-stamp

Introduction

From 1926, when he was appointed the Sir Ernest Cassel Reader in Economic Geography at the Joint School of Geography at King's College, London (KCL) and the London School of Economics (LSE), until August 1966, forty years later, when he abruptly died of a heart attack in a hotel elevator in Mexico City, Professor Sir Laurence Dudley Stamp (1898–1966) unremittingly and enthusiastically devoted his life and work to championing Geography – one obituarist called him "one of the happiest geographers" (Anon, 1968, 186).[1]

Geography's calling, Stamp believed, was as an applied discipline, able to solve practical problems.[2] That required geographers, he thought, to work with or become

DOI: 10.4324/9781003343240-22

integrated within, governing and planning bodies like the local or national state or international organisations that possessed both resources and power to affect the solutions that applied geographers like himself recommended. Certainly, Stamp was integrated in his professional work. George Kimble (1967, 247) wrote, "no other British geographer has equaled and few have approached [Stamp's level] … of public service". In the first half of his career Stamp applied geography to solve problems for the UK national state around rural land use. In the second half, while continuing to serve the British government, advising on agriculture and conservation, he proffered geographical solutions to the problem of world development, attending branches of the United Nations (UN) and participating in the International Geographical Union (IGU).

Consequently, "geography was, to him, more than academic" (Wise 1966, 594). He was never interested in theory.[3] Theory's abstractions, Stamp believed, took him away from the world, while he desperately wanted to be in the thick of it. The discipline of Geography offered precisely that opportunity. That is why Stamp was such a tireless promoter, such a disciplinary fan. It explains his decision in 1921 to sit the undergraduate degree finals in Geography the first time they were offered at the University of London even though he already had three other degrees under his belt – Kimble (1967, 246) called it a Geography degree "by 'moonlighting'". He was awarded a first-class degree without attending a single Geography course lecture.[4] He could not get enough of the discipline. Fittingly, his last mortal breath was at an international Geography meeting.

Geography enabled Stamp to be a man of action, a doer; to apply to national and global problems his formidable geographical knowledge. Much of that knowledge was acquired from his land-use studies. The first was in the UK, the Land Utilisation Survey (LUS), covering the entire country. Its origins were in the late 1920s, but it took even the indefatigable Stamp nearly 20 years to finish. Later, he tried to survey the whole world, although this was never completed. In both cases, his studies were not simply descriptive, but constitutive, in a sense creating spaces not known before, and, following the French philosopher-cum-intellectual historian Michel Foucault, providing opportunities for their regulation and governance.[5] Once Stamp entered public service, first at the level of the national state, later as participant in international organisations, that is what he did. He shaped the governance of those spaces. He brought a particular form of physical planning to the geography of the UK, formalised by the 1947 Town and Country Planning Act, and internationally he encouraged policy that promoted 'development'.

That Stamp could do so much was in part because of his capacious talents: a photographic memory; unflagging energy and drive; honed organisational skill; a knack for absorbing large amounts of complex information along with an ability for incisive simplification; sharpness, fluency and persuasiveness on his feet; and the skill to write clearly, lucidly, winningly, and perhaps most importantly, with swiftness and punctuality. It was said that by the end of a five-day transatlantic voyage he would have ready his next book, and he always "completed to time" (Wise 1988b, 181). In his unpublished autobiography, he reflected on his writing method, which he said entailed "saturating myself with the material, writing a book in my head (often laying on bed) and then proving it out on paper or in dictation on a Dictaphone" (quoted in Wise 1988b, 263). Not mentioned, though, but also vital, was his personal assistant, Mrs Audrey Clark, who took those tapes and hand-written papers "for further editing and typing" (Wise 1988b, 177).

There was a second, related strand to Stamp's geographical project stemming from his unconditional faith in the power of geography: pedagogical. For his project to succeed, students needed to sign up. For that, geographical teaching techniques and materials should inspire and enliven, not as was often the case, bore and deaden. To that end, Stamp and his wife Elsa (née Rea, 1900–1962), another geographer – he was her undergraduate tutor at the Joint School of Geography – joined to become a Geography textbook publishing machine, writing over 50 titles, many going into several editions (Chibnall 1968).[6] They started their own profitable publishing company, Geography Publications Ltd., producing geographical volumes to educate, edify and excite, to make new citizens of geography. The 1955 *Times Educational Supplement* reckoned that along with "[Thomas] Arnold, Archimedes, [and] Aristotle", "L. Dudley Stamp ... shaped our schooling" (quoted in Wise 1988b, 176). The Stamps' texts served not only the UK's domestic market, but before World War II were sold throughout the British Empire, and after, all over the British Commonwealth.

The remainder of this chapter is divided into two main parts. The first is about the extraordinarily full life of Stamp. He never wasted a minute, holding to a "relentless work ethic" (Tobin 2021). His life moved with the rhythms of the 20th century. His professional life began in 1921 as an agent of British colonialism in the Far East. After he returned home, he helped shape a wartime and early post-war planned and managed national state. And post war he took on, among other roles, that of the globe-trotting geographical expert. The second focuses on Stamp's work both as he gathered geographical knowledge within Britain – the LUS – and later fashioned government policy once he was enmeshed within the apparatus of the national state. While that policy was partly economic, about raising efficiency and productiveness, it was also deeply geographical, defined by what he thought was an ideal relationship between nature and society rooted in the land. Stamp was a geographer to the core.

Life and times

Early years: World War I and Empire

Growing up upper middle-class at the turn of the twentieth century in a London suburb, Stamp missed long stretches of schooling because of childhood sickness. That never adversely affected him, though. It may have been an asset. Inculcated at home by the belief, as he put it in his autobiography, that "success ... [w]as the natural reward of clean living and hard work" (quoted in Wise 2004), Stamp taught himself the natural sciences. It was good enough that at "the tender age of fifteen years" he took the entrance examination at KCL and was admitted (Anon, 1966). Awarded a First in Geology in 1917, the following year Stamp published his first academic paper, the beginning of a prolific publication record (Chibnall 1968). Stamp said later, "if you believe in your own work, publish and be damned" (quoted by Tobin 2021). Stamp was a believer, and published prodigiously, although he was rarely damned.[7]

In 1918, commissioned as a Lieutenant in the Royal Engineers, Stamp served in Northern France during the last two months of World War I. Remarkably in his time off he managed to complete research that in the same year earned him an MSc in Geology from King's – he was given special leave while still on active duty for the exam (Wise 1988b, 185). Demobbed in September 1919, he immediately entered King's doctoral programme, receiving a doctorate in Geology two years later. Rather than staying in the academy, however, Stamp took a job as an oil geologist in Burma (now Myanmar), then a British colony. He worked first, for the Yomah Oil Company (1920–1923) (which folded), and then very briefly in 1923, for Indo-Burma Petroleum. Neither satisfied. In

1923, he returned to academia, becoming the inaugural Professor of Geology and Geography at the University of Rangoon (established in 1920, and now the University of Yangon). It was there with his new wife, Elsa – they married in 1923 – that he began writing their aids and textbooks for teaching school Geography. An early version of their school blockbuster, *The World* – it ran to 17 editions – was written in Burma and published in 1927.

Middle years: The Land Use Survey, World War II and government

In 1926, Stamp returned to the heart of Empire, London, offered the Cassel Readership at the Joint School of Geography at KCL and LSE. While the position was to teach and research in the field of economic geography, Stamp had little intellectual sympathy with economists, especially those at LSE.[8] This was because of his conception of Geography. Given his training in the natural sciences, Stamp keenly emphasised the importance of the physical environment for the discipline, albeit a physical environment holistically integrated with the social world and most fully expressed in land use. That gave Geography a groundedness – Geography as earth writing (its literal translation from the Greek) – but absent in economics given its disciplinary aspiration for theoretical abstraction. As Wise (1988b, 180) put it, Stamp "was a 'real world' man, his ideas rooted in soil, vegetation and people working the land, not in abstract theory". Stamp's favourite precept was, "Face the world" (quoted by Wise 1988b, 180–181).

It was precisely that view that led Stamp shortly after he returned to the UK to instigate a land use survey for Great Britain. Land use was exactly where "soil, vegetation and people working the land" converged. The last comprehensive land survey of the country was *The Domesday Book* (1085–1086) completed 850 years earlier. That "Great Survey" undertaken for William the Conqueror who invaded the country in 1066 recorded all his new holdings within England and taxes owed. Domesday here meant the last word or final judgement. That was Stamp's aim too in his own "modern *Domesday Book*" (Towles 2023).

The details of Stamp's LUS are discussed in detail in the next section. It involved cartographically recording on very large-scale maps – six inches to the mile – the land use of every "acre of land", field by field, in the British countryside. It was a mammoth undertaking. Stamp estimated in total it took 44,000 days of labour (over 120 years; Stamp 1948, 36). Even Stamp's seeming inexhaustible determination and resolve were tested as he travelled day and night on boats, trains and cars throughout the country. With Elsa as chauffeur, Stamp (1948) recalled "often standing up on the front seat of my car with my head through the sunshine roof and a roll of six-inch maps in front of me". In the first year, Stamp calculated he travelled 2,000 miles (3600 kms) a month (Towles 2023). He did so to check the accuracy of map recordings, to organise local teams of surveyors, to give pep talks, and to speak to and advise national and local politicians, planners and bureaucrats at both public and private meetings. Invitations to speak were a temptation he could never resist, including on the relatively new BBC radio service (inaugurated in 1922; Wise 1968, 262). And all this while carrying out his day job of giving undergraduate lectures, supervising graduate students, attending School meetings, writing textbooks, reports, academic papers, and engaging in a full mailbag of correspondence.

The importance of the LUS, and the consequent elevation of Stamp from everyday geographer to a member of Britain's governing inner circle, was triggered by the outbreak of World War II. Or, more precisely, by a larger changed sensibility around governance precipitated by the war and defined by a shift towards expert-based national and local planning (Tanner 1987, 31). It was the moment for LUS and Stamp to shine. And it did, and he did too.

During the 1930s the combination of plentiful, cheap food imports – around 70% of the UK's food products were imported – and the economic effects of the Great Depression, devastated the British agricultural sector. The result, as Whitehead, Jones and Jones (1997, 100) wrote, was that "large swathes of the British countryside … [were] left uncultivated and turned to rough brush". Stamp called it "the tragic age of British farming" (quoted by Towles 2023). That neglect came home to bite from 1939 after World War II began. A tight German blockade around Britain enforced by packs of Nazi U-boats prevented food imports into the country. With consequent food shortages and the resulting rationing, measures to make British agriculture efficient and productive became a national imperative. Here LUS and Stamp were indispensable. They provided detailed information about land already used for food production and, even more importantly, about land that could be used for food production. With that knowledge, land was 'mobilised' for war, becoming 'a weapon of war'. It allowed a newly recruited Land Army to 'dig for victory' in what was called the "plough-up campaign" (Matless 1998).

It was then that Stamp was enrolled in government as an expert, and part of much broader moves by the British state to establish a central planning apparatus to fight the war. In Stamp's case, his specialised geographical knowledge of land use was crucial to formulating a national plan to meet both immediate needs in the UK– the plough-up campaign – but also the longer-term relation between towns and countryside. In 1941, Stamp was appointed Vice-Chair of Lord Justice Scott's Committee on Land Utilisation in Rural Areas. That committee, and the resulting 1942 Scott Report, regulated land use during the remaining War years (Stamp became Chief Adviser on rural land in the Ministry of Agriculture after the Report was published), but also shaped future British state planning of land for the next 45 years or so.

The 1942 Scott Report, formalised in 1947 as The Town and Country Planning Act with its own newly minted government ministry, curbed urban development incursions into the British countryside, preserving, indeed, encouraging, traditional farming practices and their associated land use. For Stamp (1943, 16) the Report and subsequent legislation, were "based on geographical principles" – *his* principles of "soil, vegetation, and people working the land". Before the war, the countryside was in danger of eradication because of both the decline of British agriculture and wanton urban-fringe development. The 1947 Act ensured it would now be preserved. Stamp's LUS created a space to be governed and realised by the general shift to national central planning and specific recommendations of the Scott Report.

Late years: Professor Stamp goes global

In 1945, while continuing to serve government – he received a CBE for his wartime service that same year[9] – Stamp was promoted to Professor and made Departmental Head. Added to all his other responsibilities was now running LSE's Geography Department. It did not seem to slow him down. He became even busier, now travelling the globe.

Stamp was always an internationalist. But during the last third of his life, from the very late 1940s until his death, his internationalism became more directed. Partly, that was a result of his growing involvement with the IGU. Founded in 1871, the International Geographical Congress, as it was then called, originated to serve European exploration, colonialism and militarism. The first signs of change were during the inter-war period when its meetings became less about exploration and colonial management and more about international scholarly exchange and debate around ideas, methods, techniques and uses of academic Geography (Kish 1972). Given Stamp's conviction about the power of geography, the IGU's global mission of spreading the geographical word greatly appealed. In 1949, Stamp was elected IGU's Vice President, and in 1952 he became President with a four-year term. He took early retirement from LSE in 1958 but continued to work just as hard. In 1964, he became Chair of the Organizing Committee for the July London meeting of the IGU. The Post Office issued four commemorative stamps for the occasion – especially appreciated by Stamp as a keen philatelist (remarkably he had time for hobbies). The Queen, as the *Western Mail* reported, wearing a "blue shimmering dress", and accompanied by Prince Phillip, officially opened the IGU at London's Royal Albert Hall.[10] The Queen gave a speech about Geography during which, it was also reported, the Duke of Edinburgh appeared "deep in thought". Stamp followed the Queen, giving his version of Geography as "the study of Earth as the home of man [sic]" (quoted by Wise 1988b, 178). Neither Her Majesty's nor her Prince Consort's reaction was recorded, however. Stamp continued his involvement with the IGU literally until the end. It was at a Latin American IGU regional conference that Stamp suffered his fatal heart attack.

The other form Stamp's internationalism took was scaling up his land-use survey to the world as a whole. In 1949, he helped launch the World Land Use Survey with UNESCO Boston, which later opened an office in London. Two years later he became Director of its operations. "The objective" of the survey, as Stamp put it in a 1963 BBC radio interview, was "to record on maps the use of land according to a common system for countries all over the world".[11] Again, like the LUS in Britain, behind the survey was a planning objective, in this case to promote, in the language of the time, the "development" of "underdeveloped" countries. As Stamp continued, "we've got this problem all over the world of population increasing, and so we have a pressure on land resources". It was this that led to "underdevelopment". By carrying out a land-use survey worldwide, thought Stamp, it would be possible to provide advice on how to use "land to produce more and more efficiently than at present", and hence "freeing the world from hunger". Stamp recognised that inefficient land use was also a problem of developed countries. Nevertheless, and showing his Western centrism, by comparing how land was used between "temperate" developed regions and "tropical" underdeveloped regions, he said the "best answer for the World as a whole" was the agricultural model found in "countries of North-West Europe". Consequently, Western land use became the benchmark to which "underdeveloped" countries should aspire, which experts like Stamp (1953, 1960a, 1960b) would "help" them to achieve through various international organisations.

The Land Utilisation Survey of Great Britain

Shortly after he returned to the UK from Burma Stamp had the idea for the LUS, of mapping in detail the land use of the entire countryside of Great Britain using field survey techniques (Stamp 1948; Board 1968). Even for a relatively small country – although not that small given it was just over 94,000 square miles (243,500 square kms) – it was an ambitious undertaking. Peter Hall (2003, 548) later called it an "audacious step". Nothing of that scope had been done before at a national level anywhere (Board 1968, 29–30). The *Domesday Book* was far less wide-reaching, covering only England and parts of Wales, with no maps and less accurate.

Map making

The intellectual impetus for the LUS was the regional survey tradition (Stamp 1932, 710; Matless 1992). That began in nineteenth-century France with the engineer Frédéric le Play. In turn, his work was taken up during the 1920s and 1930s in Britain, especially by sociologists and geographers who formed the Le Play Society (Stamp was a participant). As an approach, it emphasised research on rural areas, detailed first-hand empirical study by means of place-based surveys, cartographic representation, and a connection between physical and social processes. Consequently, it echoed very much Stamp's conception of Geography. He was leery, though, of the "mystical environment-society dialect" that was invoked sometimes within that tradition (Rycroft and Cosgrove 1995, fn. 12, 104). Instead, Stamp's approach "was more technocratic", a "pragmatic rendering" (Rycroft and Cosgrove 1995, fn. 12, 104). He wanted to stick with the facts.

That said, underlying Stamp's LUS was also a deep-seated normative judgement. He believed the current land-use configuration of Great Britain was disordered and unbalanced, a result of "uncontrolled *laissez faire* economics and industrial exploitation of the land's resources during the nineteenth century" (Rycroft and Cosgrove 1995, 100). That produced significant loss of agricultural land, "waste" unfit for cultivation and undermining what he believed was the previous harmonious relation between nature and society (Stamp 1960a, 1960b, 9; Rycroft and Cosgrove 1999, 123). Stamp strongly believed land use in Britain *should* return to its formerly ordered and balanced relation, with society and nature made whole again. That normative stance was reflected in the title of his 1948 book summarising the history, methods and achievements of LUS: *The Land of Britain: Its Use and Misuse*. He could not have been clearer. Bad land use ("misuse") *should* be eradicated, and replaced by good land use. The LUS was for him the first step in land's normative makeover. As Whitehead, Jones and Jones (1997, 90–91) write, "it is wrong to think of land-use maps as purely maps of how land is being used. Land-use maps have always provided chronicles of how land is being under-used, or even misused, and indications of how it could be used more effectively in the future."

Before the UK could replace bad land use with good land use it was necessary to know the existing land use. The was the purpose of LUS, to provide up-to-date, comprehensive cartographic data. Those geographical facts, in turn, brought into view, made visible, a space that now could be controlled and regulated, governed by the national state. That visibility allowed state planning to get to work, to make improvements by eradicating waste, to improve efficiency, to restore balance and equilibrium between society and nature. LUS was a deliberate geographical intervention. It was never impartial, never only a data-gathering exercise. As Whitehead, Jones and Jones (1997, 90) again write:

> land-use mapping is an exercise in making … a nation's land into an actually knowable spatial entity…. [It] involves a significant amount of time, political coordination, and expertise, but it can result in the … expanded governmental reach of the state over its territorial natures. Accordingly, we do not see the mapping of land as simply a neutral recording process, but as a strategy with particular political goals and aims.

LUS definitely involved a significant amount of time, coordination and expertise. The first major issue faced was staffing. LUS's completion required an army of surveyors. In 1929, Stamp had a light-bulb moment when he heard Mr E. E. Field speak about his land-use study for the single county of Northamptonshire in the English Midlands (Stamp

1931, 44). With assistance from the Local Education Committee, Field had enlisted local school children to carry out the surveying. Stamp would do the same, with himself, KCL/LSE staff and graduate students acting as supervisors and fact checkers.

In the end, roughly a quarter of a million, hopefully keenly observant British schoolchildren, including Scouts, Girl Guides and Rovers, from 10,000 schools were signed up by Stamp as on-the-ground land-use surveyors (Rycroft and Cosgrove 1995, 91). They became Stamp's "mobile eyes" (Whitehead, Jones, and Jones 2007, 106). Each surveying group received from the British mapping agency, the Ordnance Survey (OS), two 6 inch-to-the-mile quarter-base maps for their district representing about 6 square miles.[12] The children would fan out into the countryside and record land uses (their first map was for "rough field work", the second for "finished copy"; Stamp 1931, 47). Under teacher supervision, students were given "two days or three days" to complete the assignment (Stamp 1931, 46). The land use of any given land parcel in the district was designated using one of seven broad land utilisation categories: forest; meadowland; arable; heathland; gardens; wasteland; and ponds. Each category was marked on the map either by the first letter of the land-use type (e.g., 'F' for forest), or by a standardised and suggestive colour (e.g., dark green for forest).[13] Stamp "left to the discretion of the surveyor whether each piece of land is marked by the initial letter ... or lightly coloured by pencil" (Stamp 1931, 44) (Figure 19.1).

There was also the issue of money. LUS was hugely costly. The bill for the base maps alone – 44,000 were required – was enormous. Although the OS sold them to Stamp at the discounted price of 1s./6d. each (7.5p), the total bill came to £3,300 (UK £285,000 in 2023 prices; US$367,000). The Rockefeller Foundation gave an initial £500, but clearly that did not go far. The result was that the survey was in "almost permanent financial difficulty" (Hall 2003, 548). From 1936, Stamp (1948, 12) took "on complete financial responsibility for the finances ... of the survey" with the LUS becoming a subsidiary of his and Elsa's publishing company, relying on income earned from school textbooks they wrote (Southal, Brown and Burton 2003, 4). Fortunately for LUS, the pair were exceedingly productive, their textbooks best sellers and highly profitable. Further, their company negotiated all contracts and received all royalties (Wise 1968, 263).

The bulk of the LUS research was undertaken between 1931 and 1934, although the last survey was not finally completed until 1941 (Stamp 1948). "Finished copy" maps from each district were sent to the LSE, a "centre of calculation", and collated. They were further "translated" into 169 1-inch-to-10-miles county maps (Southall, Brown and Burton 2003, 2–5). That last work was only done in eight weeks by, as Stamp put it, "girls with a good school record that included Geography" (quoted in Towles 2023). In turn, those maps also formed the basis for 92 separate county reports, each between 40 and 200 pages long, and variously written by Stamp, University of London Geography lecturers or graduate students.

Producing cartographic knowledge on paper maps could be fraught, however. Office records, printing plates and maps, both the 1 inch to 10 miles and the 6 inches to 1 mile, were destroyed in two separate fires. One in May 1941 occurred during the London Blitz of World War II, and the other was the result of student protests at LSE in January 1969 (Southall, Brown and Burton 2003, 4–5).

Policy making

While Stamp undertook LUS independently, not explicitly for purposes of the national state, the state was always present in one form or another. Most immediately, Stamp believed that enrolling school children as surveyors would contribute to their education not only as geographers, furthering the interests of the discipline, but as national citizens. The

Figure 19.1 A portion of a six-inches-to-a-mile OS map marked according to the 'initial letter' of Stamp's seven land utilisation categories.

Source: Stamp, L. Dudley. "The Land Utilization Survey of Britain." *The Geographical Journal*, vol. 78, no. 1, 1931, p. 44.

grounded practices of surveying, mapping and analysing local land, he believed, gave them knowledge about the nation, showed them what was at stake in its land-use change, and, by providing them with information about "land and its workings", made them conscious of the importance of the British countryside and its need for protection using national planning (Rycroft and Cosgrove 1995, 95; Whitehead, Jones and Jones 1997, 102–103).

After the outbreak of World War II, the national state loomed ever larger as the machinery of central planning moved into gear, with LUS one of its component parts. Shortly after the declaration of war, Stamp remembered he was inundated with

> receipt of several telegrams asking to borrow sets of original [LUS] maps as a basis for planning the ploughing up campaign … the next two or three months were occupied in sending out thousands of our original six-inch sheets on loan to County War Agricultural Committees.
>
> (quoted Whitehead, Jones and Jones 1997, 109)

Figure 19.2 Dudley Stamp (R) having a laugh with the urban planner, Patrick Abercrombie (L), 1957.

Source: Royal Institute of British Architects, https://www.ribapix.com/sir-patrick-abercrombie-with-geographer-sir-dudley-stamp-on-the-right_riba44971#

The plough-up campaign began in 1940, continuing through to the next year. Stamp supported it "vociferously", immensely "pleased [by the resulting] … revived agriculture" (Rycroft and Cosgrove 1995, 99), sending as much material from LUS to facilitate its success as he could. By 1941 it was not just his maps that were in demand, but himself.

By then, he and his work on LUS were known to some important pre-war figures in government policy. He had shared the stage at meetings with perhaps the country's most important urban and regional planner, Patrick Abercrombie, who later re-planned London and other bombed British cities (Figure 19.2). The Conservative politician Montague Barlow had also drawn on LUS for his influential 1940 Royal Commission Report on the geographical distribution of the industrial population. In this sense, Stamp was already vetted as a potential participant in national planning. It was no surprise, then, that in early 1941 he was invited by Lord Reith, another government insider, former Director General of the BBC and now Minister of Works and Buildings, to be on his Consultative Panel. That job did not last long before Stamp became Vice-Chair of the Scott Committee, but that was not the point. Stamp was now officially a government policy insider, serving as an expert adviser. He remained there for the rest of his life, applying geography to national state policy problems.

As Tanner (1987, 34) reflects specifically about Stamp, "one of the main characteristics of the policy community is that there is regular interaction between its members most of who knew each other very well". That was Stamp's life. Even early on Stamp "knew people in public life" in part because of his closeness to his famous elder brother, Josiah, a long-time key financial adviser to the British government and who travelled in impressive political circles (Tanner 1987, 35). Later, interaction was fostered through Stamp's heavy involvement and multiple leadership roles in professional associations, societies, panels, boards and committees – if he was not President or Chair, he was a member. Further, those roles gave him a mouthpiece to disseminate his views. After all, he even got to lecture the Queen. Stamp's multiple involvements nurtured exactly the right kind of environment for him to participate in policy discussions.

Stamp's 1941 appointment as Vice-Chair of Lord Justice Scott's committee "to consider the conditions which should govern buildings and other constructional development in country areas" resulted in his greatest impact as an applied geographer compared to anything else he did (Stamp 1943, 16). While the terms of reference of the Scott Committee might have sounded dull, the committee's recommendations embodied in its Report were not at all. As Stamp (1943, 16) said, they were a "blueprint ... for a future Britain ... based on geographical principles" (Stamp1943, 16). Those principles were primarily Stamp's.[14] They were a belief in the geographical harmony of society and nature, and best found as agricultural practices and associated land uses. That geography in the UK had come undone, as LUS showed. The misuse of land – waste – piled up, while 'good' use shrank. Farms were abandoned, producing a wasteland of "brambles and scrub", while urban infringement produced a different kind of wasteland, "building estates of hundreds of acres of unsightly derelict 'plots'" (Stamp 1943, 17–18). Stamp believed misuse must stop, and good use must be encouraged. This could only be done, said the Scott Report, by a Central Planning Authority (Stamp 1943, 18). Later enacted in post-war legislation and government reorganisation – the 1947 Town and Country Planning Act and the Ministry of Town and Country Planning – rural areas were revitalised by encouraging farming and village life, while development intrusions were held at bay by implementing green belts, restrictive planning permission, and highly regulated new town construction. For Stamp, the earlier unbuckling of nature and society provoked a disordered, even ugly, British countryside. But with applied geography, and under the aegis of the state, that could be reversed in his view: the country returned to order and beauty.

Conclusion

During his time, Dudley Stamp was the most well-known British geographer. He moved within the higher echelons of British society and the academy. He received a knighthood the year before he died (some said it was long overdue); had his photo officially taken for the National Portrait Gallery; received countless academic awards and gold medals (Wise 1988b, 185–187, details the fully tally); and was a guest on such BBC popular programmes – when there was only the BBC, so it attracted a large public audience – as *The Brains Trust* and *Desert Island Discs*. But Stamp is little remembered today and rarely taught.

Partly he became lost because of the nature of Geography as a discipline. It has no canon unlike, for example, economics, that vigorously upholds and retells its history, including the lives of important economists, to justify its canonical present. Geography, in contrast, is always moving on to the next new thing. In the process, its past, and those who populated it, are erased, forgotten, become spectral, names without substance. In this chapter, I've tried to bring back to life, to re-corporealise, Dudley Stamp. Certainly, he was a man of his times, which partly accounts for, say, his casual sexism or willingness to serve Empire. He is more than this caricature, though. He had a clear intellectual geographical project crystallised in his belief in a sympathetic relationship between humans and their physical environment, and best manifest in rural land use. His applied geography was the means to realise that project and best done, he believed, by harnessing the power of the state. That is why he became a state insider. Of course, as a strategy it is always fraught. It is so easy to be co-opted, with one's fundamental beliefs diluted or washed away in favour of the state's interest. But at least during the immediate post-War period Stamp's strategy appeared to succeed. It is a reason to remember his life in Geography.

Notes

1 Going to his importance for British geography, the Institute of British Geographers organised a posthumous *Festschrift* for Stamp. He was a founding member and later President of that organisation. Chapters by his friends R. Ogilvie Buchanan (1968) and Michael Wise (1968) provided biographical details. Wise (1966, 1988a, 1988b, 2004) also wrote an obituary and two other biographical essays.

2 Stamp's 1945 inaugural lecture as Professor at LSE was titled 'Applied Geography'. He reused the title for perhaps his most popular book for a general lay audience published in 1960 by Penguin, marketed at an affordable 3s./6d. (UK 17.5p. or US 50¢ in 1960 prices).

3 '*The Times*' obituary put it delicately, saying Stamp "was not a profound master of theory" (Anon, 1966).

4 A 'First' (always capitalized) in the British degree marking system is the top grade. In the past in many English universities, the degree mark was determined by only a single set of exams – 'finals' – sat at the very end of the degree programme.

5 Studies of governmentality were pursued by the French philosopher Michel Foucault (1926–1984) during the later part of his life in the late 1970s and early 1980s. They represented his continuing interest in the exercise of power. Surveying and drawing maps was one technique for doing so often, although not exclusively, by the state. Once defined cartographically, activities within space, like land use, could be subject to state power, managed, controlled and disciplined.

6 Stamp's prolificness earned him the nickname "Deadly Stamp" (Beaver 1966, 388). Apart from contributing to geographical pedagogy through his books, both his maps and the plaster-of-paris physical terrain models he used to illustrate earth processes were creatively crafted – "veritable works of art" according to *The Scottish Geographer* – firing the geographical imagination of students (Tobin, Lorimer and Naylor 2021).

7 See Chapter 4 in which Stamp's antagonism to geography's quantitative revolution and to Peter Haggett's work in particular is discussed. While certainly not damning, Peter Haggett reputedly said that Stamp gave him "the pip". It was hard to damn someone, as *The Times* said of Stamp, whose "generosity of spirit ... [included] love of helping lame dogs over stiles" (Anon, 1966).

8 LSE economists at the time included Lionel Robbins (1935, 16) who wrote the classic definition of orthodox (neoclassical) economics as "the science which studies human behaviour as a relationship between ends and scarce means which have alternative uses". Orthodox economics, which continues to dominate the field, remained historically the most abstract version of the discipline, the farthest removed from Stamp's 'Face the world' aphorism.

9 CBE stands for Commander of the British Empire, an award given by the British monarch to an individual for public service.

10 The quotes about the opening of the London IGU Congress in this paragraph come from *The Western Mail* and were provided to the author by Mike Heffernan; email to the author, July 24th, 2023.

11 All of the quotes from Dudley Stamp in this paragraph come from a conversation about his work with Roy Plomley on the long-running BBC radio programme (and now podcast), *Desert Island Discs*, BBC Home Service, May 6th, 1963. A three-and-a-half minute extract from that programme can be heard at: https://www.bbc.co.uk/programmes/p009y5gf.

12 It took 22,000 6-inch-to-the-mile OS base maps to cover the entire countryside of Great Britain.

13 The colours chosen for rural land use categories resonated with interwar British landscape painting, from "chalky toned greens to light browns" (Rycroft and Cosgrove 1995, 95). In contrast, urban and industrial land – "carboniferous squalor" in Stamp's mind – were depicted in disturbingly vibrant scarlet (Rycroft and Cosgrove 1995, 96). Tobin (2021) likens Stamp's maps to "decorative wall hangings". Colour reproductions are found in Towles (2023).

14 Originally there were 14 members of the Scott Committee. One "died tragically in a black-out accident", one resigned, and of the remaining 12, only Stamp was a geographer. The idea that the Report rested on "geographical principles" surely came from him. There was one other academic on the committee, but they were the only member to vote against the proposal: "a Professor of Economics trained in the *laissez faire* school", i.e., a Lionel Robbins neoclassical economist (see fn. 8) (Stamp 1943, 16).

Study task

Variations on this question can be set on an unseen examination, an open-book examination or as a term paper assignment.

Dudley Stamp believed that Geography best proved itself as an applied discipline. Do you agree with him? Provide reasons and examples either for or against his view.

Starter reading list

Child, E. and Barnes, T. J. 2020. State and territory. In M. Domosh, M. Heffernan and C. W. J. Withers (Eds) *The Sage Companion to Historical Geography Volume 2*. London: SAGE, pp. 468–496.
Matless, D. 1992. Regional surveys and local knowledges: The geographical imagination in Britain, 1918–39. *Transactions of the Institute of British Geographers* 17, 464–480.
Rycroft, S. and Cosgrove, D. 1995. Mapping the modern nation: Dudley Stamp and the Land Utilisation Survey. *History Workshop Journal* 40(1), 91–105.
Stamp, L. D. 1948. *The Land of Britain: Its Use and Misuse*. London: Longmans Green.
Whitehead, M., Jones R. and Jones, M. 2007. *The Nature of the State: Excavating the Political Ecologies of the Modern State*. Oxford: Oxford University Press.
Wise, M. J. 1988. Laurence Dudley Stamp: 1898–1966. In T. W. Freeman (Ed.) *Geographers Biobibliographical Studies*. London: Mansell Publishing Limited, pp. 175–188.

References

Anon. 1966. Obituary: Professor Sir Dudley Stamp: Outstanding British Geographer. *The Times*, August, 10.
Anon. 1968. Sir Dudley Stamp. *Transactions of the Institute of British Geographers*, 44(1), 185–186.
Beaver, S. H. 1966. Obituaries Sir Dudley Stamp, C.B.E., D.Sc, D.Lit. An Appreciation by the President. *Geography* 51(4), 388–391.
Board, C. 1968. Land use surveys: principles and practice. In *Land Use and Resources: Studies in Applied Geography – A Memorial Volume to Sir Dudley Stamp*. Oxford: Alden Press, pp. 29–42.
Buchanan, R. O. 1968. The man and his work. In *Land Use and Resources: Studies in Applied Geography – A Memorial Volume to Sir Dudley Stamp*. Oxford: Alden Press, pp. 3–11.
Chibnall, J. 1968. Bibliography of publications by Sir Dudley Stamp. In *Land Use and Resources: Studies in Applied Geography – A Memorial Volume to Sir Dudley Stamp*. Oxford: Alden Press, pp. 71–81.
Hall, P. 2003. Geographers in the urban century. In R. J. Johnston and M. J. Williams, (Eds), *A Century of British Geography*. Oxford: Oxford University Press, pp. 545–562.
Kimble, G. H. T. 1967. Obituary: Laurence Dudley Stamp 1898–1966. *Geographical Review* 57(2), 246–249.
Kish, G. 1972. The participants. *International Geographical Union: Commission on History of Geographical Thought: Geography: Through a Century of International Congresses*. Paris: UNESCO, pp. 35–49.
Matless, D. 1992. Regional surveys and local knowledges: The geographical imagination in Britain, 1918–39. *Transactions of the Institute of British Geographers* 17, 464–480.
Matless, D. 1998. *Landscape and Englishness*. London: Reaktion.

Robbins L. 1935. *An Essay on the Nature and Significance of Economic Science*. London: MacMillan and Co.
Rycroft, S. and Cosgrove, D. 1995. Mapping the modern nation: Dudley Stamp and the Land Utilisation Survey. *History Workshop Journal* 40(1), 91–105.
Rycroft, S. and Cosgrove, D. 1999. Regional knowledge for pedagogy and planning: Dudley Stamp and the Land Utilisation Survey. In A. Buttimer, S. D. Brunn, and U. Wardenga, (Eds), *Text and Image: Social Construction and Regional Knowledge*. Leipzig: Institut für Länderkunde, pp. 122–128.
Southall, H., Brown, N. and Burton, N. 2003. *Digitising the Inter-War Land Use Survey of Great Britain: A Pilot Project*: Report to the Environmental Agency. https://researchportal.port.ac.uk/portal/en/publications/digitising-the-interwar-land-use-survey-of-great-britain-a-pilot-project(1df4d95b-8f2f-4b1c-a7ee-07dd85e3afcf).html
Stamp, L. D. 1931. The land utilisation survey of Britain. *Geography* 16, 44–51.
Stamp, L. D. 1932. Land utilisation survey. *Nature* 129, 709–711.
Stamp, L. D. 1943. The Scott Report. *The Geographical Journal* 101(1), 16–20.
Stamp, L. D. 1948. *The Land of Britain: Its Use and Misuse*. London: Longmans Green.
Stamp, L. D. 1953. *Our Underdeveloped World*. London: Faber and Faber.
Stamp, L. D. 1960a. *Applied Geography*. Harmondsworth: Penguin.
Stamp, L. D. 1960b. *Our Developing World*. London: Faber and Faber.
Tanner, M. F. 1987. Sir Dudley Stamp 1898–1966. *Planning History Bulletin* 9(2), 30–35.
Tobin, G. 2021. Be inspired: The relief models of Sir L. Dudley Stamp, November 22nd. https://www.rgs.org/geography/online-lectures/be-inspired-the-relief-models/
Tobin, G., Lorimer, H. and Naylor, S. 2021. The art of earth-building: Placing relief models in the culture of modern geography. Paper given at the one-day symposium, "40 Years of HPGRG – Looking back and looking forward", September 7th.
Towles, A. 2023. "A modern Domesday Book": Sir Dudley Stamp and the Land Utilisation Survey maps. https://blogs.lse.ac.uk/lsehistory/2023/06/09/a-modern-domesday-book/
Whitehead, M., Jones R. and Jones, M. 2007. *The Nature of the State: Excavating the Political Ecologies of the Modern State*. Oxford: Oxford University Press.
Wise, M. J. 1966. Obituary: Professor Sir Dudley Stamp. *The Geographical Journal* 132(4), 591–594.
Wise, M. J. 1988a. Sir Dudley Stamp: His life and times. In *Land Use and Resources: Studies in Applied Geography – A Memorial Volume to Sir Dudley Stamp*. Oxford: Alden Press, pp. 261–271.
Wise, M. J. 1988b. Laurence Dudley Stamp: 1898–1966. In T. W. Freeman, (Ed.), *Geographers Biobibliographical Studies*. London: Mansell Publishing Limited, pp. 175–188.
Wise, M. J. 2004. Stamp, Sir (Laurence) Dudley (1896–1966). *Oxford Dictionary of National Biography*. https://www.oxforddnb.com/display/10.1093/ref:odnb/9780198614128.001.0001/odnb-9780198614128-e-36238?rskey=i3L9Sr&result=1

20 Geographically empowering the marginalized

Bill Bunge, expeditions, and maps

Luke Bergmann and Trevor Barnes

Bill Bunge (1928–2013)

"We must make geography relevant. We do make geography Black." — Yvonne S. Colvert (1971, preface)

"Everyone but geographers knows what geographers do: they explore and map" — Bill Bunge (1983, 209)

Introduction

During the tumultuous long decade of the 1960s, America was shaken to its core.[1] There was no respite. To blink would be to have missed something momentous. Things fell apart; the centre no longer held. One of the constant provocations, and broadcast daily

DOI: 10.4324/9781003343240-23

in full colour into the living rooms of the nation, was the Vietnam War and its daunting, mounting statistics – over 58,000 US soldiers and around 3 million Vietnamese dead by the war's end. Protests were incessant and everywhere. On university campuses they sometimes turned tragic as at Kent State, Ohio, when in 1970, four unarmed undergraduate War protestors were shot dead by the US National Guard. Earlier, in 1963, 250,000 marched on Washington, DC to hear Martin Luther King in his "I had a dream" speech demand civil rights and the end of racism. Civil rights were a long-standing issue in America, but they became increasingly urgent and visible from the mid-to-late 1950s, associated with non-violent acts of civil disobedience including boycotts, sit-ins, and marches. Their purpose was to stop legalized racial discrimination and segregation, particularly in the US South, but also in large, industrialized US Northern cities where racial minorities faced harsh prejudice in employment and housing markets. The assassination in April 1968 of Martin Luther King in a Memphis motel incited a storm of inner-city riots, and mass arson in major US cities. It was the greatest civil unrest since the American Civil War, with disturbances in over 100 cities. Moreover, other inner cities had already gone up in flames earlier, most famously in 1965 Watts in Los Angeles, and in 1967 in a city especially germane to our chapter, Detroit (Barnes and Sheppard 2019).[2]

Shaken to his core too, especially by what happened in Detroit, but also politically energized especially around issues of civil rights and race, was the subject of this chapter, the American geographer, William Wheeler Bunge (1928–2013). Already politicized by the late 1950s when he was a graduate student at the University of Washington in Seattle – he joined the Communist Party USA – he was further impelled, revolutionized, and transformed by America's upheavals of the 1960s. So was his practice of Geography, which later resonated more widely within the discipline, pushing it to become more participatory, activist, embedded within grassroot issues of the community studied. It was geography at ground level not as a God's-eye view. Through intense study from the mid-1960s of his own one-square-mile urban neighbourhood of Fitzgerald in Detroit, Michigan, Bunge pioneered a new geographical approach that turned on the idea of the expedition – initially he dubbed his work and those that contributed to it, the "Detroit Geographical Expedition" (DGE). It involved neighbourhood engagement, participation, and organization. Later, with more resident involvement and with specific educational goals, it was formalized as the Detroit Expedition and Geographical Institute (DGEI). Overtly political, directed towards progressive change for predominantly Black residents, it often drew on pointed and novel forms of cartography, which Bunge called "oughtness maps" (Figure 20.1). The purpose of Bunge's work was always to make a better world for those who were oppressed, exploited, discriminated against and marginalized. It was never merely to describe. Riffing on Marx, Bunge said: "it is not the function of geographers to merely map the earth, but to change it."

One stream of Bunge's work in Detroit was published as the book, *Fitzgerald: Geography of a Revolution* (1971). Note the subtitle. It was not only about documenting and precipitating on-the-ground radical change within an urban neighbourhood, but also radicalizing the practice of the discipline of Geography, to "make it groan and protest" (Foucault 1980, 54), in order to effect that change. There never has been a book quite like it in the history of Anglo-American academic Geography. It was defiantly unrespectable: tortured, controversial, angry, partial, with leaps of logic and under-substantiated claims. It even looked different: oversized, more like a coffee-table book, full of striking black-and-white photos of the neighbourhood and its mostly Black residents, with confrontational maps and diagrams (Figure 20.2), and pull-out textboxes including about Bunge,

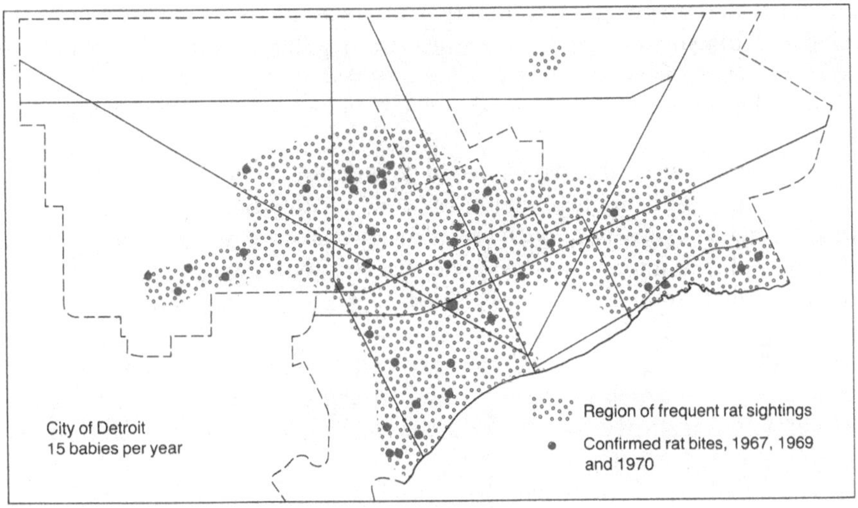

Map 3.16 Region of rat-bitten babies
If a middle-class suburbanite sees a rat in a slum ghetto he may feel disgust, but a slum mother feels horror. The rats – predacious carnivores – are looking for any food to eat, be it garbage or her infant's toes, fingers and nose. In New York City there are an estimated 4 million rats, almost one to every person. But in slum ghettos like Harlem there are 20 rats per baby, hunting, hunting.
Source: Detroit Department of Health, 1967, 1969 and 1970.

Figure 20.1 'Region of rat-bitten babies', which appears here as redrawn years later within Bunge's (1988, 167) *Nuclear War Atlas*, but originally drafted for the Detroit Geographical Expedition.

Figure 20.2 Bunge at a block club meeting in his Fitzgerald [neighborhood] home (Bunge 1971, 137).

his wife, and their families. It was the polar opposite of the then traditional academic geographical scholarship defined by neutrality, measured judgement, comprehensiveness, balance, precision, transparent logic, and step-by-step documentation. It was the former qualities not the latter that accounted for the book's creativity and lasting impact, however.

One of *Fitzgerald's* several ends was to do Geography differently. Bunge pressed geographers to become politically involved in the concerns of communities that they studied and to take up those issues as activists. Arm's-length study was unacceptable. There was an ethical and political imperative to engage directly. With typical bravado Bunge (1969, 5) exhorted, "armchair geographers of the world arise, you have nothing to lose except your middle-aged flab".

One critical form of geographical involvement, thought Bunge, was the expedition. Not expedition as it is usually conceived, however, that is, bound up with colonial expansion and extractivism. Instead, expedition should be "stood on its head" to become a geographical practice of collective liberation (Bunge 1969, 4). In the first of four *Field Notes* – occasional typed mimeographs written by Bunge and Fitzgerald community members[3] – Bunge (1969, 4) wrote:

> The purpose of the Expedition is to help the human species most directly. It is not a 'nice' geography or a status quo geography .... It is also a democratic, as opposed to an elitist expedition. Local people are to be incorporated as students and as professors. They are not to be further exploited. Their point of view is given first place.

This wasn't merely a matter of whose knowledge was prioritized, but who had control, intellectually and financially. As Bunge continued:

> Granted that unlike earlier expeditions, so many of which were exploitative, (indeed the word root of 'expedition' is the same as that of 'exploitation,') human explorations are 'contributive,' (resource contributing instead of resource taking,) how does this affect the character of the exploration? Totally. Priorities are totally reversed. The world of geography is stood on its head.... [This means a] reversal of the priorities of the definition of the research problem from campus-career oriented to community-people oriented.... [Further, there is] a second reversal: ... if the explorer is dependent on the community for his [sic] life, his [sic] power relations are exactly reversed from the normal academic ones.... And most importantly, the power of the expedition itself, who hires and fires, who writes checks and so forth must be in the hands of the people being explored, risky as that sounds to academics. The idea of big outside white money is absolutely ruinous.
>
> (Bunge 1969, 4, 6–7, emphasis original)

Within this manifesto of radical expeditions, Bunge invites us to be attuned to the challenges, contradictions, and limits of academics participating in these sorts of coalitions and activities. Expeditions are not utopian exercises seeking political purity.

More generally, Bunge believed geographers should live in the complicated worlds they help make. The Expedition, Bunge (1969, 4) wrote, "is democratic also in that if planning work results, and that is one of the main purposes of the Expedition, then the planners, the geographers, are expected to live in the mess they create". Further, the impetus to change comes not from above but from below: "The frustration of the planner is that his [sic] dreams...run into the complacency of the well-off who essentially wish to keep things the same. Therefore, active cooperation with the outs, especially the poor, allows full creativity and generates the necessary political power for [progressive change]" (Bunge 1969, 4). The reappropriated language of "expedition" was subsequently taken up and practised elsewhere: in Toronto and Vancouver, Canada, Sydney in Australia, and New York in the United States (Bunge and Bordessa 1975; Katz 2012; Blomley and McCann 2019). Furthermore, even though this use of 'expedition' may be less common, its vision and practice continue to resonate and echo through the field.

Bunge was certainly no saint, though. There is the basic issue of whether he should have been involved in community activism and struggle in Fitzgerald, given he was not Black, nor a long-time resident, nor socially or economically marginalized. Of course, he worked with residents and activists including the then Black teenage organizer, Gwendolyn Warren, whose influence over the DGEI and Bunge's own thinking was substantial.[4] But the relation between Bunge and Warren was often fraught. Warren later recalled Bunge as mostly unsaintly: insensitive, arrogant, heteronormative, and on occasion, homophobic, sexist and a bully (Warren et al. 2019). Bunge believed he was practising a liberatory racial politics arrayed against White hegemony. Warren and others thought, however, that Bunge's concepts and approaches to Blackness and Whiteness contained their own unacceptable hegemony. That said, Bunge ultimately knew that non-academics like Warren and others, and not him, must control the DGEI if it was to acquire deeper and broader relevance. The quote by Yvonne Colvard that opens this chapter, drawn from her writings in the DGEI's *Field Notes*, reflects that realization.

The remainder of the chapter is chronologically organized and reviews Bunge's rich but often chaotic life. He lost jobs and university appointments at an alarming rate, along the way alienating friends and colleagues. Bunge always believed he was hounded and tormented by various authorities, from Department of Geography Chairs to the CIA. Despite such turmoil, he completed four linked major intellectual geographical projects that were about trying to change the world for the better: the first on formal geographical theory, the basis of his doctoral dissertation at the University of Washington, Seattle, during the late 1950s (Bunge 1966); the second *Fitzgerald* and his involvement in the DGEI in Detroit during the second half of the 1960s; the third was the Toronto Geographical Expedition and its associated book with Ron Bordessa, *The Canadian Alternative*, written during the first half of the 1970s after he left the United States effectively as a political refugee to live in Ontario, Canada; and the last was the *Nuclear War Atlas* completed during the 1980s after he left Ontario for Quebec.

Bill Bunge: The early years

Born in 1928, Bunge grew up in La Crosse, Wisconsin, a river town on the east bank of the Mississippi. In a biographical text box from *Fitzgerald*, he wrote that "Bunge generations alternate between money-making and cause-serving" (Bunge 1971, 135). His father made money as a mortgage broker, while his son vigorously served many causes both political and intellectual.

After attending a private preparatory school, in 1946 Bunge enrolled as an undergraduate in Beloit College in southern Wisconsin. His family had long-standing connections – one of the College's theatres is still named after the family. Graduating in 1950, but with the outbreak of the Korean War the same year, Bunge was drafted into the US Fifth Army, stationed at Fort McCoy, Tomah, Wisconsin. Within the camp was the Chemical, Biological and Radiological Wartime School. Ironically, given Bunge's (1988) last book, the *Nuclear War Atlas* (1988), he was assigned to "teach atomic war" at the School (Bunge 1988, xi). While at Fort McCoy he also enrolled in a Geography extension course offered by the University of Wisconsin, Madison. Faculty from there would travel to various outposts of the university to give classes (Bunge 1988, xi). Fatefully, Bunge's class was taught, as he put it, by "the first professional geographer [I met] ... in my life, ... Richard Hartshorne" (Bunge 1988, xi). At that time, Hartshorne (1899–1991) was likely America's most famous living geographer. He was author of *The Nature of Geography* (1939), which meticulously explicated, rigorously justified, and genealogically fixed the discipline like no other English language volume before it. Arguing for regional geography, Hartshorne contended that regions were singular and exceptional entities that could be only described and not causally explained. In the language used to categorize different academic disciplines, Geography was an ideographic science, concerned with studying unique objects, and not a nomothetic science that drew on universal causal laws (e.g., like physics and its use of the law of gravity). "Regional geography ... is essentially a descriptive science concerned with the description and interpretation of unique cases", wrote Hartshorne (1939, 449). Bunge wrestled for decades with Hartshorne's designation of Geography as ideographic, but most immediately, he had to contend with Hartshorne as a teacher and graduate student supervisor.

Hartshorne was no charismatic teacher, but Bunge was riveted by his extension course lectures, wanting more. In 1954, Bunge entered the Geography Master's programme at the University of Wisconsin, Madison, moving to the PhD programme in 1956. That's when problems began. Partly because of the influence of non-geographers at Wisconsin, and partly because of his sympathy, later obsession, with Kurt Schaefer's 1953 paper that had attacked Hartshorne's ideographic position and offered instead the view of Geography as a universal law-seeking (nomothetic) science, Bunge fell out intellectually, and in most other ways too, with Hartshorne (Barnes 2016; Barnes and van Meeteren 2023). He miserably failed his preliminary doctoral exams at Wisconsin, with Hartshorne casting a negative vote.

By the time Bunge received that exam result, he was already planning to visit the University of Washington, Seattle. He knew that some in the Department there were carrying out the kind of "nascent mathematical geography" he wanted to pursue and allied with Schaefer's nomothetic geography (Bergmann and Morrill 2018, 291). After Bunge arrived in Seattle, he was "allowed to stay [on] .... for his doctorate" (Bergmann and Morrill 2018, 291). He joined the "space cadets", the name originally coined as a form of ridicule, but later worn as a badge of honour, to describe a group of mathematically minded graduate students who serendipitously all arrived to study Geography in Seattle between 1955 and 1958. Exclusively male, the group's members were talented, ambitious, energetic, intensely competitive, but frustrated by the woeful intellectual pedestrianism of descriptive regional geography. They wanted to overturn it, with Richard Hartshorne their principal enemy. Their end was a nomothetic scientific geography, spatial science.

Bunge's doctoral supervisor in Seattle was the young professor, Bill Garrison, who served as mentor, muse, paymaster, and protector for the larger group of space cadets (Bunge 1988, xi, instead called them with resonances of both armed resistance and revolution "Garrison's Raiders"). In a 1959 gloating letter to Hartshorne, Bunge described the culture of inquiry at Washington:

> [It] is an intellectual market place … Students barge into offices, sharpshoot during lectures, quarrel openly and sometimes personally (although seldom feud) with staff. Here men [sic] sink or swim on their intellectual vigor and this is one of those rare places in the world or time, a department enjoying a golden period. It is such an atmospheric thing that I am always afraid something will upset its rare metabolism.[5]

Bunge's 1960 PhD thesis was likely the most glittering of the achievements of that 'golden period'. Titled simply 'Theoretical Geography', the dissertation made both a philosophical and a practical argument, showcasing a set of techniques and mathematical constructs drawn especially from geometry in support of Geography as a science-based, nomothetic discipline. As he crowed to Hartshorne, "*We are achieving universality at the theoretical level*.... We are theoretical or fundamental geographers" (emphasis in the original).[6]

An amended version of the thesis was published in 1962 (followed by a further extended but only lightly revised second edition in 1966) by C. W. K. Gleerup Publishers of Lund, Sweden. Torsten Hägerstrand, likely Sweden's best-known geographer who met Bunge when he was a visiting professor in Seattle in 1959, arranged publication. From his prior encounters, Hägerstrand believed Bunge "had a stroke of genius".[7] William Bunge Sr., the father, and fortunately in the money-making half of the family, still needed to cut a cheque for $1800 to cover printing costs. He did. The presses rolled, and a Geography classic was born. With evangelical gusto it celebrated Geography as a science of spatial relations with its foundation the mathematics of space, geometry. It became as Kevin Cox (2001, 71) wrote, "perhaps the seminal text of the spatial-quantitative revolution". It changed the world of Geography.

The Detroit and DGEI decade

Following a disastrous one-year teaching post at the University of Iowa, Iowa City, 1960–1961, when Bunge accused the Chair of the Geography Department of muzzling him – unsurprisingly, Bunge's contract was not renewed – he was offered an Assistant Professor position at Wayne State University in Detroit. Deciding to live in Detroit's Fitzgerald neighbourhood, Bunge's (1969, 8) family "piled in the car and went up and down every street … like a milkman" until they found a house to rent. Historically working-class, Fitzgerald during the 1960s was transformed from a white to a Black neighbourhood. It was not where white middle-class professors typically lived, but Bunge was no typical white middle-class professor.

At least early on in his time at Wayne State he continued to focus on the intellectual agenda of *Theoretical Geography*. To that end, he helped to launch the Michigan Interuniversity Community of Mathematical Geographers (MICMOG). Meeting once a month in the back room of a tavern in Brighton, Michigan, the participants, several of them former space cadets, talked mathematical symbols and the axioms of geometry. The next year Bunge began working with another pioneer of scientific geography, Bill Warntz, to write a successor to *Theoretical Geography*, *Geography: The Innocent Science*. It continued in the earlier evangelical spirit. In June 1963, Bunge wrote to Warntz as they were finishing the prospectus for the book: "Was it in your ear that God whispered that you were supposed to reorganize the profession? I hope so. Maybe if I stand on a mountain the outline will come as a tablet with a burst of lightning" (quoted in Bunge 2001, 76). Presumably it didn't. And neither did the longer text that Bunge continued working on at least into the 1990s, apparently never completing.

What intervened, distracting Bunge, was "the Crime" (Bunge 1979, 170), and more generally, the turmoil of the 1960s with which this chapter began. The crimes for Bunge were racism and the flagrant abuses of civil rights. He threw himself into every protest going. "I went to Selma. I went to everything. Peace demonstrations in New York, in Washington, Civil Rights demonstrations in Jackson, Mississippi" (Bunge 1979, p. 170).[8] It also led him to become involved in local community politics, to start undertaking research differently. By 1964, he was already sending his Wayne State students to survey the poorest inner-city districts on Detroit's east side, as well as shadowing "new poverty program workers" in the same neighbourhoods (Bunge 1969, 9–10). In 1966 he had the first germ of the idea for *Fitzgerald*, although he imagined it initially as simply a "background study for Federal aid in the Great Society 'boomlet'" (Bunge, 1969, 10).[9] The 1967 Detroit riots (or "rebellion" as Bunge and others called it), which lasted over five days, July 23–28, and resulted in 43 deaths, with 1,000 injuries, more than 7,000 arrests, and close to 3,000 buildings burnt down, changed all that. As he put it, during "the smoke of [that 1967] revolution ... I lived in everyone's definition of freedom – No state ... [It] had been driven out ... I was free to think freely, so I did, I wrote a peace book, *Fitzgerald*" (Bunge 1988, xix).

1968 was famously a momentous year for America, but it also was for Bunge personally. First, Bunge was fired from Wayne State. The reasons are not entirely clear, but rumours include him swearing in class, complaints from parents about the material he was teaching, and sending students out to map in communities some viewed as dangerous. No longer having a job, however, gave him more time to engage in community organization and activism.

To those ends, also in 1968, Bunge along with Gwendolyn Warren, already a community activist in Fitzgerald, conceived at a meeting in Bunge's living room what would eventually become the Detroit Geographical Expedition and Institute (DGEI) (a community meeting in Bunge's home is shown in Figure 20.2). In part, the 'Institute' dimension of the DGEI came about as a response to Warren's rebuke to Bunge that he had not given back anything to the community. The DGEI took on both community education and research, both of which emphasized transformative types of, and roles for, cartography.

With Bunge acting as a go-between, DGEI set up free university courses for Detroit inner-city residents. Initially they were offered at the University of Michigan (held at Wayne State) for 40 students and then, in 1970, at Michigan State University for 400 students (Horvath 1971). The courses included subjects like urban planning and design, along with cartography, which would enable residents to use academic knowledge for their own purposes. For Heyman (2017, 2), this represented a "radical move" in pedagogy, away from academics as distant experts, to playing the role of civic educators "who help people become better equipped to organize and take control over their communities".

The research was carried out by the community for the community using, in part, techniques learned in university classes. As Knudson (2017, 3) puts it, Fitzgerald's "residents [became] contributors to knowledge rather than sources of knowledge to be appropriated". That research would take different forms, but it always involved being there, "talking, listening, arguing, befriending, and [even] making enemies", as Bunge (1969, 6) wrote. The result could be community flyers, advocacy reports, position papers, submissions to city council, or maps. Especially maps. But not maps as "a god's eye view of the world ... as austere and authentic" (Thatcher 2017, 3) but instead representations of the "lived, daily experience of a community ... made legible through mapping". Those maps would be, as Phil Cohen (2017) labelled them, "life maps, maps that tell stories in a way that brings data to human life".

Both DGEI's operations and research along with the community maps represented an approach to academic geography never tried before: activist, community based, politically uncompromising. Brought into conversation with historical geographic method, these approaches were central to the completion of *Fitzgerald*, which was published in 1971. The year before, with no job, and continuing to be harassed by the state – in 1970 the Chairman of the US House of Representatives Internal Security Committee publicly listed Bunge among the country's 65 "radicals and/or revolutionary speakers" on campus (Rosenbaum 1970) – he left Detroit, left America, exiling himself to Canada.

Canada and the Toronto Geographical Expedition

Initially, Bunge accepted a one-year visiting professor position at the University of Western Ontario in London ["Western"], Ontario (1970–1971). The post allowed Bunge both to retain a connection to Detroit, which was just over a two-hour drive away, and to begin to think how projects deploying the idea of the expedition might be extended. His undergraduate lectures about what he had done in Detroit captured the imagination of at least some of the students at Western. Tom Scanlan was galvanized: Bunge "made it sound like we were going to the North Pole, like it was an expedition".[10] Scanlan signed up. He didn't go to the North Pole, however, but to Toronto.

In Fall 1972, Bunge had been appointed to another one-year, but later extended to two years, visiting professorship, this one at the Department of Geography, York University, Toronto. It began well enough with Bunge starting a new project, the Toronto Geographical Expedition (TGE). There were overlaps with DGEI. It involved intense on-the-ground neighbourhood research, original cartography, collaborations with community groups, and meetings with City Hall politicians and bureaucrats. The focus was Christie Pits, an older, mixed-ethnic and mostly working-class inner-city neighbourhood northwest of downtown along the central east–west transportation corridor, Bloor Street. That choice was serendipitous. Bunge originally looked for "a black ghetto" to study, not believing Bryn Greer Wooten, his friend who drove him around Toronto scouting for sites, that there wasn't one. Having almost finished the tour, Wooten turned on to Bloor Street, and Bunge said, "OK, somewhere around here".[11]

Bunge set up his "base" as he called it for the TGE at a nearby rented house at 283 Brunswick Avenue (Figure 20.3). A "big house with big rooms", it was where Bunge held his York undergraduate class on urban expeditions, gave public lectures, held hundreds of TGE project meetings, planned research expeditions, and continuously wrote.[12] Scanlan observed, "Bunge just typed all the time.... [He had] to get his ideas down as he had them."[13] The Brunswick house was also where several of the TGE student leaders like Tom Scanlan lived and worked. This marks a major difference with DGEI. Much of the research for the TGE was carried out by unpaid university students from York, the University of Toronto, or, like Scanlan, from Western (if they lived at the Brunswick house, however, they paid no rent). These students, called "explorers", were responsible for specific substantive research topics like transportation or housing (Stephenson 1974, 98). Residents and interested members of the public, "folk geographers" in Bunge's lexicon, also contributed to the research but they were not involved in setting the agenda as in Detroit. Neither was there any equivalent to a Gwendolyn Warren,[14] nor were free university courses on offer for neighbourhood residents (although they could join Bunge's own class for free).

Figure 20.3 The outside of 283 Brunswick Avenue, Toronto, May, 1974. That day the house was used as a meeting place for the inauguration of the Union of Socialist Geographers, a relatively short-lived organization of radical geographers, 1974–1982. Many of North America's radical geographers were there including Bill Bunge, top row, centre left.

Photo by Colm Regan.

Theoretically, though, the TGE was more ambitious than the DGEI. That is best seen in the volume produced from it, *The Canadian Alternative*, published in 1975. While Bunge and Bordessa were listed as twin authors, Bordessa later said the book's "underlying way of viewing the world, understanding the world, interpreting the world" was Bunge's alone.[15] The *Canadian Alternative*, Bordessa said, was "a window ... on the complex panoramic panoply of issues and connections ... that existed in Bill [Bunge's] mind".[16] Not that it was easy to write them down. Bunge and Bordessa "spent day-after-day talking through things ..., typing up pieces of paper ..., [putting them] in piles". Then "at some point", Bordessa said, "we began to have enough of these conversations and enough piles of paper ... [and] gradually began ... to find coherence".[17]

The coherence is set out in Table 20.1, taken from *The Canadian Alternative*. The central theme is human survival. That's clear from the book's first sentence: "The monograph addresses the question of the survival of mankind [sic]" (Bunge and Bordessa 1975, i). In turn, survival is examined in *The Canadian Alternative* at five different geographical scales – neighbourhood, city, nation, continent, and planet – and in relation to three different types of environments – human ("Mankind" [sic]), technological ["Machinekind"], and ecological ["Nature"]. The detailed field work ("expeditions") of the TGE most directly bore on these three different environments at the neighbourhood and city scales, although on occasion it became relevant at broader scales. Holding everything together was a larger argument that tied the three different environments together. It was that capitalist technology [Machinekind] at every scale posed a destructive threat for the survival of both humans and the ecological environment. As such, Table 20.1 served as a 3x5 grid in which to insert for each cell an appropriate empirical

Table 20.1 The matrix of scales and spaces used to organize material in *The Canadian Alternative* (from Bunge and Bordessa 1985, ii)

Spaces/Scale	*I* *Christie Pits*	*II* *Toronto*	*III* *Canada*	*IV* *North America*	*V* *Earth*
A: Mankind	Hidden Landscapes	People Defined Regions	Urban Nationalism	1. Detroit Humanly Viewed 2. The Black Salvation of Canada	One World
B: Machine-kind	Traffic, Parks and High-Rises	Recreational Geography: Two Views	Federal Prejudice Against Big Cities	Toronto-Detroit: A Tale of Two Countries	Children and Machines
C: Nature	The Upper Gerrard Ravine	Restoration of the Urban Environment	The Natural City	The Indian Plan for North America	Survival

case that helped demonstrate and integrate the theory. Some of those cases at the narrower scales would come directly from field work, the urban expedition, other cases at broader scales from different sources.

TGE emphasized finding strategies to curb machinic destructive tendencies at the neighbourhood and city scales. For example, TGE provided City Hall planners and politicians with road maps of vehicle traffic flows compared to routes children walked to schools and parks within Christie Pits. Those along with maps of traffic accidents involving children convinced the City to reroute traffic within the neighbourhood using new one-way systems and vehicle-calming measures.

Two student "explorers" heavily involved in the Christie Pits mapping, Tom Scanlan, along with another Western undergraduate who enrolled in Bunge's Western class, Derek Stephenson, later drew on TGE to help advance a small visionary Toronto community organization, "Is Five". Involved in numerous urban and environmental campaigns, it pioneered the "blue box" for recycling that is now used extensively across North America. Is Five Press also published several books including *Neighbourhood Geography* (Scanlan 1978) that adapted and popularized the approach of the TGE and *The Canadian Alternative* for use in schools. Even Toronto's schoolchildren have in them a bit of Bunge.

More generally, TGE and the earlier DGEI along with the expeditions in other cities that they prompted were central to producing an increasingly participatory, community-based radical human geography, a geography of the street and less of the academy. In addition, the use of innovative cartography in both the DGEI and TGE was a catalyst for later radical counter-cartography (see Wood 2010). As the geographical expeditions showed, maps could make powerful arguments about injustice, inequity, and social marginalization, becoming guides to political action and alternative geographical imaginaries. Aspects of critical GIS, critical cartography, and qualitative GIS all draw from this tradition that originated with Bunge's expedition, and starkly different from the technocratic, detached approach that otherwise remains pervasive in GIS. What if Geography's cartography and GIS classes and practices had prioritized 'oughtness maps' as part of a movement not to merely describe what is but to realize what should be?

Quebec and the *Nuclear War Atlas*

Although a faculty colleague at York, Gerald Walker, thought that "Bill ... has been functioning very well here", [18] that was not the majority opinion of the Department. Bunge's contract was not renewed. Indeed, despite advertising within the Department, only three people turned up at Bunge's swan-song presentation about TGE. Not that it deterred Bunge, who stood "at the lectern ... [speaking] at full blast".[19] Yet, Bunge was not at a loss for ways to continue to engage the world, which included working as a Toronto taxi-driver. He told one interviewer that if you "drive a taxicab ... you really learn about the city. You will know more, if you have driven a cab, than a man sitting there with a factorial ecological printout" (Bunge 1976).

Bunge found other ways of organizing his activities. While in Detroit in 1968, Bunge had founded an organization, The Society for Human Exploration, "for the purposes of providing formal recognition of the need to reorient geographical research in directions of direct human concern" (Bunge 1969, Preface). After moving to Quebec from Ontario in the second half of the 1970s, he resumed work under that banner, except that it was also rendered in French, *Société Pour l'Exploration Humaine*. Specifically, he pursued the theme of human survival set out in *The Canadian Alternative*. But this was planetary human survival rather than at the level of the neighbourhood and worked out in his last widely published project, the *Nuclear War Atlas* (1988).

The nuclear bomb had been a menacing presence throughout Bunge's adult life. He was 17 when the American Air Force dropped atomic bombs on the Japanese cities of Hiroshima and Nagasaki; 22 when he taught "nuclear war" at Camp McCoy, Wisconsin; 34 during the Cuban missile crisis when over the course of 13 days thermonuclear war between the US and the Soviet Union appeared imminent; and 55 when President Ronald Reagan announced his "Star Wars" plan that had US nuclear missiles orbiting earth ready to destroy any weapon launched by the Soviets. For Bunge the nuclear bomb was the ultimate machine of destruction, the greatest threat to planetary human survival. He had been geographically theorizing nuclear war for decades, but the circumstances of the early 1980s compelled him to engage nuclear war and human survival directly.

In 1982, under the auspices of the Society, a large multi-folded pamphlet called the *Nuclear War Atlas* containing 28 maps showing the geography of nuclear Armageddon was printed. Bunge sold the pamphlets for $5 each, including door to door in select residential neighbourhoods in both Montreal and Toronto. Blackwell Publishers then went ahead with a full-length book version containing both a substantial text and the addition of another 29 maps (making 57 maps in total). It was not for the faint of heart, though. As Fraser MacDonald (2006, 58) wrote, "Bunge's *Atlas* maps out a post-apocalyptic terrain, without any attempt to 'soften the theme of unremitting and sense-numbing disaster.' Few geographers have offered their readers such a bleak cartography."

The bleakness had a purpose, though – in fact, the same purpose as the maps in *Fitzgerald*, the *Field Notes*, and *The Canadian Alternative*. As Bergmann and Morrill (2018, 292) write, they were "motivated by an urgency to have geography analyse and respond to human crises of the times, responses that relied on analysis with creative spatial methods". That creativity is found, for example, in the map titled "The closest neighbour ever – the Soviet Union and the United States" (Figure 20.4). It is not how world maps 'should be', with seemingly missing boundaries (in this case, those of the Soviet Union and the USA), but the borders of the superpowers in fact are there; indeed, they are everywhere as the legend and red-hash markings that cover the entire map

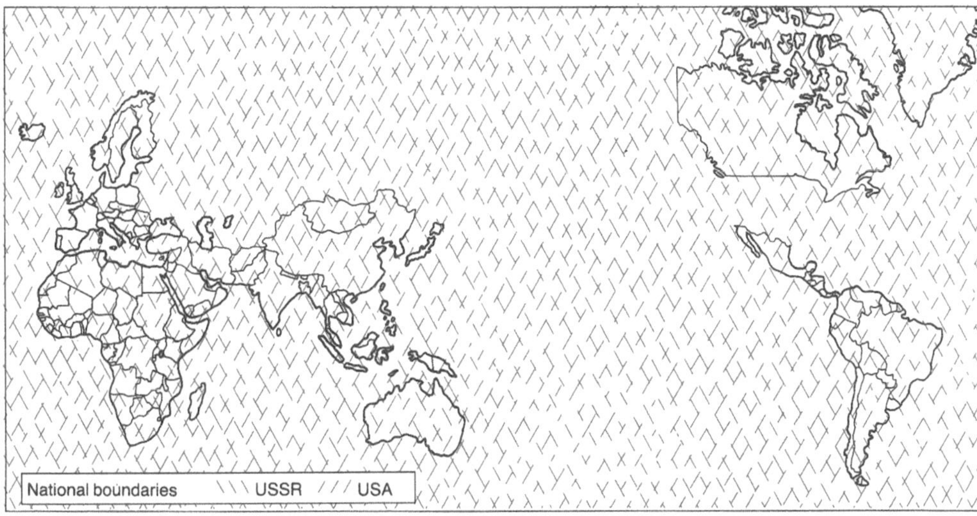

Figure 20.4 "The Closest Neighbours Ever: The Soviet Union and the United States"

Bunge [1988, 93], used with permission, with legend emphasized for clarity.

indicate. The point, as Timothy Barney (2015, 200) suggests, is to show how nuclear armed inter-continental ballistic missiles (ICBMs) have produced a world of "radical proximity", where the entire surface of the globe shares the same nuclear apocalyptic space. There is no hiding, no escaping outside – no outside. The survival of everyone on the planet is in peril. By rendering that possibility so starkly, Bunge hoped it would provoke enough people to be sufficiently engaged that such a possibility would never occur.

The last couple of decades of Bunge's life were lived at a greater distance from academic Geography in the sense that Bunge published little within the formal literature. He lived in Quebec, continuing to think, to engage politically, and to write extensively. Many of those who Bunge knew or wished to be in contact with received his letters, calls, and copies of items published by The Society for Human Exploration. He passed away in 2013 (Bergmann and Morrill 2018). The Society of Human Exploration, which Bunge wished to be an umbrella for a reinvigorated geographical approach to expeditions (see also Bunge 1983), went dormant with Bunge's own passing.

Conclusion

While earlier in his life Bunge struggled to choose between either an ideographic or a nomothetic geography, over time, he began to see their relation as in a productive tension rather than as mutual exclusivity. Geography could be both a mathematical science and a qualitative and engaged field subject (Bunge 1983). Likewise, physical geography and human geography might appear to be separate but trying to disentangle them was very hard. So, why try? Or again, while the local and the global appeared at opposite scalar poles, to study one meant studying the other. Consequently, over his life, Bunge accepted that there were not opposing Geographies but only one Geography. Much more important for him, and remaining constant throughout his life, were society's ingrained inequalities and the precariousness they brought to marginalized social groups, but which nonetheless could be ameliorated by political action using the tools and methods of Geography.

Partly caused by his personality, but also by his intellectual and political commitments, Bunge's life was frequently difficult, wrought by conflict, struggle, and failure. Few have been willing to pay the price Bunge paid for his life. But in our own moments of crisis and promise – ecological, social, political, intellectual – a remarkable number of geographers have found lessons and inspirations in learning from Bunge and the many other compelling people and currents that his life passed by, pushed apart, and pulled together.

Notes

1 The long decade of the 1960s is inexact and depends on the particular issue at hand, but generally it begins roughly in the mid-50s and continues some time to the early-to-mid-1970s. It marks an extended period of political, social, and cultural change, particularly in the US. The term was used by Frederic Jameson (1984) in his influential essay, "Periodizing the 60s."

2 While the Vietnam War and civil rights likely did most to unsettle America during the 1960s, there were other concurrent mass political movements aiming to rouse fundamental social change including Women's Liberation associated with second-wave feminism and environmentalism – the first Earth Day was in 1970, bringing on to the street 22 million people.

3 For discussion on the Field Notes including their origin as well as copies of the Field Notes themselves see: https://antipodeonline.org/2017/02/23/dgei-field-notes/.

4 Bunge credits his concern with children and their suffering in a built environment filled with dangerous machinery, such as cars, to Warren (Horvath 1974; Bunge and Bordessa 1975; Wood 2010, p. 170).

5 W. Bunge to R. Hartshorne, Tuesday [June 1, 1959]. American Geographical Society Library, University of Wisconsin, Milwaukee, Papers of Richard Hartshorne, "Correspondence William Bunge," File F, Box 194.

6 W. Bunge to R. Hartshorne, Tuesday [June 1, 1959]. American Geographical Society Library, University of Wisconsin, Milwaukee, Papers of Richard Hartshorne, "Correspondence William Bunge," File F, Box 194.

7 Torsten Hägerstand to Edward L Ullman, 22 June 1962, Posthumous Papers of Torsten Hägerstrand, Vol. 6, Lund University, Lund, Sweden.

8 One of the most well-known civil rights protests was the Selma to Montgomery march in 1965 in Alabama, a state with a deeply entrenched history of racist policies and deadly violence. The 54-mile march took three days, with civil rights demonstrators protected from white vigilante groups and local state enforcement officers by Federal National Guard troops.

9 The term Great Society was first used by President Lydon Johnson in a 1964 speech. It began to take legislative force from 1965 as a series of Federal programmes to eliminate poverty and racial injustice.

10 Tom Scanlan interviewed by Luke Bergmann, Toronto, ON, June, 2023.

11 The quotations in this paragraph come from Bryn Greer Wooten, interviewed by Trevor Barnes, Toronto, ON, July, 2021.

12 Bryn Greer Wooten, interviewed by Trevor Barnes, Toronto, ON, July, 2021.

13 Tom Scanlan interviewed by Luke Bergmann, Toronto, ON, June, 2023.

14 While there was no Gwendolyn Warren, Bunge did have Ron Bordessa, Chair of the Geography Department at Atkinson College, York University. Atkinson College's Department of Geography was separate from York's Faculty of Arts Geography Department that hired Bunge. Nonetheless, it was Bordessa, a long-time admirer of Bunge going back to his reading of *Theoretical Geography* he "happened upon" as an undergraduate, who mostly funded TGE. He paid for renting the Brunswick Avenue house, office supplies, photocopying, and perhaps most spectacular cost, the telephone. The quotation is from Ron Bordessa, interviewed by Trevor Barnes and Luke Bergmann, Victoria, BC, August, 2023.

15 Ron Bordessa interviewed by Trevor Barnes and Luke Bergmann, Victoria, BC, August, 2023.

16 Ron Bordessa, interviewed by Trevor Barnes and Luke Bergmann, Victoria, BC, August, 2023.

17 Ron Bordessa, interviewed by Trevor Barnes and Luke Bergmann, Victoria, BC, August, 2023.

18 Letter from Gerald Walker to Clark Akatiff, November 13, 1973, Series T1, Box 09, Folder 377, William Bunge Archive, Department of Geography, University of British Columbia.

19 John Warkentin interviewed by Trevor Barnes, Toronto, ON, July, 2021.

Study task

Variations on this question can be set on an unseen traditional examination, an open book examination or as a term paper assignment.

What sorts of "oughtness maps" can you work with others in your communities to create and to achieve what kind of end?

Starter reading list

Barnes, T. J. 2018. A marginal man and his central contributions: the creative spaces of William ("Wild Bill") Bunge and American geography. *Environment and Planning A* 50 (8), 1697–1715.

Bergmann, Luke and Richard Morrill. 2018. William Wheeler Bunge: Radical geographer (1928–2013). *Annals of the American Association of Geographers* 108 (1), 291–300.

Bunge, W. 1977. The first years of the Detroit Geographical Expedition Institute: A personal report. In R. Peet, Ed., *Radical geography: Alternative viewpoints on contemporary social issues.* Chicago: Maaroufa Press, pp. 31–39.

Bunge, W. 1979. Perspective on *Theoretical Geography. Annals of the Association of American Geographers* 69 (1), 169–174.

Merrifield, A. 1995. Situated knowledge through exploration: Reflections on Bunge's geographical expeditions. *Antipode* 27 (1), 49–70.

Warren, G. C., Katz, C. and Heynen, N. 2019. Myths, cults, memories, and revisions in radical geographic history: Revisiting the Detroit Geographical Expedition. In T. J. Barnes and E. Sheppard, Eds, *Spatial histories of radical geography: North America and beyond.* Chichester: John Wiley & Sons Ltd., pp. 59–85.

References

Barnes, T. J. 2016. The odd couple: Richard Hartshorne and William Bunge. *The Canadian Geographer* 60 (4), 459–465.

Barnes, T. J. and Sheppard, E. 2019. Introduction. In T. J. Barnes and E. Sheppard, Eds, *Spatial histories of radical geography: North America and beyond.* Chichester: John Wiley & Sons Ltd., pp. 1–36.

Barnes, T. J. and van Meeteren, M. 2023. The great debate in mid-twentieth-century American geography: Fred K. Schaefer vs. Richard Hartshorne. In S. Lovell, S. Coen and M. Rosenberg, Eds, *The Routledge handbook of methodologies in human geography.* Routledge: Milton Park and New York, pp. 9–23.

Barney, T. 2015. *Mapping the Cold War: Cartography and the framing of America's international power.* Chapel Hill, NC: UNC Press.

Bergmann, Luke and Richard Morrill. 2018. William Wheeler Bunge: Radical geographer (1928–2013). *Annals of the American Association of Geographers* 108 (1), 291–300.

Blomley, N. and McCann, E. 2019. "A necessary stop on the circuit": Radical geography at Simon Fraser University. In T. J. Barnes and E. Sheppard, Eds, *Spatial histories of radical geography: North America and beyond.* Chichester: John Wiley & Sons Ltd., pp. 117–147.

Bunge, W. 1966. *Theoretical geography.* 2nd ed. Lund, Sweden: C W K Gleerup.

Bunge, W. 1969. The first years of the Detroit Geographical Expedition Institute: A personal report. In R. J. Horvath and E. J. Vander Velde, Eds, *Field Notes, the Detroit geography. expedition, a series dedicated to the human exploration of our planet, discussion paper no. 1, The Detroit Geographical Expedition Institute.* East Lansing, MI: Michigan State University, pp. 1–30.

Bunge, W. 1971. *Fitzgerald: Geography of a revolution.* Cambridge, MA: Schenkman.

Bunge, W. 1973. The geography. *The Professional Geographer* 25 (4), 331–337.

Bunge, W. 1976. Geographers on film interview with William Bunge. *Library of Congress, Geographers on Film.* https://www.loc.gov/item/mbrs01844974/

Bunge, W. 1979. Perspective on *Theoretical Geography. Annals of the Association of American Geographers* 69 (1), 169–174.

Bunge, W. 1983. Geography is a field subject. *Area* 15 (3), 208–210.
Bunge, W. W. 1988. *Nuclear war atlas*. Oxford: Blackwell.
Bunge W. W. 2001. Classics in human geography revisited: Bunge, W., *Theoretical geography*. Author's response. *Progress in Human Geography* 25(1), 75–77.
Bunge, W. and Bordessa, R. 1975. *The Canadian alternative: Survival, expeditions and urban change*. York University Geographical Monographs, No. 2. Toronto: York University.
Cohen, P. 2017. William Bunge: Expeditionary geography: A commentary on six iconic maps for "Fitzgeralds's geography of a revolution". *Mapworks* # 2 Spring, 1–5. https://livingmaps.review/journal/index.php/LMR/article/download/59/99 (accessed 23 June 2018).
Colvert, Y. 1971. Preface. *The Geography of the children of Detroit. Field notes: A series dedicated to the human exploration of our planet, discussion paper no. 3*. Detroit, MI: Detroit Geographical Expedition and Institute.
Cox, K. R. 2001. Classics in human geography revisited: Bunge, W., *Theoretical geography*. Commentary 1. *Progress in Human Geography* 25 (1), 71–73.
Foucault, M. 1980. *Power/knowledge*. New York: Vantage.
Hartshorne, R. 1939. *The nature of geography: A critical survey of current thought in the light of the past*. Lancaster, PA: Association of American Geographers.
Heyman, R. 2017. Knowledge production, political action and pedagogy in Trumbull, Antipode Foundation Symposium. https://radicalantipode.files.wordpress.com/2017/02/dgei-field-notes_rich-heyman.pdf (accessed 23 June 2018).
Horvath, R. J. 1971. The "Detroit Geographical Expedition and Institute" experience. *Antipode* 3 (1), 73–85.
Horvath, R.J. 1974. Machine space. *Geographical Review* 64 (2), 167–188.
Jameson, F. 1984. Periodizing the 60s. *Social Text* 9(10), 178–209.
Katz, C. 2012. NYC Geographic Expedition and Institute: Liberation education for geographic inquiry. 2012–13 grant from Antipode Foundation.
Knudson, C. 2017. Detroit, geography, expedition, institute: Unpacking the history and structure of the DGEI. Antipode Foundation Symposium. https://radicalantipode.files.wordpress.com/2017/02/dgei-field-notes_chris-knudson.pdf (accessed 23 June 2018).
MacDonald, F. 2006. Geopolitics and "the vision thing": Regarding Britain and America's first nuclear missile. *Transactions of the Institute of British Geographers*, 31, 53–71.
Rosenbaum, D. E. 1970. Congressman lists 65 "radicals" despite injunction. *The New York Times*, October 15, 1970.
Scanlan, T. 1978. *Neighbourhood geography*. Toronto: Is Five Foundation Press.
Schaefer, F. K. 1953. Exceptionalism in geography: A methodological introduction. *Annals of the Association of American Geographers* 43(3), 226–249.
Stephenson, D. 1974. The Toronto Geographical Expedition. *Antipode* 6(2), 98–101.
Thatcher, J. 2017. Looking back to Detroit for a (counter-mapping) path forward. Antipode Foundation Symposium. https://radicalantipode.files.wordpress.com/2017/02/dgei-field-notes_jim-thatcher.pdf (accessed 23 June 2018).
Warren, G. C., Katz, C. and Heynen, N. 2019. Myths, cults, memories, and revisions in radical geographic history: Revisiting the Detroit Geographical Expedition. In T. J. Barnes and E. Sheppard, Eds, *Spatial histories of radical geography: North America and beyond*. Chichester: John Wiley & Sons Ltd., pp. 59–85.
Wood, D. 2010. *Rethinking the power of maps*. New York: Guilford Press.

21 Making other economies possible

Geographies of ethical action

Katherine Gibson

Katherine Gibson

Introduction

How did two Geography graduate students who met in 1977 and co-wrote their second semester paper about plant closures in Fitchburg, Massachusetts end up writing, two decades later, a book (*The End of Capitalism (As We Knew It): A Feminist Critique of Political Economy*, 1996) that was to launch a new way of thinking about the economy? Why did Julie Graham and Katherine Gibson become the combined authorial subject—J.K. Gibson-Graham? And how has their work inspired people to 'take back the economy as a site of ethical action'? This chapter tells the story of how a collective 'we'

DOI: 10.4324/9781003343240-24

developed the Diverse Economies Research Project—our way of making Geography matter. This way has allowed us and others to see all manner of current and future-possible geographies that would otherwise be unjustly ignored.

To start, I'll present some scenes that identify the kind of Geography that a diverse economy approach makes possible. Then I'll move back and forth in time to focus on key moments in the evolution of this research agenda. My aim is to show how the discipline of Geography provides an intellectual environment in which to imagine alternative ways of living, and more excitingly, to enact non-capitalist community economies.

Scene 1: Community meeting hall, Springfield Massachusetts, 2000

A group of community researchers have been enrolled in Julie Graham's action research project "Rethinking Economy: Envisioning Alternative Regional Futures", in the Pioneer Valley of Massachusetts. It's a motley crew of people who are usually seen as marginal to the economy and certainly never consulted when it comes to regional development strategies. There's a local artisan, high school students, intermittently or self-employed individuals, retired people, a high school dropout, a maintenance worker, social service providers who worked with drug users and recent immigrants, and people involved in volunteering. Each reports back on interviews they've conducted with people in their networks about what activities they did over the past week. A picture of economic diversity emerges—people have been gifting, volunteering, bartering, caring, donating, sharing, making, selling, cooking, tending, exchanging. When asked to identify which they associate with the capitalist economy on a grid of paid and unpaid activities, market, and non-market transactions, only four out of 50 are singled out (Figure 21.1).

From this exercise the image of a floating iceberg emerges (Figure 21.2). In the tip above the water line are activities associated with the capitalist economy (paid waged labour, commodity exchange, for-profit capitalist business) which are the subject of economics theory, and in the region below the waterline are the plethora of other activities associated with making a living (Community Economies Collective, 2001).

The iceberg image is a powerful communicator that has been used in countless conversations with people who rarely see themselves as 'acting subjects' in the economy. The effect of inventorying the underwater transactions and activities is to redefine what an

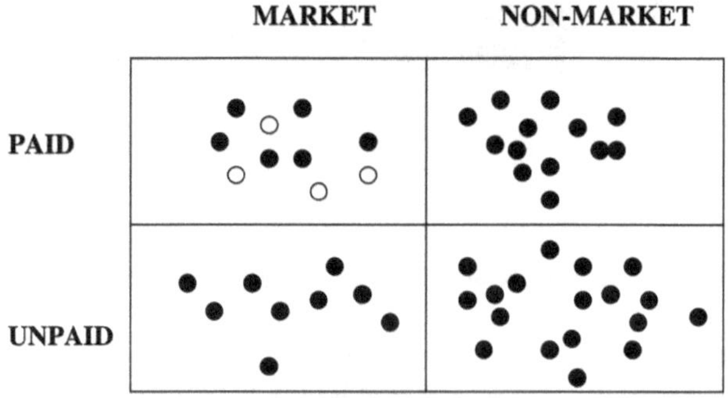

Figure 21.1 Distribution of activities from community researchers' interviews with the white dots representing capitalist activities.

Source: Gibson-Graham, 2002

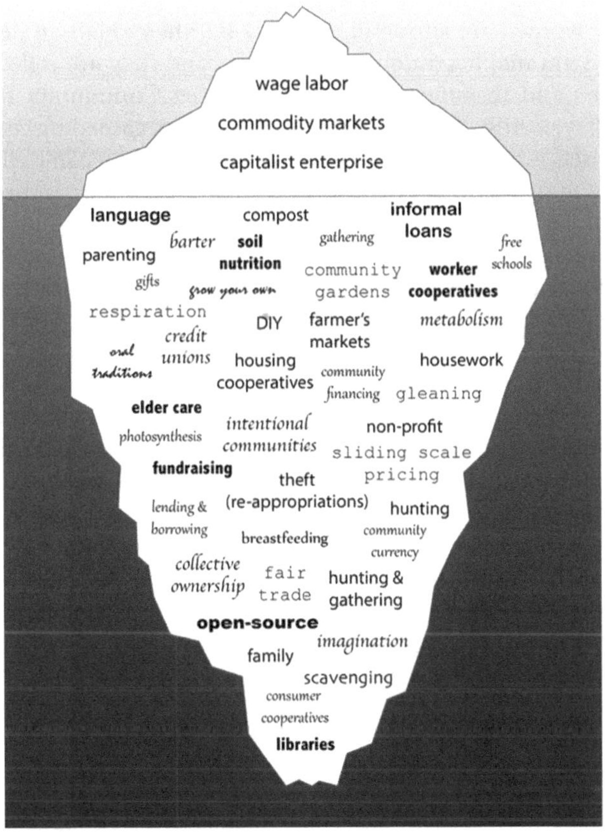

Figure 21.2 The diverse economies iceberg.

Source: Diverse Economies Iceberg by Community Economies Collective is licensed under a Creative Commons Attribution-ShareAlike 4.0 International License. https://www.communityeconomies.org/resources/diverse-economies-iceberg

economy *is*, that is, to represent it as always diverse.[1] In our action research approach this move acts as a prelude to brainstorming interventions that take back the economy as a site of ethical action around the wellbeing of people and planet.

The iceberg has evolved over time to include more than human transactions that support livelihoods—the exchanges with earth others for instance, such as soil nutrition, respiration and photosynthesis. It has also taken other forms more appropriate to context. For research projects in the Pacific and Southeast Asia a floating coconut is the template on which activities are recorded. The floating iceberg/coconut has turned up in art galleries where it is used to map out the diverse economy of the art world, in student papers where it has been used to identify the diverse economy of seed production, in farmer workshops where it serves to highlight the gendered nature of the division of labour and access to cash.

Place matters in the study of Geography. The iceberg image records the geography of economic difference, that is, the place-based specificity of economies. At the same time, it is a reminder that so much of what contributes to livelihoods on the ground is ignored by conventional economic theory (or economic geography).

Scene 2: Community meeting hall, Jagna, Bohol, Philippines 2009

A group of elderly women are sitting in the Laca Barangay Hall all decked out in white overalls and caps. I am making a documentary film and they are reflecting on becoming social entrepreneurs and forming the Laca Ginger Tea Community Enterprise (Gibson with Hill and Maclay, 2009). They talk about the benefits gained by coming together and mobilizing local assets—the plentiful ginger that farmers often can't sell, their traditional knowledge of how to grind ginger, mix it with sugar and water and reduce it over heat into a powder that can be stored and used to make a nourishing drink with medicinal properties, and how their association of women farmers who previously banded together to raise funds for local activities has fostered this new business. Maureen, the community researcher recruited from the community, outlines how *hungus*, a traditional practice of reciprocal labour exchange, has been modified to support this enterprise, so that if any of the women are unable to attend the weekly collective workday, another family member attends in her place. The women talk about now having some cash to purchase new spectacles or pay for medical tests. They talk of their pride in a community enterprise that younger women can participate in and maybe then be less inclined to leave the local area to work as domestic helpers in far-flung countries, and of their surprise in having found a voice in local affairs.

The film I'm making is documenting a range of social enterprises that have emerged from an action research project, 'Negotiating alternative economic strategies for regional development in Indonesia and the Philippines', and from the work of the NGO Unlad Kabayan Migrant Services Foundation with whom I have worked for many years (Gibson, Law and McKay, 2001). Our approach to regional development focuses on strengthening *community economies*, that is those that build on assets, service local needs and are less susceptible to the vagaries of global markets (see www.communitypartnering.info).

Some 20 years later, the Laca enterprise is still thriving. The inventory of traditional economic practices that we began to document in the Philippines (Gibson, Cahill and McKay, 2010) has been extended to encompass many more across Asia (Gibson et al., 2018).

Geography is not only the study of what is, but also what can be, so it's important that we attend to what our research is making more real and what kinds of possibility it fosters. In the case of this project in the Philippines, our action research increased interest in community-led social enterprise development and demonstrated the legitimate role for place-based, asset-building strategies in local economic policy agendas.

Scene 3: Project Report launch "Beyond Business as Usual: A 21st-Century Culture of Manufacturing in Australia" at Western Sydney University, 2020

Up on the stage are four federal and state politicians (Green, Labor, National and Independent). They are discussing the findings of research on Australian manufacturing that showcases the ways ten enterprises are conducting business with an eye to economic justice and environmental sustainability *as well as* financial viability (Gibson et al., 2019). Company representatives from the studied firms are also present. Over the two years of the project, they have met the research team many times, and each other at a one-day workshop. They had been surprised to find how much they have in common across their differences in terms of product (mattresses, milk, meat, furniture, carpet tiles, bespoke engineering equipment, high end clothing, recycled electrical goods and circuit boards) and enterprise organization (global multinational public companies, national-scale family business, producer cooperatives and social enterprises).

This project was prompted by a national debate about the future of manufacturing in Australia that took place as the last multinational car manufacturers closed their assembly plants (Gibson-Graham et al., 2019). We resisted the sensationalist claim that 'manufacturing in Australia is dead' and set out to explore the diversity of operating manufacturers, including not only capitalist businesses, but a range of other enterprises involved in the sector. We were keen to explore diversity in the tip of the iceberg.

We deliberately chose businesses performing aspects of "just sustainability" (Agyeman et al., 2003). Rather than keeping to Milton Friedman's narrow dictate that the only social responsibility of business is to increase its profits and satisfy the shareholders, the companies we sought out accepted that their social licence to operate involved responsibilities not only to workers and their families, but also to the next generation of owners, to communities in which firms are located, to the environment that provides inputs and suffers the detrimental effects of industrial production *and* to maintaining financial viability.

Economic geography has long favoured critical studies of the capitalist 'beast', priding itself on exposing the landscapes of unevenness, exploitation and plunder it thrives on. Important as this work is, it's just as important for geographers to work on finding or creating other landscapes of care, cooperation and repair, even within the belly of the beast.

Scene 4: Report back time in my Women's Studies undergraduate class on Feminist Ecologies and Economies at Harvard University, 2022

Between the weekly seminars students are doing the exercises from *Take Back the Economy: An Ethical Guide for Transforming our Communities* (Gibson-Graham, Cameron and Healy, 2013) to prepare for in-class discussions. *Take Back* was written for a popular audience and was designed to be read in groups, with people doing the exercises and confronting the ethical negotiations that community economies are built upon. We have reached the part of the course that is concerned with Economy, something that most of the students feel uncomfortable about or ill-prepared to discuss, this being a class in Women's Studies. They have been asked to research a non-capitalist enterprise in the Boston region and come to class prepared to present a 'people's account' of the business as laid out in *Take Back*. The account should attempt to distinguish necessary payments to workers, surplus generation and distributions of surplus.

Each student introduces the enterprise they have found with as much information as they can gather about organizational structure, employment, wages, financial distributions and ethos. The mood in the class is buoyant. The students are excited to share their discoveries. Surprisingly, among the 15 students there is no duplication. One presents the case of a landscaping business that operates as a social enterprise, another a cooperative restaurant that serves organically grown food. One student has tracked down the Harvard graduate who initiated TUNIQ, a North African fashion label and artisan's collective that operates according to Islamic values. Another reports on her visit to the Dudley Co-op, a housing cooperative that has operated on the Harvard campus since 1958.

The shift in affect since the beginning of the course is notable. At the outset there had been much discussion of the oppressive nature of the capitalist patriarchal system and the distributed power dynamics that made the students feel hopeless about engaging with economic activism. In this class there's a sense of hope and welcome surprise that others in the community are working in contexts where exploitation and profit making at all cost are not the norm. The exercise opened new options for involvement by the students who were keen to step outside the Harvard bubble and be more integrated into life in the city.

Geography is a subject that invites engagement with the world around us. As teachers we have an opportunity to direct students' attention to economic experiments with non-capitalist forms, including both newly emerging and longstanding cases, especially in a world where the options for transformation often appear to be closing down.

As I hope these vignettes have demonstrated, the Diverse Economies Project is both an academic research programme and a set of strategic practices enacted by researchers, artists, community organizations and policy makers. Importantly it involves a politicized form of knowledge production that attends to the power of *language* to shape the world, the affective disposition of *subjects* who can harness their power to make a difference in the world, and where *collective action* is directed to building 'other' worlds. The project has many descriptions ranging from the more scholarly—such as reading against the grain of capitalocentrism,[2] to the more practical and overtly political—such as imagining and enacting postcapitalist practices. In what follows I'll situate this relatively new sub-discipline of Geography within the trajectory of J.K. Gibson-Graham's 'life'. Of the many key moments that energized this project I'll focus on a few personal/political and contextual/place-specific moments, moving in time non-sequentially to build up the story.

The birth of J.K. Gibson-Graham in a humble dorm room at Rutgers University, New Brunswick, 1992

It's a steamy summer and Associate Professor Julie Graham (Department of Geology and Geography, University of Massachusetts Amherst USA) (Figure 21.3), and Dr Katherine Gibson (Senior Lecturer, Department of Geography and Environmental Science and Director of the Centre for Women's Studies, Monash University, Melbourne, Australia)

Figure 21.3 Katherine with Julie Graham (1945–2010).

are lounging around preparing a presentation for a workshop organized by the IGU Commission on Geographical Perspectives on Women. It is based on Gibson's field work with women in the remote mining communities of Central Queensland where union politics and feminist concerns about gender equity are clashing around the introduction of longer work weeks for open-cut coal miners, more overtime pay for men, and the destruction of a family-centred weekend (Gibson, 1992). We have used this material to apply our newly acquired tools of poststructuralist feminist theory to think together about decentred subjectivities and a feminist-inspired economic politics that challenges the centrality of the capitalist wage worker as a revolutionary subject. The paper is an amalgam of our two intellects, observations and expressions. It represents a major step forward in our thinking and it seems strange to attach one name to the paper, or even two in a sequence that will be read in terms of the binaries that structure academic valuation (the primary vs subordinate researcher/the theorist vs the data collector, etc.).

In a heady mood that we serious, tenured 'adults' could often slip into—a condition of hilarity that transported us back to our younger selves, those school girls on different continents, rolling in mirth around the playground or on the walk home from school with friends, unable to stop laughing, each new utterance pushing the merriment and the cackling to a higher scale—the idea emerged of blending our names together and attributing the paper to a single author. The amalgam *J.K. Gibson-Graham* had a certain appeal—it was gender neutral, it rolled off the tongue, it resonated obliquely with the liberal economist John Kenneth Galbraith (this was long before J.K. Rowling was on the scene). By adopting this moniker, we followed in the footsteps of many other women who were keen to be read as writers first, not as *women* writers (Gibson-Graham, 2022). And what started as a bit of fun, as with the joke or the Freudian slip, the birth of J.K. Gibson-Graham (JKGG from here on) signified much more than met the eye. Julie and I went on to publish under this name for the next 30 years, whenever the writing was part of our collective project, even after Julie's death in 2010.

The very idea of JKGG challenged the masculinism of Marxian political economy and radical economic geography, the disciplines we trained in as graduate students at Clark University. But our critique was less motivated by exposing the dominance of male voices in these fields, though with some notable exceptions like Doreen Massey and Susan Christopherson in economic geography, it was pretty much a male bastion. It was more directed to the style of thinking that had become the norm in economic analysis.

Emboldened by Jacques Derrida and Luce Irigaray and their work of deconstruction, we had become increasingly uncomfortable with the 'phallogocentrism' of political economy, that is, the way it privileged the masculine (the symbolic phallus) in understanding meaning and, related to this, the way in which 'woman' figured as an emptiness or lack against the positivity of master terms. In economic geography, phallogocentrism manifested in the superior agency and determining dynamics attributed to capitalist production (the engine of the economy), corporate decision makers and the 'productive' industrial working class, and the relegation of urban households, women's work and agrarian livelihoods to a peripheral status as the 'acted upon'. Our attraction to deconstruction extended out beyond 'the economy', our subject of study, to ourselves and the very meaning of authorship. JKGG offered a way to queer authorship—the master term 'author' was multiplied in such a way that no-one could try and differentiate the theoretician, the practical field worker, the writer, the editor, the conceptualizer, the describer, the American, the Australian.

Liberated by philosophers Richard Rorty's pragmatism and Judith Butler's concept of performativity, we had come to reject thought as an accurate reflection of reality—it was no less important and effective, but thinking and 'the real' was to be experimented with. JKGG allowed us to enact a playful feminist strategy of obfuscation, it became a way of entering economic debates that took themselves very seriously, a way of laughing off the ridicule offered by senior male scholars in economic geography who taunted us to walk through the wall of the conference room if we thought reality was all in the mind. And it was a way of getting invited to amazing (usually male-only) events under the guise of 'being a man'.

Above all, however, it was the freedom associated with becoming JKGG and the power it unleashed to step outside our comfort zones and begin to think big thoughts that fuelled the generation of the Diverse Economies Research Program. As JKGG, our various strengths were combined and amplified, and our doubts and weaknesses were diminished or dealt with. In 1992 we were both in tenured positions in the academy so, to those who queried, but what about getting tenure? (this was a bit before the academic obsession with performance indicators, when the question would have been, but what about your H-index?), we were able to thumb our noses. JKGG was ever a space where we explored making an academic life that was not only bearable but pleasurable, not always playing strictly by the rules, but making up new rules where a collective subject found a place to breath and create. Geography provided an accommodating home.

Geography is a discipline that is rather undisciplined. Perhaps this is because it is one of the few fields of study that straddles the science/social studies divide and finds it hard to police its practitioners. While there have been fierce paradigmatic battles between approaches within Geography, there is considerable tolerance for a diversity of approaches, and a history of experimentation with ways of 'doing' Geography. Think, for example, of the Detroit Geographical Expedition, or the work of the Bawaka Collective in Australia.

A geographer from Sydney and a literary critic from Detroit

The two people who met at graduate school and started researching regional restructuring in the deindustrializing Northeast of the USA brought very different training and experience to their collaboration and friendship. Before moving on to discuss the project we embarked upon, it is worth looking back into what shaped us as individual scholars and what we thus brought to the collective persona we were to inhabit for many decades.

Katherine

I had grown up with a geographic grounding. My grandmother had studied Geology at the University of Sydney at the turn of the 20th century. My mother was a keen lapidarist who took the family on field trips to distant sites to search for sapphires, gold, jasper and agate. The shared knowledge of these women meant that landscape and strata were presences in my childhood world. To this geo-spatial sensibility was added an interest in the inter-connectedness of the globe. Growing up in Australia, at the edge of the world for some, I always had an interest in what lay beyond. As a youngster, the USA loomed large. My parents embarked on a three-month trip to the USA in 1957—Dad to visit all the fish canning factories around the country to study the latest in production technology, Mum to visit her oldest friend who had travelled to Los Angeles as a war bride. As a preschooler left in the care of my maternal grandparents, I eagerly awaited the postcards of Disneyland, the Grand Canyon, New Orleans, the Statue of Liberty, the Maine coast. The world was large, interesting and within reach.

Geography was one of my favourite subjects at school and I pursued its study at Sydney University as part of a science degree, with a joint major in Geology and a minor in Social Theory. In my third year, I was introduced to the idea of epistemology by Professor John Rutherford, a quietly spoken unassuming agricultural geographer who asked searching questions, and my intellectual world exploded. That there were different ways to understand the world was mind-blowing. And that narrative played a big role in theory making was a shock. It gave me a way of stepping back from the realist representation of what 'science' did. In my geology studies the new field of plate tectonics was one big story that was organizing a plethora of empirical facts. My fellow students (and bar one, they all were fellows) were destined to find jobs in the burgeoning mining industry. I just couldn't see myself spinning stories to help the mining corporations locate more gold or uranium in the Australian outback. I came to a fork in the road of my studies and left geology behind to start on an Honours research project in human geography.

In 1975 one couldn't help but become politicized. Gough Whitlam, Australia's first Labor Prime Minister after 23 years of conservative rule, was summarily sacked after presiding over the most exciting three years of policy experimentation. The nation was awash with social democratic programmes—universal healthcare, free university education, an end to military conscription, legal aid, regional development. In the inner-city neighbourhood where I lived next to the university, the new federal government had stepped in to purchase run-down 19th-century housing owned by the Anglican Church with the aim of preserving the working-class community and maintaining low-income social housing close to the city. The Glebe Estate was also in the pathway of a controversial expressway that the conservative NSW government planned to build. This context provided me with a topic to research and I jumped into a 'humanistic study' of long-term residents in Glebe to understand what the 'community' of Glebe was all about. My sociocultural study was framed by phenomenology (a sense of place) and perception studies, but the stories I collected told of a stranded working-class generation whose children had moved to outer suburbs to follow jobs no longer located in industrial areas around Sydney Harbour. Gentrification by new white-collar professionals migrating in from the suburbs was on the rise. Purchase of the Glebe Estate ensured that market forces were kept at bay and low-income social housing was preserved in the area, while all around new residents renovated the run-down urban fabric, converting rooming houses back into single family mansions, revaluing the Victorian-era terraces (row houses) and replacing green painted concrete front yards with unwieldy, bark-dropping native plants and trees.

Upon finishing my thesis, I came under the spell of the powerful explanatory force of Marxist theory, as introduced into Geography by David Harvey's 1973 book *Social Justice in the City*.[3] For many years my Glebe study remained an embarrassment to me. Who needed to talk to people to understand the workings of capitalist investment and disinvestment in the city? I could have just looked at the rent books and financial records to see that capitalist restructuring was at the root of everything I had observed. Talking to old people was irrelevant. I needed to know more—so off I went to graduate school in the USA to deepen my education.

Julie

In 1964 Julie Graham arrived in Northampton, Massachusetts to study English Literature at the prestigious women-only Smith College. She had spent a childhood in Corning, New York State and in the suburbs of Detroit, Michigan—both major centres of manufacturing where her father had been employed as an engineer with the glass maker Corning before moving west to establish his own small machine-tool business serving the auto industry. We never made much of it, but it was a curious coincidence, given what we later chose as our focus of research, that both our fathers had been manufacturers riding the post-war long boom. Our comfortable childhoods were underpinned by the largesse this economic upturn distributed to managers and workers alike, that is, until it didn't. By the late 1970s corporate restructuring under pressure from globalization made life in the manufacturing sector much more precarious and both our fathers experienced this from the management side.

Smith College had a reputation for turning out well-educated ladies who would move seamlessly upon graduation into corporate wifehood. Of course, there were outliers, such as Betty Friedan, Sylvia Plath and Julie Child, to name some of the more well-known graduates. In the mid-1960s Northampton became a centre of second-wave feminism and Smith College was not immune to the questioning that was taking place about the role of women in society. Meanwhile Julie's education in English literature at Smith was strongly influenced by the New Criticism—a movement that studied texts as objects divorced from social context or intentions of the author. Close reading was to focus on formal structures internal to the text, and any attempt to contextualize the meaning of a poem or novel beyond the text itself was discouraged. Not surprisingly, this sat uncomfortably with a world which (once more) was opening to the realization that women writers, artists, musicians and theoreticians were absent from the curricula, or judged as definitively inferior. Julie graduated from Smith *magna cum laude* and was on track to begin a PhD at Yale, but the attractions of continuing to toe the New Criticism line were dimming and she declined the offer of a place. What remained with her from this education was an incredible training in close reading and precise expression. Also, a sensibility, that she was to draw on later, gained from one of her teachers (the renowned literary critic Helen Vendler) who encouraged her to have a tasting, rather than a judging, mind.

By the end of her degree Julie had fallen in love with the Connecticut River Valley where she lived for the rest of her life. She delighted in its rewilded woodlands, remaining farms and charming 18th-century hill towns with their white wooden structures, its network of 19th-century industrial mill towns where immense red-brick factories, now abandoned, cling to the banks of rapidly flowing streams. She immersed herself in feminist activism, teaching writing at Smith to make ends meet. As radical separatist feminism began to take over the local movement, she gravitated towards the environmental movement. In the early 1970s at the height of the back-to-the-land movement, Julie and friends scraped together funds to purchase land adjoining the watershed of the mighty Quabbin Reservoir, the water supply for Boston Massachusetts that had been created in the late 1930s by drowning a valley and its towns. They built their own houses, learning carpentry, bricklaying, well-digging and more, and unofficially took on the name of a road (Cooleyville) that no longer led anywhere.

Eager to learn more about the environment and politics, Julie enrolled in some courses in Geography at the University of Massachusetts Amherst, and after completing a few subjects applied to enter the PhD programme in the Graduate School of Geography at

Clark University. Driving her second-hand yellow VW Beetle with an "Eat the Rich" bumper sticker, she travelled some 50 miles from the home she had built to start a PhD programme that did not require an undergraduate degree in Geography. Meanwhile, I travelled halfway across the world to the programme I had selected because of its unique mix of non-mainstream schools of thought—Marxist geography and humanistic geography.

The Worcester that we encountered in 1977 when we rocked up to Clark University was decidedly in decline. The city's role at the centre of the American industrial revolution had long passed. Little did we know of its 'firsts'—the monkey wrench, the rickshaw, the typewriter, the liquid fuel rocket, shredded wheat, the birth control pill. What met us was a run-down CBD, abandoned factories and streets of three-story walk-up houses built for immigrant families, one on each floor, that could be rented by graduate students reasonably cheaply. We joined the small cohort of graduate students interested in bringing Marxist theory to bear on the study of Geography.

In the first years of graduate training, we read Marx's *Capital* Volumes 1, 2 and 3 and studied political economic analyses of 20th-century capitalist development. All around we saw evidence of uneven economic development. We were eager to understand what was driving these geographical relocations. Fortunately, Clark had just employed Don Shakow, a Marxian econometrician who had returned to academia keen to theorize and rethink political action, having spent five years post PhD establishing and running Seattle's first employee-run enterprise at the Pike Place Market. Under his tutelage we applied the Marxian labour theory of value to understanding capitalist crisis and theorized geographical restructuring as fuelled by the tendency for sectoral rates of profit to fall. We became adept at matrix algebra and manipulating National Income and Product Accounts using input-output analysis to calculate industry sector profit rates in value terms. Holy Moly! From bus trips with old age pensioners in Glebe and close readings of the poetry of Yates and Eliot, to complex mathematical transformations and statistical programmes! In a few short years Julie and I retrained as Marxian political economists and were engaged in the value debates at the highest level (Graham, Gibson, Horvath and Shakow, 1988; Gibson, Graham and Shakow 1989; Shakow, Graham and Gibson, 1992). We had acquired a new language—that of Marxian political economy—and, reading political theory on the side, we were keen to insert ourselves in debates about collective action.

Geography, like any school of knowledge production, is not politically neutral. At Clark where Antipode *was born, we asked: Could economic geography serve the interests of the exploited, marginalized and abandoned? And not the business sector eager to know where to locate the next supermarket to maximize returns?*

Transforming economic geography with growing dissatisfaction

In 1980, still as graduate students, we helped to organize an international workshop on regional restructuring in Luxemburg. Those gathered were researchers committed to 'changing the world, not only understanding it'. We all agreed that a newly emerging form of global capital was reshaping the world's economic geography and posing great challenges for regional wellbeing. State action informed by nascent neoliberal policy was becoming part of the story of restructuring. The Brits in the group were reeling from the first round of Thatcher's attack on working-class communities. En route to the workshop I had travelled via Newcastle-on-Tyne and attended a mass demonstration of retrenched steelworkers in Consett, County Durham, where the nationalized British Steel Corporation's plant and mines had been summarily closed. I remember sheltering from the cold in the concrete bunker on the edge of the football stadium listening to the enraged speeches by union leaders. I was filled with hope that our theoretical work on understanding the economics behind this closure would lead to more informed class action.

Drawing on Marxian theories of accumulation and crisis, geographers increasingly led the way in analyzing the changing production landscape. By the end of the 1980s it was no longer a career-threatening move to adopt a Marxist approach. Countless studies were being done on the geography of global production, from the maquiladoras of Mexico to the export-processing zones of Asia, from the union-busting sunbelt strategies of auto companies to the rise of a contract labour migration to supply the global construction industry. Each new volume of Peter Dicken's *Global Shift* (first edition in 1986, 7th in 2014) provided up-to-date studies of sectors undergoing change as a new geography of capital accumulation transformed space. But, we wondered, how was our scholarship contributing new strategies for addressing exploitation and marginalization?

The politics of radical analysis began to have a hollow ring. I remember a clarifying moment. I was in Australia post PhD seeking to bring my academic research and political activism together. I went to an address by Laurie Carmichael, a firebrand speaker, a communist and secretary of the Amalgamated Metal Workers' Union. His critical analysis of the global political economic context aligned with our findings and criticism of capitalist restructuring. His strategy for Australian workers was to be part of the new World Car order, to buckle under, accept the destruction of any kind of national protection, take up a role in the global production chain, ensure the workers received their piece of the pie in terms of wages and conditions, and let the changes rip. There was no alternative, it seemed.

I left that meeting shaken by the glimpse of rank self-interest on the part of relatively privileged male workers and disappointed by my lack of a comeback. The disconnect between what we were theorizing and what we could say to those affected by disinvestment and capital flight was great. It looked as if all my work, motivated by the thought that it would contribute to change for the better, had come to nothing. It was leading nowhere. If the revolution was to come, we had a long time to wait.

Paradigm shifts in Economic Geography have seen what was once 'radical' move to the mainstream and become unremarkable. How then, does the discipline remain creative and experimental? One way is by engaging with a wide range of thinkers 'outside the tent'.

Feminism (and anti-essentialism) to the rescue!

Growing dissatisfaction with studying the inventive geographies of capitalist restructuring pushed us towards new horizons of thought. At her first job post PhD at the University of Massachusetts Amherst Julie encountered the work of anti-essentialist Marxists Richard Wolff and Stephen Resnick in the nearby Economics Department. Their critique of economic determinism fell on fertile ears as our commitment to classical Marxist value analysis began to wane (Resnick and Wolff, 1987). It was their approach to class as the process of producing, appropriating and distributing surplus labour (as opposed to an identity, ranking or structural location) that sparked our interest in pursuing the study of economic difference. Not only could we theorize the coexistence of different forms of capitalist business, but we could also analyse the coexistence of capitalist, feudal, communal and self-appropriating class processes. Importantly, it was their theorization of an economic subject who could simultaneously occupy multiple class processes that excited our interest in the decentred self, a topic of much discussion within feminist theory (Gibson-Graham, Resnick and Wolff, 2000, 2001).

In Australia, a move to Monash University to take up the Directorship of the Centre for Women's Studies had brought me in touch with world-leading feminist theoreticians trained in poststructuralist feminist theory. Teaching in Women's Studies involved getting on top of a whole new field of feminist thought which at the time was rife with debate about essentialized identities, political standpoints, subjectivity and performativity. Together Julie and I extended our philosophical knowledge, reading complex texts of French theory and bringing our newly acquired concepts and approaches to bear on questions of economic identity and determination. We began to ponder the question "Why can feminists have revolution now, while Marxists have to wait?"

It seemed that under the influence of poststructuralist theory, the political subject and the social totality had been ripped apart and theorized as continually under construction, as decentred, as constituted by antagonisms, as discursively, as well as socially, constructed, so why was it, we wondered, that Capitalism/The Economy remained immune to reconceptualization? Something was standing in the way of action to build a different, non-capitalist world. We came to see that in fact, it was us and what we were doing as thinkers.

When Capitalism was represented as the model of Economy, placed at the pinnacle of the development ladder, seen as ever creative, able to bounce back reinvigorated after crisis, there could be no possibility of imagining and enacting 'other' economies. We needed to deconstruct Capitalism, to queer capitalist identity and to liberate the radical heterogeneity of a diverse economy.

A set of questions guided the thinking laid out in our 1996 book, *The End of Capitalism (As We Knew It): A Feminist Critique of Political Economy*. What if we stepped back to see our Marxian analyses as performing economic dominance? as producing a fantasy of wholeness that operated to obscure diversity and disunity? as strategically painting ourselves into a corner where no project of socialist or non-capitalist construction could possibly flourish? Furthermore, what if we were to recognize the volume and value of non-market transactions both within and without the household, or the many ways of producing, appropriating and distributing surplus labour in coexisting class processes?

Feminism gave us the impetus to smash (the discourse of) Capitalism and see it in a thousand pieces. We started to liberate economic diversity from the capitalocentric discourse within which it was imprisoned. It was in this wreckage that we glimpsed new openings for the Diverse Economies Research Project.

In conclusion: geographies of ethical action

At the outset, *The End of Capitalism (As We Knew It)* upset many who were committed to their masterful analyses of capitalist transformation (Leyshon, 2010), but a growing number of scholars, geographers included, were being drawn to study economic diversity, spurred on by the World Social Forum's movement of movements (Gibson-Graham, 2006). Fast forward a few more decades and *The Handbook of Diverse Economies* showcases the blossoming of scholarship since releasing the iron grip of capitalocentrism.

Three forms of politics have stoked this proliferation of work:

1 a politics of language to create a new language of the diverse economy,
2 a politics of the subject to identify a broad range of subjects of the economy other than and in addition to the capitalist and working-class subjects, and
3 a politics of collective action to build ethically oriented economies, what we called *community economies*.

The first is a politics of representation and language. Scene 1, where the chapter starts, is an exercise in challenging mainstream representations of the economy. In a theoretical move central to the method of feminist deconstruction, it does the work of detaching ourselves from the *A/not A* structuring of meaning, exposing the harm done when capitalist activities (the dominant A) at the tip of the iceberg are seen as the only legitimate ones that make up an economy. Documenting all the *not A/non-capitalist practices* in place is a first step towards capturing and naming economic diversity. Scene 2 gives one small example of this politics—a recognition and valuing of the term *hungus*, denoting a place-specific form of reciprocal labour exchange. The second step is to make the previously dominant term *A/capitalist practice* also different from itself, so we can identify a set of diverse economic practices on *both* sides of the divide. Scene 3 presents an example of doing just this—proliferating our understanding of what business is—producing a new language to capture the possibility of doing business with ethical commitments to well-being, equity, sustainability *and* collectively or privately appropriated profit. To capture this heterogeneity, the language of a diverse economy is ever evolving.

The second is a politics of the subject. As this chapter has tracked, our training as radical critical thinkers introduced us to the pleasures of mastery and the commanding powers of explanation. But pleasure was soon replaced by political paralysis fuelled by debilitating doubts about the viability of non-capitalist practices and experiments. We had become thinking subjects of capital, acted upon, denied any agency other than resistance. As we argued in *A Postcapitalist Politics*, we needed to work on ourselves to liberate our thinking from capitalocentrism. This is not to ignore the power of capitalist business or the financial sector that panders to it, or to underplay the devastating effects on people and communities of capitalist exploitation shored up by sexism, colonialism and white supremacy. It is to acknowledge these realities, but *also* to employ an anti-essentialist lens that counters economic determinism and liberates possibility, that queers the assumed alignments between capitalist identity, practices and hegemonic power. Through the process of documenting the incredible variety of non-capitalist practices and experiments, it is difficult not to be affected (Gibson-Graham and Roelvink, 2009). Scene 4 illustrates a small example of the resubjectivation that takes place as familiarity with economic difference grows. The next step along the road from inventory is enactment—that is, creating new economic experiments, negotiating ethical commitments to living well. As Scene 2 shows, it is through this process that community economies can be activated through the becoming of new subjects of a community economy committed to surviving well, distributing surplus, encountering others responsibly, caring for commons and safeguarding the interests of future generations.

The third politics is that of collective action. Here action is not only conceived of as resistance to capitalist exploitation and oppression. Our call to collective action focuses *also* on the positive project of building 'other' economies, ones that draw on those practices within the diverse economy that attempt to promote equity and wellbeing, racial justice, environmental stewardship, and distributive mechanisms that spread wealth and address inequality. Scenes 2, 3 and 4 all present glimpses of collective actions that are building alternative ways of living together.

The study of Geography can be everything to everyone, but what is its purpose? I started out being drawn to this discipline because it puts things in context, from rocks and skyscrapers to conflicts and perceptions. As I moved in a more scholarly direction, I wanted Geography to offer explanations for why things were as they were. Through human geography I was introduced to the powerful explanatory tools of Marx's theory of capitalist development. In this chapter I have given some glimpses into how J.K. Gibson-Graham came to reframe Geography as a place where we can imagine and enact different, less exploitative, less extractive ways of living, ones oriented towards an ethic of care for people and the planet.

Notes

1 See Gibson-Graham and Dombroski, 2020: 9-17 for the most evolved inventory of the Diverse Economy.

2 Capitalocentrism is the dominant economic discourse whereby the Economy is modelled on the capitalist economy. Within this discourse 'other' forms of economy are denied any independent identity or dynamism. They are only ever understood in reference to capitalism as either the same as, modelled upon, a precursor to, a substandard version of, a complement to, or existing within its orbit (Gibson-Graham, 1996: 6, 35).

3 And to me by Ron Horvath, an American radical social urban geographer who came to work at Sydney University.

Study task

This can be set as a term paper exercise or an exam question:

What are some of the ways that Geography can be practised so as to make diverse economies more visible and community economies more possible?

Starter reading list

Cameron, J. and S. Wright, 2014. 'Researching diverse food initiatives: From backyard and community gardens to international markets', *Local Environment: The International Journal of Justice and Sustainability* 19(1), 1–9.

Cameron, J. (with Craig Manhood and Jamie Pomfrett), 2011. 'Bodily learning for a (Climate) changing world: Registering differences through performative and collective research', *Local Environment*, 16(6), 493–508.

Cameron, J., and K. Gibson, 2020. 'Action research for diverse economies' in J.K. Gibson-Graham and K. Dombroski, eds, *The Handbook of Diverse Economies*, pp. 511–519. Cheltenham, UK: Edward Elgar Publishing.

Gibson-Graham, J.K., 2020. 'Reading for economic difference' in J.K. Gibson Graham and K. Dombroski, eds, *The Handbook of Diverse Economies*, pp.476–485. Cheltenham, UK: Edward Elgar Publishing.

Gibson-Graham, J.K., 2014. 'Rethinking the economy with thick description and weak theory', *Current Anthropology* 55 S9, Crisis, Value, and Hope: Rethinking the Economy, August: S147–S153.

Gibson-Graham, J.K. and G. Roelvink, 2009. 'An economic ethics for the Anthropocene', 40th Anniversary Issue of *Antipode: A Radical Journal of Geography* 41, 320–346.

Harris, E., 2009. 'Neoliberal subjectivities or a politics of the possible: Reading for difference in alternative food networks', *Area* 41(1), 55–63.

References

Agyeman, J., Bullard, R., Evans, B. (eds) (2003) *Just Sustainabilities: Development in an Unequal World.* Massachusetts, USA: MIT Press.

Community Economies Collective. (2001) 'Imagining and enacting noncapitalist futures', *Socialist Review*, 28 (3+4), pp. 93–153.

Dicken, P. (1986) *Global Shift.* SAGE Publications.

Gibson, K. (1992) '"Hewers of cake and drawers of tea": Women, industrial restructuring and class processes on the coalfields of central Queensland', *Rethinking Marxism* 5 (4), pp. 29–56.

Gibson, K., Cameron, J., Healy, S. and McNeill, J. (2019) *Beyond Business as Usual: A 21st Century Culture of Manufacturing in Australia.* Sydney, Australia: Institute for Culture and Society, Western Sydney University.

Gibson, K., Astuti, R., Carnegie, M., Chalernphon, A., Dombroski, K., Haryani, A.A.R., Hill, A., Kehi, B., Law, L., Lyne, I., McGregor, A., McKinnon, K., McWilliam, A., Miller, F., Ngin, C., Occeña-Gutierrez, D., Palmer, L., Placino, P., Rampengan, M., Than, W.L.L. (2018) 'Community economies in Monsoon Asia: Keywords and key reflections', *Asia Pacific Viewpoint*, 59(1), pp. 3–16.

Gibson, K., Cahill, A. and McKay, D. (2010) 'Rethinking the dynamics of rural transformation: Performing different development pathways in a Philippines municipality', *Transactions of the Institute of British Geographers*, 35 (2), pp. 237–255.

Gibson, K. with Hill, A. and Maclay, P. (2009) 'Building Social Enterprises: Strategies for Local Development in the Philippines', A 50minute DVD Human Geography, Research School of Pacific and Asian Studies, The Australian National University. https://www.youtube.com/watch?v=ut8Np7l24Yg&list=PL8nyPQdAvhRDDWG3gyephFXCpmJ3XLH18&index=2

Gibson, K., Law, L. and McKay, D. (2001) 'Beyond heroes and victims: Filipina contract migrants, economic activism and class transformations', *International Feminist Journal of Politics* 3 (3), pp. 365–386.

Gibson, K., Graham, J. and Shakow, D. (1989) 'Calculating economic indicators in value terms for the Australian economy and for individual industrial sectors, 1974–75 and 1978–79: Method and findings', *Journal of Australian Political Economy* 25, pp. 45–73.

Gibson-Graham, J.K. (1996) *The End of Capitalism (As We Knew It): A Feminist Critique of Political Economy with a New Introduction.* Oxford: Blackwell Publishers.

Gibson-Graham, J.K. (2002) 'Beyond global vs local: Economic politics outside the binary frame', in Herod, A. and Wright, M. (eds) *Geographies of Power: Placing Scale.* Oxford: Blackwell Publishers, pp. 25–60.

Gibson-Graham, J.K. (2006) *A Postcapitalist Politics.* Minneapolis: University of Minnesota Press.

Gibson-Graham, J.K. (2022) 'Collective authorship—"not mine and thine but ours"', *Left Cultures*, 1, p. xx.

Gibson-Graham, J.K. and Dombroski, K. (eds.) (2020) *The Handbook of Diverse Economies.* Cheltenham UK: Edward Elgar Publishing.

Gibson-Graham, J.K. and Roelvink, G. (2009) 'An economic ethics for the Anthropocene', 40th Anniversary Issue of *Antipode: A Radical Journal of Geography*, 41, pp. 320–346.

Gibson-Graham, J.K., Cameron, J., Healy, S. and McNeill, J. (2019) 'Roepke Lecture in Economic Geography—Economic Geography, manufacturing, and ethical action in the Anthropocene', *Economic Geography*, 95(1), pp. 1–21.

Gibson-Graham, J.K., Cameron, J. and Healy, S. (2013) *Take Back the Economy: An Ethical Guide for Transforming Our Communities.* Minneapolis: University of Minnesota Press.

Gibson-Graham, J.K., Resnick, S. and Wolff, R. (eds.) (2000) *Class and Its Others.* Minneapolis: University of Minnesota Press.

Gibson-Graham, J.K., Resnick, S. and Wolff, R. (eds.) (2001) *Re/presenting Class: Essays in Postmodern Marxism.* Durham, NC: Duke University Press.

Graham, J., Gibson, K., Horvath, R. and Shakow, D. (1988) 'Restructuring of U.S. manufacturing: the decline of monopoly capitalism', *Annals of the Association of American Geographers*, 78 (3), pp. 473–490.

Leyshon, A. (2010) 'Classics in human geography revisited: Gibson-Graham, J.K. 1996: *The End of Capitalism (As We Knew It): A Feminist Critique of Political Economy*. Oxford: Blackwell', Commentary 2, *Progress in Human Geography*, 34 (1), pp. 120–122.

Resnick, S. and Wolff, R. (1987) *Knowledge and Class: A Marxian Critique of Political Economy*. Chicago and London: University of Chicago Press.

Shakow, D., Graham, J. and Gibson, K. (1992) 'Industrial restructuring in the U.S. economy: A value analysis', *Capital and Class*, 47, pp. 35–66.

22 Speaking truth to power

Microplastics and the sewage scandal from the rivers of Manchester to Westminster

Jamie Woodward

Jamie Woodward (1965–)

Introduction

Plastic pollution is a wicked problem and one of the defining environmental issues of the 21st century. It has been described as a planetary threat because it impacts nearly every marine and freshwater ecosystem (Borrelle et al., 2020). More plastic has been produced in the last decade than in the previous century and huge volumes of plastic waste continue to end up in marine and freshwater environments (Williams and Rangel-Buitrago, 2022). This chapter stems from landmark research by a small team of physical geographers at the University of Manchester investigating microplastic contamination in the rivers of their city region. This research began as a curiosity-driven piece of river catchment science (Hurley, Woodward and Rothwell, 2017b, 2018). It quickly gained visibility beyond the academy to become part of a national debate about the state of Britain's rivers, the environmental impact of sewage-dumping private water companies, and years

DOI: 10.4324/9781003343240-25

of failure to monitor and regulate water company activities. This research has shown that wastewater from water company assets is the main supplier of microplastics to UK rivers (Woodward et al., 2021). Challenging the environmental performance of a £120 billion water industry is not for the faint-hearted.

The microplastics research at Manchester has generated widespread media coverage and led to a range of policy-shaping impacts at Westminster. It's taken my research in new directions, presented opportunities to publish in high-impact research journals, and led to engagement with new audiences to explain its significance. This work has featured in documentaries produced by independent film makers and the BBC.[1] The last five years have certainly deepened my knowledge of news media, environmental campaigning, the power of social media, corporate behaviour (including disinformation strategies) and the workings of national government policy development. This chapter sets out why our microplastics research matters and why a geographical perspective matters. It explains not only why the water industry attempted to undermine one of our key publications, but why the conclusions of that paper were taken up so readily by policy makers and campaigners seeking to tackle those who pollute our rivers and coastal waters.

Microplastics are plastic particles smaller than 5 mm. They are present in the environment in a wide range of shapes, sizes, and polymer types and have been found in all parts of the globe from the highest mountains to the depths of the ocean. They've even been detected in Antarctic snow (Aves et al., 2022). Many researchers are investigating the presence and impact of microplastics in fish (McGoran et al., 2017, 2018) and other aquatic fauna (Covernton et al., 2019). The potential effects of microplastics on humans is a rapidly growing area of research (Blackburn and Green, 2022). Primary microplastics (e.g., microbeads) are manufactured for specific applications, while secondary microplastics (e.g., microfibres and fragments) are breakdown products from the degradation of larger plastic items. When we began this work most research had focused on microplastics in the oceans, but awareness has grown rapidly around the threat that microplastic contamination poses to the health of freshwater ecosystems. As a result, the study of microplastics in rivers has grown exponentially over the last five years or so (e.g., Rowley et al., 2020). I'll also explain below how our work on the catchments around Manchester has helped to show that rivers are the major supplier of microplastics to the global ocean. This is a good demonstration of how local fieldwork matters and how it can become big-picture physical geography.

A geographical perspective

Before we can begin to tackle plastic pollution and clean up the environment, we need to understand more fully where microplastics are found, how they enter and move through the environment – in this case in river catchments from source to ocean – and where on that journey they accumulate to form an ecological hazard. We need better information on the geography of microplastic contamination, including data on the location of microplastic hotspots and the processes involved in hotspot formation and dispersal. We also need to make sure we are collecting the right kinds of samples. Should we sample river water, sediment, or the guts of fish, for example? Our research has shown that sampling microplastics in river water tells only a tiny part of the story and can miss much of the contamination in rivers; in many urban and sub-urban catchments, most of the microplastics are found stored on river channel beds and not in the water column (Woodward et al., 2021). The political significance of channel-bed microplastics will be explored later.

The first phase of our microplastics work formed part of a PhD project in the Department of Geography at Manchester.[2] Rachel Hurley's doctorate started out looking at heavy metal pollution in the rivers of Greater Manchester (Hurley, Rothwell and Woodward, 2017a). One evening in October 2015, after watching David Attenborough on TV talking about microplastics in the oceans, I began to think about where all the marine microplastics were coming from and how large cities like Manchester – with lots of people and industry – must be important sources of all kinds of microplastics. We'd already collected 40 samples of fine bed sediment (in spring and early summer 2015) from ten rivers around Manchester for Rachel's PhD project. Were those samples full of microplastics? I suggested to Rachel that she find a method to extract microplastics from those bed sediments. In just a few weeks we had our first microplastics data and Rachel's PhD – early in her second year – took on a whole new dimension.

A distinctive feature of our ongoing work is a focus on the river *bed* environment (Figure 22.1). River beds are important habitats and a key source of nutrition for the entire aquatic ecosystem. Countless creatures live, feed, and reproduce in this environment; fish and bird species, for example, forage on organic detritus and invertebrates by sifting through fine bed sediments. If the river-bed habitat is contaminated – whether with heavy metals, microplastics, or other potentially harmful substances – these creatures are vulnerable to the ingestion of contaminants. In the case of fish, microplastic particles and other pollutants can enter the aquatic food chain through direct ingestion during feeding and drinking or via preying on river-bed invertebrates that have previously ingested microplastics. Ecologists call this latter pathway *trophic transfer*. Researchers have reported a range of negative impacts associated with the ingestion of microplastics by fish (Lei et al., 2018). Adverse effects can include physical damage to the gastro-intestinal tract and the leaching of polymers and chemical additives into the fish as the microplastics degrade. The impacts depend on several factors, including how long the microplastic particles are retained, the level of accumulation, and the extent of

Figure 22.1 Use of an aluminium cylinder to collect samples of fine bed sediment for microplastic analysis. We isolate a portion of the channel bed and disturb the gravels with a metal trowel. Fine sediment and microplastics are brought into suspension. Bulk samples (15–20 litres) of turbid river water are returned to the laboratory where the microplastics are recovered, identified, and counted.

penetration into remote organs and tissues. Microplastic particles can also be a carrier for pathogens (Viršek et al., 2017). We have recently reviewed the multiple impacts of microplastic ingestion on freshwater fish (Sulaiman, Woodward and Shiels, 2023).

Mapping the geography of microplastic contamination on river beds at the catchment scale is therefore important – not least because the quality of this habitat affects the ecological status of the entire riverine ecosystem, but also because mapping helps us to estimate the ecological risk from those microplastics and provides clues about the sources of those microplastics and how we might shut them down. A geographical perspective is absolutely fundamental to understanding and tackling this problem.

Mapping microplastics in space and time

Figure 22.2 shows our first set of microplastic data from ten Greater Manchester rivers across a catchment area of some 1500 km.[2] Manchester city centre is where the rivers Irwell, Irk, and Medlock converge. Rachel Hurley quickly mastered new laboratory methods to produce these data. This map shows the geography of microplastic contamination in spring/early summer 2015 at 40 locations across a range of land uses: rural headwater streams, lowland agricultural landscapes, Manchester's suburbs, and the dense urban city centre. The microplastics recovered from these river-bed sites included a mix of primary and secondary microplastics. They were grouped into four categories: fragments, microbeads, fibres, and other. The last category was quantitively unimportant (~1%) in this survey and included a mix of microplastic types including glitter, foam, and plastic film. The vast majority of the channel-bed microplastics shown here are smaller than 1 mm across, falling within the range 60 to 500 µm, equivalent in size to fine and medium sand. In this first survey, microplastics were detected at 39 out of 40 sites; the

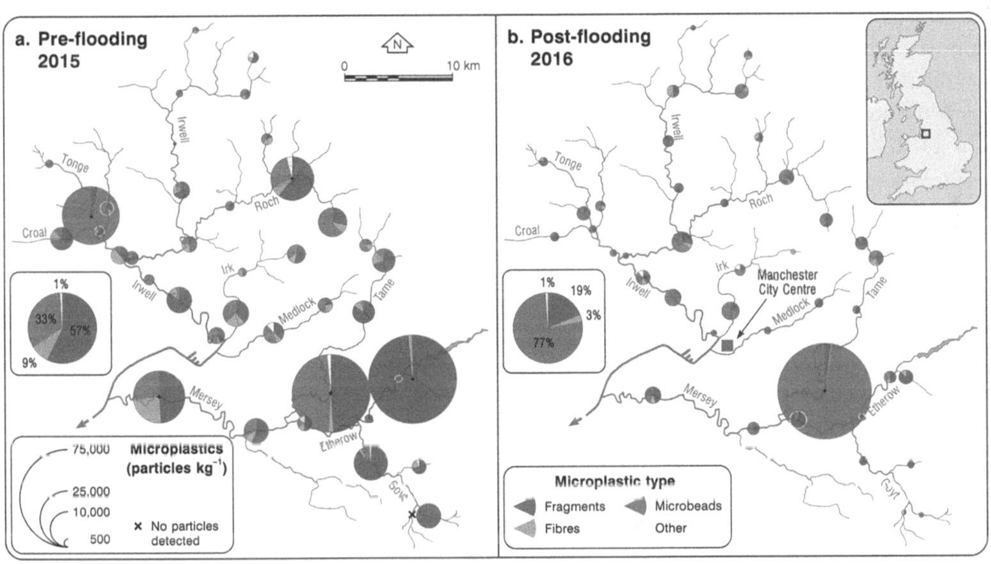

Figure 22.2 Microplastic concentrations and assemblages on river channel beds across Greater Manchester in 2015 (a) and 2016 (b) before and after a sustained period of winter flooding (after Hurley, Woodward and Rothwell, 2018). Note that microplastics were recorded at all 40 sites in the 2016 survey.

rural headwaters in all catchments show low levels of microplastic contamination (Figure 22.2a). The largest circles on the map show the highest concentrations. In the 2015 survey, five sites were labelled as microplastic contamination *hotspots* where the channel bed concentrations exceeded 15,000 microplastic particles per kg of fine bed sediment. This equates to >15 microplastic particles in every gram of fine bed sediment. Each hotspot displayed a different microplastic assemblage.

This study was the first to show that microplastic contamination was widespread on UK river beds, from rural headwater catchments to heavily urbanised reaches in large centres of population. It's still one of the largest surveys of microplastic contamination in freshwaters. The spatial variability in both concentrations and assemblages is notable. Some sites are dominated by microbeads or fragments, for example; others record an even mix of the three main microplastic types. While the rural headwaters show low concentrations of microplastics, there are no consistent downstream trends in the urban rivers. We knew immediately that these data were important, but they also raised lots of questions. An interesting map wasn't enough. As the Christmas break approached in late 2015, we started to delve into the literature to establish the wider significance of these microplastic concentrations, but there were few studies of microplastics in rivers at that time and it was often difficult to compare data when different field methods were employed. We also wanted to better understand the processes involved in microplastic accumulation and transport. We had no idea, for example, how stable the microplastic hotspots were, or how much time was needed for them to accumulate.

The winter of 2015/2016 saw a period of sustained rainfall and flooding across many parts of the UK with several named storms moving across the country. In December 2015 Storm Eva produced record flooding in the North of England. On Boxing Day (26 December) the River Irwell recorded the largest flood in 70 years of monitoring and Manchester city centre was placed on flood alert. In contrast, south of the city the Upper Mersey and its tributaries saw relatively modest flooding. This provided an excellent opportunity to test some ideas about microplastic storage and hotspot formation. When we returned to campus after the Christmas break we decided to resample all 40 sites to see how much of the microplastic load had been flushed from the beds of the ten rivers. The post-flooding dataset for spring and early summer 2016 is shown in Figure 22.2b. Note how the circles have decreased in size at most sites showing a catchment-wide fall in microplastic concentration. A key outcome of this research was the demonstration that the microplastic load on these river beds could be flushed away during flood events. The rivers clean themselves. This process had not been previously documented.

The microplastic hotspots recorded in the 2015 survey were all removed during the flooding of winter 2015/2016, when 28 sites recorded a significant fall in microplastic concentrations. Eighteen of those sites saw their microplastic load decrease by an order of magnitude. We also observed the removal and formation of a hotspot on the River Tame that became dominated (98%) by microbeads with a concentration of some 72,400 microplastic particles per kg of fine bed sediment (Figure 22.2b). This became known as the Denton hotspot – the most heavily contaminated site known on any river at that time.[3] We didn't think this was the most important feature of our study, but it attracted a great deal of media attention.

The global context

Next we worked out how to *quantitively* estimate microplastic loss from the river beds across the Upper Mersey and Irwell channel network. We used channel-bed area and microplastic concentrations to estimate the entire microplastic load stored on these river beds in the catchment before and after the flooding. We subtracted the 2016 post-flood total from the 2015 pre-flood total to estimate the quantity of microplastics lost from channel beds across the entire network. This was the first empirical determination of catchment-wide microplastic flux.[4] The winter flooding had exported approximately 70% of the microplastic load stored on the river beds (equivalent to 0.85 ± 0.27 tonnes or 43 ± 14 billion microplastic particles). We then stuck our necks out and put this figure in global context. A recent paper had estimated the total floating microplastic burden in the global ocean at 4.85 trillion particles (van Sebille et al., 2015). Our flux estimate of 43 billion was just over 1% of that total. In a single winter flood season the small rivers around Manchester could not supply 1% of the microplastic load floating in the world's oceans. One of these big numbers must be wrong, but we were confident that our microplastic flux estimate was robust. All of this meant that the total microplastic load in the world's oceans must be far higher than previously estimated and that was an important conclusion of our 2018 paper.[5] We were relieved that this finding was welcomed by Eric van Sebille when he commented on our work in *The Guardian*: "This work supports a much higher estimate of global microplastic pollution in the oceans. I'm not surprised by that conclusion."[6] We published this work in *Nature Geoscience* in March 2018 (Hurley, Woodward and Rothwell, 2018).

New questions

The first phase of our research demonstrated that the microplastic contamination of river beds could have a rapid rate of turnover. In other words, the residence time of the microplastic bed load is low and largely controlled by flood frequency. While the Irwell catchment saw extreme flooding on Boxing Day 2015, the Upper Mersey catchments (Tame, Etherow, and Goyt rivers) experienced flood events of more typical magnitude, yet still saw widespread microplastic flushing from channel beds. This was significant because it revealed that even moderate flooding could flush microplastic loads from channel beds. Since the flow regimes of these rivers are regularly punctuated by flood events capable of flushing microplastics from channel beds, this observation raised new and important questions. Where are the microplastics coming from and under what conditions do microplastics enter the river channel so that they can accumulate on the bed in high concentrations? Many polymers are less than or close to the density of water so they should be easily washed downstream.

If the microplastic load stored on these river beds is transient, this contaminant must be regularly replenished, otherwise microplastic hotspots would not form. It is well known from studies in many cities around the world that urban wastewater can contain high concentrations of microplastics (Murphy et al., 2016; Gündoğdu et al., 2018; Blair et al., 2019; Long et al., 2019). These microplastics come from a range of sources including domestic wastewater, industrial effluents, and urban runoff. There is a large literature on microplastics in wastewater and on the effectiveness of wastewater treatment works (WWTW) at removing those microplastics (see Iyare, Ouki, and Bond 2020 for a review). Microplastics are suspended solids and studies have shown that conventional tertiary treatment can remove between 95 and 99% of the microplastic load in wastewater (Carr et al., 2016; Talvitie et al., 2017; Simon et al., 2018). WWTWs in the UK are only supposed to release untreated wastewater to rivers under 'exceptional circumstances' during very heavy rainfall when rivers are in flood. If this is the case, how do river beds become heavily contaminated with microplastics when we have shown that even moderate floods will flush channel-bed microplastics away? This puzzled us for a while, so we designed a field experiment in summer 2019 on the River Tame to better understand the conditions that promote microplastic accumulation. We sampled the river at higher spatial resolution than before immediately after a very large summer flood[7] so that we could observe the geography and rate of microplastic hotspot formation. After a sustained period of low flows we collected channel bed and floodplain sediment and water samples at 14 sites. We targeted several point sources of wastewater including WWTWs and CSOs (Figure 22.3). Figure 22.4 shows the data from these sites along the River Tame.

We found that treated wastewater contained only very small concentrations of microplastics, predominantly fibres (Figure 22.4a and d). We were also able to track and trace microplastic sources by comparing the microplastic assemblages in wastewater from outfalls with the assemblages on the channel bed immediately downstream. The Tame study found that microplastic hotspots were much more common in urban rivers than we had previously documented (Figure 22.4c). Most importantly we showed that these hotspots had formed under low flow conditions within weeks of the big July flood that had effectively 'reset' the channel microplastic load. These hotspots had distinctive microplastic assemblages that mirrored the microplastic content of the nearest upstream point source of wastewater. This was typically a wastewater treatment works outfall or a combined sewer overflow (Figure 22.3). There was no evidence of microplastic mixing between sites. A key conclusion of this work was that hotspots of microplastic contamination could only form on channel beds if *untreated* or inadequately treated wastewater was discharged to rivers when the rivers were at low flow and therefore unable to transport those microplastics downstream (Woodward et al., 2021). This work very quickly became political.

Figure 22.3 Collecting a sample of untreated wastewater and urban runoff from a combined sewer overflow (CSO) outlet that empties into the River Tame in Greater Manchester in summer 2019. The grill is blocked with wet wipes and sanitary products from the sewerage network after a large flood in late July.

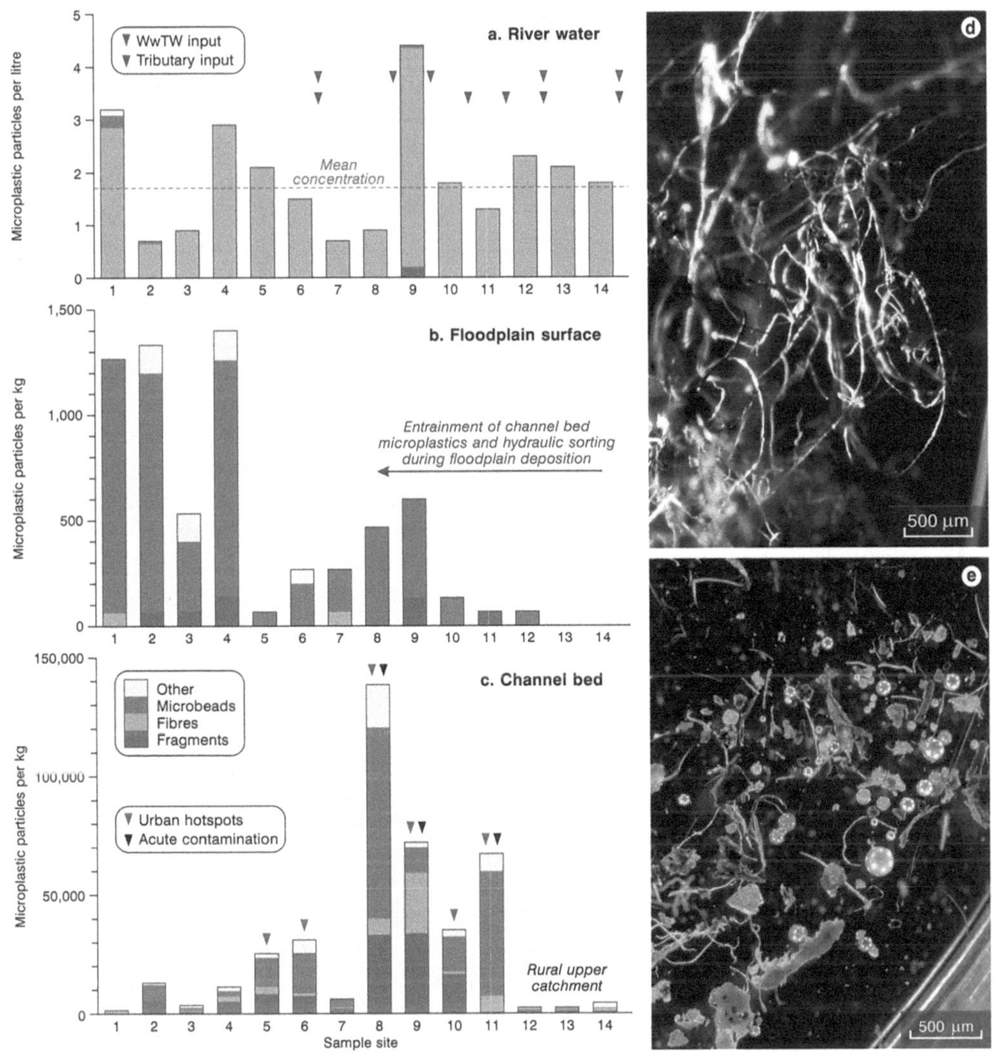

Figure 22.4 The uneven geography of microplastic contamination in the River Tame (after Woodward et al., 2021). This diagram shows 14 sites. Note how samples of river water (a) show low concentrations of microplastics, yet the channel bed (c) at the same location can contain very high levels of microplastic contamination. The floodplain surface (b) was dominated by microbeads which typically are the highest density microplastics in these assemblages. The photographs show microplastic assemblages in river water at low flow (d) and on the channel bed (e).

Microplastic pollution and water company behaviour

Microplastic particles are highly mobile. The presence of microplastics in high concentrations on river channel beds is a clear sign that wastewater treatment is not happening as it should. The geography of contamination in the urban river environment revealed a clear relationship between wastewater infrastructure and the highest levels of microplastic contamination. The microplastic assemblages shown in Figure 22.4c contain large numbers of microbeads. This is significant because microbeads are *primary* microplastic particles. They provide an incontrovertible link to the wastewater system because they cannot be the product of in-channel degradation of larger plastic items. We published the River Tame study in the journal *Nature Sustainability* in May 2021 under the title 'Acute riverine microplastic contamination due to avoidable releases of untreated wastewater'.[8] This paper became highly political because not only did it argue that water company assets were the main conduit of this microplastic contamination, but it also linked microplastic pollution of rivers with the emerging national scandal of water companies dumping sewage and untreated wastewater into rivers and coastal waters *outside* of the permitted conditions.[9] In the abstract to that paper we wrote:

> Here we show how untreated wastewater laced with microplastics and raw sewage is routinely discharged into UK river flows that are too low to disperse the microplastics downstream. These 'dry weather' spills lead to acute microplastic contamination of river bed habitats. Many aquatic fauna feed in the benthic zone, the quality of which affects the entire riverine ecosystem. All microplastic types accumulate to high concentrations on the channel bed until flushed downstream by floods. These findings pose fundamental questions about the sustainable management of urban wastewater. Treating the wastewater would shut down the major source of microplastic fragments and microbeads in such rivers and prevent their transport to the ocean.

Our microplastics research began with the aims of quantifying the extent of the microplastic contamination problem in UK rivers and of better understanding the key processes involved. It was also driven by a concern to assess the ecological threat posed by these particles. As this research progressed we were able to show that river-bed microplastic surveys are a key tool for riverine ecosystem quality mapping and for the assessment of microplastic risk. The storage of microplastics on the channel bed presents the greatest ecological threat because many creatures live, feed, and reproduce in that habitat. What we didn't fully anticipate as we began this work was the extent to which our findings would challenge corporate interests and become highly political in demonstrating the utility of bed sediment microplastic data as a tool for monitoring poor (and often illegal) wastewater practice. The fact that hotspots appear to be limited in spatial extent makes them useful for identifying input locations and poor wastewater practice.

What began as a curiosity-driven piece of environmental science took us into new territory slap-bang in the middle of an emerging sewage scandal that had begun to engulf the private water companies in England from the middle of 2021.[10] In March 2021 Professor Peter Hammond[11] published a paper showing that water companies were routinely underreporting discharges of sewage and inadequately treated wastewater to rivers (Hammond et al., 2021). This was a key step in uncovering the sewage scandal. Peter Hammond contacted me as soon as our *Nature Sustainability* paper was published (May 2021) and we talked via Zoom. He told me that he "jumped for joy" when he read our paper! That's easily the best feedback I have ever had on a piece of academic research. This conversation was important because we'd reached the same conclusion about water company behaviour completely independently using very different methodologies. Peter Hammond had carried out forensic analysis of flow data time series from wastewater treatment works. We had studied microplastic assemblages on river channel beds. Both approaches provided hard evidence for the routine discharge of untreated wastewater to rivers outside of permitted conditions. Both showed that regulation of the water industry and environmental stewardship were failing.

The sewage scandal

The microplastics research at Manchester immediately attracted media and political attention because it connected the microplastic pollution of our rivers with the national scandal of water companies dumping untreated sewage.[12,13] The scourge of plastic and microplastic pollution had been in the public consciousness for several years at this point and the UK government introduced a range of policies to tackle aspects of the problem. For example, in January 2018 when Michael Gove was Secretary of State for Environment, Food and Rural Affairs, he introduced a ban on the use of microbeads in personal care products. The last thing the water companies needed was more adverse publicity and another piece of academic research that placed the responsibility for another high-profile form of environmental pollution squarely at their front door. Water companies cannot be blamed for the microplastics that enter the sewerage system, but it is their duty to ensure that wastewater treatment takes place and suspended solids are removed. When our paper was published on 13 May 2021, United Utilities released a statement[14] that sought to undermine its findings. Jo Harrison, Director of Environmental Planning and Innovation at UU, said:

> As a society, it is vitally important that we increase our understanding of microplastics pollution, allowing us to develop meaningful solutions to tackle this significant issue – and United Utilities is already working with others to address the problem. It's therefore disappointing to see a number of flaws in the University of Manchester's research, which takes a narrow view of the issue and makes several assumptions without evidence.

Our paper went through rigorous peer review at *Nature Sustainability* and our field sampling methodology has been objectively evaluated against a large number of studies and ranked first for its rigour (Lofty et al., 2023).[15] United Utilities did not present any evidence to counter our conclusions. This is a classic corporate disinformation strategy. At the time of writing, almost three years after the publication of this research, United Utilities has failed to provide an alternative explanation for the microplastics data from the River Tame.

A key aim of the next phase of our research was to raise awareness of the significance of our findings to interested parties including the public, regulators, and policy makers. This ambition has been aided by extensive national and international media coverage, by briefing policymakers, and by social media, primarily Twitter/X. We also created a whiteboard animation that sets out the wider significance of our microplastics findings in the context of the sewage dumping scandal[16] (Figure 22.5). In summer 2021 with the help of Policy at Manchester[17] I gave presentations on our findings to various bodies including the UK Government's Department for Environment Food & Rural Affairs (DEFRA), the water industry regulator (Ofwat), the All-Party Parliamentary Group (APPG) for Microplastics, the Rt Hon. Philip Dunne's office, and the UK Parliament's Environmental Audit Committee. In 2022 I presented to the new Office for Environmental Protection (OEP) that was set up by the 2021 Environment Act with the aim of protecting and improving the environment by holding government and other public authorities to account. Our findings raised a few eyebrows at DEFRA, but were favourably received in all cases. In all of this, by speaking truth to power, I've worked to change the discourse around water pollution in Britain's rivers by ensuring that microplastic pollution was included in discussions around policies for environmental protection. This has contributed to a new level of scrutiny of the water companies and a wider debate about the effectiveness of those charged with environmental stewardship.

Figure 22.5 The whiteboard animation produced by We are Cognitive showing the journey from field science to policy impact at Westminster.

Tim Farron, the environment spokesman for the Liberal Democrats, whose constituency falls within the United Utilities region, has been an outspoken critic of water industry pollution and executive salary excesses. His comment from June 2023 captured the public mood:

> The system is clearly broken when water bosses are rewarded with millions of pounds for destroying the environment. This is a slap in the face for people who are seeing their water bills rise and our treasured rivers and lakes polluted with sewage by United Utilities. The Environment Secretary needs to wake up and stop the shameless profiteering of water companies.[18]

In 2022 United Utilities released raw sewage into rivers and coastal waters for some 425,491 hours (Figure 22.6). These data do not tell the full story, since Peter Hammond has shown that there is widespread underreporting and water companies are not required to release data on the *volumes* of sewage and wastewater they discharge to the environment. We are not told if a discharge is 100 litres or 1 billion. I have argued that data on discharge *volumes* are needed if we are to have any confidence in the assessment of industry and environmental improvement (Woodward, 2023).

One of the most remarkable features of the sewage scandal was the fact that it was exposed by campaigners and university researchers and not by Ofwat or the Environment Agency who are responsible for the regulation of the water industry and environmental stewardship, respectively, in England. The major political parties in the UK were all rather slow to appreciate the depth of public outrage around this issue. The state of the UK's rivers and the behaviour of sewage-dumping water companies in England was one of the defining issue in the General Election in July 2024.[19]

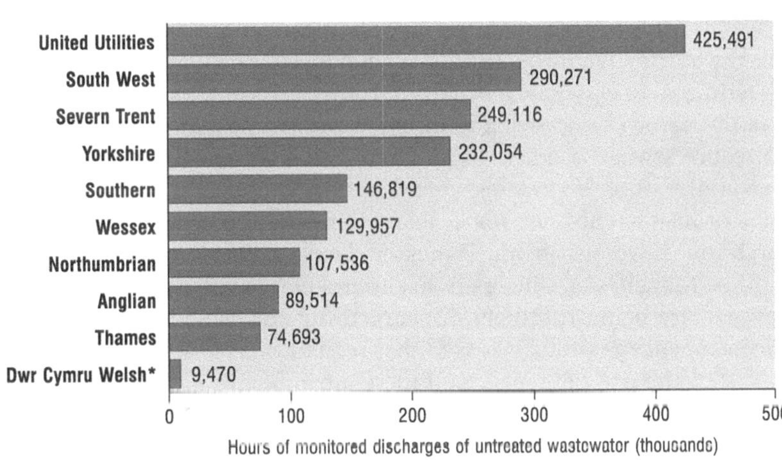

Figure 22.6 Discharges of untreated sewage and wastewater to rivers and seas in England from water company overflows in 2022. The total for 2022 across the industry was almost 1.8 million hours of discharges. These data are self-reported by the water companies and collated by the Environment Agency. They are published at the end of March for the previous year. United Utilities also topped the charts for the number of discharges in 2022. *The value for Dwr Cymru is for discharges into rivers in England only.

Policy impact at Westminster

In 2021 our findings around microplastics in rivers influenced policy development at Westminster in several ways. The 2021 Environment Act included a wide range of provisions around air and water quality and the protection of wildlife and biodiversity. It also included provisions to reduce plastic waste. The Act formed part of the post-Brexit legislative framework to enhance and protect the environment. The bill saw several delays, and the debate around sewage pollution and how to enshrine in law stronger powers to tackle polluting water companies was especially protracted and widely covered in the media. The government was put under intense pressure to tackle the issue. As the bill passed into law on 9 November 2021, Environment Secretary George Eustice said:

> The Environment Act will deliver the most ambitious environmental programme of any country on Earth. It will halt the decline of species by 2030, clean up our air and protect the health of our rivers, reform the way in which we deal with waste and tackle deforestation overseas. We are setting an example for the rest of the world to follow.[20]

The Act was vigorously debated in Parliament in September and October 2021. Those parts of the bill intended to tackle sewage and wastewater discharges by the water companies generated lengthy discussions and members of the House of Lords tried to strengthen this part of the legislation. The government proposed the establishment of a Storm Overflows Taskforce to report on this issue. Widespread and graphic media coverage of sewage dumping into rivers and coastal waters around Britain heaped pressure on the government and social media went into overdrive. The former pop star Feargal Sharkey has been in the vanguard of this campaign to highlight the environmental impacts of sewage dumping and the multiple failures of regulation.

During the discussion of the Environment Bill in the House of Lords on 13 September 2021, our work on microplastics was cited by Baroness Altmann.[21]

> I welcome the Storm Overflows Taskforce announced last August, which "has agreed to set a long term goal to eliminate harm from storm overflows". This, too, is most welcome but, so far, this involves improving monitoring and transparency rather than meaningful action to reduce sewage overflows into rivers and waterways. So far, the Environment Agency has clearly struggled to assess compliance with discharge rules and impose enforcement action or fines to galvanise noticeable action and stop or reduce these overflows.
>
> Research on sewerage from Professor Peter Hammond and Professor Jamie Woodward of Manchester University has found clear evidence that untreated sewage or wastewater are being routinely discharged outside the conditions allowed by the Environment Agency permits. It is vital that regulation of discharges of untreated sewage and wastewater are tightened, and these amendments would assist in this regard. The Government's plan is to set targets on reducing pollution from wastewater, agriculture and so on, but setting targets is not an active reduction of this pollutant.

On the same day the Earl of Caithness[21] put the following question to Lord Goldsmith, who represented the government in the debate:

> Has he been briefed on the latest research from the University of Manchester, which has demonstrated a direct link between poor wastewater management and high levels of microplastic pollution in the United Kingdom? When we have these overflows, the

> microplastics go out into the water system—not only the rivers, but the sea, thus negating a whole lot of good that the Government have been trying to do in reducing microplastics. If this were not enough of a bad situation before, it is now really bad.

When our 2021 paper was published, the APPG on Microplastics was already collaborating with The National Women's Institute to make the case for new legislation to tackle the problem of microplastics from clothing being released into the aquatic environment. This APPG published its first report on 21 September 2021, 'Microplastic Policies for the Government'.[22]The report recommended the fitting of microfibre filters in all new domestic and commercial washing machines from 2025 to eliminate the release of microplastic fibres from washing machines into British rivers and coastal waters. I gave evidence to this APPG and was pleased to see that the riverine microplastic problem (as well as marine microplastics) that we had described featured prominently in the report.

The APPG on Microplastics was chaired by Alberto Costa, MP who introduced a private members' bill on 30 November 2021. Costa cited our microplastics data from the River Tame (published in 2018) in his opening statement to the House of Commons:[23]

> Microplastics can then escape from the wastewater treatment works, polluting our rivers and seas. These microplastics, which can contain chemicals, can be ingested by fish and other small aquatic creatures and travel up the food chain, and we then enjoy them on our plate when we eat our fish fingers or other aquatic produce. We are ingesting these microfibre plastics. Research led by the University of Manchester has found that the River Tame in Greater Manchester contains such enormous levels of microplastics that it is one of the most polluted waterway systems in the world. Investigators found 517,000 plastic particles per square metre of sediment. That figure is extremely concerning for all of us, the environment and aquatic life.

He went on:

> Microfibre plastic pollution is a huge problem, and the Government should explore all avenues to tackle the different types of pervasive plastic, in addition to the already excellent work that the Government are doing generally to tackle other sources of plastic pollution.

Throughout 2021 the UK government's Environmental Audit Committee (EAC) was conducting an Inquiry into the state of river water quality in England. The Committee and Inquiry were chaired by the Rt Hon. Philip Dunne, MP who had been a long-time campaigner for tighter controls on sewage discharges into rivers. He introduced a private members' bill in October 2020 "to place a duty on water companies to ensure that untreated sewage is not discharged into rivers and other inland waters".[24] That bill ran out of parliamentary time but many of the issues were covered by the Inquiry. The Inquiry took evidence from many sources including water company executives and the leadership of the Environment Agency. I gave oral and written evidence to the Inquiry.[25]

In July 2021 the EAC visited the River Windrush on a fact-finding exercise where they met members of the Windrush Against Sewage Pollution campaign.[26] Peter Hammond lives on the banks of the River Windrush and had become involved in campaigning and water company practices as he witnessed first-hand a steady deterioration in its water quality and aquatic ecosystem. We both presented our findings to the EAC

Figure 22.7 Demonstrating bed sediment sampling and discussing the significance of channel-bed microplastics to members of the Environmental Audit Committee in the Windrush catchment in July 2021. The Rt Hon. Philip Dunne (Chair of the EAC) is holding the sample container. Photo credit: Adrian Arbib.

delegation (Figure 22.7). The Committee published its report in January 2022 (Environmental Audit Committee, 2022). In relation to the state of rivers in England and the impact of water company behaviour, the following extracts from its opening remarks are salient here:

> Getting a complete overview of the health of our rivers and the pollution affecting them is hampered by outdated, underfunded and inadequate monitoring regimes. It is clear, however, that rivers in England are in a mess. A 'chemical cocktail' of sewage, agricultural waste, and plastic is polluting the waters of many of the country's rivers. Water companies appear to be dumping untreated or partially treated sewage in rivers on a regular basis, often breaching the terms of permits that on paper only allow them to do this in exceptional circumstances.
>
> The sewerage system is overloaded and unable to cope with the increasing pressures of housing development, the impact of heavier rainfall, and a profusion of plastic and other non-biodegradable waste clogging up the system. Successive governments, water companies, and regulators have grown complacent and seem resigned to maintaining pre-Victorian practices of dumping sewage in rivers. There has been investment in the network since privatisation, but underlying problems have not been resolved and capital investment has not kept pace with housing and other development pressures on the drainage and treatment network.

The report lists 82 conclusions and recommendations. The following recommendation was prominent and is especially relevant here:

> We recommend that the Secretary of State for Environment, Food and Rural Affairs commission, in conjunction with the devolved administrations, a UK-wide survey of

> emerging pollutants and microplastic pollution of river environments, including an assessment of their potential impact on aquatic ecology.

The Environment Agency saw its monitoring budget slashed in half in the government's 2015 spending review so that it effectively lost the capacity to assess the impact of water company discharges on rivers and coastal waters. Other changes meant that it ceded control of wastewater monitoring to the water companies, which ended up effectively marking their own homework. The water companies in England collect their own samples and provide data on wastewater discharges to the EA. Peter Hammond's work has shown that these data are subject to very limited scrutiny. One of the key recommendations of the EAC report *Water Quality in Rivers* is now being implemented by the Environment Agency. The EA is now developing its strategy for establishing a microplastics monitoring framework for rivers in England.[27]

Criminal investigations into water company behaviour and a public apology

On 18 November 2021, barely a week after the Environment Act passed into law, the Environment Agency and Ofwat announced separate major investigations into potential widespread non-compliance by water and sewerage companies at wastewater treatment works. At the time of writing the EA was "investigating more than 2,200 WWTWs that discharge into English waters. This relates to all water and sewage companies." The outcomes of these investigations will not be known until 2024 or 2025 but the EA's initial assessment "indicates that there may have been widespread and serious non-compliance with the relevant regulations by all water and sewerage companies". At the same time the new OEP is investigating both the Environment Agency and Ofwat for their failure to hold the water companies to account.

On 18 May 2023, after two years of public outrage and unprecedented criticism from the media, campaigners, university researchers, and politicians, Water UK[28] issued the following statement:[29]

> Water and sewage companies in England have today apologised for not acting quickly enough on sewage spills. To put things right, industry plans to make the largest ever investment in storm overflows as part of a major programme to reduce spills into rivers and seas. In addition, a new national environmental hub with information on all 15,000 overflows in the country will increase transparency and allow the public to hold companies to account.

Ruth Kelly, the newly appointed Chair of Water UK, said:

> The message from the water and sewage industry today is clear: we are sorry. More should have been done to address the issue of spillages sooner and the public is right to be upset about the current quality of our rivers and beaches. We have listened and have an unprecedented plan to start to put it right. This problem cannot be fixed overnight, but we are determined to do everything we can to transform our rivers and seas in the way we all want to see.

The criminal investigation into the water companies by the Environment Agency and Ofwat was still in progress in summer 2024. It is expected to be completed by the end of 2025.

Closing reflections

A key part of our approach as geographers has been to focus on deepening understanding of microplastic contamination patterns and how these patterns are controlled in space and time by both human activity and river catchment processes. We have explored microplastic contamination at different scales and placed the data from Manchester's rivers in global context. Most importantly, each stage of this research has generated new questions and therefore new opportunities. Perhaps the most important finding was recognising that the very presence of microplastic assemblages in high concentrations on river beds was a clear indication that wastewater treatment was often not taking place as it should, and sewage (untreated wastewater) discharges were taking place into low river flows *outside* of permitted conditions. The timing of the 2021 paper was critical as our work on microplastic contamination in rivers became part of a nationwide sewage scandal and the policy response. Our expertise, and our aim to let the hard-won evidence do the talking, allowed us to shape key elements of public and governmental discussions relevant to public policy.

All of this was achieved without a big research grant. What mattered most was the formulation of interesting research questions, a motivated team, and a determination to target the very best journals to bolster our case. Most importantly, through a carefully crafted programme of field and laboratory investigations, we were able to better understand the processes and timescales involved in the creation of microplastic contamination patterns and shine a light on potentially illegal water company practices.

There is a widespread perception of how science can or should influence wider society. It's sometimes called 'the linear model', where it is presumed that rigorous objective science will provide evidence and/or technologies that society will pay close attention to if the benefits are immediately obvious. Our experiences show this model is only partly accurate. Certainly, our fundamental river catchment science attracted attention outside the University of Manchester: the state of a nation's rivers, after all, is a 'hot button' issue for most societies. But it takes time for governments to respond to science, even when its insights attract widespread public support. The intensity of public outrage at the sewage scandal in England certainly had the government of the day on the back foot. Governments, of course, juggle multiple demands, and laws and regulations can't usually be altered overnight, but the passage of the Environment Bill in 2021 became a focal point for protest and political lobbying. In these so-called 'post-truth' times, governments sometimes ignore science or seek to keep scientists quiet. 'Speaking truth to power' can be both arduous and frustrating. It is often unsuccessful. At the time of writing, we are collaborating with the UK Environment Agency to design a national monitoring strategy for microplastics in rivers. Our research has played a part in policy development and holding polluting water companies to account. We are hopeful that it will lead to improvements in river water quality in the UK.

Notes

1 *Plastic Warriors* was released in September 2022 by Outpost Pictures http://www.outpost-pictures.com/documentary and *Our Troubled Rivers* with Paul Whitehouse, screened on BBC2 in March 2023, attracted over 1 million viewers https://www.bbc.co.uk/programmes/m001jw74.

2 Rachel Hurley's PhD (2014–2017) was supervised by Jamie Woodward and James Rothwell at the University of Manchester. Dr Rachel Hurley is now a research scientist at the Norwegian Institute for Water Research (NIVA) in Oslo specialising in microplastics in soils and freshwaters (https://www.niva.no/en). Hurley, R.R. (2017) *Metals and microplastics in the first*

industrial city: Fluvial sediment contamination in the upper Mersey and Irwell catchments, UK. PhD Thesis, The University of Manchester https://research.manchester.ac.uk/en/studentTheses/metals-and-microplastics-in-the-first-industrial-city-fluvial-sed.

3 Microplastic concentrations for the Denton hotspot on the River Tame actually increased between the 2015 and 2016 sampling periods from 48,300 to 72,400 microplastic particles per kg of fine bed sediment respectively. The pre-flooding assemblage was dominated by microbeads and fragments in roughly equal quantities; the post-flooding bed assemblage was 98% microbeads. This shift in composition demonstrated that channel-bed flushing and hotspot removal did take place in this reach, but also that it was quickly followed by an influx of microbead-rich wastewater from a local point source. When this work was published in 2018 this was the most heavily contaminated site known on any river in the world and it gained extensive national and international media attention. The news media that covered the implications of this paper can be viewed here https://nature.altmetric.com/details/34247539/news.

4 Note that this is a minimum estimate of total microplastic flux from these catchments over this period since it does not include sources such as CSOs and urban runoff that operate during the flood events.

5 The BBC covered this work on radio and TV https://www.bbc.co.uk/news/science-environment-43363545.

6 'Microplastic pollution in oceans is far worse than feared, say scientists', *The Guardian* 12 March 2018:

"A study reveals highest microplastic pollution levels ever recorded in a river in Manchester, UK and shows that billions of particles flooded into the sea from rivers in the area in just one year" https://www.theguardian.com/environment/2018/mar/12/microplastic-pollution-in-oceans-is-far-greater-than-thought-say-scientists.

7 The flood that took place on 29 July 2019 was the 4th largest on record with a discharge of 99.665 cumecs. The channel-bed gravels were extensively reworked and there was extensive overbank deposition of sands. The flow record at this site began in 1943 https://nrfa.ceh.ac.uk/data/station/info/69027.

8 This paper was written in early 2020 during the first year of the Covid pandemic. The fieldwork was carried out in summer 2019 by Jamie Woodward, James Rothwell and Jiawei Li. Jiawei completed her PhD at Manchester in 2022 supervised by JW and JR: Li, J. (2022) *Urban floodplain and river channel records of microplastic contamination*. PhD Thesis, The University of Manchester.

9 The Sewage Scandal: 'The Worst We Have Seen in Years' – Why British Bathing Water is the Dirtiest in Europe *Byline Times* 6 December 2022 https://bylinetimes.com/2022/12/06/the-sewage-scandal-the-worst-we-have-seen-in-years-why-british-bathing-water-is-the-dirtiest-in-europe/.

10 The water industry in England was privatised in 1989, leaving Scotland and Northern Ireland under public ownership with Wales as a not-for-profit company. In England nine private water and sewerage companies (WaSCs) were formed alongside 13 much smaller companies providing drinking water services only. At the time of writing, the water industry in England is mostly (70%) owned by overseas holdings and has paid out some £76 billion to shareholders since privatisation. There has been intense criticism of water company behaviour and the industry's failure to divert more of its profits to improving wastewater treatment infrastructure. Sewage discharges to rivers and the sea have become routine. Dumping sewage has formed part of a hugely profitable business model where executives are richly rewarded. One water company executive, Steve Mogford, who was the CEO of United Utilities from 2011 to 2022, made some £30 million while at the helm of Britain's biggest dumper of sewage.

11 Professor Peter Hammond is a key member of the Windrush Against Sewage Pollution (WASP) campaign and has played a key role in uncovering widespread sewage dumping by water companies in England https://www.windrushwasp.org/.

12 *The Guardian*: Water firms are main source of microplastics in UK rivers, study says https://www.theguardian.com/environment/2021/may/13/water-firms-are-main-source-of-microplastics-in-uk-rivers-study-says.

BBC: Wastewater is 'polluting rivers with microplastic' https://www.bbc.co.uk/news/science-environment-57092135.

13 It is important to note that untreated wastewater contains a wide range of contaminants as well as human sewage even though the latter has attracted most attention in the media.

14 United Utilities statement on microplastics research https://www.unitedutilities.com/corporate/newsroom/latest-news/united-utilities-statement-on-microplastics-research/ United Utilities is responsible for water and wastewater services in the North-West of England. Of all the water companies in England they have the worst record for dumping sewage in the environment by both numbers of sewage discharge events and total hours.

15 In a recent critical evaluation (using 10 objective criteria) of the sampling procedures employed in 36 field-based river studies around the world that quantified microplastic occurrence presence in the water column and/or bed sediment, our methodology scored highest for rigour and reliability (Lofty *et al.* 2023).

16 Animation highlights importance of microplastics research in driving water company investigations (June 2022) https://www.manchester.ac.uk/discover/news/animation-highlights-importance-of-microplastics-research/ JW discussed the making of this animation and the wider significance of the microplastics research with the whiteboard animation specialists We are Cognitive: *Microplastics in UK Rivers: the small particles pointing to a big problem* https://www.wearecognitive.com/blog/microplastics-rivers-whiteboard-animation.

17 Policy at Manchester is the University of Manchester's policy engagement institute that connects researchers with policymakers. Find out more here: https://www.policy.manchester.ac.uk/.

18 https://www.thisismoney.co.uk/money/markets/article-12229679/Water-boss-makes-30m-Britains-biggest-polluter-United-Utilities.html.

19 The 18 Tory seats identified as most at risk at the next general election due to the sewage scandal https://inews.co.uk/news/politics/tory-seats-at-risk-next-general-election-sewage-scandal-2157409.

20 The Environment Act was a major piece of legislation that followed the UK's departure from the European Union https://www.gov.uk/government/news/world-leading-environment-act-becomes-law.

21 Hansard: Volume 814: debated on Monday 13 September 2021. https://hansard.parliament.uk/lords/2021-09-13/debates/D1F7DC1B-0DF4-4086-A4F2-32910E340F53/EnvironmentBill.

22 On 21 September 2021, the APPG on Microplastics published its first report: *Microplastic Policies for the Government* https://www.thewi.org.uk/campaigns/key-and-current-campaigns/end-plastic-soup/all-party-parliamentary-group-on-microplastics.

23 Hansard: https://hansard.parliament.uk/commons/2021-11-30/debates/D58EF879-FAC4-4723-AF9C-8B43214BCDA3/MicroplasticFilters(WashingMachines).

24 Sewage (Inland Waters) Bill https://bills.parliament.uk/bills/2625.

25 JW submitted written evidence to the Environmental Audit Committee Inquiry into Water Quality in Rivers in September 2021 highlighting some of our key findings on microplastic pollution in UK. https://committees.parliament.uk/writtenevidence/38790/pdf/.

26 Windrush Against Sewage Pollution is a Charitable Trust led by Ashley Smith that campaigns for an end to sewage pollution and for improved treated effluent standards with the ultimate aim of making river ecosystems safe for all life https://www.windrushwasp.org/.

27 The River Tame in Greater Manchester is in the frame to be one of two exemplar catchments in England where this monitoring strategy is being developed.

28 Water UK is a membership body representing the water industry. Its members include all water and wastewater companies across England, Wales, Scotland and Northern Ireland https://www.water.org.uk/.

29 Full statement here: https://www.water.org.uk/news-views-publications/news/water-and-sewage-companies-england-apologise-sewage-spills-and-launch.

Study task

Variations on this question can be set as an unseen examination, an open-book examination, or as a term-paper assignment:

Focusing on the recent riverine microplastics research at The University of Manchester, examine the merits and challenges of making geography relevant to public policy.

Starter reading list

Beaumont, Justin, Maarten Loopmans, and Justus Uitermark. "Politicization of research and the relevance of geography: Some experiences and reflections for an ongoing debate." *Area* 37.2 (2005): 118–126.

Brewer, Jennifer F. "Toward a publicly engaged geography: Polycentric and iterated research." *Southeastern Geographer* 53.3 (2013): 328–347.

Demeritt, David. "Harnessing science and securing societal impacts from publicly funded research: Reflections on UK science policy." *Environment and Planning A* 42.3 (2010): 515–523.

Dorling, Danny, and Mary Shaw. "Geographies of the agenda: Public policy, the discipline and its (re)'turns'." *Progress in Human Geography* 26.5 (2002): 629–641.

Eden, S. "Green, gold and grey geography: Legitimating academic and policy expertise." *Transactions of the Institute of British Geographers* 30 (2005) : 282–286.

Harvey, David. "What kind of geography for what kind of public policy?" *Transactions of the Institute of British Geographers* 63 (1974): 18–24.

Kass, Gary, Alice M. Milner, and Klaus Dodds. "The 'borderlands' of the science–policy interface." *The Geographical Journal* 188.4 (2022): 591–599.

Martin, Ron. "Geography and public policy: The case of the missing agenda." *Progress in Human Geography* 25.2 (2001): 189–210.

Massey, D. "Geography on the agenda." *Progress in Human Geography* 25.1 (2001): 5–17.

Owens, Susan. "Making a difference? Some perspectives on environmental research and policy." *Transactions of the Institute of British Geographers* 30.3 (2005): 287–292.

Peck, Jamie. "Grey geography?" *Transactions of the Institute of British Geographers* 24.2 (1999): 131–135.

References

Aves, A.R., Revell, L.E., Gaw, S., Ruffell, H., Schuddeboom, A., Wotherspoon, N.E., LaRue, M., and McDonald, A.J. (2022) First evidence of microplastics in Antarctic snow, *The Cryosphere*, 16, 2127–2145.

Blackburn, K., Green, D. (2022) The potential effects of microplastics on human health: What is known and what is unknown. *Ambio* 51, 518–530.

Blair, R. M., Waldron, S., and Gauchotte-Lindsay, C. (2019) Average daily flow of microplastics through a tertiary wastewater treatment plant over a ten-month period. *Water Research*, 163, 114909.

Borrelle, S.B. et al. (2020) Predicted growth in plastic waste exceeds efforts to mitigate plastic pollution. *Science*, 369, 1515–1518.

Carr, S.A., Liu, J., and Tesoro, A.G. (2016) Transport and fate of microplastic particles in wastewater treatment plants. *Water Research*, 91, 174–182.

Covernton, G.A., Collicutt, B., Gurney-Smith, H.J., Pearce, C.M., Dower, J.F., Ross, P.S., and Dudas, S.E. (2019) Microplastics in bivalves and their habitat in relation to shellfish aquaculture proximity in coastal British Columbia, Canada. *Aquaculture Environment Interactions*, 11, 357–374.

Environmental Audit Committee (2022) *Water Quality in Rivers, Fourth Report of Session 2021–22*. London, UK, House of Commons, 137 pp. https://committees.parliament.uk/publications/8460/documents/88412/default/

Gündoğdu, S., Çevik, C., Güzel, E., and Kilercioğlu, S. (2018) Microplastics in municipal wastewater treatment plants in Turkey: A comparison of the influent and secondary effluent concentrations. *Environmental Monitoring and Assessment*, 190, 626.

Hammond, P. et al. (2021) Detection of untreated sewage discharges to watercourses using machine learning. *NPJ Clean Water*, 4, 18.

Hurley, R.R., Rothwell, J.J., and Woodward, J.C. (2017a) Metal contamination of bed sediments in the Irwell and Upper Mersey catchments, northwest England: Exploring the legacy of industry and urban growth. *Journal of Soils Sediments*, 17, 2648–2665.

Hurley, R.R., Woodward, J.C., and Rothwell, J.J. (2017b) Ingestion of microplastics by freshwater tubifex worms. *Environmental Science and Technology*, 51, 12844–12851.

Hurley, R.R., Woodward, J.C., and Rothwell, J.J. (2018) Microplastic contamination of river beds significantly reduced by catchment-wide flooding. *Nature Geoscience* 11, 251–257.

Iyare, P.U., Ouki, S.K., and Bond, T. (2020). Microplastics removal in wastewater treatment plants: A critical review. *Environmental Science: Water Research and Technology*, 6(10), 2664–2675.

Lei, L., Wu, S., Lu, S., Liu, M., Song, Y., Fu, Z., Shi, H., Raley-Susman, K.M., and He, D. (2018). Microplastic particles cause intestinal damage and other adverse effects in zebrafish Danio rerio and nematode *Caenorhabditis elegans*. *Science of the Total Environment*, 619–620, 1–8.

Lofty, J., Ouro, P., and Wilson, C.A.M.E. (2023) Microplastics in the riverine environment: Meta-analysis and quality criteria for developing robust field sampling procedures, *Science of The Total Environment*, 863, 160893.

Long, Z. et al. (2019) Microplastic abundance, characteristics, and removal in wastewater treatment plants in a coastal city of China. *Water Research*, 155, 255–265.

McGoran, A.R., Clark, P.F., and Morritt, D. (2017) Presence of microplastic in the digestive tracts of European flounder, *Platichthys flesus*, and European smelt, *Osmerus eperlanus*, from the River Thames. *Environmental Pollution*, 220, 744–751.

McGoran, A.R., Cowie, P.R., McEvoy, J.P., Clark, P.F., and Morritt, D. (2018) Ingestion of plastic by fish: A comparison of Thames Estuary and Firth of Clyde populations. *Marine Pollution Bulletin*, 137, 12–23.

Murphy, F., Ewins, C., Carbonnier, F., and Quinn, B. (2016) Wastewater treatment works (WWTW) as a source of microplastics in the aquatic environment. *Environmental Science and Technology*, 50, 5800–5808.

Rowley, K.H., Cucknell, A-C., Smith, B.D., Clark, P.F., and Morritt, D. (2020) London's river of plastic: High levels of microplastics in the Thames water column. *Science of the Total Environment*, 740, 140018.

Simon, M., van Alst, N., and Vollertsen, J. (2018) Quantification of microplastic mass and removal rates at wastewater treatment plants applying Focal Plane Array (FPA)-based Fourier Transform Infrared (FT-IR) imaging. *Water Research*, 142, 1–9.

Sulaiman, B., Woodward, J.C., and Shiels, H.A. (2023) Riverine microplastics and their interaction with freshwater fish. *Water Biology and Security*, 2 (4) 100192.

Talvitie, J., Mikola, A., Koistinen, A., and Setälä, O. (2017) Solutions to microplastic pollution–Removal of microplastics from wastewater effluent with advanced wastewater treatment technologies. *Water Research*, 2017, 123, 401–407.

Van Sebille, E. et al. (2015) A global inventory of small floating plastic debris. *Environmental Research Letters*, 10, 124006.

Viršek, M.K., Lovšin, M.N., Koren, S., Kržan, A., and Peterlin, M. (2017). Microplastics as a vector for the transport of the bacterial fish pathogen species *Aeromonas salmonicida*. *Marine Pollution Bulletin*, 125, 301–309.

Williams, A.T. and Rangel-Buitrago, N. (2022) The past, present, and future of plastic pollution. *Marine Pollution Bulletin*, 176, 113429.

Woodward, J.C., Li, J., Rothwell, J.J., and Hurley, R. (2021) Acute riverine microplastic contamination due to avoidable releases of untreated wastewater. *Nature Sustainability* 4, 793–802.

Woodward, J.C. (2023) To clean up England's rivers we need to know how much sewage is dumped – but water firms won't tell us. The Conversation, 23 February, 2023 https://theconversation.com/to-clean-up-englands-rivers-we-need-to-know-how-much-sewage-is-dumped-but-water-firms-wont-tell-us-200322

23 Talking geography in the public realm

Danny Dorling

Danny Dorling (1968–)

Introduction

I was once a 'normal' academic researcher and teacher – I mostly published papers in academic journals and gave lectures to university students but not the public – and then, somewhat to my own surprise, my work took a 'public turn'. This became central to my *modus operandi*, and continues to be in many ways. The public domain is the space beyond the pay-walls that prevent most people, should they even want to, from accessing most academic publications. It is the 'place' where citizens are able to share information, ideas, and viewpoints of relevance to many of them – sitting between the domains of the

DOI: 10.4324/9781003343240-26

private realm, of business, academia and the state. The public domain, if it works well, can be a vital support to those political systems called democracies. However, we can overplay its importance and underplay the extent to which there is also a great deal of public debate in those countries we think of as undemocratic. To enter the public domain, academics' contributions must be easily accessed, the language used be clear, and topics explored be of interest to a wider group. Although popular books still come at a price, that price is usually much cheaper than an academic book. And, of course, there are now plenty of other non-print ways to communicate.

Public scholarship is valuable because we don't live in a top-down world where experts have all the answers and a wider public only needs to occasionally be informed about what decisions are being taken from on high. The to-and-fro of ideas is needed across all sections of a society if a society is to function well. Putting your ideas into the public domain invites far more criticism than even the most scrupulous of academic journal referees can muster – and some of it is very telling. You can learn much, as an academic, when you share your ideas more widely.

It is geographical insights that I have communicated most often. Geography often has an image problem compared to more prestigious subjects like Economics. Yet many issues of major public relevance – such as concentrated poverty or uneven flows of international investment – are deeply geographical. Every topic I've discussed in public is geographical in nature. Some of the most useful insights into the effects of inequalities are geographical comparisons between countries. Geography is, of course, much wider than such simple factual observation, but there is a useful place for empiricism – emphasizing usually quantitative evidence. It could also be argued that much of the more complex theorizing that human geographers engage in will have a far shorter shelf-life and be deemed in future as having little relevance, other than in telling us what geographers did then.

Background

My becoming a 'public geographer' was unusual but not entirely unique. In the UK, for instance, the physical geographers Mike Hulme, Mark Maslin, and Nick Middleton enjoy some public visibility, as did the late Harm de Blij in the USA. Male academics appear much more often than female ones to be in the public eye and it is hard to know to what extent that is due to them being put forward more, or their views being taken more seriously by a sexist media (Jones, 2013). My route to public geography began through producing a large number of research papers, later republished in several edited collections (the last two of which were Dorling, 2018 and 2024). These detailed the rise in geographical inequalities in housing, health, education, well-being, and more across the UK, Europe, and worldwide. They offered suggestions as to why this was occurring. I came to believe that when spatial divisions were growing so wide, geographers should not simply write about the world for students and each other.

When asked to define their discipline, academics geographers often say 'geography is what geographers do'. I prefer 'geography is what geographers *are*' if I'm ever asked to define the subject. I then explain that my academic colleagues and I are highly influenced by the places that we have come from, where we grew up, what schools and universities we attended, and where we now live – our lived experience. For example, the academic subject of Geography internationally is very much about just the tiny number of countries worldwide in which the majority of academic geographers happen to live and work. We cannot enter a different space of academic discourse where personal positionality and experience does not influence our understanding greatly.

By 2023 I was writing that Britain was 'broken', that 'middle England' had become a patchwork of haves and have-nots, with many people now much more rarely mixing with others outside a certain milieu. I explained that before the pandemic of 2020, life expectancy dropped as a result of poverty and cut-backs. These were the first large falls seen against what might have been expected since the 1930s. I tried to explain why it was that 22 of the (now) 27 European Union countries in recent years have better mortality rates for new-borns (lower infant mortality). Fifty years ago the UK had been world leading in child health. I provided analysis that showed that no other European country has such miserly unemployment benefits; schools that do so poorly when ranked on their ability to encourage flourishing imagination; university fees so high; housing so unaffordable; or a government economically so far to the right – including to the right of the USA (Dorling, 2023a and Figure 23.1).

My public role goes beyond highlighting geographical issues of public relevance. I support campaigns, such as the one to introduce 20 mph vehicle speed limits in towns and villages. At one level, my geographical research on who's been most harmed by cars informs this, but it is also an issue of social justice. Below I discuss some of the challenges of 'going public', and why so few academics – never mind geographers – do this sort of thing. Public scholarship includes listening to, and working out how to communicate with, a very wide audience.

So why do it? There's value in using one's expertise to shape wider understanding, both analytically and politically. If you've learnt or discovered anything of value, then conveying that more widely will itself have value. Public scholarship also acts as a reality test: is anyone interested? Is there a major flaw in your argument that will be exposed if it's subject to enough scrutiny? Can you make a little bit more of a difference than you might otherwise have done? But, as I will explain towards the end of this chapter, in some countries the public domain has become very rancorous, uncivil and even at times vicious. This can take a toll on the would-be public geographer.

The case for 'Public Scholarship'

In 2005 an article was published by leading American sociology professor Michael Burawoy. It made the argument that 'public sociologies should not be left out in the cold, but should be brought into the framework of our discipline'. In its very first paragraph it claimed: 'Today, at the dawn of the 21st century, although communism has dissolved and fascism is a haunting memory, the debris continues to grow skyward'. As Burawoy saw it, the 'debris' in question was the harm caused by neoliberal capitalism. A 'public sociology' would seek to actively shape public understandings of social issues, such as the increased use of illegal drugs among young people. It would do so by making sociological evidence and argument more widely known (Burawoy, 2005).

Within 15 years, it was being reported that China had become a more important trading partner for the majority of the world's nations than the USA. As recently as 2005 that had been unimaginable (*The Economist*, 2021). China is still very much a communist country, even if its communism has evolved to the point where its best-off 1% receive an almost identically large share of national income to the 'top 1%' in the UK, but still a little less than in the USA (Chancel et al., 2022). In 2017, Xi Jinping, the President of China, had written that the two greatest evils facing that country were now greed and indifference. These had replaced feudalism and colonialism, the two great evils that the first communist leader, Chairman Mao, had identified in 1942. Rather than communism having dissolved away, as Burawoy suggested, it had re-grouped and transformed

(Milanović, 2022; Mattick, 2013). Unlike Mao, Xi Jinping appeared more concerned about public debates within China. These public debates in China concerned recent and home-grown problems.

Fascism is no longer a haunting memory either. There is a fear of fascism rising 'at home' in the USA today, and in how far to the political right most politicians in the UK have moved. In the USA the storming of the Capitol on January 6, 2021 focussed attention. The rightwards political march of the Republican Party long predated that, resulting in the election of President Trump and, great fear when he won again in 2024. In the early 2020s, I was worrying more about what political direction the country I lived in, England, was taking after Brexit (Dorling and Tomlinson, 2019), the pandemic, growing social divisions over who could afford the most basic necessaries in life, and whether this might also result in a further lurch towards authoritarianism, dictatorial leadership, more centralized autocracy, more militarism, suppression of opposition and the right to strike, the further promotion of a belief in a natural social hierarchy, and more national rhetoric concerning the perceived innate good of the British (or English) as a nation.

In September 2022, three elections in Europe resulted in new leaders coming to power. An 'ex'-neo-Nazi movement in Sweden gained enough votes to be part of a wider coalition that would topple the government there, but not enough to choose the Prime Minister (Sands, 2022). In Italy the party most closely linked to past fascism, The Brothers of Italy, gained enough votes to see their leader elected as Prime Minister (Kirby, 2022). And in the UK the elevation of Liz Truss to Prime Minister (chosen by 21% of MPs and less than 0.16% of the adult population) resulted in the British Conservative Party becoming the most right-wing of all political parties in both the rich world and in what the *Financial Times* called 'emergent market countries' (Burn-Murdoch, 2022). Truss was in power for only 49 days. When Truss was replaced by Rishi Sunak and then Keir Starmer as Prime Minister there was very little shift back (see Figure 23.1 – it is very geographical – each circle is of a political party within a particular nation-state).

The *Financial Times* used the work of hundreds of political scientists to create Figure 23.1. It's a good example of unsung public social science – where the work of many is used to inform a wider population. Some would not put the US Democrats so far to the left, suggesting that others may have done so because American political scientists were a little mesmerized by Biden, after the shock of Trump. But it is interesting to think that worldwide, Geography as a discipline is still dominated by journals and publications produced by scholars living in the two states (the USA and UK) with some of the most far-right politics of the democratic world.

I began my geographical career 30 years ago by writing the computer algorithm that allows diagrams such as that shown in Figure 23.1 to be drawn. Although it's very likely that others did the same at the same time, it's possible that my code was utilized. I was at first best known in Geography for drawing cartograms that looked a little like Figure 23.1. Graphs are largely two-dimensional. I was interested in making maps out of circles, showing relationship as well as detail (Dorling, 1992, Figure 23.2). Solving the mathematical problem of ensuring that a series of circles don't overlap, with all just touching each other, and all being placed where they best illustrate how everything is connected, is an odd beginning to becoming a public geographer. But there's no obvious road to becoming one and, I discovered, it is quite a lonely road. Geography, in comparison to sociology, economics, or politics, is a small discipline worldwide. As a result, there are few 'public geographers' and we tend not to be that well known.

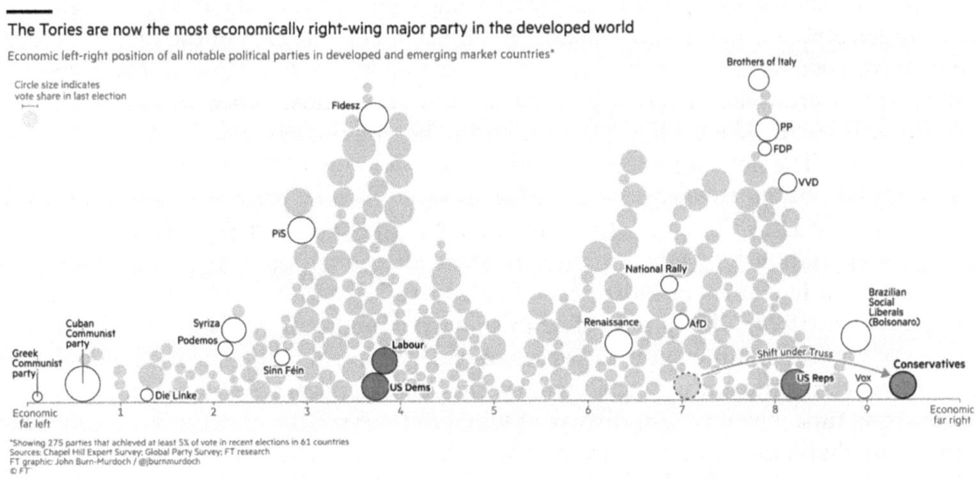

Figure 23.1 Economic position of all major political parties in rich and middle-income countries in 2022.

Source: *Financial Times* (Burn-Murdoch, 2022).

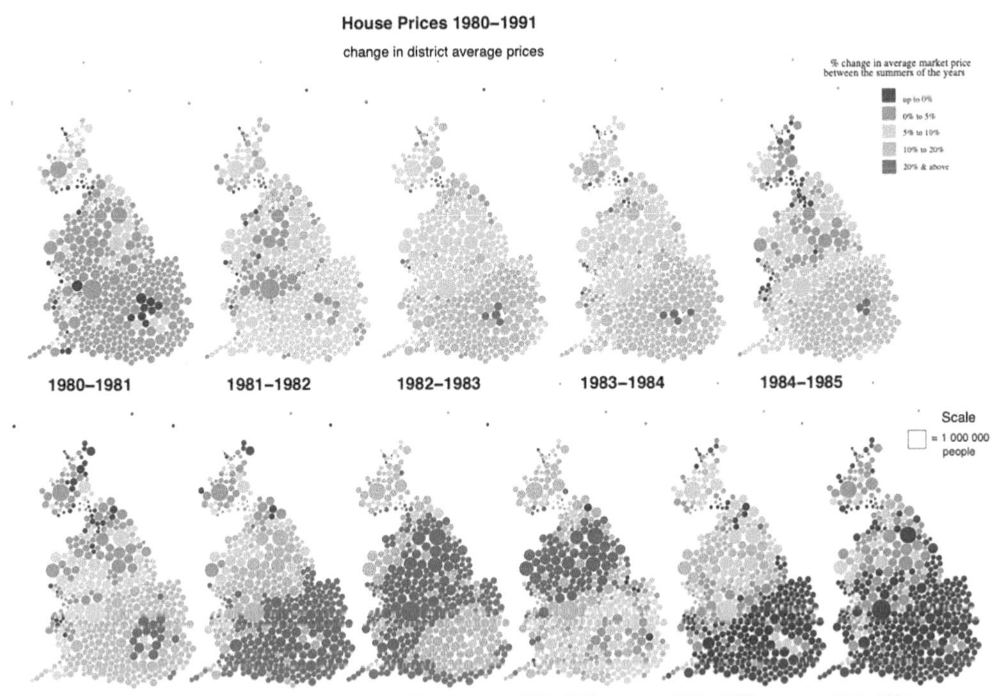

Figure 23.2 House-price change in England, Wales, and Scotland, 1980–1991.

Source: Dorling, D. (1995) *A New Social Atlas of Britain*, Chichester: John Wiley, p.127.

Figure 23.2 shows one simple example of using geographical data and visualization to better understand a geographical phenomenon. Each circle in each of the maps is a local authority district drawn with its area in proportion to its population. The largest is Birmingham. I drew each circle so it most often touches those other areas it shares the longest border with. The circles are coloured by the rate of change in house prices there in the last year. The housing boom that began in London in 1982 can be seen, and how it grew and spread northwards. Figure 23.2 also reveals how price rise slowed down in London as early as 1987, long before the famous market crash of September 1989. The circles reveal relationships whereby nearby areas affect each other the most, as well as detail such as a few areas of North-East England still experiencing house-price rises as late as 1991. In 1995, when I published this image in an academic book, I did not even imagine trying to explain what it showed to a wider public. On the one hand population cartograms were quite a novel idea then, on the other hand people who bought a home at the wrong time may have angrily asked why I did not see the crash coming and warn them (doing that itself could have exacerbated the crash, of course).

By 'public Sociology' Michael Burawoy meant sociologists who write books '...read beyond the academy, [books that] become the vehicle of a public discussion about the nature of American society – the nature of its values, the gap between its promise and its reality, its malaise, its tendencies'. The North America-centric focus in that sentence stands out today; and perhaps it helps illustrate how much progress there's been in such a short time by moving beyond seeing the countries that we live in as all-important. However, that sociological change was not primarily driven by academics. We mostly write books and papers that few people outside universities read. Very few people read my 1995 book. Almost all who did were other academics. Academia likes to think it leads public understanding and propels sea-changes in views. Looking back at the examples given in Burawoy's paper almost two decades on, it's clear that academics *don't* usually have the effects or insight they thought they had (he was wrong about both the future of communism and fascism) – but he was correct in suggesting that the public do notice the few academics who write in the opinion pages of national newspapers. I still work on housing today, and much more publicly than I did in the early 1990s; but I'm glad I had that experience of studying a housing crash outside the glare of publicity.

Talking Geography

For the last decade I've worked at the University of Oxford in the UK. As a result I get asked to talk publicly more than I did before. I have given over 1,200 public lectures and seminars in the last 30 years. I probably like the sound of my own voice too much; but if you are a university lecturer, then being given the opportunity to talk to people who've actually chosen to listen to you is an honour. You know when they're not interested, when you're becoming too boring or less relevant, when others stop asking you to come and talk. You also learn far more from the questions that school children or members of the public ask, than often you do from what your colleagues ask (because you become used to academic questions).

People invited me to come and talk because I had written books that were sometimes aimed at and purchased by the general public. I have published, with colleagues, over 50 books. In the last decade these included joint authored books on *Geography* (2016), two *Social Atlases of Europe* (2014, 2017), a *21st-century Atlas of the UK* (2016), a book on *Why Demography Matters* (2018), and *Finntopia* (2020), a book on Finland (Europe's most equitable and happiest country). Alone, but as always actually with help, I wrote *The 32 Stops* (2013), an account of lives along London's Central Line, *Population 10 Billion* (2013) on the implications of world population heading toward ten billion, *All that is Solid* (2014) on the housing crisis in Britain, a book on *A Better Politics* (2016), one on *The Equality Effect* (2017), another on *Peak Inequality* (2018), one titled *Inequality and the 1%* (2014), and another on *Slowdown* (2020), concerning how most global trends appeared to be decelerating, one titled *Shattered Nation* (2023) – about the UK and what it had become, and most recently *Seven Children* (2024) – part fiction, mostly fact The latest in this series is a more global book, The Next Crisis (2025).

I began writing about the UK in the late 1980s at the point when it was only just beginning to become clear it was heading to be ever more riven politically, economically, and socially. It was not at all obvious to me at the time I started work on the human geography of the UK that this state (made up of several countries) would later become such an outlier in Europe, or that inequality would become my main focus. But by 2014 the British Conservative party had left the 'Conservative bloc' of European political parties and joined a group which included the German far-right party, *Alternative für Deutschland.* As Figure 23.1 shows, it had also moved its economic policies to the far right. It was these political changes that made what I was writing and talking about much more concerning. The rise in economic and social inequality within the UK became more interesting to a wider audience than any public lecture on how you could see the world differently if you first reprojected the base map to show the country as a series of circles slowly and slightly changing their positions over time as district grew and shrunk, coloured something like house-price changes (Figure 23.2).

In hindsight, the kernel of what I now concentrate most on was part of the work I began in the 1980s and 1990s. Initially I had been most interested in what very large datasets (such as those from UK census information or hundreds of thousands of mortgage records) could show us about how the UK was changing. In my PhD thesis, which I worked on between 1989 and 1991, I found new ways to see changes between the 1971 and 1981 UK censuses (by reprojecting the base map). Initially, I'd absolutely no idea that in 1971 the UK had been amongst the most equitable of all European countries. I was not alone, I vividly remember reading political essays about how unequal Britain was, written in the late 1960s and early 1970s, but devoid of facts. How wrong they were! But that was not my focus. Instead, I was looking for the variations between places, not the similarities. And I was interested in new Geographical Information System (GIS) techniques, not contemporary or past political or social theory. I constructed numerous segregation indices, and (with others) produced atlases of each census. Some forms of geographical segregation appeared to be falling over time, especially by ethnicity, between 1991 and 2001; other forms, such as the segregation of university students away from the rest of the population (and between social classes), were rising.

It can be the smallest things that catch the public imagination. You can't predict what might stick. In 2005 I wrote an article in a Sunday newspaper about a prominent commentator being wrong to say that ethnic minority ghettos were forming in the UK (Dorling, 2005). Just one observation caught peoples' imagination: 'Cut Britain up horizontally rather than by neighbourhood, and you do find minority-majority areas. For example, above the fifth floor of all housing in England and Wales a minority of children are white. Most children growing up in the tower blocks of London and Birmingham – the majority of children "living in the sky" in Britain – are black.' Imagery matters: a single statistic from the national census report can have as much impact as the most complex of analyses. Complexity matters too and needs to be communicated publicly in an accessible way. However, the complex academic debate about 'ghettos' in the UK, USA, Australia, and New Zealand may have missed the key point. When the Grenfell fire occurred in 2017, that point became more obvious.

I'm not going to provide a long list of references to my own work – you can look up the books and articles in Google Scholar if you want and a few are listed above. One advantage of talking in the public realm is that you can more easily be Googled. I try to put a copy of all the papers I write on my website, and for every book I put all the data used, and all the graphics drawn from that data, on the web (see www.dannydorling.org). This site is aimed for public use.

Instead of talking more about my ideas, I'll instead now touch on the criticism an academic writing in public can expect. This illustrates one reason why (in the UK today) fewer academics now put their heads above the proverbial parapet – and why so many who do quickly put their heads down again when they are targeted by a small group of people who don't want the issues that I work on more widely discussed. Often they say that wider discussion would be unhelpful, for instance, that discussing publicly who it is that is most likely to study for a geography degree might put poorer students off applying to study geography. Or they suggest that an academic viewpoint has become too dominant, as the UK Prime Minster did in September 2022 when during her very brief tenure, she claimed: 'The economic debate for the past 20 years has been dominated by discussions about redistribution. And what has happened is, we have had relatively low growth ... and that has been holding our country back' (Harris, 2022).

For the last 30 years, as I've become more and more involved in the debate about inequality and redistribution, I knew that the very small number of academics involved in this publicly had not dominated economic debate – after nine years of the so-called New Labour government it felt as if no one was listening. We pointed out that the UK had become the most unequal country in Western Europe under New Labour. I was asked why I was criticizing Labour. Later, at first, only a few noticed the significance when the UK became the second most unequal country in all of Europe in 2019, after nine years of Conservative and Coalition governments. Today, that is much more widely known than it was when it was first recorded and I repeatedly reported the fact. It's common now to read that: 'The OECD ranks the UK 8th most unequal out of 40 major economies, more unequal than every EU country except Bulgaria' (Hildyard, 2022).

Talking publicly

As the evidence that the UK was going in the wrong direction mounted, those people who believed that greater inequality was good, or at the very least that inequality wasn't a problem, began to become angrier about what some university academics were finding (e.g., Niemietz 2014, 2022; Snowdon 2015, 2017; Grant 2018; Butcher 2019). One wrote: 'It's no coincidence that the most vocal academics in the inequality debate – Richard Wilkinson, Thomas Piketty, Danny Dorling et al – are men of the left' (Snowdon, 2016). Most of the popular books and papers written by academics on inequality are only read by a few people, but collectively they can at times have great effect, as the criticism I received from right-wing pundits implied. In that list of references above you'll find three of my books referred to as 'the worst contribution' or 'worst book' written so far (they cannot all be the worst ever book, unless each was successively worse than the last!). There was a suggestion that I portrayed neoliberals as a crossover between Ebenezer Scrooge and Hannibal Lecter. Another, that I didn't understand statistics on inequality. Another, that I (and the charity Oxfam) were advocating what one termed 'Karl Marx's wretched immiseration theory'. I'll leave you to read their criticisms and what I actually said, if you wish. However, after autumn 2022 I noticed that something may have changed and the criticism suddenly became far less frequent. This coincided with the disastrous 49 days of the Liz Truss premiership in which the ideas of my main critics were tried and seen, most spectacularly, to have failed (Dorling, 2023b).

The criticism that I was receiving as a public academic had come to a head four years earlier, in March 2019. It was then that the editor of *The Spectator* magazine wrote an article trying to encourage university students to take up internships at his workplace that summer (Nelson, 2019a). On the face of it the aim appeared laudable: to try to recruit outside of the usual circles which that magazine attracted. One of the questions it asked potential candidates to answer might be now seen as especially prescient – and misguided: 'List why 2019 will likely be the best year in human history'. However, it was the question asked directly beneath that one which caught my attention, for obvious reasons (see Figure 23.3).

I asked Fraser Nelson if he could let me know what the applicants found. He replied 'Happy to update you as entries come in. But we're disallowing the easiest hits: i.e. your publishing a book called "peak inequality" at a time when income inequality was at a 30yr low' (Nelson, 2019b). Nelson claimed that inequality was highest in '1986' – my critics tend to believe each other's 'facts'. He had picked this 'fact' up from another critic (Snowdon, 2017). What I learnt through this experience was how small and tight the

THE SPECTATOR

SUBSCRIBE

Internships at The Spectator for summer 2019; no CVs (or names!) please

Fraser Nelson

Since we abolished CVs for *The Spectator*'s internship scheme, it has acquired quite a reputation. There are fewer than two dozen journalists here in 22 Old Queen St and we recruit people rarely – but when

Fraser Nelson
31 March 2019

THE SPECTATOR

RESEARCH TASKS

- We have a large data library, but all in Excel sheets. We'd like to make this publicly available, perhaps showing a menu of charts and then providing access to the data behind it. What technology should we use?
- Find two factual errors made by a writer or politician of your choice.
- List five respects in which 2019 will likely be the best year in human history. Support your answers with data.
- Find three factual inaccuracies in articles or comments by Oxford's Prof. Danny Dorling
- Are home ownership costs rising or falling for millennials? Find data to support your answer.
- Outline a means of automatically ranking recent tweets by click through or engagement rate.
- Suggest a new piece of software/app/technological development that could benefit *The Spectator*

Figure 23.3 When you know you have touched a nerve – *The Spectator* advertisement.

Source: A tweet of part of the article: https://twitter.com/geoviews/status/1113940557116649472

circle was of people who consistently attack academics who speak in public. You can weather one-off attacks of course – most of the attacks that I received were from racists annoyed about what I had written elsewhere on immigration. But I was more surprised when a concerted and sustained criticism began to come more and more relentlessly from a small coterie of right-wing writers. These were people who read and often echoed each other's criticisms of what they saw as 'lefty academics'. Nelson never updated me, which was a pity as surely some of the applicants must have found some genuine mistakes I'd made. However, one should not assume those you're debating with are actually interested in how the world truly is – instead, they may have another agenda about how they think it should be.

It's hard to try to get into the minds of the people you make angry. It's not something most academics have to do much of if they're out of the public eye. Something about what I was writing, and especially the graphs, maps, and statistics I was producing, had touched a nerve with Fraser Nelson and his colleagues. Looking back, a few years after this episode, I think it illustrates just how wedded some people were to the idea that everything was getting better and those who suggested it was not, like me, were simply wrong; and any fact that counteracts their view must also be wrong. They really thought that '2019 will likely be the best year in human history'.

Once you're in the public eye people search out anything you write, even if printed in the most obscure of journals. Later, in 2019, a piece I'd written in an academic journal few people would read (Dorling, 2019) was picked up by *The Daily Mail* (a tabloid) and numerous other newspapers. Under a headline including the words 'posh and dim' (see Figure 23.4), a writer in *The Telegraph* newspaper summed it up: 'Writing in the journal *Emotion, Space and Society*, Dorling posits that Geography has become, as many have long suspected, "a soft option for those who come from upper middle class families"' (Hodges, 2019). I never used the phrase 'posh and dim'. There is no evidence that Geography undergraduates are especially non-intelligent. I published a follow-up academic paper showing that, of every single university degree subject, not only was UK Geography still taking the highest proportion of undergraduates from the most privileged geographic locations, but that during the last decade it also rose up to first place in terms of being the subject you were least likely to study at university if you'd ever qualified for a free school meal in the UK. Geography, across all its UK university departments, was taking the smallest proportion of young people who'd ever experienced poverty (Dorling, 2022). There is no Geography department in the UK that's especially diverse, equal, or

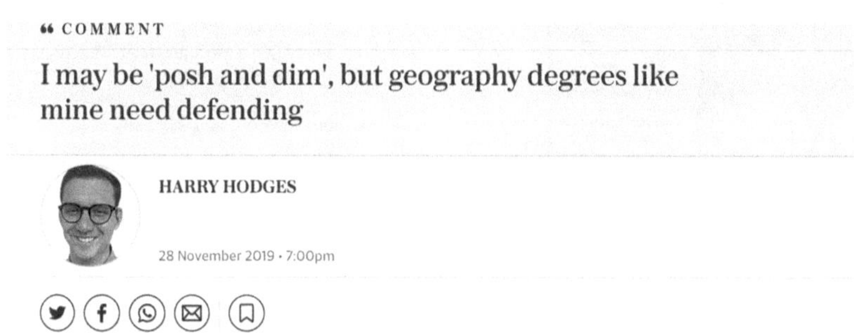

“ COMMENT

I may be 'posh and dim', but geography degrees like mine need defending

HARRY HODGES

28 November 2019 · 7:00pm

Figure 23.4 How *The Telegraph* newspaper headlined and illustrated a story on Geography.

Source: Hodges 2019, with permission from *The Telegraph*

inclusive when it comes to those who are poorer. This rise in how exclusive the academic subject of Geography had become is extremely likely to be linked to the rise in economic inequality within the UK over recent decades.

To me what matters most is that academic geographers themselves, let alone their students, are so unaware of this. In that follow-up paper I quoted 89 geographers, a group that included 59 heads of department (the majority of all university departments in the UK), who wrote a letter published in the *Times Higher Education* magazine titled 'Geography degrees are preparing disadvantaged students for relevant careers'. I think these academics honestly believed that they were. But the letter would have been more accurately titled 'Around one, or less than one, disadvantaged student per year studies in each Geography department in the UK'. Since that debacle, several people have written to me who were Geography students in the recent past and from a poor background describing how strange their experience of studying Geography at university was, and how alienating it felt to them – how alone and unusual they were. They will also have studied when it was slightly more likely for them to have been admitted than it is now. For someone from a normal or poor background, studying Geography at a UK university may be even more alienating today because it is now more rare.

I'm not complaining about my colleagues. I too was completely unaware of just how unusual Geography had become as an academic subject in the UK until I stumbled upon those statistics from an analysis of all UK students going to all UK universities in 2019. However, Geography has a long history of being elitist and targeting young people, especially young men, from posher backgrounds.

Cecil Rhodes and Halford Mackinder

You may not know who either Cecil Rhodes or Halford Mackinder were, but it shouldn't take you very long to find out and it's very much worth your while doing so. The post I hold at the University of Oxford is named after Mackinder. In 2021 a university committee recommended that the statue of Cecil Rhodes be removed from the highest plinth on Oxford's High Street. Statues in public places raise questions about signification: what do they stand for and are attempts to keep them in place partly about putting others in their place and trying to maintain an old hierarchy? A part of the wider definition of fascism is a suppression of opposition, belief in a natural social hierarchy, subordination of individual interests for the perceived good of the nation and race, and strong regimentation of society and the economy. Statues are often there to be looked up to, to put you in your place (below and beneath them), to present idealized images of the nation and race, and to give a sense of regimentation and order.

The recommendation to remove the statue was ignored by Oriel College's governing body (Rhodes attended Oriel), despite the governing body having asked for the establishment of that committee and appearing to be committed to following its recommendations. I, along with hundreds of other academics, objected to the recommendation being ignored. The petition calling for an academic boycott of that college was leaked very shortly after people had begun to sign it online. I was one of the signatories. I had not organized the petition, but being more in the public eye – and working in one of the UK academic disciplines that awards so many degrees to the children of better-off people – I was singled out. How could a geographer support this boycott – wasn't I being a hypocrite? After all, some might say, Rhodes was one of the 'great adventurers' of the 19th century. Who was I to object to the statue of Cecil being kept high up there towering above us all?

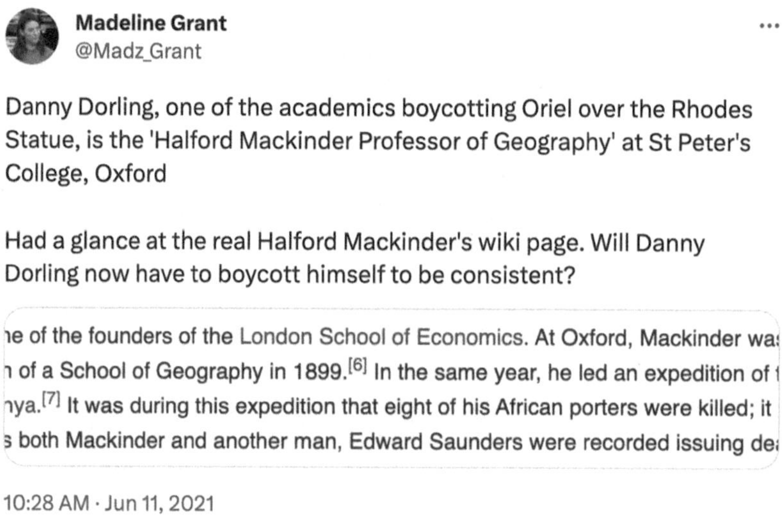

Madeline Grant
@Madz_Grant

Danny Dorling, one of the academics boycotting Oriel over the Rhodes Statue, is the 'Halford Mackinder Professor of Geography' at St Peter's College, Oxford

Had a glance at the real Halford Mackinder's wiki page. Will Danny Dorling now have to boycott himself to be consistent?

ne of the founders of the London School of Economics. At Oxford, Mackinder wa
n of a School of Geography in 1899.[6] In the same year, he led an expedition of
nya.[7] It was during this expedition that eight of his African porters were killed; it
s both Mackinder and another man, Edward Saunders were recorded issuing de

10:28 AM · Jun 11, 2021

Figure 23.5 An extract from Halford Mackinder's Wikipedia page – tweeted by Madeline Grant.

Source: Grant, 2021, quoting in turn from: https://en.wikipedia.org/wiki/Halford_Mackinder

Madeline Grant, the Institute of Economic Affairs contributor who describes herself on her Twitter page as '*Telegraph* columnist and Parliamentary sketch-writer', wrote this: 'Danny Dorling, one of the academics boycotting Oriel over the Rhodes statue, is the 'Halford Mackinder Professor of Geography' at St Peter's College, Oxford. Had a glance at the real Halford Mackinder's wiki page. Will Danny Dorling now have to boycott himself to be consistent?' (Grant, 2021). That tweet was posted with the accompanying image (Figure 23.5).

I would not object if there were a push to change the title of my job so that people hired to the post in future don't have deal with that legacy. However, for me it serves as a reminder of who I work for – a university that contains colleges which still implicitly endorse the British Empire. While Mackinder does not compare to Rhodes in the damage he inflicted on others, it helps to remember what Mackinder really did, including the death threats he made to his black porters in Africa, which were then acted on (Kearns, 2009; see Chapter 2, this volume). It makes me less complacent. When I was a child growing up in Oxford, one of the 'houses' of my school was named after Cecil Rhodes. After I left that school, in the late 1980s, the children refused to be part of that house and its name was changed to Martin Luther King House. I supported the Oriel boycott because the students of that college were trying to do the same as the children I attended school with – the Oxford University students were just unsuccessful, this time.

Grant's colleagues at *The Telegraph* newspaper, when discussing the Oriel boycott and why I'd signed the petition, wrote that 'Mackinder dedicated his life's work to the renewal of the British Empire, which he saw as viable in the aftermath of the Second World War' (Turner et al., 2021). A small group of people still harbour the idea of Britain rising up again in the world order. For some of them, the Rhodes statue is emblematic, a symbol of what could be again.

Early in 2021, the British government had proposed a new Policing Bill which recommended up to ten years' imprisonment for defacing a statue, but only five for rape. After the bill was passed it was proposed to change the law again so that the ten years' sentence could be given even if the damage done to any statue was minimal, costing less than £5,000 to repair. The reason given was as follows: 'Concern has been voiced in Parliament and society that the law focuses too heavily on the monetary value of the damage with insufficient consideration given to the emotional or wider distress caused by this type of offending, and as a result, the punishments do not fit the crime. We intend, where there is damage to a memorial, to remove the consideration of monetary damage, which would otherwise, in some cases, determine venue and limit sentencing powers' (Home Office, 2022).

So you could now go to prison for ten years simply for painting a single slogan on a statue, or even for just using chalk. A few people really care about keeping these statues – they do find the idea that they may be removed, defaced, or even just discussed and objected to distressing. To them, even showing you do not support the statue remaining where it is can be horrifying. And some of them do not understand that others might experience emotional or wider distress by knowing they have to live in a society where the statues of racists are specially protected.

The Spectator magazine listed the names of all the academics who supported the Rhodes statue boycott (Steerpike, 2021). The anonymous writer of that piece decided to pick my name out of the list of 150 and include it first. Male, white, middle-class, middle-aged academics (like me) can at least sometimes absorb some of the vitriol that's almost always more fiercely directed at others. Perhaps *The Spectator* writer picked on me first because they might otherwise be criticized for racism if they'd instead singled out one of my black colleagues; or bullying, if one of my female colleges had been chosen? The world is slowly changing and what is acceptable is changing – it is even changing how people working for *The Spectator* magazine behave.

Conclusion

Being in the public eye can result in pot-shots written in the most obscure of places. I was recently sent a published article which said: 'Danny Dorling's interesting analyses of inequality data were useful but lacked theoretical explanation and critique' (Sayer and Morgan, 2022). I was grateful to the person who sent it, as I wouldn't have found this out otherwise. I read the rest of the paper and learnt that another somewhat more capable academic than me, the hugely influential late Michel Foucault, was apparently also a disappointment because 'his theoretical reasoning is often confused'.

Being compared at times to far more famous academics such as Richard Wilkinson, Thomas Piketty, Ha-Joon Chang, Karl Marx, and Michel Foucault is flattering, but it's also rather ridiculous. I became a public geographer largely by accident, at first by drawing weird maps (Figure 23.3), and then propelled along more recently by the ire of commentators who are mostly to the far right of British politics, and the extreme far right of European politics. I probably wouldn't have the job I have now at the University of Oxford if enough people hadn't so disliked the arguments I was making that, by criticizing me, they drew so much more attention to them! My first widely read book, *Injustice: Why Social Inequality Persists* (Dorling, 2010) was disliked by many people. It doesn't look as radical or unreasonable if you read it today as when it was published in 2010, but the reception to it is probably why I was appointed to the Mackinder Chair in 2013.

It's easier to speak out when you are older, and if you have a secure job. However, if you work, as I do, at one of Britain's elite universities you'll be constantly told that 'the donors do not like what you are saying' – as if that mattered (although to the person telling me that, it really does). It will be easier to speak out in future if more geographers are vocal. I think I ended up being a public geographer at a point of extreme inequality in the UK precisely *because of* that inequality and the fact that I happened to study, analyze, and describe it. I hope that soon we'll be on the downward slope of that, at the start of an era where those forces trying to propel us to ever greater inequality are on the back foot. All kinds of public geography can be worthwhile. One day the UK discipline of Geography will not be quite what it has become, not as posh, not as elitist, and neither will the UK be in that same, sad, position.

Study task

Variations on this question can be set on an unseen examination, an open-book examination, or as a term-paper assignment:

Geographers 'going public' with their research is a nice idea, but in practice too difficult to be worth the effort involved. Discuss, using the public engagements of Danny Dorling as a focus.

His personal website is here: https://www.dannydorling.org. It's a treasure trove of information and includes recordings of some of his presentations.

(Note that, in the USA, the efforts of the late Harm de Blij could also make for a good focus for this study task if instructors can curate his many public-facing activities for the benefit of students.)

Starter reading list

Burawoy, Michael. "Public sociologies: Contradictions, dilemmas, and possibilities." *Social Forces* 82.4 (2004): 1603–1618.

Castree, Noel. "The struggle over geography: Prospects for advancing public pedagogy." *Dialogues in Human Geography* 3.2 (2013): 163–166.

Fuller, Duncan. "Public geographies: Taking stock." *Progress in Human Geography* 32.6 (2008): 834–844.

Murphy, Alexander B. "Enhancing geography's role in public debate." *Annals of the Association of American Geographers* 96.1 (2006): 1–13.

Murphy, Alexander B. et al. "The role of geography in public debate." *Progress in Human Geography* 29.2 (2005): 165–193.

Nickel, Patricia M. "Public Sociology and the Public Turn in the Social Sciences." *Sociology Compass* 4.9 (2010): 694–704.

Turner, B. L. "Geography's profile in public debate 'inside the beltway' and the national academies." *The Professional Geographer* 57.3 (2005): 462–467.

Worth, Nancy. "Public geographies and the gendered experience of saying 'yes' to the media." *The Professional Geographer* 72.4 (2020): 547–555.

References

Burawoy, M. (2005) For public sociology, *American Sociological Review*, 70, February, 4–28.

Burn-Murdoch, J. (2022) The Tories have become unmoored from the British people, *The Financial Times*, 30 September, https://www.ft.com/content/d5f1d564-8c08-4711-b11d-9c6c7759f2b8

Butcher, J. (2019) Is this the worst book written about Brexit? *Spiked Magazine*, 28 March, https://www.spiked-online.com/2019/03/28/rule-britannia-the-worst-book-written-about-brexit/

Chancel, L., Piketty, T., Saez, E., and Zucman, G. (2022) *World Inequality Report 2022*, World Inequality Lab, online: https://wir2022.wid.world/

Dorling, D. (1992) Visualizing people in space and time, *Environment and Planning B*, 19, 613–637.
Dorling, D. (2005) Why Trevor is wrong about race ghettos, *The Observer*, 25 September, pp. 14–15, https://www.theguardian.com/uk/2005/sep/25/communities.politics
Dorling, D. (2010) *Injustice: Why social inequality persists*, Bristol: Policy Press.
Dorling, D. (2018) *Peak Inequality: Britain's Ticking Time Bomb*. Bristol: Policy Press.
Dorling, D. (2019) Kindness: A new kind of rigour for British Geographers, *Emotion, Space and Society*, 33, November 2019, 100630.
Dorling, D. and Tomlinson, S. (2019) *Rule Britannia: Brexit and the End of Empire*. London: Biteback, https://www.dannydorling.org/books/rulebritannia/
Dorling, D. (2022) Where are we heading? The example of generational change in British academic Geography. *Environment and Planning F*, 1, 1–18.
Dorling, D. (2023a) *Shattered Nation: Inequality, and the Geography of a Failing State*. London: Verso.
Dorling, D. (2023b) A brief lull in the fighting? *Public Sector Focus Magazine*, January/February issue.
Dorling, D. (2024) *Peak injustice: Solving Britain's Inequality Crisis*. Bristol: Policy Press.
Grant, M. (2018) Ricardo's ideas are as indispensable today as they were 200 years ago, 19 April, https://iea.org.uk/ricardos-ideas-are-as-indispensable-today-as-they-were-200-years-ago/
Grant, M. (2021) Tweet posted at 10.28am on 11 June, https://mobile.twitter.com/Madz_Grant/status/1403283098650005514
Harris, J. (2022) Forget 'levelling up' – Liz Truss isn't even pretending to care about inequality, *The Guardian*, 18 September, https://www.theguardian.com/commentisfree/2022/sep/18/levelling-up-liz-truss-boris-johnson-britain-inequalities
Hildyard, L. (2022) Why Liz Truss is wrong on redistribution of wealth, *High Pay Centre*, 21 September, https://highpaycentre.org/why-liz-truss-is-wrong-on-redistribution-of-wealth/
Hodges, H. (2019) I may be 'posh and dim', but geography degrees like mine need defending, *The Telegraph*, 28 November, https://www.telegraph.co.uk/environment/2019/11/28/may-posh-dim-geography-degrees-like-mine-need-defending/
Home Office (2022) Criminal damage to memorials: Police, Crime, Sentencing and Courts Act 2022 factsheet, *Her Majesty's Government*, 20 August, https://www.gov.uk/government/publications/police-crime-sentencing-and-courts-bill-2021-factsheets/police-crime-sentencing-and-courts-bill-2021-criminal-damage-to-memorials-factsheet
Jones, S. (2013) The 'star' academics are so often white and male, *The Guardian*, 22 April, https://www.theguardian.com/education/2013/apr/22/university-jobs-not-being-advertised
Kearns, G. (2009) *Geopolitics and Empire: The Legacy of Halford Mackinder*. Oxford: OUP.
Kirby, P. (2022) Italian elections: Meloni gets to work on picking right-wing government, *BBC News*, 27 September, https://www.bbc.co.uk/news/world-europe-63044973
Mattick, P. (2013) *Marxism--Last Refuge of the Bourgeoisie?* Abingdon: Routledge.
Milanović, B. (2022) Socialism with Chinese characteristics for the young person, *Brave New Europe Posting*, 13 August, http://glineq.blogspot.com/2022/08/socialism-with-chinese-characteristics.html
Nelson, F. (2019a) Internships at The Spectator for summer 2019; no CVs (or names!) please, *The Spectator*, 31 March, https://www.spectator.co.uk/article/internships-at-the-spectator-for-summer-2019-no-cvs-or-names-please
Nelson, F. (2019b) Tweet posted early in the morning, 1.03am on 5 April 2019: https://twitter.com/FraserNelson/status/1113955563308769283
Niemietz, K. (2014) Danny Dorling's 'All that is solid': The worst book on the housing crisis so far, *Institute for Economic Affairs*, 4 July, https://web.archive.org/web/20140722180408/https://iea.org.uk/blog/danny-dorling%E2%80%99s-%E2%80%98all-that-is-solid%E2%80%99-the-worst-book-on-the-housing-crisis-so-far
Niemietz, K. (2022) Politically active left-wingers are the equivalent of the tribal football fan, *1828 blog*, 31 August, https://www.1828.org.uk/2022/08/31/politically-active-left-wingers-are-the-equivalent-of-the-tribal-football-fan/
Sands, L. (2022) Sweden election: How an ex neo-Nazi movement became kingmakers, *BBC News*, 15 September, https://www.bbc.co.uk/news/world-europe-62913356

Sayer, A. and Morgan, J. (2022) A realist journey through social theory and political economy: an interview with Andrew Sayer, *Journal of Critical Realism*, 21, 4, 434–470, https://www.tandfonline.com/doi/full/10.1080/14767430.2022.2049078

Snowdon, C. (2015) The rich versus the super-rich, *Institute for Economic Affairs*, 16 February, https://iea.org.uk/blog/the-rich-versus-the-super-rich

Snowdon, C. (2016) Is envy of the super-rich by the rest of the rich driving the inequality debate? *Institute for Economic Affairs*, 25 May, https://iea.org.uk/blog/is-envy-of-the-super-rich-by-the-rest-of-the-rich-driving-the-inequality-debate

Snowdon, C. (2017) How Danny Dorling (and Oxfam) recycle Karl Marx's wretched 'immiseration theory', *Institute for Economic Affairs*, 20 January, https://iea.org.uk/from-karl-marx-to-danny-dorling-and-oxfam-why-the-immiseration-theory-just-wont-go-away/

Steerpike (2021) Meet the academics behind the Rhodes boycott, *The Spectator*, 2.37pm on 11 June, https://www.spectator.co.uk/article/meet-the-academics-behind-the-rhodes-boycott

The Economist (2021) Joe Biden is determined that China should not displace America, *The Economist*, 17 July, https://www.economist.com/briefing/2021/07/17/joe-biden-is-determined-that-china-should-not-displace-america

Turner, C., Penna, D., and Heslop, L. (2021) Oxford dons boycotting Oriel are themselves funded by imperialists, *The Telegraph*, 11 June, 8.46pm, https://www.telegraph.co.uk/news/2021/06/11/oxford-dons-boycotting-oriel-funded-imperialists/

Glossary

Action research a philosophy and methodology of research employed in parts of the social sciences, including human geography. It seeks transformative change through the simultaneous processes of doing research and trying to make it count practically in the world, which are linked together by critical reflection and usually through engagement with non-academic organisations (e.g., NGOs). It involves the researcher and the researched in a collaborative process of exploration with the aim of co-creating new discourses and practices that create change, often for marginalised or relatively powerless social actors.

Anthropocene the name given to the current period of geological time during which human activities have had a very significant impact on the surface of the Earth. The Anthropocene is often said to have begun around 1950. The term is not yet an official part of geologists' vocabulary but is widely used in the geosciences, social sciences, humanities and parts of the arts.

Anthropogeographie named after the work of German geographer Friedrich Ratzel, it understands the relation among migration, settlement patterns and the physical environment as determined by the impact of nature on people.

Anti-essentialism is the belief that there are no fundamental, fixed and enduring traits or qualities that make a given entity, social or physical, what it is. As a philosophical practice it dislodges reductionism, the attempt to base causal explanation on what are believed to be foundational entities, for example, atoms in matter or individuals in society. Anti-essentialism suggests the geographical world's constituent parts are complex and changeable, malleable and contingent.

Anti-capitalism (see capitalism)

Anti-colonialism (see colonial)

Anti-imperialism (see imperialism)

Anti-modernism (see modernism)

Berkeley School a network of scholars directly or indirectly influenced by Carl Sauer's cultural landscape research from the 1930s onwards. Members of the School undertook detailed field work and archival research in order to trace landscape change over time in different regions.

Cadastral from cadastres, the comprehensive cartographic records of land holdings and boundaries of different pieces of property. Some of the earliest cadastral maps were produced under the Roman Empire, allowing state-owned lands to be mapped and delineated to prevent their occupation by private land owners.

Capitalism a mode of production of goods and services characterised by profit-seeking, employment of wage workers, competition among private producers, ceaseless innovation in production processes and products, and cycles of economic growth, stagnation and crisis. Under capitalism, capitalists own the means of production, commanding the majority of societal wealth, while workers, employed by capitalists, are said by many critics to be exploited, often oppressed, and far less well.

Capitalocentrism the idea that capitalist economic relations are necessarily and naturally dominant. In contrast, all non-capitalist economic relations are designated subordinate, 'inferior', underdeveloped, marginal or dependent.

Cartography the representation of a portion or the totality of the Earth on a map. Maps denote the features (human and physical) significant to the creator. Cartographers often argue that maps are 'neutral' scientific representations of the world, but they are cultural products reflecting the world views of their makers. As such, maps are texts that can be deconstructed (that is, examined for hidden meanings that are perhaps not immediately obvious), in much the same way that a Shakespearean play or a novel can be deconstructed to reveal the author's view of the world.

Circulation temporary and repetitive movement, such as commuting to work or the regular shipping or receipt of goods.

Central place theory put forward by Walter Christaller in 1933, which understands the spatial distribution of settlements as based on the ability to provide economic services to surrounding areas.

Choropleth maps maps that are coloured or shaded to represent significant spatial variations of a given variable across the surface of the Earth.

Class a system of ordering people and societies based on the unequal distributions of wealth and power (see Capitalism). Class differences are found in most societies, but the exact class groupings and causes of class differentiation may vary.

Coexistence the co-occupation of landscapes or territories by otherwise different people, and/or different species and non-human entities.

Colonialism a process where one country controls and appropriates the land and resources of another, imposing its own laws, rules, norms and customs. The native population is subjected to the control of an external national power.

Commercial geography branch of geography that studied the geographical flows of trade, including characteristics of the nature and pathways of traded goods, and the causes and consequences of uneven spatial patterns of trade. The term has now limited use, with 'economic geography' much more common.

Commodification the transformation of things such as goods, services, ideas, nature or information into objects traded between buyers and sellers.

Community economies the production of goods and services in which people and agents negotiate ethical actions around joint wellbeing, distributing surplus to support life, responsibly encountering those near and far, sharing common resources and investing in liveable futures.

Country (with an upper-case C) a word in Aboriginal English that includes not just the territorial, land-based notion of homeland, but encompasses humans as well as waters, seas and all that is tangible and non-tangible. It describes the totality of past and present reality, both seen and unseen.

Critical human geography a diverse set of research and teaching practices centred on opposition to repressive and inequitable power relations in any given society or context. Critical geographers stress the role of dominance and opposition in the production and reproduction of landscape, place, territory, nation, common resources and so on. Critical geography marries critique with proposals for change so as to achieve greater justice in the world. Critical human geography replaced the earlier term 'radical geography' current in the 1970s and 1980s.

Critical physical geography a branch of geographical knowledge combining detailed study of biophysical systems with an end of environmental and/or social transformation. It is sensitive especially to the role of social power in the creation and actioning of that knowledge. It acknowledges that scientific knowledge can be employed in unjust ways, disadvantaging particular people or parts of the non-human world.

Customary law the systems and practices prevalent among pre-industrial or present-day Indigenous peoples that developed over time from accepted moral norms. They regulate human behaviour, mandate sanctions for non-compliance, and connect people with the land, water and with each other.

Cultural geography a branch of geographical inquiry that studies the relationships between culture, landscape and place. It includes the study of values, beliefs, norms, religious practices, tastes and their material expression at the levels of a home, neighbourhood, city, countryside and so on.

Cultural imperialism the practice of one culture dominating another and rendering it secondary, 'exotic' or 'inferior'. European colonisation of overseas territories was normally accompanied by cultural imperialism (e.g., the British in colonial India).

Cultural landscape a blending of human and non-human components to create a particular set of material landscape forms and processes that are then interpreted and valued.

Darwinian evolution a well-evidenced theory based on the ideas of British naturalist Charles Darwin. It suggests that all species descended from a common ancestor, developing physical and behavioural traits over time that made them fit to live in specific environments, allowing survival and species reproduction. For Darwinians, all organisms undergo slow change as they seek to adapt to environments that also change.

De-colonial / decolonised (see Colonialism)

Decennial census a national effort to count the country's population and tabulate other demographic and socioeconomic information. Results are collected for reporting zones such as provinces, counties, cities and city blocks. In many countries a census is conducted every ten years (the meaning of decennial).

Deconstruction as developed by philosopher Jacques Derrida, an intellectual approach highlighting the unfixity and contestability of meaning, calling into question fundamental axioms of Western thought that divided the world into supposedly secure 'ontological' components such as 'society' and 'nature'.

Diverse economies a term used to refer to the heterogeneity of economic practices – from the standard capitalist form to co-ops, social enterprises, Islamic finance, and fair-trade partnerships – many of which are invisible in representations of 'the economy'.

Disembodiment lack of connection to the physical body or reality. It is often used to mean de personalised, abstract knowledge.

Electoral bias the extent to which one or more political parties gain an unfair advantage over their rivals due to the operation of a country's electoral system.

Electorate all adults in a country who are registered and eligible to vote in that country's national and local elections.

Emancipation a process which results in social, legal or political freedom for a person or a social group.

Environmental determinism the idea that the conditions of the non-human environment (climate, physical features) in a particular place determine the character of human development in that place.

Environmental flow water released from a dam or weir to maintain downstream river health, and the health of the environment.

Environmental governance the ensemble of organisations, rules and practices that together govern how the non-human world is used by people in a given region, country or set of countries. Governance extends beyond government and involves other actors beyond the state, such as businesses, charities and NGOs.

Environmentalism the beliefs, knowledges, political arguments and practical actions directed towards supporting, protecting, understanding and enhancing the environment. It comprises various social, cultural and political movements that aim to conserve and protect the more-than-human environment by inducing change in human behaviours (e.g., towards nature conservation). Environmentalists are 'green' in an ethical sense.

Epistemology a set of beliefs or axioms about how we know the world, including ourselves, that's taken to be correct, truthful or otherwise accurate.

Ethno-geomorphology a transdisciplinary branch of geomorphology which includes the study of the ways people interpret, value and manage the land to create biophysical-cultural landscapes. It focuses on a relational understanding of people and place within the context of diverse cultural worldviews in decision making.

Fascism a political system in which a dictator or ruling party seeks to control all aspects of social, cultural, economic and political life. Fascist governments seek to limit what is sayable and doable, often stigmatising critics as 'heretics', 'enemies of the people' and so on. Fascists try to unify a population around a limited number of often nationalistic beliefs and aspirations.

Fellow a person who holds a position in a professional academic society or at a university.

Feminist an advocate of equal rights for men and women; a person who's critical of values, norms and practices that privilege men.

Feminism a broad range of views that assert the needs, rights and entitlements of women in the economic, social, cultural and political spheres.

Foundationalism a belief that there is a deep-seated single, indisputable truth that allows understanding of the world.

Gender the physical and behaviour traits conventionally associated with men and women respectively. Transgender people identify with some blend of these traits that unsettle the man–woman binary.

Geodesy way of measuring and representing the Earth based on its shape, location in space and gravitational fields, and their change with time.

Geographic Information Science branch of geography using computational approaches to collate, order and visualise information about our worlds, and to model and interpret spatial patterns in three dimensions.

Geographical imagination the ideas or images of a person or a whole society about the world's peoples, places and environments. Imaginations are formed over time through processes of school education, news reporting and so on. They are often simplistic depictions.

Geohumanities study of place, spatial patterns and landscape from the specific perspective of the arts and the humanities (and not the social sciences). This perspective emphasises different ways people can invest the world with meaning. The 'same' world is thus rendered 'different' depending on a person's interpretation of reality.

Geopoetics both a practice and disciplinary field which rejects the notion of a universal, objective understanding of the world in favour of knowing the worlds around us using frames of perception from literary to art to creative imagination.

Geopolitics the study of the political relationship between a country or a block of countries and other countries.

Gerrymandering the process of deliberately manipulating the boundaries of electoral districts so as to make it harder to elect politicians for rival parties and easier to elect politicians from one's own party.

GIScience geographic information science searches for general principles governing information about geography, that is, information about the distribution of physical and social phenomena over and near the surface of the Earth. These general principles can be used in the design of technologies such as geographic information systems.

GIS geographic information systems emerged in the late 1960s as computer applications for the handling of geographical information. GIS has since expanded to an international industry serving the needs of applications that intersect with virtually all aspects of human activity.

Global North those countries currently possessed of significant monetary, economic, military and technical resources. Not limited to countries in the Northern Hemisphere but dominated by those in Europe and North America. Also includes Southern Hemisphere nations such as Australia and New Zealand.

Governance system of rules, policies and practices which determine how a community, town, city, region, and so on are run and involving both governmental and non-governmental actors.

GPS the Global Positioning System is a collection of satellites, ground control stations and receivers capable of accurate measurement of a receiver's position on or near the Earth's surface. Originally developed by and for military applications, it is now widely used for a host of civilian purposes.

Gramscian following the beliefs of influential Italian Marxist thinker Antonio Gramsci. Gramsci argued that capitalist elites secure power and authority not only through command of money but also by shaping the ideas of ordinary people so as to create 'hegemonic' beliefs. Gramsci argued that left-wing political parties needed to counter these dominant ideas, not only focus on wealth differentials.

Heuristic a mental shortcut for solving problems that delivers a result that suffices given limited time or other constraints.

Historical geography branch of geography exploring past societies, regions, places and environments.

Humanistic geography an approach to geography drawing on philosophies of meaning, phenomenology, and existentialism, which places human experience and subjectivity at the centre of understanding our world.

Hydro-social cycle a socio-natural process by which water and society make and remake each other over space and through time.

Jurisdiction (i) a legal-political unit, (ii) the official power to make decisions and administer justice in or over that unit.

Institution an enduring structure of rules and practices, or an organisation which structures human behaviour.

Ideology system of ideas and beliefs about how the world should be interpreted and constructed. It shapes understanding of behaviours and actions and may be consciously or subconsciously held and expressed.

International Geographical Union international society that promotes the study of geography through coordination of activities (including teaching and research) via conferences, resourcing, networking and publication.

Indigenous culture rich, diverse set of beliefs, knowledges and practices that define non-Western, traditional social and cultural groups through their connection and relationships with ancestral lands.

Imperialism assertion of military, economic, cultural and political power by one nation over one or more other nations.

Keynesian economics founded by the English economist John Maynard Keynes in the 1930s, it was a response to the inability of conventional economics to understand or cure the then Great Depression. Keynes demonstrated that the problem was the lack of effective demand, the inability of consumers to buy goods. His solution was massive state involvement in the economy, investment by the state in both business and social welfare schemes, e.g., old-age pensions. In both cases, money was provided to consumers who previously lacked it, thereby raising effective demand, and lifting the economy out of recession.

Land Council in Australia, a statutory organisation that assists Aboriginal people to acquire and manage their traditional lands and seas.

Landsat a system of Earth-observing satellites first established in 1972, and widely used for mapping and land resource management.

Landscape a polysemic term indicating a portion of land or territory which the eye can comprehend in a single view, as well as representing a scene (e.g., photographically).

Lifeworld the real and virtual environment as perceived by a person. A lifeworld shapes a person's sense of self, other and the wider world.

Latourian from the French philosopher-sociologist Bruno Latour and his pioneering work in science studies. His actor–network theory conceived knowledge as the product of shifting connections (networks) among very different entities (actors) that could be both human and non-human, with neither privileged. Nothing exists outside the networks and the changing linkages they represent.

Lefebvrian following the beliefs of Henri Lefebvre on space as a social construction which affects and is shaped by both human practice and perception. Terrestrial space is seen as more than a set of phenomena arrayed across the surface of the Earth. For Lefebvre, space is socially created and affects society.

Location theory often an abstract, mathematically informed approach to explaining why economic activities are located in the geographical sites they are. Originating in Germany in the 19th century, it was most popular within Anglophone economic geography during the 1960s and 1970s. Interest waned with the rise of radical and then critical geography.

Malapportionment an electoral abuse which involves manipulating the size of the electorate in different electoral districts in order to make it easier to elect politicians from one party than from its rivals.

Marxism a body of analysis and political practices derived from Karl Marx (and to some extent Frederik Engels). Communism, Marx's alternative to capitalism, is predicated on broad equality among all people. Marxism shaped critical human geography significantly since the mid-1970s, initially as radical geography.

Mercator projection a means of representing the spherical globe on a flat piece of paper using a cylindrical projection which retains the relative location of north and south as well as local land shape. It has the effect of exaggerating the size of the land surface at the poles to infinity and shrinking land near the equator.

Metaphysics literally transcending the physical world. It is a study of things that cannot be analysed by a scientific examination of the material external world.

Microplastic contamination particles of plastic which are less than 5mm in length that are found polluting environments. They are typically the result of erosion, weathering or decay of plastics, either during use or when discarded. They may also be intentionally fabricated for retail in products such as cosmetics and then released into the environment. Microplastics comprise many different compounds and little is known about their distribution, toxicity or persistence.

Migration more permanent and less frequent movements where the intent is to leave one place and arrive in another to stay.

Mobility movement across space at all scales. Mobility has become central to work in the 'new mobilities paradigm' or 'mobility turn'. It can also defined as a combination of the fact of movement, meaning assigned to movement and the practice of movement, understood in the context of power relations.

Modifiable Areal Unit Problem a problem occurring when changing the boundaries of areas in which the value of a variable is reported. Boundary changes significantly alter the results of inferential statistical analysis carried out on those variables.

Modernisation transformation of an existing society to one dominated by industrial development and complex technologies, often secular and urban.

More-than-human a description of the non-human world that avoids the nature/culture binary, thereby facilitating different imaginations of environmental development emphasising human dependence on (rather than dominance of) the worlds around us.

Native Title the recognition that Indigenous people have rights and interests to land and waters according to their traditional law and customs. Law courts in settler-colonies like Australia are currently considering native title claims lodged by various Indigenous groups.

Naturalism the view that there's a single external natural world comprehensible through science.

Neogeography because of technological advances and reduced costs, a geographic technology that is accessible to consumers and produced by a wide range of individuals.

Neighbourhood effect in elections, the tendency for people living in an area to change their political views to fit in with the local majority view.

Neoclassical theory the dominant school of economics originating in the late nineteenth century. Its methods and theories are defined by the school's overriding concern to maximise the greatest social and economic benefit from using a minimal amount of resources: the most for the least.

Neoliberalism a social and economic philosophy prioritising the freedom of individuals, families, communities and firms. In the USA, the UK and elsewhere, neoliberal ideas became prominent from the early 1980s. Critics argue that this has reduced social solidarity among citizens, engendered selfishness and fuelled soaring economic inequality, with national governments withdrawing social support.

Nomadic metaphysics an alternative to a 'sedentarist' metaphysics that centres human mobility in our understanding of society, culture and politics.

Ofwat the UK Water Services Regulation Authority.

Orthodox traditional ideas, beliefs and practices accepted as normal within a social group.

Ontology concerned with the nature of existence and the fundamental structure of reality. Ontological statements focus on what are taken to be the building blocks of the world, with reality and its component parts.

Pedagogy the theory and practice of how knowledge and skills are taught. Pedagogic refers to an idea or practice which is related to teaching.

Peters projection a means to represent a spherical Earth on a flat piece of paper that retains the area of a land mass over its shape. Europe becomes a much smaller area on a Peters projection map compared to the Mercator projection.

Place (i) a unique point on the Earth's surface, (ii) an arena for daily physical activity, or (iii) a locus of identity and belonging for people who inhabit places for extended periods of time.

Political economy emerging in Western Europe in the 18th century, the study of the relationships between politics, particularly in the form of the state, and the economy. It is the older name for what became Economics.

Possibilism, environmental the idea that societies are limited or constrained by their environments, but not fully determined by them. Consequently, similar environments are associated with different forms of society and culture.

Positivism an **epistemology** that claims real or genuine knowledge is true either by definition, such as mathematical statements, or based on empirical facts that are proved by the scientific method. All other sources of knowledge – intuition, introspection, religious revelation – are spurious, meaningless.

Post-colonialism a critical academic study of the political, cultural and economic effects of colonialism and imperialism on both colonies, typically of Western European countries, and the colonial powers themselves.

Post-Brexit the time period after Britain left the European Union on 31 January 2020.

Poststructuralism a philosophical movement originating in France during the 1960s that suggested that the meaning of words was radically unstable, escaping the control of an author. Meaning was ultimately secured by multiple systems of power spiralling through society.

Post-truth idea that objective facts, scientific reason and evidence no longer have sway, especially in contemporary politics. Instead, the post-truth world is shaped by emotion and subjective personal opinion buttressed by a massive media apparatus.

Production of scale the construction through political and economic struggles of the various spatial resolutions at which social life is organised. Geographical scales like the urban, the regional, the national and the global are neither given ontologically, nor simply revealed by the investigations of researchers, but are instead actively made by social actors.

Presentism the doctrine that reality only exists in the 'now'. It results in a bias towards the importance of current events over those in the past or future.

Public geography geography as communicated to non-academic audiences by academic geographers. For example, geographical ideas presented through the media or non-specialist books or talks.

Quantitative geography a subfield of geography that uses measurement, calculation and mathematical models to represent and analyse geographical phenomena in space and time.

Racialised: a person or group characterised according to 'racial' categories supposedly distinguishing them from other people of other 'races'. In the past, 'race' was taken as an essentialist category, fixed and biologically given. It is now taken as socially constructed, used to discipline, control and exclude particular groups of people.

Radical geography a set of approaches to the study of human geography developed from the early 1970s. These approaches came from different radical political traditions – from Marxism to anarchism – taking issue with various perceived social injustices.

Regional geography a branch of geography that emphasises the study of unique regions, taken as fundament geographical units for the organisation of social, cultural and more-than-human entities.

Riverscape all the features of the landscape that can be found along a river path, both human and more-than-human.

River Styles Framework a set of tools and principles to explain and predict river patterns, and used in transdisciplinary decision making.

Royal Charter a traditional document issued by the ruling monarch that gives power or rights to a particular organisation.

Sauerian geography an approach to the inter-disciplinary study of landscape following the work of Carl Sauer and the Berkley School of cultural geography.

Sedentarist metaphysics pervasive patterns of thought that see a fixed and bounded nation-state (or other kind of place or territory) as a moral container of identity, culture and politics.

Sentient capable of having or experiencing feelings and sensations.

Settler-colonialism a type of colonialism in which foreign settlers emigrate and permanently reside on land already inhabited by Indigenous peoples.

Situatedness the idea that individual experience and social context affects the ways in which we 'know' (understand and interpret) the worlds around us. Consequently, the meanings we give to events, processes and patterns cannot be objective and independent of the observer and their context.

Socialism a political system in which economic wealth is shared among all citizens relatively equally.

Space The humanly created infrastructure and material geographical environments that enable and constrain human activity.

Spatial data handling the management of spatial data usually using Geographical Information Systems (GIS). GIS applies a geographic tag to all data points for storage, visualisation and processing purposes.

Spatial history A branch of history concerned with the roles of geographical location and context.

Spatial science branch of geography that emphasises the scientific visualising, processing and modelling of spatial data.

Socio-geomorphology a study of the reflexive relationship between social processes and land features and the ways in which we interpret and assign meaning to the landscape.

Social Darwinism the application of Darwinian principles of natural selection to the evolution of societies. It presumes humanity's sociological characteristics are akin to their natural/physiological ones. Historically it was used to justify the domination of a people through imperialism based on racial prejudice.

Structuralist social theory any theory of human behaviour that highlights the rules, norms and institutions of society (structure) within which a person is socialised and currently lives. It is the social structure that determines a person's position and their mode of life including form of consciousness.

Territory a clearly defined and bounded space that delimits the scale and scope of human action, both within it and between its inside and outside. It is controlled space, usually overseen by a government and a legal system.

Topography the shape and arrangement of phenomena at the surface of the Earth, and their study and representation. A topographical view sees regions as different absolute spaces that are clearly distinguishable from one another, with the spaces within any given regional boundary assumed to be more similar than those outside it – as when one crosses from, say, an industrial region into an agricultural region.

Topology an understanding of the world through networks and how they connect places together. Under this approach spatial scales like the local and the global are not viewed as discrete entities at opposite ends of a scalar spectrum but instead as a distinction between shorter and less connected networks (local) from longer and more connected ones (global).

Traditional owners any group of Aboriginal people who have 'primary spiritual responsibility' for sacred sites on a piece of land, and who are entitled by Aboriginal tradition to hunt and gather on that land.

Transgression literally 'crossing a boundary'. The boundary can be geographical or socio-cultural but is often both.

Turnout In an election, the number of people who actually cast a vote in an election expressed as a percentage of the total number of people eligible to vote: a measure of levels of political participation.

Ward an electoral district used in elections for sub-national local governments in the United Kingdom and in some other countries.

Water governance the practices of coordination and decision making of different actors around the distribution of water among various users and locations.

Whig history a form of history-telling that highlights progress over time, identifying key people, events or organisations supposedly improving society, or a social group or a place.

White supremacy the belief that white people are superior and should dominate other 'races' and ethnic groups.

Wicked problem a multi-faceted social, environmental or cultural problem caused by complex, inter-connected human and more-than-human processes that cannot be resolved using single or multi-parameter optimisation approaches because of incomplete, contradictory or changing evidence. Part of the wickedness is that the problems in question are perceived multiply, meaning different sets of social values are used to define what is at stake.

Zeitgeist from the German meaning the culture and values of a particular period in time and space.

Index

Pages in *italics* refer to figures and pages in **bold** refer to tables.

Kolibri 360 GmbH
Ferdinand-Jühlke-Straße 7
99095 Erfurt, Deutschland
produktsicherheit@kolibri360.de